Mössbauer Spectroscopy Applied to Magnetism and Materials Science

Volume 1

MODERN INORGANIC CHEMISTRY

Series Editor: John P Fackler, Jr.
Texas A&M University

CARBON-FUNCTIONAL ORGANOSILICON COMPOUNDS
Edited by Václav Chvalovský and Jon M. Bellama

GAS PHASE INORGANIC CHEMISTRY
Edited by David H. Russell

HOMOGENEOUS CATALYSIS
WITH METAL PHOSPHINE COMPLEXES
Edited by Louis H. Pignolet

INORGANOMETALLIC CHEMISTRY
Edited by Thomas P. Fehlner

THE JAHN-TELLER EFFECT AND
VIBRONIC INTERACTIONS IN MODERN CHEMISTRY
I. B. Bersuker

METAL INTERACTIONS WITH BORON CLUSTERS
Edited by Russell N. Grimes

MÖSSBAUER SPECTROSCOPY
APPLIED TO INORGANIC CHEMISTRY
Volumes 1 and 2 • Edited by Gary J. Long
Volume 3 • Edited by Gary J. Long and Fernande Grandjean

MÖSSBAUER SPECTROSCOPY
APPLIED TO MAGNETISM AND MATERIALS SCIENCE
Volume 1 • Edited by Gary J. Long and Fernande Grandjean

ORGANOMETALLIC CHEMISTRY OF THE
TRANSITION ELEMENTS
Florian P. Pruchnik
Translated from Polish by Stan A. Duraj

A Continuation Order Plan is available for this series. A continuation order will bring delivery of each new volume immediately upon publication. Volumes are billed only upon actual shipment. For further information please contact the publisher.

Mössbauer Spectroscopy Applied to Magnetism and Materials Science

Volume 1

Edited by

Gary J. Long

University of Missouri–Rolla
Rolla, Missouri

and

Fernande Grandjean

University of Liège
Sart-Tilman, Belgium

PLENUM PRESS • NEW YORK AND LONDON

Library of Congress Cataloging-in-Publication Data

Mössbauer spectroscopy applied to magnetism and materials science / edited by Gary J. Long and Fernande Grandjean.
p. cm. -- (Modern inorganic chemistry)
Includes bibliographical references and index.
ISBN 0-306-44447-X (v. 1)
1. Mössbauer spectroscopy. 2. Magnetism. 3. Materials science.
I. Long, Gary J., 1941- . II. Grandjean, Fernande, 1947- .
III. Series.
QC491.M623 1993
620.1'1297--dc20 93-14059
CIP

ISBN 0-306-44447-X

A Division of Plenum Publishing Corporation
233 Spring Street, New York, N.Y. 10013

Printed in the United States of America

Contributors

Lawrence H. Bowen • Department of Chemistry, North Carolina State University, Raleigh, North Carolina 27695-8204

S. J. Campbell • Department of Physics, University College, The University of New South Wales, Australian Defense Force Academy, Canberra, ACT 2600, Australia

Gordon Czjzek • Kernforschungszentrum Karlsruhe, INFP, W-7500 Karlsruhe 1, Germany

Eddy De Grave • Laboratory of Magnetism, University of Gent, B-9000 Gent, Belgium

Brent Fultz • Division of Engineering and Applied Science, California Institute of Technology, Pasadena, California 91125

H. Gleiter • Werkstoffwissenschaften, University of Saarland, D-6600 Saarbrücken, Germany

J. M. Grenèche • Equipe de Physique de l'Etat Condensé, URA CNRS No. 807, Université du Maine, F-72017 Le Mans Cedex, France

Q. A. Pankhurst • Department of Physics, University of Liverpool, Liverpool L69 3BX, England

R. J. Pollard • Department of Physics, Monash University, Clayton, Victoria 3168, Australia

W. Potzel • Department of Physics, Technical University of Munich, D-85747 Garching, Germany

G. Principi • Materials Section, Department of Mechanical Engineering, University of Padua, I-35131 Padua, Italy

William M. Reiff • Department of Chemistry, Northeastern University, Boston, Massachusetts 02115

Robert E. Vandenberghe • Laboratory of Magnetism, University of Gent, B-9000 Gent, Belgium

F. Varret • Département de Recherches Physiques, URA CNRS No. 71, Université Pierre et Marie Curie, F-75252 Paris Cedex 05, France

Preface

During the past 30 years materials science has developed into a full-fledged field for basic and applied scientific enquiry. Indeed, materials scientists have devoted their efforts to creating new materials with improved electronic, magnetic, thermal, mechanical, and optical properties. Often unnoticed, these new materials are rapidly invading our homes and automobiles, and may be found in our utensils, electronic equipment, textiles, home appliances, and electric motors. Even though they may go unnoticed, these new materials have either improved the efficiency and lifetime of these items or have reduced their weight or cost. In particular, magnetically ordered materials are useful in various applications, such as motors, magnetic imaging, magnetic recording, and magnetic levitation. Hence, much effort has been devoted to the development of better hard magnetic materials, magnetic thin films, and molecular magnets.

During the same period of time, Mössbauer-effect spectroscopy has grown from a laboratory curiosity to a mature spectroscopic technique, a technique that probes solid-state materials at specific atomic sites and yields microscopic information on the magnetic and electronic properties of these materials. Iron-57 is the most commonly and easily used Mössbauer-effect isotope and, of course, is particularly relevant for the study of magnetic materials. Various applications of Mössbauer spectroscopy to magnetic materials are discussed in the first six chapters of this volume. Other isotopes such as zinc-67 and gadolinium-155 have recently been used to study the electronic properties of zinc compounds and the electronic and magnetic properties of rare-earth transition metal compounds. The latter compounds include high-temperature superconductors and hard permanent magnetic materials, two groups of materials that have potentially important practical applications.

The mechanical and magnetic properties of nanostructured materials are of current interest to engineers, materials scientists, and theoretical physicists. The use of Mössbauer-effect spectroscopy to investigate the microscopic electronic and magnetic properties of these nanostructured materials is discussed in Chapter 7.

As we finish this volume, we would like to thank all the authors who have worked diligently through a tight schedule to include in their manuscripts the most up-to-date literature references and to present the scientific results in a comprehensive, readable fashion. We hope that this book will be useful to both the Mössbauer-effect and the materials science research communities.

Finally, we would like to thank Amelia McNamara and Christopher Curioli of Plenum Press for their editorial and technical assistance during the preparation of this volume.

Gary J. Long
Fernande Grandjean

Rolla, Missouri

Contents

6. Mössbauer Spectroscopy Studies of Slow Paramagnetic Relaxation and Magnetic Interactions in Ferrocenium-Based Molecular Magnetic Materials

William M. Reiff

8. Zinc-67 Mössbauer Spectroscopy

W. Potzel

9. Mössbauer Spectroscopy of New Materials Containing Gadolinium

Gordon Czjzek

1

Chemical Systematics of Iron-57 Hyperfine Magnetic Field Distributions in Iron Alloys

Brent Fultz

1. INTRODUCTION

There have been several excellent reviews of applications of iron-57 Mössbauer spectrometry in physical metallurgy and metals physics.[1-9] These reviews typically demonstrate applications of the four measurable quantities in Mössbauer spectroscopy: recoil-free fraction, isomer shift, quadrupole effect, and hyperfine magnetic field (HMF). The present chapter has a more focused goal. It shows how local, short-range chemical information can be obtained from the HMF distribution of iron-57 nuclei in ferromagnetic bcc alloys.

At first thought, this goal may seem excessively specialized. While the recoil-free fraction and isomer shift can be measured for most materials containing a Mössbauer isotope, significant HMFs exist only for magnetic materials. The methods described here are yet further specialized to distributions of HMFs in ferromagnetic bcc iron alloys. This limits significantly the materials suitable for study. Nevertheless, the reward for studying these materials is the richness of local chemical detail that can be resolved in their Mössbauer spectra; the observable level of atomic detail is much greater than for isomer shifts or quadrupole splittings. In favorable alloys, the HMF

Brent Fultz • Division of Engineering and Applied Science, California Institute of Technology, Pasadena, California 91125.

Mössbauer Spectroscopy Applied to Magnetism and Materials Science, Volume 1, edited by Gary J. Long and Fernande Grandjean. Plenum Press, New York, 1993.

distribution provides not just the average solute occupation of the first-nearest-neighbor (1nn) shell of iron-57 atoms. It can also provide the fractions of iron-57 atoms with each different solute occupancy. This level of detail can, at least for polycrystalline materials, surpass the capabilities of modern X-ray and neutron scattering experiments.

There are several challenges in using HMF distributions for determining local chemical environments. There is the practical problem of extracting reliable HMF distributions from the Mössbauer spectra. Several techniques have been developed for this task,[10–16] and a software package based on one of these methods is almost a necessity. These methods will not be discussed here, however. This chapter is devoted to understanding how the shapes of HMF distributions originate with the solute environments around the iron-57 atoms.

This chapter begins by describing an old but popular approach to modeling the HMF distribution in an alloy. Each solute atom in each nearest-neighbor shell of the iron-57 atom is assumed to cause a characteristic HMF perturbation. The HMF is assumed proportional to the number of solute atoms in each nearest-neighbor shell, so this model is termed the "additive perturbation model." The chemical systematics of these solute-induced HMF perturbations are then examined. These systematics can be explained well with a more detailed model of how the conduction and core electrons respond to the magnetic moments in the alloy. We call this more detailed model the "magnetic polarization model." When the solute is a simple magnetic hole in a bcc iron crystal, the magnetic polarization model reduces to the additive perturbation model. For those solutes that induce changes in the magnetic moments at the host iron atoms, however, the models become inequivalent, and the additive perturbation model becomes unreliable. This chapter ends with a discussion of how different electronic phenomena are included in the magnetic polarization model, and suggestions are made for improving the accuracy and realism of the model.

This chapter presents only three examples from metals research of how HMF distributions have been used to obtain local chemical information. The examples are not intended as a review, but to indicate the nature of the information that can be obtained. Many more examples exist in the literature. I apologize to both the reader and my colleagues for neglecting so much excellent work.

2. THE ADDITIVE PERTURBATION MODEL

A useful, phenomenological model of how solute atoms perturb the HMF at iron-57 atoms, here termed the "additive perturbation model," has a long history. As early as 1961, Flinn and Ruby[17] observed satellite sextets in Mössbauer spectra of Fe–Al alloys, and they correctly attributed these satellites to iron-57 nuclei that have aluminum atoms as nearest neighbors. In a widely cited paper, Wertheim *et al.*[18] reported results from their study of many dilute Fe–X alloys. They showed how their Mössbauer spectra could be explained with a simple systematics where each nearest-neighbor solute atom caused a perturbation of the iron-57 HMF, and the combined effects of several solute atoms were approximately additive. Many solutes cause a significant perturbation of the iron-57 HMF when they are in either the first- or second-nearest-neighbor shell of an iron-57 atom. (Statistical analysis with the χ^2 test rarely supports the assignment of HMF perturbations to solutes beyond the second-nearest-neighbor shell, however.[44]) As a specific example, a first-nearest-neighbor vanadium atom causes a decrease in the magnitude of the iron-57 HMF of +30 kG. This HMF perturbation is denoted ΔH_1^{V}, and that for a second-nearest-neighbor vanadium solute atom, which is about +20 kG, is denoted ΔH_2^{V}.[19] The HMF perturbations of the more distant neighbors, which are usually much weaker, are included by considering only their average concentration dependence, denoted κ. The HMF at a specific iron-57 atom having n_1 first-nearest-neighbor solute atoms (of species X) and n_2 second-nearest-neighbor solute atoms in an alloy with a solute concentration of c will be:

$$H = H_0 + \Delta H \tag{1}$$

where

$$\Delta H = n_1\,\Delta H_1^{\mathrm{X}} + n_2\,\Delta H_2^{\mathrm{X}} + \kappa c \tag{2}$$

Here, H_0 is the HMF at an iron-57 nucleus in pure iron metal (−330 kG at 300 K, −339 kG at 4 K). To calculate the HMF distribution for an alloy, it is necessary to know the probabilities for the different n_1 and n_2. If the alloy is a random solid solution with no chemical order, the probability of finding an iron-57 atom with n_1 and n_2 solutes in its first- and second-nearest-neighbor shells is the product of the binomial probabilities:

$$P(n_1, n_2) = P(8, n_1, c)P(6, n_2, c) \tag{3}$$

where the individual binomial probabilities are:

$$P(N_{\mathrm{j}}, n_{\mathrm{j}}, c) = \frac{N_{\mathrm{j}}!}{(N_{\mathrm{j}} - n_{\mathrm{j}})!\,n_{\mathrm{j}}!}\,c^{n_{\mathrm{j}}}(1 - c)^{N_{\mathrm{j}}-n_{\mathrm{j}}} \tag{4}$$

The Mössbauer spectra of many dilute alloys show distinct individual sextets from the different first- and second-nearest-neighbor chemical environments. An example is presented in Fig. 1, where these sextets are seen to contribute clear satellite peaks to the outer peaks of the main sextet. Tests of the systematics of Eqs (2) and (3) have been either direct or indirect goals of much research with Mössbauer spectrometry and NMR spectrometry.[18–57] The most complete set of measurements of $\{\Delta H_{\mathrm{j}}^{X}\}$ parameters was performed in the doctoral work of I. Vincze, published with I. A. Campbell in 1973.[44] There are also reports of $\{\Delta H_{\mathrm{j}}^{X}\}$ parameters for solutes in bcc Fe–Ni[58–62] or Fe–Co[63,64] matrices, and in fcc[65–68] and hcp[69] alloys.

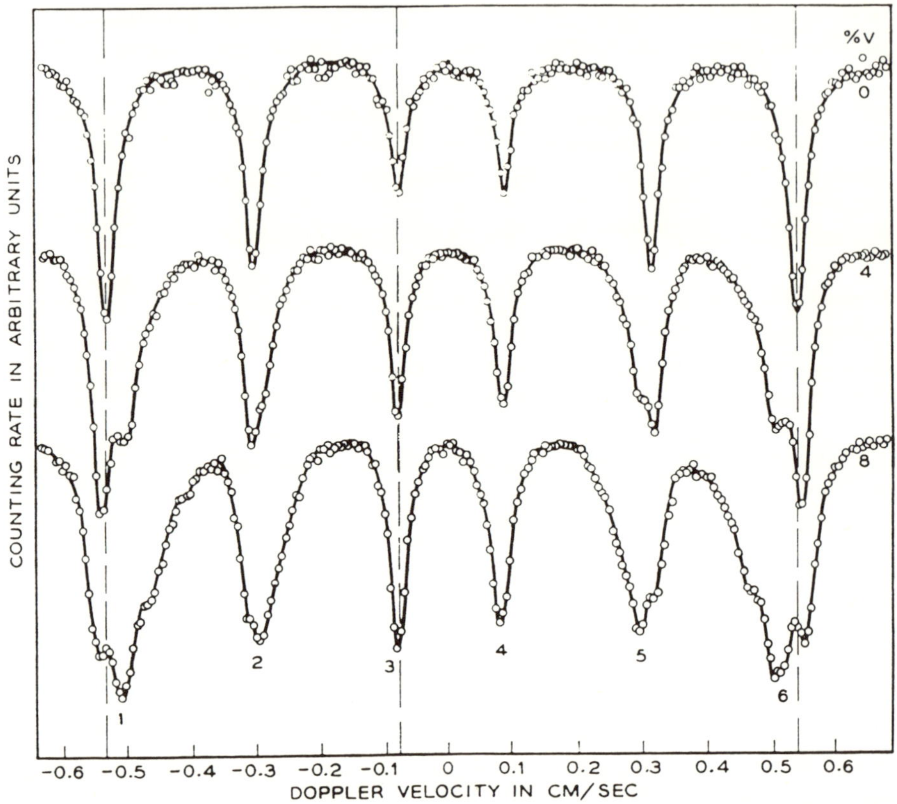

Figure 1. Mössbauer spectra from pure Fe (top), Fe–4% V (middle), and Fe–8% V (bottom), showing structure on the insides of the peaks of the sextet.[18] The structure originates with 1nn and 2nn V atoms, which cause similar HMF perturbations at iron-57 nuclei. [From G. K. Wertheim *et al., Phys. Res. Lett.* **12,** 24 (1964).]

Many alloys do not, however, show well-resolved satellite peaks in their Mössbauer spectra, and show only a shift and broadening of the main peaks in the sextet. This shift and broadening can be analyzed with Eqs. (2) and (3) by calculating the mean and variance of the HMF. For the mean, $\langle \Delta H \rangle$:

$$\langle \Delta H \rangle = \sum_{n_1=0}^{n_1=8} \sum_{n_2=0}^{n_2=6} P(8, n_1, c) P(6, n_2, c)[n_1 \, \Delta H_1^X + n_2 \, \Delta H_2^X + \kappa c] \quad (5)$$

$$\langle \Delta H \rangle = c[8 \, \Delta H_1^X + 6 \, \Delta H_2^X + \kappa] \quad (6)$$

and for the square of the variance, $\langle \Delta H^2 \rangle - \langle \Delta H \rangle^2$:

$$\langle \Delta H^2 \rangle - \langle \Delta H \rangle^2 = c(1-c)[8 \, \Delta H_1^{X^2} + 6 \, \Delta H_2^{X^2}] \quad (7)$$

The mean and the variance of the HMF distribution, Eqs. (6) and (7), allow the extraction of ΔH_1^X and ΔH_2^X from a Mössbauer spectrum that has no clear features in its spectral lines. If ΔH_1^X and ΔH_2^X are constants in a fixed

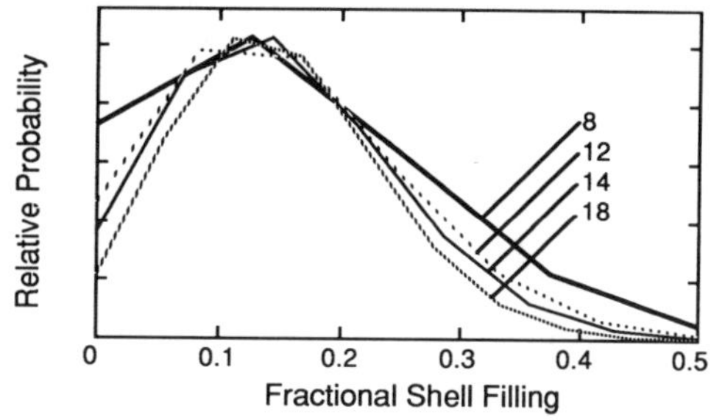

Figure 2. Binomial probability distributions, $P(N, n, c)$, for n successes in N trials, where the probability of success $c = 0.15$. $N =$ number of sites in shell; $n =$ number of solutes; $c =$ solute concentration; fractional shell occupancy $= n/N$. Curves are labeled by N.

ratio, Eqs. (6) and (7) show that both the center and the width of the HMF distribution will increase in proportion to ΔH_1^X. It may then be possible to use the different proportionality constants for the mean and the width for determining approximately the number of nearest neighbors that perturb the iron-57 HMF. This was recently attempted for fcc Fe–Ni alloys,[68] where the concentration dependence of the mean HMF was used to extract a ΔH_1^X. The width of the iron-57 HMF distribution was then obtained satisfactorily by assuming that only the 12 first-nearest-neighbor atoms in the fcc structure caused significant perturbations of the iron-57 HMF.

This approach is only a modestly sensitive way to obtain the number of effective nearest-neighbor shells, however. The same mean, $\langle \Delta H \rangle$, can be predicted with either many sites having small ΔH_j^X, or few sites having large ΔH_j^X. The width of the HMF distribution will, however, decrease as the inverse square root of the number of effective nearest-neighbor sites. This is illustrated in Fig. 2, which presents typical cases for bcc and fcc structures. Graphed in Fig. 2 are binomial probabilities. With the additive perturbation model, the HMF distribution will be the same shape, and related by a simple scaling of the horizontal axis. The sharpness of the distribution is noticeably increased when more nearest-neighbor sites are effective, but such changes are difficult to resolve experimentally.

3. APPLICATIONS OF THE ADDITIVE PERTURBATION MODEL

For nearly 20 years, the additive perturbation model of Eqs. (2) and (3) has been used to obtain microscopic chemical information from ferromagnetic iron alloys. The cumulative body of work is too large to review here, but the reader is directed to several reviews (Refs. 1–8, especially 3). It is not hard to use the additive perturbation model [Eqs. (2) and (3)] for routine chemical analyses of ferromagnetic bcc iron alloys, but the HMF distribution can be used to obtain more information than just the average chemical composition. Below are two examples where the additive perturbation model was able to provide the full distribution of local chemical environments, and this information was essential for understanding two types of phase transformations in physical metallurgy.

3.1. Spinodal Decomposition of Fe–Cr–Co Alloys

When an alloy is unstable against infinitesimally small fluctuations in composition, chemical segregation may occur by spinodal decomposition. A distinct nucleus of new phase does not form, but the composition develops heterogeneities that intensify with time. Most studies of spinodal decomposition use diffraction methods, from which it is possible to study the amplitude and periodicity of the chemical heterogeneities. Real space imaging techniques, such as transmission electron microscopy, are also used. When Mössbauer spectrometry can measure the distribution of local chemical environments, it provides a different, but equally useful viewpoint for studying spinodal decomposition.

Mössbauer spectrometry has been used in several studies of spinodal decomposition in Fe–Cr alloys.[70–75] The work by Chandra and Schwartz[3,71] was the first to use the additive perturbation model, but our example is taken from a later study. The alloy Fe–33% Cr–21% Co, sometimes known as "chromindur," was developed for permanent magnet applications.[76–79] Much as for binary Fe–Cr alloys, a solid solution of 33% Cr in a bcc Fe–Co matrix is unstable. In equilibrium we expect most of the chromium to segregate to a σ-phase, but the formation of the σ-phase is sluggish. The alloy instead undergoes the process of spinodal decomposition as the chromium segregates to chromium-rich regions.

Figure 3 shows a sequence of Mössbauer spectra taken from an alloy of Fe–33% Cr–21% Co studied by Tahara *et al.*[64] This sequence represents snapshots of the alloy as it undergoes the early stages of spinodal decomposition. As will be described in Section 4, cobalt atoms cause only a small change in the iron-57 HMF, so most of the structure in the HMF distribution is caused by chromium atoms. The broad sextet of peaks at the top of Fig. 3 was best fit with the additive perturbation model of Eqs (2) and (3) by using $\Delta H_1 = +20\,\text{kG}$ and $\Delta H_2 = +41\,\text{kG}$. This is a dubious ratio for first- and second-nearest-neighbor HMF perturbations, but the average of +29 kG for all 14 atoms in the first and second neighbor shells is reasonable.

Once spinodal decomposition begins, the alloy is no longer a homogeneous random solid solution, so the probability distribution of Eq. (3) is no longer applicable. A new probability distribution for the local chromium environments must be calculated. Mössbauer spectrometry is certainly sensitive to the variance of the composition distribution, but less so to the precise shape of the distribution. (Kuwano *et al.*[75] report some sensitivity to the shape, however.) The HMF distribution of the disordered alloy is approximately proportional to a binomial probability distribution of local chromium environ-

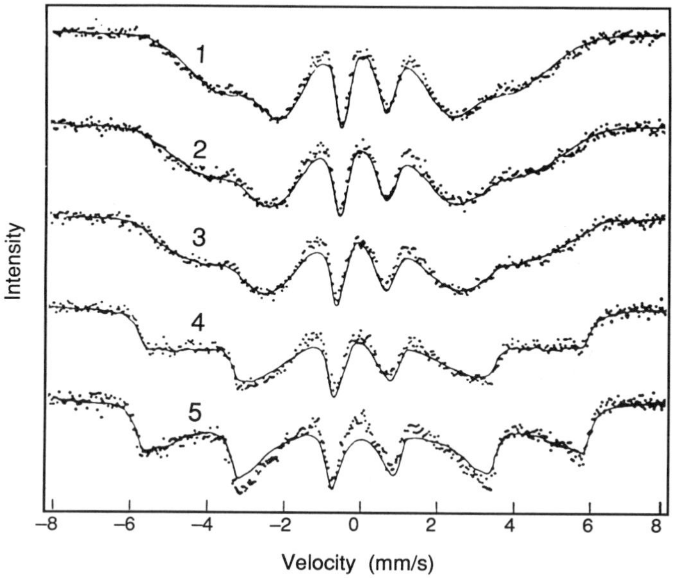

Figure 3. Mössbauer spectra from an alloy of Fe–33% Cr–21.5% Co, quenched and annealed at 560°C for various times: (1) as-quenched, (2) 6 s, (3) 10 s, (4) 30 s, (5) 200 s.[64] (Illustration courtesy of Y. Nakamura.)

ments, $P(14, n, 0.33)$. This distribution gradually changes to a broad distribution that includes many iron-57 atoms in chromium-depleted regions, and fewer iron-57 atoms in chromium-rich regions. (The sum of sine functions used by Tahara *et al.*[64] is certainly an oversimplification, but it does contain the essential behavior.) The HMF distribution evolves continuously toward a stronger sharp peak characteristic of bcc Fe–Co, superimposed on a weaker, wide distribution characteristic of an iron alloy with higher chromium concentration. This is seen in the shapes of the first and sixth peaks of Fig. 3, located near Doppler velocities of −5.5 and +5.5 mm/s. With annealing (progression from spectrum 1 to spectrum 5), and further decomposition of the alloy, these peaks develop a sharp edge at higher Doppler velocities, and a wide tail on their lower Doppler velocity sides.

Mössbauer spectrometry is not so sensitive to the wavelength of the compositional fluctuations. This is not surprising because composition fluctuations in spinodal decomposition are on the order of 100 Å, very large compared with the range of HMF perturbations. It is perhaps more surprising that the HMF distribution of an alloy of Fe–33% Cr–21% Co can be interpreted without consideration of the cobalt. Understanding the different iron-57 HMF perturbations caused by chromium atoms and cobalt atoms is the subject of Section 4.

3.2. Short-Range Order in Fe–Al Alloys

When atoms have an energetic preference for neighbors of unlike type, an alloy may develop an ordered structure at low temperature. Mössbauer spectrometry can be useful for monitoring the evolution of local chemical environments in ordering alloys, and the additive perturbation model of Eq. (2) can be useful for analysis of the data. Again, as for the case with spinodal decomposition, the distribution of solute atoms undergoes strong changes during ordering. Once ordering has begun, the probability distribution of Eq. (3) is no longer applicable. In favorable cases, the structure of the HMF distribution permits measurement of the full near-neighbor probability distribution. Such information is valuable for modern studies of the kinetics of ordering.

The clarity of structure in the iron-57 HMF distribution of Fe–Al alloys has encouraged many investigations of chemical order in Fe–Al, especially near the Fe_3Al composition.[52,77–98] The low-temperature equilibrium phase of Fe_3Al has the $D0_3$ ordered structure, depicted in Fig. 4. There are two iron sites in this ordered structure; the majority iron site has four first-nearest-neighbor aluminum atoms, and the minority iron site has zero first-nearest-neighbor aluminum sites. It is therefore not surprising that the Mössbauer spectrum from a fully ordered $D0_3$ structure has two sextets of peaks: a strong sextet with an HMF of −220 kG (majority site), and a weaker sextet with an HMF of −310 kG (minority site). Starting with a disordered solid solution of Fe–25% Al, a low-temperature annealing was performed to induce ordering. Some HMF distributions extracted from Mössbauer spectra at intermediate times of annealing are presented in Fig. 5, and the early growth of these two peaks in the HMF distribution is clearly visible. It is interesting that the position of the peak from the minority site is essentially unchanged during

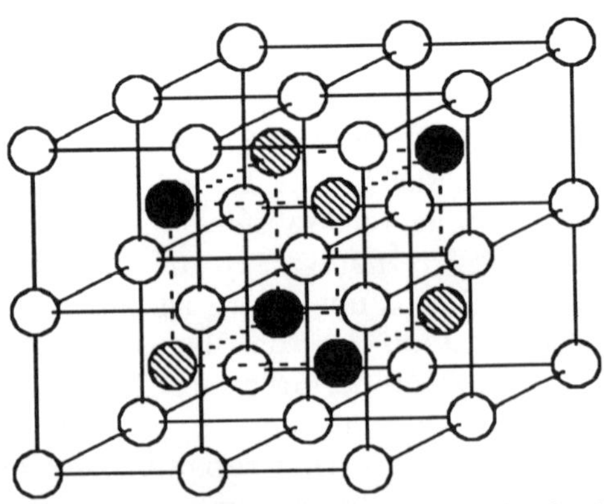

Figure 4. The $D0_3$ ordered structure. Al atoms are black, minority Fe atoms are gray, and majority Fe atoms are white.

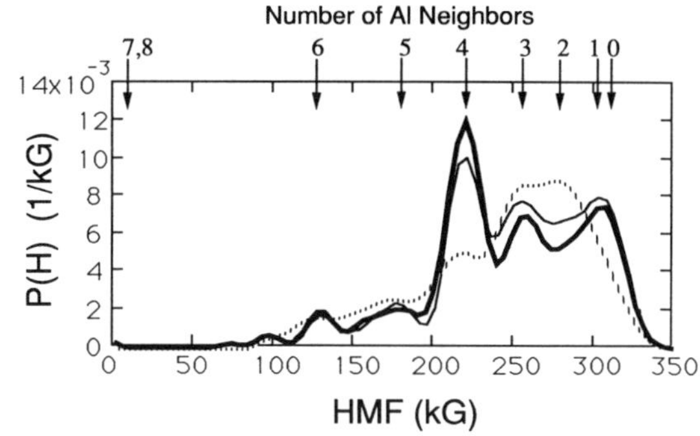

Figure 5. Experimental HMF distributions from Fe–25% Al alloy.[96,102] Dashed curve; piston–anvil quenched, disordered solid solution. Thin solid curve: after 6 h of annealing at 185°C. Thick solid curve: after 47 h of annealing at 300°C.

ordering. This is inconsistent with strong second-nearest-neighbor HMF perturbations in the alloy, because the number of second-nearest-neighbor aluminum atoms for the minority site changes from 1.5 to 6 during ordering. With a $\Delta H_2^{\mathrm{Al}} = +7.8\,\mathrm{kG}$, obtained from dilute Fe–Al alloys, this change in second-nearest-neighbor occupancy would cause a substantial change of HMF of about $+35\,\mathrm{kG}$ during ordering. Evidently the second-nearest-neighbor interactions become weaker with aluminum concentration, or they become weaker during ordering.[98]

The peaks in the experimental HMF distributions in Fig. 5 are labeled at the top with the number of aluminum atoms in the first-nearest-neighbor shell of the iron-57 atom. Note the clarity with these local environments are resolved. This resolution provides much information about the short-range order in the alloy, and how it changes during a disorder → $D0_3$ order transformation. In particular, it is possible to ask how the different local environments of the $D0_3$ structure evolve individually as ordering occurs. The differences in the two HMF distributions for the partially ordered alloys (Fig. 5) indicate that as ordering occurs at lower temperatures, the minority site for iron atoms grows relatively faster than the majority site. This is consistent with X-ray observations that $B2$ order tends to evolve more quickly than $D0_3$ order at lower temperatures.[99] This result is not expected from the thermodynamics of Fe–Al alloys, and probably requires an explanation based on kinetics.[100–102]

4. THE MAGNETIC POLARIZATION MODEL

The additive perturbation model of iron-57 HMF perturbations, described in Section 2, can be a workhorse technique in physical metallurgy, as shown in Section 3. This model uses the parameters ΔH_1^X, ΔH_2^X, and κ, but makes no attempt to justify their chemical origin. A deeper understanding of the origin of the HMF perturbations, presented here, is of fundamental interest. It is also helpful for identifying the limitations of the additive perturbation model. A

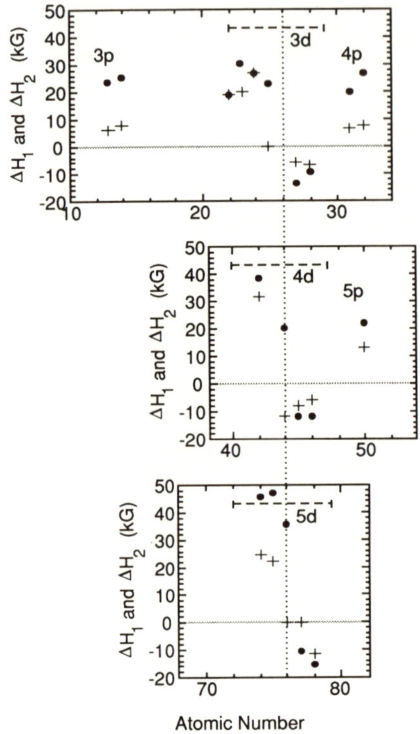

Figure 6. The parameters ΔH_1^X (dots) and ΔH_2^X (crosses), obtained from dilute bcc Fe–X alloys, as a function of the atomic number of the solute. Vertical line denotes iron column in the periodic table.

compilation of some reliable experimental measurements of ΔH_1^X and ΔH_2^X[44,47,55–57] is presented in Fig. 6. The trends to observe are:

1. Most solutes cause a positive ΔH_1^X and ΔH_2^X, with $\Delta H_1^X > \Delta H_2^X$.
2. Small, negative ΔH_1^X and ΔH_2^X occur for the late transition metals, located to the right of iron in the periodic table.
3. The positive values of ΔH_1^X and ΔH_2^X tend to be larger for solutes with larger atomic volumes.

It may seem desirable to obtain the iron-57 HMF distribution from *ab initio* electronic structure calculations. Not many such calculations have been performed, however, and accurate *ab initio* electronic structure calculations for magnetic alloys are certainly state-of-the-art.[103–106] Electronic structure calculations have been important for identifying the individual mechanisms of iron-57 HMF perturbations, however, and some of this work is described in Section 8.

There has been a long effort to use chemical systematics to test the candidate mechanisms for HMF perturbations. Such comparisons have tended to emphasize the HMFs at the solute atoms themselves. Nevertheless, early workers developed a picture of how the iron-57 HMF could originate with mechanisms of core electron polarization and conduction electron polarization.[107,108] The model presented here was developed through work by Stearns[38,46,51] and Vincze and Campbell[44,109] during the late 1960s and 1970s. Further development was performed by Fultz and Morris for Fe–Ni alloys[59,110] and Hamdeh, Fultz, and Pearson[111–113] for Fe–Co alloys. In essence, conduction electrons and core electrons are spin-polarized by exchange interactions with the unpaired $3d$ electrons in magnetic iron alloys. The conduction and core electrons with s-like character have some probability of being at the iron-57 nucleus, and thus affect the iron-57 HMF through the Fermi contact interaction. Specifics of these interactions are deferred to Section 8. Here we require only the fact that the conduction and core electron polarization mechanisms depend on the unpaired electron densities at atomic sites, and therefore reflect the atomic magnetic moments in the alloy.

The HMF at the iron-57 nucleus is obtained as the sum of two components. The first component is H_L, and is caused by the unpaired $3d$ electrons local to the iron-57 atom. These unpaired $3d$ electrons polarize, through exchange interactions, all s-like electrons in their vicinity. This includes the core polarization (CP) of the $1s$, $2s$, relativistic $2p$, and $3s$ electrons of the iron-57 atom, plus the conduction electron polarization (CEP) of the $4s$ electrons. Changes in H_L are proportional to changes in the local magnetic moment at the iron-57 atom, $\Delta\mu(0)$, induced by neighboring solute atoms:

$$\Delta H_L = (\alpha_{CP} + \alpha_{CEP})\,\Delta\mu(0) \tag{8}$$

$$\Delta H_L = -86\left(\frac{\text{kG}}{\mu_B}\right)\Delta\mu(0) \tag{9}$$

where α_{CP} and α_{CEP} are constants of proportionality between ΔH_L and $\Delta\mu(0)$, caused by the core polarization and conduction electron polarization mechanisms, respectively. We reference all HMF perturbations to the HMF of pure iron metal, and all changes in the individual components of the HMF perturbations (e.g., ΔH_L and ΔH_{NL}) to the strengths of these components in pure iron.

Each solute neighbor out to the fifth or sixth nearest-neighbor shell can alter the magnetic moment at the iron-57 atom.[114–121] To accommodate environments with multiple solute neighbors, we assume that the changes in $\Delta\mu(0)$ from solute neighbors are additive. (Nonlinearities with multiple solute atoms are discussed in Section 6.)

$$\Delta H_L = -86\left(\frac{\text{kG}}{\mu_B}\right)\sum_{\mathbf{r}\neq 0}\delta(\mathbf{r})\,g_X^{\text{Fe}}(r) \tag{10}$$

The sum is performed over the vectors, $\mathbf{r}$, connecting the iron-57 atom to all of its neighboring lattice sites. The Kroneker delta function equals 1 if the site is occupied by a solute atom, and equals 0 if it is occupied by an iron atom. The variable $g_X^Y(r)$ is the change in magnetic moment of a Y atom when it has an X solute in its coordination sphere of radius $|\mathbf{r}|$. (We assume that the magnetic disturbances are the same for each site in a nearest-neighbor shell.) All perturbations of iron magnetic moments are referenced to μ_{Fe}^0, the magnetic moment of an iron atom in pure iron.

The second component is a transferred HMF, H_{NL}. It arises from spin polarizations of nonlocal 4s electrons at the iron-57 nucleus, caused by exchange interactions with spins at neighboring atoms. For a solute atom at $\mathbf{r}$ having a magnetic moment, $\mu(\mathbf{r})$, differing from μ_{Fe}^0, the change in H_{NL} is

$$\Delta H_{\text{NL}} = \alpha_{\text{CEP}} \sum_{\mathbf{r}(\neq 0)} f(r)[\mu(\mathbf{r}) - \mu_{\text{Fe}}^0] \tag{11}$$

where $f(r)$ is the fraction of conduction electron polarization at the iron-57 nucleus produced by a change in magnetic moment at r, with respect to the conduction electron polarization produced by the same change in magnetic moment at $r = 0$.

We separate ΔH_{NL} into two terms: ΔH_{DNL} and ΔH_{INL}. The direct nonlocal term, ΔH_{DNL}, comprises the contributions from those lattice sites occupied by solute atoms. The indirect nonlocal term, ΔH_{INL}, comprises the contributions from those lattice sites occupied by iron atoms, but whose magnetic moments are perturbed by nearby solute atoms:

$$\Delta H_{\text{NL}} = \Delta H_{\text{DNL}} + \Delta H_{\text{INL}} \tag{12}$$

$$\Delta H_{\text{DNL}} = \alpha_{\text{CEP}} \sum_{\mathbf{r}(\neq 0)} \delta(\mathbf{r}) f(r) \left[\left(\mu_X^0 + \sum_{\mathbf{r}' \neq \mathbf{r}} \delta(\mathbf{r}') g_X^X(|\mathbf{r}' - \mathbf{r}|) \right) - \mu_{\text{Fe}}^0 \right] \tag{13}$$

$$\Delta H_{\text{INL}} = \alpha_{\text{CEP}} \sum_{\mathbf{r}(\neq 0)} (1 - \delta(\mathbf{r})) f(r) \sum_{\mathbf{r}' \neq \mathbf{r}} \delta(\mathbf{r}') g_X^{\text{Fe}}(|\mathbf{r}' - \mathbf{r}|) \tag{14}$$

All perturbations of solute magnetic moments are referenced to the magnetic moment of an isolated solute atom in an iron matrix, μ_X^0.

Using the same $\alpha_{\text{CEP}} f(r)$ parameters in both Eqs. (13) and (14) implicitly assumes the same amount of conduction electron polarization per magnetic moment for either an iron atom or a solute atom. This assumption may be incorrect if unpaired electrons at the solute atom have a significantly different form factor from those of iron. The assumption was found to be nearly true for Fe–Ni and Fe–Co alloys[59,111] and is a moot point for nonmagnetic solute atoms. The set of $\{\alpha_{\text{CEP}} f(r)\}$ parameters can be obtained from experimental measurements with solutes that act as magnetic holes in an iron lattice, and do not disturb the magnetic moments of host iron atoms. Neutron diffuse magnetic scattering[117,118] and other Mössbauer spectrometry work[38,51,55] show that 3p and 4p solutes are suitable magnetic holes. Figure 6 shows that these solutes cause approximately the same HMF perturbations, as do many other nonmagnetic solutes. The HMF perturbations from aluminum and silicon solutes were also measured when these solutes were dilute impurities in matrices of Fe–Ni and Fe–Co.[59,111] This was done by comparing experimental Mössbauer spectra from Fe–Co–Al and Fe–Co–Si alloys to spectra from Fe–Co, and similarly for Fe–Ni–Si and Fe–Ni.

"Calibration" experiments with magnetic holes suggest a set $\{\alpha_{\text{CEP}}f(1), \alpha_{\text{CEP}}f(2), \alpha_{\text{CEP}}f(3)\}$ of: $\{-11.5, -3.5, +2.5\}\text{kG}/\mu_{\text{B}}$, as given by Fultz and Morris.[59] This set predicts the observed shapes of spin echo NMR spectra of Fe–Al by Budnick and Skalski,[122] Fe–Si by Dean *et al.*, and continuous wave NMR spectra of Fe–Sn by Mendis and Anderson,[26] although these authors[26,36,122] offer other peak assignments. Another reasonable set of $\{\alpha_{\text{CEP}}f(r)\}$ parameters is: $\{-12.1, -2.7, +2.4\}\text{kG}/\mu_{\text{B}}$.[46] Usually these two sets do not predict substantially different HMF distributions.

Neutron diffuse magnetic scattering provides information on the solute magnetic moments in bcc iron, and it also provides information on the $\{g_X^{\text{Fe}}(r)\}$ parameters. In general, solutes to the left of iron in the periodic table cause only small perturbations of the magnetic moments at host iron atoms, but transition metal solutes to the right cause large perturbations.[118] These host moment disturbances extend to the 5nn shell of the solute atom or farther, but have generally similar shapes. For nickel solutes a consistent set of parameters is: $\{g_{\text{Ni}}^{\text{Fe}}(r)\} = \{0.070, 0.056, 0.025, 0.018, 0.015\}\mu_{\text{B}}$[115,116,119] and another set is: $\{g_{\text{Ni}}^{\text{Fe}}(r)\} = \{0.093, 0.062, 0.022, 0.013, 0.012\}\mu_{\text{B}}$.[46] For other transition element solutes to the right of iron in the periodic table, it is probably adequate to use the same r-dependence of these parameters. Their magnitude, however, must be scaled so that the magnetizations of dilute alloys are consistent with experimental data. For cobalt and nickel solutes, the $\{g_X^{\text{Fe}}(r)\}$ parameters are so large that in calculations for nondilute alloys, the iron magnetic moment will rise to unrealistically high values. The calculated magnetization of the alloy, obtained by summing the magnetic moments of all atoms, then rises above the experimental magnetization data. To match the experimental magnetization data, the iron magnetic moment must be allowed to saturate at some reasonable value, such as $3.0\mu_{\text{B}}$.[114] A variety of procedures to saturate the calculated iron moments are possible, but so long as they reproduce the magnetization data for nondilute alloys, they predict similar HMF distributions.[59,111]

Unlike iron atoms, the $3d$ transition metal solutes cobalt, nickel, and copper (located to the right of iron in the periodic table) all have filled $3d\uparrow$ states. The magnetic moments at these solute atoms are consequently fairly insensitive to their local environments. Increasing the magnetic moment at a nickel solute, for example, would require a loss of nickel $3d\downarrow$ electrons. A change in magnetic moment at the nickel atom therefore requires a substantial charge transfer or s–d conversion. It is therefore not surprising that the magnetic moments at cobalt[114,123,124] and nickel[114] solutes are rather constant in bcc iron. We use the sets $\{g_{\text{Co}}^{\text{Co}}(r)\} = \{0\}$, and $\{g_{\text{Ni}}^{\text{Ni}}(r)\} = \{0\}$.

5. RELATIONSHIP BETWEEN THE MAGNETIC POLARIZATION MODEL AND THE ADDITIVE PERTURBATION MODEL

Some solute atoms, when substituted for iron atoms on a bcc lattice, possess no magnetic moment themselves, and furthermore do not disturb the magnetic moments at host iron atoms. Solutes with such ideal magnetic properties are called "magnetic holes." When iron is alloyed with solutes behaving as magnetic holes, the bulk magnetization of the Fe–X alloy follows a law of simple dilution (i.e., an alloy of Fe–10% X has a bulk magnetization that is 0.9 times that of pure iron). The $3p$, $4p$ (and possibly $5p$)-series atoms tend to act as magnetic holes in bcc iron, at least for dilute alloys.

For ideal magnetic holes, the magnetic polarization model reduces exactly to the additive perturbation model, as is now shown. For less ideal magnetic holes, the equivalence between the models becomes less perfect, but may still be reasonable. For solutes that do not behave as magnetic holes, such as the late transition metal elements, the additive perturbation model is useful only for dilute alloys. Beyond concentrations of about 10%, the additive perturbation model can no longer predict the HMF associated with a particular local atomic environment.

We first consider a reasonably good magnetic hole, an aluminum atom. Because the aluminum atom does not develop a magnetic moment:

$$\mu^0_{\mathrm{Al}} = 0 \tag{15}$$

and its zero magnetic moment cannot be perturbed by surrounding aluminum atoms:

$$\sum_{\mathbf{r}' \neq \mathbf{r}} \delta(\mathbf{r}') g^{\mathrm{Al}}_{\mathrm{Al}}(|\mathbf{r}' - \mathbf{r}|) = 0 \tag{16}$$

Since the magnetic moment at the iron-57 atom is not changed by neighboring aluminum solute atoms:

$$g^{\mathrm{Fe}}_{\mathrm{Al}}(|\mathbf{r}' - \mathbf{r}|) = 0 \tag{17}$$

so Eq. (10) becomes:

$$\Delta H_{\mathrm{L}} = 0 \tag{18}$$

Using Eqs. (15) and (16), we simplify Eqs. (13) and (14):

$$\Delta H_{\mathrm{INL}} = 0 \tag{19}$$

$$\Delta H = \Delta H_{\mathrm{DNL}} = -\alpha_{\mathrm{CEP}} \sum_{\mathbf{r}(\neq 0)} \delta(\mathbf{r}) f(r) \mu^0_{\mathrm{Fe}} \tag{20}$$

Rewriting Eq. (20) in terms of the first three nearest-neighbor shells, having N_j sites per shell:

$$\Delta H = -\alpha_{\mathrm{CEP}} \sum_{j=1}^{3} \sum_{n=0}^{Nj} \delta(r_{j,n}) f(r_j) \mu^0_{\mathrm{Fe}} \tag{21}$$

where each site is now specified by the index "j" denoting its nearest-neighbor shell about the iron-57 atom, and the index "n" for the particular site within the shell.

The precise assignments of small HMF perturbations to shells beyond the second-nearest-neighbor shell are difficult and controversial. An alternative, reasonable approach is to use conduction electron response parameters $\alpha_{\text{CEP}} f(1)$ and $\alpha_{\text{CEP}} f(2)$ for the first- and second-nearest-neighbor shells, and a "band polarization" parameter to account for the third-nearest-neighbor shell and more distant neighbors.[55] For a random alloy, we transform the third-nearest-neighbor parameter, $\alpha_{\text{CEP}} f(3)$ as:

$$-\alpha_{\text{CEP}} \sum_{n=0}^{N_3=12} \delta(\mathbf{r}_3) f(r_3) \mu_{\text{Fe}}^0$$

$$= -\alpha_{\text{CEP}} \sum_{n=0}^{N_3=12} \frac{12!}{(12-n)!\,n!} c^n (1-c)^{12-n} n f(r_3) \mu_{\text{Fe}}^0 \tag{22}$$

$$= -\alpha_{\text{CEP}} f(r_3) \mu_{\text{Fe}}^0 c = \kappa^{\text{hole}} c \tag{23}$$

If we make the definitions:

$$\kappa^{\text{hole}} \equiv -12\alpha_{\text{CEP}} f(r_3) \mu_{\text{Fe}}^0 = -0.672\ \text{kG}/(\%\ \text{solute}) \tag{24}$$

$$\Delta H_1^{\text{hole}} \equiv -\alpha_{\text{CEP}} f(r_1) \mu_{\text{Fe}}^0 = +25.3\ \text{kG} \tag{25}$$

$$\Delta H_2^{\text{hole}} \equiv -\alpha_{\text{CEP}} f(r_2) \mu_{\text{Fe}}^0 = +7.8\ \text{kG} \tag{26}$$

Equations (21) and (23) become equivalent to Eq. (2) of the additive perturbation model. Our value for κ^{hole} is about 10% larger than that obtained by Dubiel and Zinn[55] for dilute alloys, but about 45% smaller than that obtained by Stearns[33] for nondilute, ordered alloys.

When the solute is not a magnetic hole, the additive perturbation model is still useful in the dilute limit when there is a low probability of having two or more solute atoms near an iron-57 atom (concentrations typically less than 2%). Transition metal solutes to the right of iron in the periodic table shift the iron-57 HMF distribution to more negative HMFs, and broaden it. Analysis of the shift and broadening [as with Eqs. (6) and (7)] provide ΔH parameters that are small and negative, but it is difficult to assign ΔH_{j} parameters to specific nearest-neighbor shells. The difference between the negative ΔH parameters for nickel and the positive ΔH parameters for a magnetic hole can be understood with the magnetic polarization model. For a nickel solute atom, which has a large magnetic moment itself, the positive contribution of ΔH_{DNL} is not so large as in Eqs. (25) and (26). Instead, the dominant contributions to ΔH are ΔH_{INL} and ΔH_{L}. Because $g_{\text{Ni}}^{\text{Fe}}(r) > 0$, these contributions are negative, and the magnitude of the average HMF is larger in bcc Fe–Ni alloys than in pure iron. The magnetic polarization model predicts that a nickel atom does not cause the largest HMF perturbation when it is in a first-nearest-neighbor shell of an iron-57 atom. The nickel atoms should be somewhat more distant to have the largest effect on ΔH_{INL}, and these more distant nickel atoms can also have large effects on ΔH_L. The magnetic polarization model predicts that nickel atoms in the third-nearest-neighbor shell will cause the largest, negative ΔH. This agrees with results from high-resolution NMR experiments of dilute Fe–Ni, for which a distinct third-nearest-neighbor satellite was resolved.[29,125]

The next section describes the implementation of the magnetic polarization model by means of Monte–Carlo calculations. Here we present some results of these calculations to demonstrate a serious problem with the additive perturbation model. Calculated HMF subdistributions from iron-57 atoms with different numbers of nickel atoms in their first- and third-nearest-neighbor shells are shown in Fig. 7a and 7b. The alloy was a random solid solution of Fe–10% Ni, and the parameters for the calculations were those of the 4 K calculations by Fultz and Morris.[59] The calculated subdistributions were fit to Gaussian functions, and these results are displayed. The mean HMF does not scale linearly with the number of first-nearest-neighbor nickel atoms (Fig. 7a), as predicted by the additive perturbation model [Eq. (6)]. The second-nearest-neighbor subdistributions showed only a weak dependence on the number of nickel atoms in the second-nearest-neighbor shell. Better linearity is found for the HMF perturbations caused by third-nearest-neighbor nickel atoms, however (Fig. 7b).

Unfortunately for the additive perturbation model, it is not possible to obtain a ΔH_1 and ΔH_3 by inspection of Figs 7a and 7b, and then calculate the HMF distribution with additive contributions from the first- and third-nearest-neighbor shells. For example, the probable environment with zero first-nearest-neighbor nickel atoms and two third-nearest-neighbor nickel atoms would have an average HMF of -375 kG, which is unreasonably large. The fundamental reason why the additive perturbation model cannot work with nondilute alloys of solutes that are not good magnetic holes is the following. There are two centers of summation: (1) the iron-57 atom itself and (2) each neighboring iron atom. Around the first type of center is a linear sum of the conduction electron responses, the $\{f(r)\}$, plus a nonlinear sum of magnetic moment perturbations, the $\{g_X^{Fe}(r)\}$. This nonlinear sum is also performed

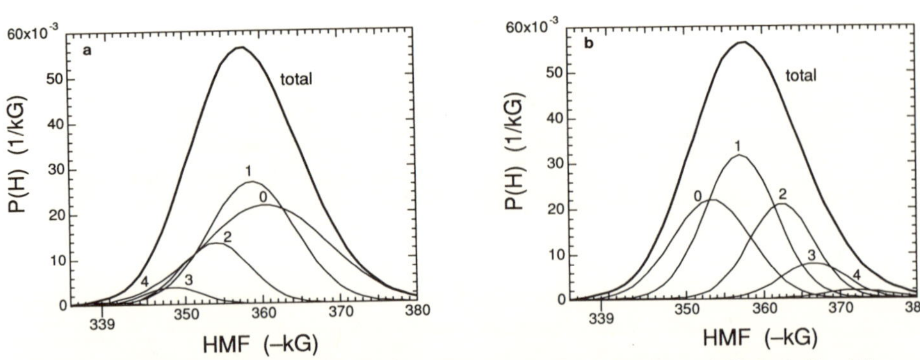

Figure 7. HMF subdistributions for different Ni occupancies of nearest-neighbor shells. Alloy was Fe–10% Ni with the parameters given in Ref. 59. (a) 1nn shell; (b) 3nn shell.

around the second center. In general, when two centers of summation are important, ΔH depends on three-body terms. Multibody effects become important even at low solute concentrations owing to the nonlinear compositional dependence of the iron magnetic moments. These effects are important when the $\{g_X^{Fe}(r)\}$ parameters are strong and long-ranged. Multibody terms are beyond the capability of the additive perturbation model, and it therefore gives unreliable predictions for nondilute alloys.

While the magnetic polarization model does treat multibody contributions to ΔH, their inclusion does not bode well for obtaining detailed information about local atomic arrangements around iron-57 atoms. There are too many multibody terms to identify from experimental HMF distributions, or from the dependence of the HMF on composition, for example. The problem is exacerbated when the HMF distributions do not have well-resolved features. In practice, detailed information about nearest-neighbor shell occupancies cannot be obtained reliably from Mössbauer spectra when iron is alloyed with transition elements to its right in the periodic table. These HMF distributions can still provide local chemical information on the length scale of the $\{g_X^{Fe}(r)\}$, however, which is around 7 Å.

6. IMPLEMENTATION OF THE MAGNETIC POLARIZATION MODEL

To obtain the iron-57 HMF distribution with the magnetic polarization model, it is necessary to know precisely the positions of all solute atoms. It is insufficient to know only the positions of the solute atoms with respect to the iron-57 atoms; it is also necessary to know the positions of the solute atoms with respect to each other. There are some cases where this knowledge is available, so the iron-57 HMF distribution can be obtained by analytical calculation. For perfectly ordered alloys, there are only a few sites for the atomic species. Calculations of the iron-57 HMFs in ordered alloys can be performed analytically for each iron-57 site by evaluating the sums in Eqs. (10), (13), and (14), using the sets of $\{\delta(\mathbf{r})\}$ appropriate for the ordered alloy. (For examples of these calculations, see Refs. 46, 51, 126.) For extremely dilute alloys it is also possible to perform analytical calculations to obtain the iron-57 HMF perturbations caused by solutes in specific nearest-neighbor shells. As the alloy becomes slightly more concentrated, it is possible to use an approximation technique that averages the multibody interactions.[60] The only mathematical trick required is the evaluation of moments of the binomial distribution.

For disordered or partially ordered alloys with solute concentrations greater than 10%, it is necessary to use a computer program to obtain HMF distributions. A flowchart for such a program is presented in Fig. 8. The program begins with a bcc crystal filled with the desired arrangement of atoms (e.g., random solid solution, partially ordered). The program then sorts

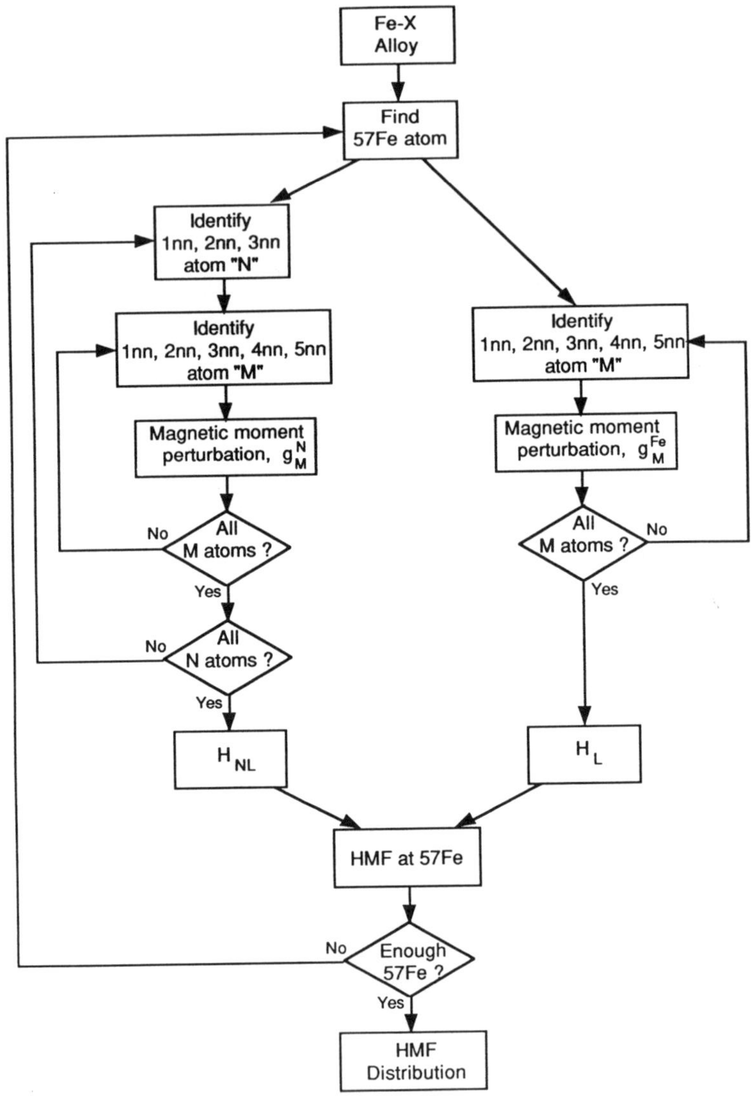

Figure 8. A flowchart for computer calculation of the iron-57 HMF distribution with the magnetic polarization model.

through the alloy, finds each iron atom, and assumes it to be an iron-57. The program next identifies the atom species in the first-, second-, and third-nearest-neighbor shells of each iron-57 atom. A scan around each first-, second-, and third-nearest-neighbor atom (denoted "N") is performed, and the effects of these surrounding solutes (denoted "M") on the magnetic moments of the atom "N" are evaluated. When the magnetic moment at each atom "N" in the first-, second-, and third-nearest-neighbor shell is determined, H_{NL} can be evaluated at the iron-57 atom. The program also determines the magnetic moment at the iron-57 atom in the same way, and evaluates H_L for that iron-57 atom. Once the total HMF of the chosen iron-57 is found, it is added to the HMF distribution. The process begins again with a different iron-57 atom.

Since the magnetic moments at most atoms in the alloy are calculated, the program can also determine the bulk magnetization of the alloy by performing a sum of individual atomic moments. As mentioned in Section 4, matching to experimental bulk magnetization data is useful for ensuring that the magnetic moments at the atoms are reasonable, and for ensuring that they have the correct dependence on solute concentration. Matching the concentration dependence in the dilute limit typically provides the parameters $\{g\}$. Assuming the alloy remains a random solid solution at higher concentrations, the dependence of the alloy magnetization, $\langle \mu \rangle$, on the solute concentration, c, is:

$$\langle \mu \rangle = c\mu_X + (1 - c)\mu_{\text{Fe}} \tag{27}$$

For alloys where only the iron magnetic moment is dependent on composition (i.e., $\{g_X^X(r)\} = 0$):

$$\langle \mu \rangle = c\mu_X + (1 - c)\left[\mu_{\text{Fe}}^0 + \sum_{\mathbf{r}\neq 0} \delta(\mathbf{r}) g_X^{\text{Fe}}(\mathbf{r})\right] \tag{28}$$

For dilute, random alloys, the distribution of the $\{\delta(\mathbf{r})\}$ for the different iron atoms will be a binomial distribution. We evaluate its first moment for the various shells:

$$\langle \mu \rangle = \mu_{\text{Fe}}^0 + c(\mu_X - \mu_{\text{Fe}}^0) + c(1 - c)G_X^{\text{Fe}} \tag{29}$$

where

$$G_X^{\text{Fe}} \equiv 8g_X^{\text{Fe}}(1) + 6g_X^{\text{Fe}}(2) + 12g_X^{\text{Fe}}(3) + 24g_X^{\text{Fe}}(4) + 8g_X^{\text{Fe}}(5) + 6g_X^{\text{Fe}}(6) + \cdots \tag{30}$$

When we know, for dilute alloys, both the composition dependence of the bulk magnetization, and the magnetic moment of the solute atom, we can obtain G_X^{Fe}. The relationship in Eq. (29) typically works only for dilute alloys, however. Deviations from this composition dependence are caused by the nonlinear compositional depenence of the iron magnetic moment. We have been able to successfully model the magnetization data at higher solute concentrations by using several different approaches to saturating the iron magnetic moments (abrupt, gradual, asymptotic, etc.). When different saturation procedures are used to match the bulk magnetization data, however, it is not obvious that these procedures will provide the same HMF distribution. Nevertheless, in cases where different methods were tried, the different methods that reproduced the experimental magnetization data also provided similar HMF distributions.[59,111]

A computer calculation of the HMF distribution can be used as a module in a larger software environment. This can be advantageous because the simulated alloy used for calculation of the HMF distribution can also be used in simulations of other experiments, such as X-ray diffractometry. Combinations of different types of experimental data generally allow for better determinations of the atomic configurations. The combination of Mössbauer spectrometry and X-ray diffractometry data, for example, can be used to obtain both the short- and long-range order of an alloy. The short- and long-range order must satisfy certain compatibility relationships, and this further defines the allowed alloy structures. The simulated alloys for the computer calculation can also be used to obtain quantities that are not measurable, but are of fundamental interest. The configurational entropy of the simulated alloy, for example, can be determined by using the entropy expressions of the cluster variation method.[127–129]

In research at the California Institute of Technology, we are trying to understand the kinetics of disorder → order transformations in iron alloys by using activated state rate theory of vacancy motion. Given a set of interatomic interactions from either first principles calculations or from ad hoc assumptions, a disordered alloy is allowed to develop order with a Monte–Carlo algorithm.[130–132] At various stages of ordering, the simulated alloy is analyzed for its HMF distribution and X-ray diffraction pattern. These results are compared with experiment to test the predictions of kinetic theory. It is difficult to know the relationship between the time scales of the simulations and the real experiments, or even if it is a linear relationship. However, when more than one microstructural variable is measured, there are often interesting features in the "kinetic path" through the space spanned by the different variables. A kink in the kinetic path through short- and long-range ordering was found in the disorder → $B2$ order transformation in FeCo.[113] This kink could be an example of a long-lived "pseudostable state" that occurs at a saddle point of a free energy function.[133,134]

7. APPLICATION OF THE MAGNETIC POLARIZATION MODEL TO Fe–Cu

Copper atoms do not act as magnetic holes in bcc iron, so Mössbauer spectrometry probably cannot provide detailed information about the copper occupancies of different nearest-neighbor shells of iron-57 atoms. The HMF distribution can nevertheless be used to measure chemical composition in regions within 7 Å around iron-57 atoms. This is an important range for chemical information in Fe–Cu alloys, which undergo clustering-type transformations. With the experimental techniques of rapid quenching, ion irradiation, laser irradiation, sputtering, and high-energy ball-milling, supersaturated bcc solid solutions of Fe–Cu have been prepared.[135–142] A bcc alloy of Fe–20% Cu, prepared by ball-milling,[142] was so far from the equilibrium solubility of copper in bcc iron that it was unclear if the alloy was really a disordered solid solution. Mössbauer spectrometry helped answer this question. The outer peaks of the sextet of Fe–Cu are presented in Fig. 9. Note the increasing shift, broadening, and skewness of these peaks with copper concentration.

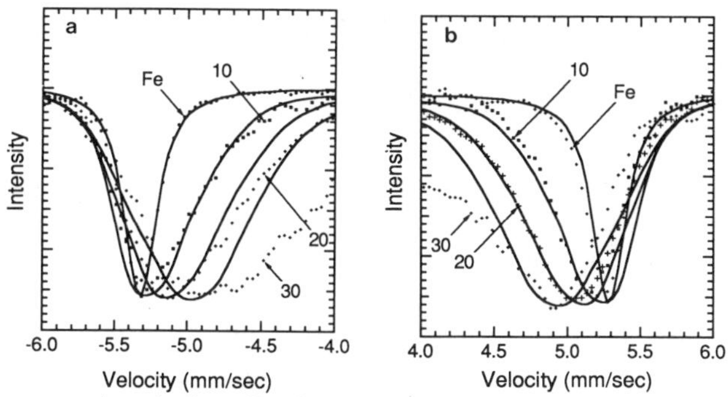

Figure 9. Mössbauer spectral peaks from Fe–Cu alloys of different compositions after high-energy ball-milling. Solid curves were calculated with the magnetic polarization model, as described in the text. (a) Low-velocity and (b) high-velocity peak of the sextet.

Calculations of the HMF distributions of Fe–Cu alloys were performed with the magnetic polarization model. Input parameters to the model were obtained from neutron diffuse magnetic scattering data.[143] The magnetic moments of the iron atoms were 2.22 μ_B and environment-dependent, the moments of the copper atoms were 0.3 μ_B and constant. The local perturbations of the magnetic moments at iron atoms near copper atoms, $\{g_{Cu}^{Fe}(r)\}$, were assumed to have the same spatial distribution as for nickel atoms,[115,116,119] but 15% of the strength. With these parameters, the calculated magnetic moment of the alloy decreased with copper concentration slower than predicted by simple dilution. Using the definition of Kneller,[144] $\beta \equiv (1/\mu_{Fe})(d\mu_s/dc)$, we calculated that $\beta = -0.75$ for alloys with compositions ranging from pure iron to Fe–20% Cu. This is in good agreement with magnetization data.[144,145] Using a pure iron spectrum for reference, the calculated HMF distributions were converted into a sextet of peaks and convolved with a Lorentzian function characteristic of the peaks from pure iron.

Calculated Mössbauer peaks are presented in Fig. 9a and 9b as solid curves. The calculated peak shapes agree well with the experimental peaks from Fe–10% Cu, Fe–15% Cu, and Fe–20% Cu alloys. The agreement is less good for the Fe–30% Cu alloy, primarily because the Fe–30% Cu spectrum contains a substantial contribution from a ferromagnetic fcc phase. Nevertheless, the spectrum from Fe–30% Cu shows that this material has considerably more than 20% Cu in its bcc phase. With empirical calibrations, it seems possible to use Mössbauer spectrometry to determine the chemical compositions of bcc Fe–Cu alloys to accuracies of a few percent.

The line shapes of the experimental spectra are consistent with the assumption of a homogeneous, random solid solution. The line shape is not very sensitive to compositional heterogeneities, however. For alloy of Fe–20% Cu, it is possible that compositional heterogeneities may be as large as ±10%.

8. ELECTRONIC ORIGIN OF HMF PERTURBATIONS

The magnetic polarization model does not present a full picture of the origin of the iron-57 HMF. There are different mechanisms responsible for the iron-57 HMF and its perturbation by solute neighbors, and the model lumps them all into two terms, H_L and H_{NL}. Here the real electronic mechanisms are discussed, and it is shown how they are included within the magnetic polarization model. The total HMF at an iron-57 nucleus is the sum of the following contributions, listed in the approximate order of decreasing importance:

$$H = H_{core} + H_{cond} + H_{ov} + H_{orb} + H_{mag} + H_{dip} \tag{31}$$

In what follows, each of these contributions is defined, and its underlying electronic mechanism is discussed.

The contributions H_{dip}, H_{mag}, and H_{orb} can be understood with classical magnetics and atomic physics. The magnetic moment at each atom near the iron-57 atom contributes an amount, H_{dip}, to the iron-57 HMF. In pure iron, this classical magnetic dipole contribution vanishes owing to cubic symmetry, but when an iron atom is replaced with a solute atom, H_{dip} will change. This contribution is weak, however, being on the order of 1 kG for nearest-neighbor distances. Such changes could be incorporated into the ΔH_{DNL} contribution to the HMF perturbation, but this requires an anisotropy of the function $f(r)$. The contribution H_{mag} is the classical magnetic field equal to $4\pi M/3$, were M is the lattice magnetization. The demagnetizing field from the return flux reduces H_{mag}. Typically, $H_{mag} < +7$ kG, and it may not change substantially with alloying. Atoms of iron retain some of their free atom character in bcc iron, and electron orbital magnetic moments remain for their $3d$ wave functions. A spin–orbit interaction, involving the nuclear spin, creates an H_{orb} of +50 to +70 kG.[146,147] If this contribution changes upon alloying, the change could be incorporated into H_L.

The other three contributions to the iron-57 HMF, H_{ov}, H_{cond}, and H_{core}, involve the Fermi contact interaction,[148,149] which is proportional to the unpaired electron spin density at the iron-57 nucleus. The Fermi contact interaction is the principal source of the iron-57 HMF and its chemical systematics. It is discussed in most references on Mössbauer spectroscopy, but the reader is also referred to some explicit calculations of the iron-57 HMF.[103–106,150–155] The Fermi contact interaction is dominated by the $1s$, $2s$, relativistic $2p$, $3s$, and $4s$ electrons, which have some finite probability of being at the iron-57 nucleus. Changes in the numbers of unpaired $1s$, $2s$, $2p$, and $3s$ electrons at the iron-57 nucleus are termed changes in the core polarization contribution to the HMF, H_{core}. They are typically induced by changing the number of unpaired $3d$ electrons at the iron-57 atom, which in turn exchange-polarize the $1s$, $2s$, $2p$, and $3d$ electrons.[150–152] Changes in the $4s$ electrons are termed changes in the conduction electron polarization contribution, H_{cond}. In the magnetic polarization model, the contributions H_{core} and H_{cond} are approximately associated with H_L and H_{NL}, respectively, but there is one difference. The 4s conduction electrons are polarized by the 3d electrons at the iron-57 atom itself and this effect is included in H_L [see Eq. (8)].

When occupied electron wave functions of the solute atom are brought next to occupied electron wave functions of the iron-57 atom, there are changes in the iron-57 HMF. The contribution H_{ov} includes the simple overlap of the tails of the solute atom wave function with the iron-57 nucleus, but an additional effect is involved.[156] It is necessary to normalize the iron-57 $3d\uparrow$ and $3d\downarrow$ wave functions to orthogonalize them to the solute atom wave functions. Because this orthogonalization is spin-dependent, the overlap of a nonmagnetic solute wave function with the iron-57 atom can change the iron-57 HMF. The H_{ov} contribution to the HMF need not scale precisely with the atomic size of the solute, but we do expect such a trend. The magnetic polarization model does not include this effect.

When the electrons from the solute atom spend some fraction of their time in unoccupied states at the iron-57 atom (or vice versa), we have covalent bonding. The covalent bonding is spin-dependent, so we expect a contribution, H_{cov}, to the iron-57 HMF.[156] If this covalent bonding involves the iron 3d electrons, we expect a change in H_{core}. The magnetic polarization model includes changes in H_{core} in Eq. (8). Changes in H_{core} are equal to $\alpha_{CP}\,\Delta\mu(0)$, and this is an important contribution to ΔH_L. When the covalent bonding involves electrons at a solute atom and the delocalized 4s electrons, it is often called "interband mixing," and its contribution to the iron-57 HMF is termed H_{hyb} (for hybridization).[51,157–163] Interband mixing occurs in the band structure of any crystal, such as pure iron, at k-vectors where two bands are nearly degenerate in energy. This mixing is altered locally by a solute atom, however, which is a local disturbance in charge and spin density. The solute atom causes both charge and spin redistributions of the 4s electrons. The redistributions of the 4s electrons perturb the iron-57 HMF, and this is included in H_{NL}.

In addition to H_{hyb}, due to chemical bonding, there are also spin polarization mechanisms of the Ruderman–Kittel–Kasuya–Yosida (RKKY) type.[157–167] These mechanisms produce spatial oscillations in electron spin. The period of these oscillations is inversely proportional to the Fermi wave vector, which cuts off abruptly the occupied conduction electron states. These disturbances in 4s electron spin density extend from neighboring solute atoms to the iron-57 atom, causing a change in the contribution to the iron-57 HMF that we term H_{rkky}. Incidentally, neither H_{hyb} nor H_{rkky} is necessarily spatially isotropic, although it has been argued that they are.[33]

Changes in the spin polarization of 4s conduction electrons are grouped together in H_{cond}, but there are several possible physical mechanisms involved,[42,51,157–173] and several ways to evaluate them.[174] It is impossible to review here the possible conduction electron polarization mechanisms and their controversies.[174–177] For a phenomenological understanding of H_{cond}, it is sufficient to consider a solute atom as a local disturbance in the charge density and in the spin density of the alloy. Mechanisms exist for the conduction electrons to carry a charge disturbance to the iron-57 nucleus (causing a local isomer shift), due to either a charge perturbation[168–173] or a spin perturbation[38,157–167] at the solute site. Mechanisms also exist for the conduction electrons to carry a spin density to the iron-57 nucleus (causing a local HMF perturbation), due to either a charge perturbation or a spin perturbation at the solute site. It was recently found that for the local isomer shifts, the chemical systematics scale with changes in the 4*s* density at the iron-57 atoms, not with indirect effects through the 3*d* electrons.[178] The observed trend is for the local isomer shifts to follow the electronegativity. This suggests that for local isomer shifts, the predominant feature of the solute atom is its number of valence electrons, not the spin polarization of its valence electrons. The systematics of the local HMF perturbations, presented in Fig. 6, are quite different. The local HMF perturbations do not follow the electronegativity, but follow instead magnetic trends, especially the trend of host magnetic moment perturbations around solute atoms.[115] The magnetic polarization model considers only contributions to H_{cond} that orginate with conduction electron spin polarizations due to spin perturbations at solute atoms.

9. BEYOND THE MAGNETIC POLARIZATION MODEL

The magnetic polarization model does not consider the size of the solute atom, so the model does not consider changes in the H_{ov} contribution to the iron-57 HMF. The importance of these solute size effects is suggested by Fig. 6. Notice that the larger 5*d* solute atoms have large values of ΔH_1 and ΔH_2. This could originate with H_{ov}. If so, an atomic volume correction term may be appropriate.

Another effect, which is less well understood, is the change in conduction electron response parameter, α_{CEP}, with alloy composition. In experiments with Fe–Si alloys of low[179] and high[33] concentrations, it was found that each additional silicon atom in the nearest-neighbour shell of an iron-57 atom creates a larger HMF perturbation by 15–30%. The nonlinear progression of the HMFs with number of aluminum neighbors seen in Fig. 5 suggests an effect somewhat greater than 20%. Corrections for this effect have not yet been attempted. If α_{CEP} depends on only the local conduction electron polarization, however, a simple correction would be to scale nonlinearly the H_{cond} contribution to the HMF by the amount seen in Fig. 5.

The magnetic polarization model assumes that the parameters $\alpha_{\text{CEP}}\{f(r)\}$ and $\{g_X^{\text{Fe}}(r)\}$ are spatially isotropic. This leads to two concerns. Conduction electron redistributions around solute atoms are probably not isotropic, but the symmetry of the crystal does not allow for any differences in the $f(r)$ or $g(r)$ parameters for the different sites in the same nearest-neighbor shell of the solute [i.e., with the crystal symmetry we can still write $f(r)$, rather than $f(\mathbf{r})$].

Comparisons of $f(r)$ or $g(r)$ parameters with different values of r will require consideration of electronic anisotropies, however, if we want to know why the parameters have their relative strengths.

Magnetic alloys provide a second type of anisotropy. A neighboring solute atom breaks the cubic symmetry at the iron-57 atom. The direction of the lattice magnetization may have different orientation with respect to the vector direction between the iron-57 atom and the solute atom. For magnetic materials we may therefore doubt the isotropy of the iron-57 HMF perturbations. Obviously H_{dip} is anisotropic. It will change its intensity at an iron-57 site, depending on the direction of the lattice magnetization. H_{dip} is small, and more important sources of anisotropic HMF perturbations are possible. The spin and the spatial coordinates of the electron are independent variables. If the lattice magnetization changes its direction (so there is a change in direction of the electron spin), we do not expect any direct effect on the spatial form of the wave functions. Nevertheless, spin–orbit interactions will couple the spatial part of the wave function to the spin part. This could cause the HMF perturbations to change with the direction of lattice magnetization. HMF perturbations that vary with the direction of magnetization are called "anisotropic HMF perturbations."

Anisotropic HMF perturbations are best observed in experiments using single crystals. The lattice magnetization is forced, by application of an external magnetic field, to lie along different crystallographic directions, and changes in the HMF distributions are measured. Moderate anisotropic HMF perturbations were found in this way for bcc alloys of Fe–Si,[24] Fe–Cr,[43] and Fe–Ni,[8] and larger anisotropic HMF perturbations were found for Fe–Mo.[3,48] In single-crystal experiments with ordered fcc Ni_3Fe alloys, however, the anisotropic HMF perturbations for iron-57 atoms on nickel sites were found to be as large as typical isotropic HMF perturbations.[8,180] The experimental systematics of the anisotropic iron-57 HMF perturbations are not yet known, largely because their proper measurement requires specimens that are single crystals. For bcc iron alloys, the anisotropic HMF perturbations are not as large as the isotropic HMF perturbations, but it is still prudent to test their importance. This can usually be done with polycrystalline samples by simply applying a saturating magnetic field to the sample, and seeing if the HMF distribution is changed appreciably. The idea is that without the applied field there will be a tendency for the magnetizations of the individual crystallites to lie along the (100) easy axis of magnetization. In the applied field, however, the lattice magnetizations will be randomized with respect to the crystalline axes. For polycrystalline alloys of bcc Fe–Ni, anisotropic effects were identified in this way[59] that were comparable to the results from single crystals.[8]

10. SUMMARY

HMF distributions can provide detailed information about the chemical environments of iron-57 atoms in bcc iron alloys. When an additive perturbation model is appropriate, the number of solute atoms in the nearest-neighbor shells of iron-57 atoms can be obtained from the HMF distribution. Examples from metallurgical studies of spinodal decomposition in Fe–Cr–Co and ordering in Fe–Al were presented. In these studies, not only the average solute environment around the iron-57 atoms was measured, but the distribution of solute environments was measured as well. The additive perturbation model is appropriate for alloys in which the solute atom behaves as a magnetic hole in a bcc iron crystal.

A different model, a magnetic polarization model, was constructed by considering explicitly how the magnetic moments at the different atom sites cause polarizations of the conduction and core electrons. Since the strengths of the magnetic responses of the conduction and core electrons have now been measured, the model has no freely adjustable parameters when the necessary data from neutron diffuse magnetic scattering and bulk magnetization measurements are available. The magnetic polarization model reduces to the additive perturbation model for solutes such as aluminum or silicon, which behave as magnetic holes. However, the magnetic polarization model also succeeds in predicting the featureless HMF distributions from nondilute alloys with cobalt or nickel, which are not magnetic holes in bcc iron. For these solutes, and other transition metal solutes to the right of iron in the periodic table, the additive perturbation model is inappropriate. The physical origins of the magnetic polarization model are discussed, as are its present limitations. It is possible, without much difficulty, to include corrections for atomic volume effects and nonlinearities of the conduction electron polarization. Corrections for the anisotropy of the HMF perturbations are less well understood, however.

The magnetic polarization model can be used for obtaining short-range chemical information about the environments of iron-57 atoms. When the solutes do not behave as magnetic holes, the chemical information originates from solutes in several nearest-neighbor shells around the iron-57 atom, not just the first. The model can still be useful for extracting chemical information from HMF distributions, however, as shown with an example of local chemical analysis of nonequilibrium bcc Fe–Cu alloys.

ACKNOWLEDGMENTS. I have been fortunate to have collaborated on Mössbauer spectrometry research with Professors J. W. Morris, Jr., H. H. Hamdeh, P. Matteazzi, G. Le Caër, H. Kuwano, and W. Keune, and with my present and former students Z. Mei, D. H. Pearson, H. Ouyang, Z. Q. Gao, J. K. Okamoto, T. A. Stephens, and L. Hong. I am grateful to the U.S. Department of Energy for their support under contract DE-FG03-86ER45270.

REFERENCES

1. F. van der Woude and G. A. Sawatzky, *Phys. Rep.* **12**, 335 (1974).
2. F. E. Fujita, in *Mössbauer Spectroscopy,* U. Gonser, ed., Vol. 5, p. 201, Springer-Verlag, Berlin (1975).

3. L. H. Schwartz, in *Applications of Mössbauer Spectroscopy,* R. L. Cohen, ed., Vol. I, p. 37, Academic Press, New York (1976).
4. U. Gonser and M. Ron, in *Applications of Mössbauer Spectroscopy,* R. L. Cohen, ed., Vol. II, p. 282, Academic Press, New York (1980).
5. M. Ron, in *Applications of Mössbauer Spectroscopy,* R. L. Cohen, ed., Vol. II, p. 329, Academic Press, New York (1980).
6. T. E. Cranshaw, in *Advances in Mössbauer Spectroscopy,* B. V. Thosar and P. K. Iyengar, eds., p. 217, Elsevier, Amsterdam (1983).
7. U. Gonser, ed., *Microscopic Methods in Metals,* Springer-Verlag, Berlin (1986).
8. T. E. Cranshaw, in *Industrial Applications of the Mössbauer Effect,* G. J. Long and J. G. Stevens, eds., p. 7, Plenum Press, New York (1986).
9. U. Gonser, M. Ackerman, H.-J. Bauer, N. Blaes, S. M. Fries, R. Gaa, and H.-G. Wagner, in *Industrial Applications of the Mössbauer Effect,* G. J. Long and J. G. Stevens, eds., p. 25, Plenum Press, New York (1986).
10. B. Window, *J. Phys. E* **4**, 401 (1971).
11. J. Hesse and A. Rübartsch, *J. Phys. E* **7**, 526 (1974).
12. G. Le Caër and J. M. Dubois, *J. Phys. E* **12**, 1083 (1979).
13. R. A. Brand and G. Le Caër, *Nucl. Instrum. Methods* **207**, 455 (1988).
14. G. Longworth, in *Mössbauer Spectroscopy Applied to Inorganic Chemistry,* Vol. 2, G. J. Long, ed., p. 289, Plenum Press, New York (1987).
15. S. J. Campbell and F. Aubertin, in *Mössbauer Spectroscopy Applied to Inorganic Chemistry,* Vol. 3, G. J. Long and F. Grandjean, eds., p. 183, Plenum Press, New York (1989).
16. D. G. Rancourt and J. Y. Ping, *Nucl. Instrum. Methods Phys. Res.* **B58**, 85 (1991).
17. P. A. Flinn and S. L. Ruby, *Phys. Rev.* **124**, 34 (1961).
18. G. K. Wertheim, V. Jaccarino, J. H. Wernick, and D. N. E. Buchanan, *Phys. Rev. Lett.* **12**, 24 (1964).
19. S. M. Dubiel and W. Zinn, *J. Magn. Mater.* **37**, 237 (1983).
20. M. B. Stearns, *Phys. Rev.* **129**, 1136 (1963).
21. M. B. Stearns, *J. Appl. Phys.* **36**, 913 (1965).
22. M. Rubenstein, G. H. Stauss, and M. B. Stearns, *J. Appl. Phys.* **37**, 1334 (1966).
23. M. B. Stearns, *Phys. Rev.* **147**, 439 (1966).
24. T. E. Cranshaw, C. E. Johnson, M. S. Ridout, and G. A. Murray, *Phys. Lett.,* **21**, 481 (1966).
25. J. I. Budnick, S. Skalski, T. J. Burch, and J. H. Wernick, *J. Appl. Phys.* **38**, 1137 (1967).
26. E. F. Mendis and L. W. Anderson, *Phys. Rev. Lett.* **19**, 1434 (1967).
27. M. Rubenstein, *Phys. Rev.,* **172**, 277 (1968).
28. G. K. Wertheim, *Phys. Rev. B* **1**, 1263 (1970).
29. J. I. Budnick, T. J. Burch, S. Skalski, and K. Raj, *Phys. Rev. Lett.* **24**, 511 (1970).
30. B. de Mayo, D. W. Forester, and S. Spooner, *J. Appl. Phys.* **41**, 1319 (1970).
31. D. Feldman, H. R. Kirchmayer, A. Schmolz, and M. Velicescu, *IEEE Trans. Magn.* **7**, 61 (1971).
32. G. H. Stauss, *Phys. Rev. B* **4**, 3106 (1971).
33. M. B. Stearns, *Phys. Rev. B* **4**, 4069 (1971).
34. P. J. Schurer, G. A. Sawatzky, and F. van der Woude, *Phys. Rev. Lett.* **27**, 586 (1971).
35. W. E. Sauer and R. J. Reynik, *J. Appl. Phys.* **42**, 1604 (1971).
36. R. H. Dean, R. J. Furley, and R. G. Scurlock, *J. Phys. F* **1**, 78 (1971).
37. L. R. Newkirk and C. C. Tsuei, *J. Appl. Phys.* **42**, 5250 (1971).
38. M. B. Stearns, *Phys. Rev. B* **9**, 3326 (1972).
39. I. Vincze and L. Cser, *Phys. Status Solidi B* **49**, K99 (1972).
40. I. Vincze and L. Cser, *Phys. Status Solidi B* **50**, 709 (1972).
41. I. Vincze and L. Grüner, *Phys. Rev. Lett.* **28**, 178 (1972).
42. G. Grüner, I. Vincze, and L. Cser, *Solid State Commun.* **10**, 347 (1972).
43. T. E. Cranshaw, *J. Phys. F* **2**, 615 (1972).
44. I. Vincze and I. A. Campbell, *J. Phys. F* **3**, 647 (1973).

45. P. J. Schurer, K. W. Maring, and F. van der Woude, *Int. J. Magn.* **4**, 291 (1973).
46. M. B. Stearns, *Phys. Rev. B* **9**, 2311 (1974).
47. I. Vincze and A. T. Aldred, *Phys. Rev. B* **9**, 3845 (1974).
48. L. H. Schwartz and A. Asano, *J. Phys.* (*Suppl. Colloq C-6*) **35**, 453 (1974).
49. A. Asano, Ph.D. dissertation, Northwestern University (1974).
50. I. Vincze and I. A. Campbell, *Solid State Commun.* **15**, 1495 (1974).
51. M. B. Stearns, *Phys. Rev. B* **13**, 1183 (1976).
52. T. E. Cranshaw, *Physica* **86–88B**, 392 (1977).
53. S. M. Dubiel, *Hyp. Interact,* **8**, 291 (1980).
54. S. M. Dubiel and J. Zukrowski, *J. Magn. Magn. Mater.* **23**, 214 (1981).
55. S. M. Dubiel and W. Zinn, *Phys. Rev. B* **26**, 1574 (1982).
56. S. M. Dubiel and W. Zinn, *Phys. Rev. B* **28**, 69 (1983).
57. S. M. Dubiel and W. Zinn, *Phys. Rev. B* **28**, 237 (1983).
58. B. Fultz and J. W. Morris, Jr., in *Nuclear and Electron Spectroscopies. Applied to Materials Science,* E. N. Kaufmann and G. K. Shenoy, eds., p. 377, North Holland, Amsterdam (1980).
59. B. Fultz and J. W. Morris, Jr., *Phys. Rev. B* **34**, 4480 (1986).
60. B. Fultz and J. W. Morris, Jr., in *Industrial Applications of the Mössbauer Effect,* G. J. Long and J. Stevens, eds., p. 237, Plenum Press, New York (1986).
61. S. S. Nagy, E. Kuzmann, A. Vertes, G. Szabo, and B. Konczos, *Nucl. Instrum. Methods Phys. Res.* **34**, 217 (1988).
62. B. Fultz, *Hyp. Interact.* **41**, 607 (1988).
63. B. Fultz and H. H. Hamdeh, *Hyp. Interact,* **45**, 55 (1989).
64. R. Tahara, Y. Nakamura, M. Inagaki, and Y. Iwama, *Phys. Status Solidi* **41**, 451 (1977).
65. A. D. Marwick, R. C. Piller, and T. E. Cranshaw, *J. Phys. F* **17**, 37 (1987).
66. T. E. Cranshaw, *J. Phys. F* **18**, 43 (1988).
67. T. E. Cranshaw, *J. Phys. Condens. Matter* **2**, 2743 (1990).
68. J. Y. Ping and D. G. Rancourt, *J. Magn. Magn. Mater.* **103**, 285 (1992).
69. S. M. Dubiel, *Phys. Rev. B* **37**, 7284 (1988).
70. G. Mima and M. Yamaguchi, *Trans. Jpn. Inst. Met.* **11**, 239 (1970).
71. D. Chandra and L. H. Schwartz, *Metall. Trans.* **2**, 511 (1971).
72. M. Shiga and Y. Nakamura, *Phys. Status Solidi* **37**, K89 (1976).
73. H. D. Solomon and L. M. Levinson, *Acta Metall.* **26**, 429 (1978).
74. M. E. Houghton, P. L. Rossiter, and P. E. Clark, *J. Mater. Sci.* **13**, 155 (1978).
75. H. Kuwano, Y. Ishikawa and T. Yoshimura, *Hyp. Interact.* **69**, 501 (1991).
76. H. Kaneko, M. Homma, and K. Nakamura, *AIP Conf. Proc.* **5**, 1088 (1972).
77. H. Kaneko, M. Homma, T. Fukunaga, and M. Okada, *IEEE Trans. Magn.* **11**, 1440 (1975).
78. S. Mahajan, E. M. Gyorgy, R. C. Sherwood, S. Jin, S. Nakahara, D. Brasen, and M. Eibshütz, *Appl. Phys. Lett.* **32**, 688 (1978).
79. M. Eibschütz, C. Y. Chin, S. Jin, and D. Brasen, *Appl. Phys. Lett.* **33**, 362 (1978).
80. L. Cser, J. Ostanevich, and L. Pal, *Phys. Status Solidi* **20**, 581 (1967).
81. M. B. Stearns, *Phys. Rev.* **168**, 588 (1968).
82. O. Brümmer, G. Dräger, and I. Mistol, *Ann. Phys.* **7**, 135 (1972).
83. S. A. Losievskaya, *Phys. Status Solidi A* **16**, 647 (1973).
84. G. A. Dorofeyev and V. S. Litvinov, *Phys. Met. Metall.* **44**, 438 (1977).
85. G. Athanassiadis, G. Le Caër, J. Foct, and L. Rimlinger, *Phys. Status Solidi A* **40**, 425 (1977).
86. K. Oki, A. Yamamura, K. Kudo, and T. Eguchi, *Trans. Jpn. Inst. Met.* **20**, 451 (1979).
87. J. E. Frackowiak, *Hyp. Interact.* **28**, 1067 (1986).
88. H. Chacham, E. Galvao da Silva, D. Guenzburger, and D. E. Ellis, *Phys. Rev. B* **35**, 1602 (1987).

89. E. Popiel, M. Tuszynski, W. Zarek, and T. Rendecki, *J. Less-Common Met.* **146**, 127 (1989).
90. J. T. T. Kumaran and C. Bansal, *Solid State Commun.* **69**, 779 (1989).
91. L. Dobrzynski, T. Giebultowicz, M. Kopcewicz, M. Piotrowski, and K. Szymanski, *Phys. Status Solidi A* **101,** 567 (1987).
92. S. Makhlouf, M. Shiga, and K. Sumiyama, *J. Phys. Soc. Jpn.* **60**, 3537 (1991).
93. J. H. Hsu and C. L. Chien, *Hyp. Interact.* **69**, 451 (1991).
94. M. Miyazaki, M. Ichikawa, T. Komatsu, K. Matusita, and K. Nakajima, *J. Appl. Phys.* **69**, 1556 (1991).
95. M. Miyazaki, M. Ichikawa, T. Komatsu, K. Matusita, and K. Nakajima, *J. Appl. Phys.* **69**, 7207 (1991).
96. B. Fultz, Z.-Q. Gao, and H. H. Hamdeh, *Hyp. Interact.* **54**, 521 (1990).
97. B. Fultz, Z.-Q. Gao, and L. Anthony, in *Alloy Phase Stability and Design,* G. M. Stocks, D. P. Pope, and A. F. Giamei, eds., p. 186, Materials Research Society (1991).
98. B. Fultz and Z.-Q. Gao, *Nucl. Instrum. Methods Phys. Res.* (in press).
99. L. Anthony and B. Fultz, *J. Mater. Res.* **4**, 1140 (1989).
100. L. Anthony and B. Fultz, *J. Mater. Res.* **4**, 1132 (1989).
101. Z.-Q. Gao and B. Fultz, in *Kinetics of Ordering Transformations in Metals,* H. Chen and V. K. Vasudevan, eds., p. 151, TMS, Warrendale (1992).
102. Z. Q. Gao and B. Fultz, *Philos. Mag.* (in press).
103. H. Ebert, H. Winter, B. L. Gyorffy, D. D. Johnson, and F. J. Pinski, *J. Phys. F* **18**, 719 (1988).
104. H. Akai, M. Akai, S. Blügel, B. Drittler, H. Ebert, K. Terakura, R. Zeller, and P. H. Dederichs, *Prog. Theor. Phys. Suppl.* **101,** 11 (1990).
105. H. Ebert, H. Winter, D. D. Johnson, and F. J. Pinski, *J. Phys. Condens. Matter* **2**, 443 (1990).
106. H. Akai, *Hyp. Interact.* **68**, 3 (1991).
107. D. A. Shirley and G. A. Westenberger, *Phys. Rev.* **138**, A170 (1965).
108. D. A. Shirley, *Phys. Lett.* **25A**, 129 (1967).
109. I. Vincze and I. A. Campbell, *Solid State Commun.* **15**, 1495 (1974).
110. B. Fultz and J. W. Morris, Jr., *Hyp. Interact.* **28**, 553 (1986).
111. H. H. Hamdeh, B. Fultz, and D. H. Pearson, *Phys. Rev. B* **39**, 11233 (1989).
112. B. Fultz and H. H. Hamdeh, *Philos. Mag. B* **60**, 601 (1989).
113. B. Fultz, *Phys. Rev. B* **44**, 9805 (1991).
114. M. F. Collins and J. B. Forsyth, *Philos. Mag.* **8**, 401 (1963).
115. M. F. Collins and G. G. Low, *Proc. Phys. Soc.* **86**, 535 (1965).
116. I. A. Campbell, *Proc. Phys. Soc.* **89**, 71 (1966).
117. T. M. Holden, J. B. Comly, and G. G. Low, *Proc. Phys. Soc.* **92**, 726 (1967).
118. G. G. Low, *J. Appl. Phys.* **39**, 1174 (1968).
119. H. R. Child and J. W. Cable, *Phys. Rev. B* **13**, 227 (1976).
120. C. G. Shull and M. K. Wilkinson, *Phys. Rev.* **97**, 304 (1955).
121. M. F. Collins, R. V. Jones, and R. D. Lowde, *J. Phys. Soc. Jpn.* **17**, Suppl. BIII, 19 (1962).
122. J. I. Budnick and S. Skalski, in *Hyperfine Interactions,* A. Freeman and R. B. Frankel, eds., p. 724, Academic Press, New York (1967).
123. N. Hamada, *J. Phys. Soc. Jpn.* **46**, 1759 (1979).
124. R. Richter and H. Eschrig, *J. Phys. F* **18**, 1813 (1988).
125. P. C. Reidi, *J. Phys. F* **8**, L210 (1978).
126. M. B. Stearns, *J. Appl. Phys.* **49**, 1555 (1978).
127. R. Kikuchi, *Phys. Rev.* **81**, 988 (1951).
128. R. Kikuchi, *J. Phys.* (*Paris*) **C7**, 307 (1977).
129. L. Anthony and B. Fultz, unpublished results.
130. B. Fultz, *J. Chem. Phys.* **87**, 1604 (1987).

131. B. Fultz, *J. Chem. Phys.* **88**, 3227 (1988).
132. H. Ouyang and B. Fultz, *J. Appl. Phys.* **66**, 4752 (1989).
133. R. Kikuchi, T. Mohri, and B. Fultz, *Mater. Res. Soc. Symp. Proc.* **205**, 387 (1992).
134. B. Fultz, *Philos. Mag.* **67B,** 253 (1993).
135. W. Klement, Jr., *Trans. AIME* **233**, 1180 (1965).
136. K. Sumiyama, T. Yoshitake, and Y. Nakamura, *J. Phys. Soc. Jpn.* **53**, 3160 (1984).
137. C. L. Chien, S. H. Liou, D. Kofalt, W. Yu, T. Egami, and T. R. McGuire, *Phys. Rev. B* **33**, 3247 (1984).
138. L. Lanotte and P. Matteazzi, *J. Magn. Magn. Mater.* **101**, 178 (1991).
139. A. Gupta, G. Principi, P. Trotta, E. Jannitti, C. Tosello, L. M. Gratton, S. Enzo, S. Lo Russo, and V. Rigato, *Surf. Coat. Technol.* **51**, 429 (1992).
140. A. R. Yavari, P. J. Desré, and T. Benameur, *Phys. Rev. Lett.* **68**, 2235 (1992).
141. K. Uenishi, K. F. Kobayashi, S. Nasu, H. Hatano, K. N. Ishihara, and P. H. Shingu, *Z. Metallk.* **83**, 132 (1992).
142. J. Eckert, J. C. Holzer, C. E. Krill, III, and W. L. Johnson, *J. Mater. Res.* **7**, 1980 (1992).
143. F. Kajzar and G. Parette, *J. Appl. Phys.* **50**, 1966 (1979).
144. E. F. Kneller, *J. Appl. Phys.* **35**, 2210 (1964).
145. A. T. Aldred, *J. Phys. C* **1**, 1103 (1968).
146. W. Marschall and C. E. Johnson, *J. Phys. Radium* **23**, 773 (1962).
147. P. D. DeCicco and A. Kitz, *Phys. Rev.* **162**, 486 (1967).
148. E. Fermi, *Z. Phys.* **60**, 320 (1930).
149. E. Fermi and E. Segré, *Z. Phys.* **82**, 729 (1933).
150. R. E. Watson and A. J. Freeman, *Phys. Rev.,* **123**, 2027 (1961).
151. A. J. Freeman and R. E. Watson, in *Treatise on Magnetism,* G. Rado and H. Suhl, eds., Vol. 2A, Academic Press, New York (1965).
152. R. E. Watson and A. J. Freeman, in *Hyperfine Interactions,* A. Freeman and R. B. Frankel, eds., Academic Press, New York (1967).
153. S. Wakoh and J. Yamashita, *J. Phys. Soc. Jpn.* **25**, 1272 (1968).
154. K. J. Duff and T. P. Das, *Phys. Rev. B* **3**, 2294 (1971).
155. T. P. Das, *Phys. Scr.* **11**, 121 (1975).
156. R. E. Watson and A. J. Freeman, in *Hyperfine Interactions,* A. Freeman and R. B. Frankel, eds., Academic Press, New York (1967).
157. R. E. Watson, S. Koide, M. Peter, and A. J. Freeman, *Phys. Rev.* **139**, A167 (1965).
158. R. E. Watson and A. J. Freeman, *Phys. Rev.* **152**, 566 (1966).
159. R. E. Watson, in *Hyperfine Interactions,* A. Freeman and R. B. Frankel, eds., Academic Press, New York (1967).
160. R. E. Watson and A. J. Freeman, *Phys. Rev.* **178**, 725 (1969).
161. R. E. Watson, A. J. Freeman, and S. Koide, *Phys. Rev.* **186**, 625 (1969).
162. M. B. Stearns, *Phys. Rev. B* **8**, 4383 (1973).
163. M. B. Stearns and L. A. Feldkamp, *Phys. Rev. B* **13**, 1198 (1976).
164. M. A. Ruderman and C. Kittel, *Phys. Rev.* **96**, 99 (1954).
165. T. Kasuya, *Prog. Theor. Phys.* **16**, 45 (1956).
166. K. Yosida, *Phys. Rev.* **106**, 893 (1957).
167. A. W. Overhauser, *J. Appl. Phys.* **34**, 1019S (1963).
168. E. Daniel and J. Friedel, *J. Phys. Chem. Solids* **24**, 1601 (1963).
169. I. A. Campbell, *J. Phys. C* **2**, 1338 (1969).
170. B. Caroli and A. Blandin, *J. Phys. Chem. Solids* **27**, 503 (1966).
171. A. Blandin and I. A. Campbell, *Phys. Rev. Lett.* **31**, 51 (1973).
172. P. Jena and D. J. W. Geldart, *Phys. Rev. B* **7**, 439 (1973).
173. P. Jena and D. J. W. Geldart, *Solid State Commun.* **15**, 139 (1974).
174. M. B. Stearns and J. M. Norbeck, *Phys. Rev. B* **20**, 3739 (1979).
175. I. A. Campbell and I. Vincze, *Phys. Rev. B* **13**, 4178 (1976).

176. M. B. Stearns, *Phys. Rev. B* **13**, 4180 (1976).
177. R. E. Watson and L. H. Bennett, *Phys. Rev. B* **15**, 502 (1977).
178. H. Akai, S. Blügel, R. Zeller, and P. H. Dederichs, *Phys. Rev. Lett.* **56**, 2407 (1986).
179. T. E. Cranshaw, C. E. Johnson, and M. S. Ridout, in *Proceedings of the International Conference on Magnetism, Nottingham, 1964,* p. 141, The Institute of Physics and The Physical Society, London (1965).
180. T. E. Cranshaw, *J. Phys. F* **17**, 967 (1987).

2

Conversion Electron Mössbauer Analysis of Radiation Effects Induced in Thin Metallic Films

G. Principi

1. INTRODUCTION

Mössbauer spectroscopy has been used in recent years as a rather unique method of providing information on an atomic scale on ion-beam-irradiated materials. An extensive review of the subject has been published by De Waard and Niesen.[1] Unfortunately, the utilization of Mössbauer spectroscopy is limited to materials containing a Mössbauer active nuclide as a constituent or as a probe. Moreover, often scientists working in the field of ion beams either do not have access to Mössbauer spectra or are familiar with the technique. The number of communications reporting Mössbauer data at the most recent International Conferences on Ion Beam Modification of Materials, held in Knoxville, Tennessee, in 1990 and in Heidelberg, Germany, in 1992 are relatively few. If Mössbauer spectroscopy had been used in all of the possible favorable cases, the frequency of utilization of the technique would have been close to that of more widely used techniques such as Rutherford backscattering spectrometry, transmission electron microscopy, and X-ray diffraction.

In the study of ion beam effects on materials, transmission and scanning electron microscopy, energy-dispersive X-ray analysis, and X-ray diffraction

G. Principi • Materials Section, Department of Mechanical Engineering, University of Padua, I-35131 Padua, Italy.

Mössbauer Spectroscopy Applied to Magnetism and Materials Science, Volume 1, edited by Gary J. Long and Fernande Grandjean. Plenum Press, New York, 1993.

usually with glancing angle geometry are the techniques most widely used to analyze the structure and morphology of irradiated surfaces. In addition to Rutherford backscattering spectrometry, but with less frequency, X-ray photoelectron spectrometry, secondary ion mass spectrometry, Auger electron spectrometry, and nuclear reaction analysis are sometimes used to obtain the depth profile of various constituents. The point is not, however, to establish the primacy of one technique over another but rather to see how useful it is to combine the information from complementary techniques.

The purpose of this chapter is to provide an up-to-date review of the combined applications of Mössbauer spectroscopy and other useful analytical techniques to the study of irradiation effects on thin metallic films. By irradiation effects we mean those effects produced on thin films deposited onto suitable substrates by energetic inert ions, ion beam mixing, or by pulsed laser beams of low energy density. This last type of surface treatment, currently referred to as laser quenching, has been used in some cases as a method to be compared with ion beam mixing in order to induce nonequilibrium phenomena at the surface of a material.

2. BRIEF SURVEY OF ION BEAM MIXING MECHANISMS

Ion beam mixing is a processing technique recently developed and used to induce, under the influence of ion irradiation, changes in the composition and structure of thin layers of material deposited onto a substrate.[2,3] The deposited species are introduced into the underlying layer, or mixed together, by means of cascade collisions and diffusive processes connected with the mobility of defects created by radiation damage. This process is an alternative to that of direct ion implantation, which can often be inadequate both because of the technical problems connected with the ion sources and because of physical limitations, such as sputtering effects. Sources of metallic or other unusual ions are less easily realized than sources of gaseous ions such as argon, helium, krypton, and xenon, which are well suited for ion beam mixing applications. Sputtering can be a severe limit to the maximum implanted ion concentration obtainable in the target material. Moreover, the thickness of the implanted layer is limited by the energy available in the bombarding ions and, in turn, by sputtering. One of the advantages of ion beam mixing is that the irradiation dose necessary to reach effects analogous to those obtainable by direct ion implantation can be as much as ten times lower.

With ion beam mixing it is possible to modify thin surface layers of materials in ways that are sometimes difficult or impossible to achieve by conventional methods. This occurs because the involved processes are generally far from thermodynamic equilibrium and can induce the formation of unusual surface structures. The process may be used for tailoring the composition, structure, and hence the properties of the material surfaces in a reproducible and controlled manner.

The most common ion beam mixing configurations are the following. The simplest one is the monolayer–bulk configuration which consists in a thin film of a material, A, deposited on a substrate, B. The two components are then mixed at an atomic level by bombardment with an ion beam, generally an inert gas, of sufficient energy such that the ion penetration range plus range straggling include the A/B interface. The bilayer configuration consists of two films deposited onto a substrate, currently oxidized silicon surfaces or suitable materials such as kapton. The monolayer–bulk and bilayer configurations are particularly useful for studying the way in which atoms are rearranged at an initially abrupt interface. Such studies can give valuable insight into the mixing mechanisms. A multilayer target is convenient for fabricating compounds in which the average composition is fixed by the thickness of each layer. With proper ion bombardment conditions the deposited film tends to become homogeneous. The energy of the ion beam used to bombard multilayers must be chosen so that the ion range plus range straggling corresponds to the position of the interfaces between the deposited materials A and B. Because the efficiency of mixing depends on many parameters, the understanding of the mixing mechanisms and the optimization of the process parameters require a systematic study with various combinations of materials.

The perturbations produced in the specimen by ion beam mixing are essentially of two classes, those resulting from *collisional or ballistic mechanisms,* in which the recoil mixing and the ion cascades are fast and athermal, and those resulting from *diffusional mechanisms,* which are connected with the thermally activated mobility of atoms and defects. The thickness of the mixed zone can be defined in several ways depending on the deposition configuration used and the temperature regime studied. Herein we will limit ourselves to definition of this parameter as the number, Q, of atoms of one of the elements embodied in the mixed layer at a given dose, ϕ. It is generally accepted, and has been verified in metal/metal and in metal/semiconductor systems, that a linear dependence of Q on ϕ indicates the occurrence of anisotropic mixing caused by primary collisions. In contrast, a square root dependence of Q on ϕ is associated with isotropic diffusion in well-developed cascades. In principle, it should be possible to describe a trend in the mixing process for a given system by a combination of a linear and a square root term,

$$Q = A\phi + B\phi^{1/2} \tag{1}$$

where A and B depend on the irradiation conditions.

The observed temperature dependence of Q at a given dose for many metal/silicon and metal/metal systems has suggested the existence of two temperature regimes. At low temperatures, $T < T_{\text{transition}}$, the ion mixing is fairly independent of temperature, whereas at high temperatures, $T > T_{\text{transition}}$, the ion mixing is strongly dependent on temperature. In terms of mixing mechanisms, the two temperature regimes are interpreted as follows.

Below $T_{transition}$ the mixing process is athermal and depends only on the atomic collisions induced by the ion beam. The parameters influencing the process are essentially the incident energy and the masses of the incident ion and of the target atoms. Above $T_{transition}$ the effects of long-distance atomic transport become predominant in the mixing process. These effects are diffusionlike processes and are strongly dependent on the defect concentration and mobility. The so-called "radiation-enhanced diffusion" plays a very important role in this temperature regime.

The ion beam mixing processes may also be classified in terms of duration. *Prompt processes,* which give rise to ballistic mixing, are athermal and occur primarily within the first picosecond after the ion collision. These are high-energy interaction processes which include linear cascades, recoil mixing, cascade mixing, and atom displacement spikes. *Delayed processes* are operative on a time scale longer than the first piceosecond and can occur for up to 1 h. The processes involved are low-energy, thermally activated processes which give rise to radiation-enhanced diffusion transport by defect fluxes and thermal spikes. The prompt and delayed processes leave the solid in a state that is not at thermal equilibrium, a situation that may persist for a long time after the irradiation. The resulting *persistent effects* may last for a time greater than an hour and can affect the physical and chemical properties of the specimen.

It is worthwhile to note that, among the various surface sensitive analytical techniques suitable for studying the behavior of ion-beam-irradiated systems, Rutherford backscattering spectrometry and conversion electron Mössbauer spectroscopy present the most interesting aspects of complementarity. Rutherford backscattering spectrometry detects elastically backscattered energetic protons or alpha particles and may be effectively applied to the study of how atoms of different mass are distributed in the near-surface region to a depth of a few hundred nanometers of a specimen.[4] Conversion electron Mössbauer spectroscopy is the best technique for investigating the bonding of particular atoms in the thin surface layers. In the case of iron, scattered electrons escape mainly from a depth of about 200 nm, a depth comparable with that analyzed by Rutherford backscattering spectrometry.[5] This depth is of the same order as the ion penetration range plus range straggling of, for example, 100-keV Ar^+ or 200-keV Kr^{2+} ions in metallic films.

3. METAL/METAL SYSTEMS

Ion beam mixing of metal/metal couples is a powerful method for studying alloy phase and phase transformations in thin films. Starting in the early 1980s, a great deal of effort has been concentrated on obtaining metastable alloy films in composition ranges generally not achievable by conventional rapid solidification methods. Effective quenching rates as high as

10^{13} to 10^{14} degrees per second can be achieved. Liu *et al.*[6] have presented a discussion of the thermodynamic factors that favor ion-induced amorphization. They have classified the binary systems into three categories, readily glass-forming, possible glass-forming, and hardly glass-forming. A prediction of the glass-forming ability should be possible on the basis of two intrinsic parameters of a system. The first parameter is the maximum possible amorphization range, defined as 100% minus the maximum equilibrium terminal solid solubility. The second parameter is ΔH_F, the heat of formation, which may be calculated by a model proposed by Miedema.[7] For large values of the maximum possible amorphization range and negative values of ΔH_F, amorphization occurs over a wide composition range. In contrast, for systems with small values of the maximum possible amorphization range and positive values of ΔH_F, little amorphization is observed. There are a number of systems with intermediate combinations of these two parameters for which amorphization is strictly composition dependent. Nevertheless, exceptions to these predictions have been experimentally verified and theoretically supported.[8] A positive ΔH_F value above 19 kJ/mol does not prohibit amorphization of Fe/Cu, as will be discussed below.

We will survey in this section the metal/metal ion beam mixing studies that most easily utilize Mössbauer spectroscopy.

3.1. Iron/Tin

Because iron and tin are the most favorable and popular Mössbauer isotopes, the iron/tin couple is very attractive for conversion electron Mössbauer studies after ion beam mixing. Moreover, iron/tin is an interesting system because of its important technological applications. For instance, thin coatings of tin on iron or steel surfaces are known to improve the corrosion resistance of the substrate. The corrosion protection ability of the tin coating can be improved by mixing the tin layer with the substrate by using energetic ions.

Electron microscopy and X-ray diffraction of ion beam mixing tin films on iron indicate the presence of fine microstructure and $FeSn_2$ whiskers that make the surface morphology highly irregular.[9] In similarly prepared samples subjected to simple heating and to the combined effect of heating and ion beam mixing by nitrogen ions, the formation of various intermetallic phases, such as $FeSn_2$, FeSn, and Fe–Sn solid solution, have been observed by Gratton *et al.*[10] Room-temperature nitrogen irradiation alone has a low efficiency of mixing as compared with irradiation and heating, probably because of the strong tendency of nitrogen-implanted ions to form iron nitrides.

Such complications can be avoided by utilizing inert ions.[11,12] The iron-57 conversion electron Mössbauer spectra and the corresponding hyperfine field distributions of as-deposited, xenon ion-irradiated (100 keV energy, 5×10^{15} ions/cm^2 dose, 2.4 μA/cm^2 current density), and annealed iron/tin bilayers

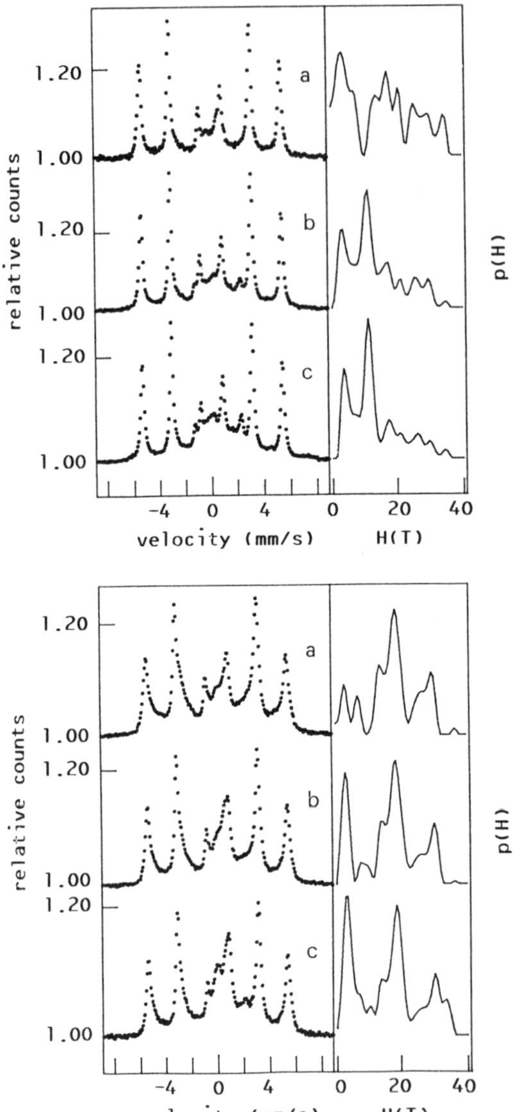

Figure 1. (Top) Conversion electron Mössbauer spectra of iron/tin sample (a) as evaporated and after vacuum annealing for 1.5 h at (b) 453 K and (c) 523 K. (Bottom) Conversion electron Mössbauer spectra of iron/tin sample after (a) xenon irradiation and subsequent vacuum annealing for 1.5 h at (b) 453 K and (c) 523 K. The corresponding hyperfine field distributions illustrate the phase evolution (metallic iron excluded) with annealing (Ref 11).

studied by Moncher *et al.*[11] are shown in Fig. 1. The field distributions were obtained by considering three different ranges of values, 0–15, 15–25, and 25–40 T, in which the isomer shift was typical of $FeSn_2$, Fe_3Sn_2, and Fe–Sn solid solution, respectively. The spectra show the presence of superparamagnetic $FeSn_2$ and ferromagnetic Fe_3Sn_2 with a hyperfine field of about 20 T. The broadening and the satellites of the iron sextet also suggest the formation of a thin solid solution with iron. It is interesting that the heat treatment of the

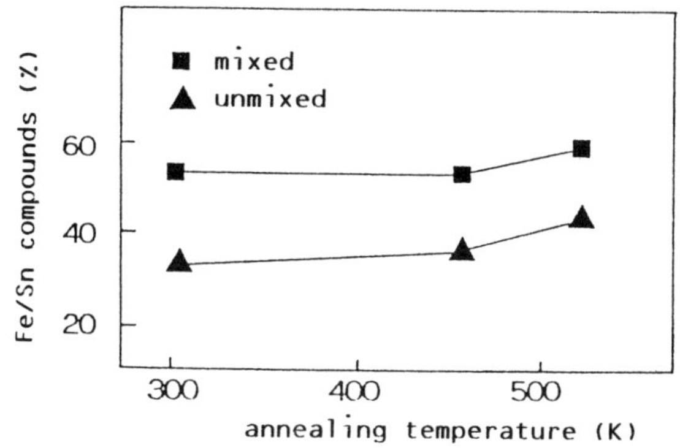

Figure 2. Evolution with annealing temperature of the relative Mössbauer spectral contributions caused by the iron–tin compounds in the iron/tin xenon ion-beam-mixed and unmixed samples (Ref. 11).

irradiated specimen results in an increase in the intensity of the superparamagnetic $FeSn_2$ component at the expense of the dissolved tin in iron component. This observation suggests that the first step in the segregation of tin solid solution is the formation of small particles of $FeSn_2$. The most obvious effect of the irradiation is a marked increase in the amount of the alloy phases, especially the Fe_3Sn_2 phase. Hence, the consistency of the intermetallic phases at the interface of iron/tin layers can be increased by ion beam irradiation and subsequent heat treatment, as shown in Fig. 2.

More detailed information can be obtained if a Mössbauer probe atom is placed at the metal/metal interface. This is shown in Fig. 3 for specimens, prepared by Sanders *et al.*,[13] having the iron/tin interface enriched with both the Mössbauer active isotopes iron-57 and tin-119. The iron-57 and tin-119

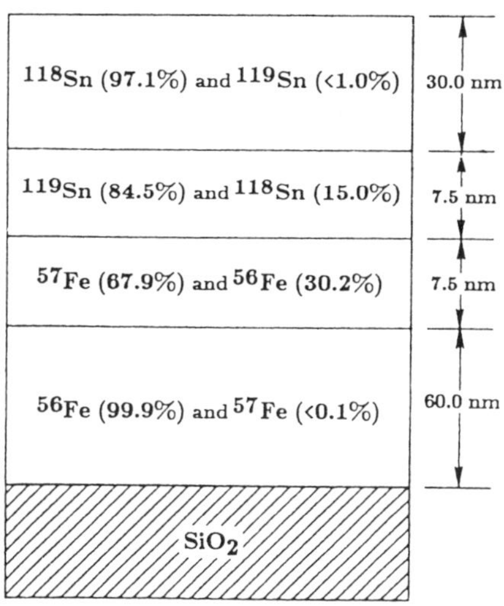

Figure 3. Iron–tin specimen after completion of sample preparation procedure (Ref. 13).

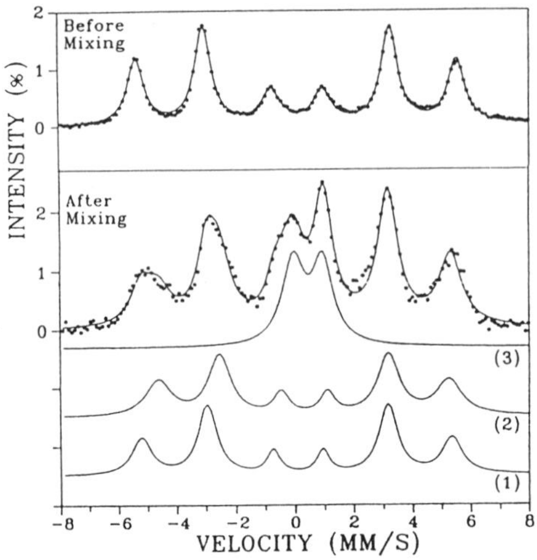

Figure 4. Iron-57 conversion electron Mössbauer spectra of sample of Fig. 3 before and after argon ion beam mixing (Ref. 13).

conversion electron Mössbauer spectra of as-deposited and argon ion (dose of $5 \times 10^{16}\,Ar^+/cm^2$ at 40 keV)-irradiated samples are shown in Figs. 4 and 5. These spectra reveal a strong effect of mixing. The iron-57 spectral profiles obtained after ion mixing have been interpreted as consisting of a solid solution of tin in iron, the two magnetic components (1) and (2), and an $FeSn_x$ $(1 < x < 2)$ alloy mixed with small magnetically relaxed iron crystallites, component (3). The corresponding tin-119 spectrum has been fit as the sum of

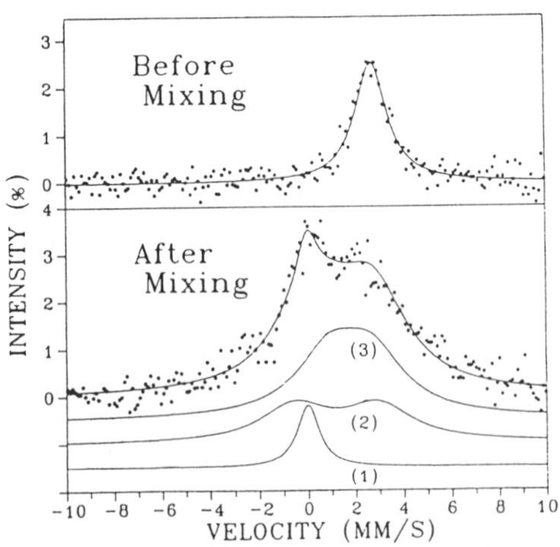

Figure 5. Tin-119 conversion electron Mössbauer spectra of sample of Fig. 3 before and after argon ion beam mixing (Ref. 13).

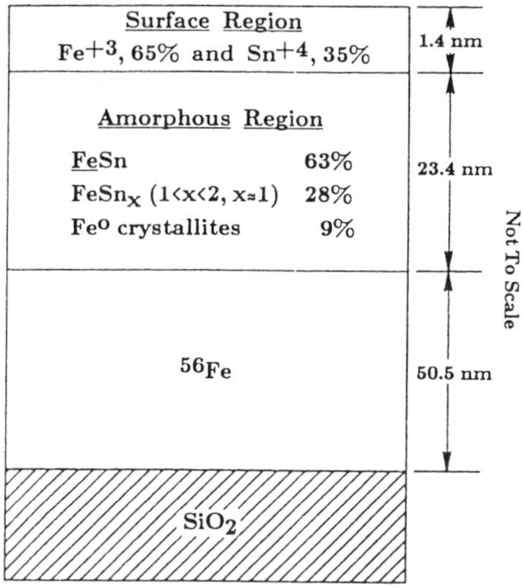

Figure 6. Overall profile of sample of Fig. 3 after argon ion beam mixing (Ref. 13).

a central single line, component (1), which provides evidence for SnO_2, two broad doublets associated with tin diluted in iron, component (2), and a tin-rich FeSn alloy, component (3). From the values of the measured hyperfine parameters and by comparison with the data of Rodmaq *et al.*,[14] amorphous character can be assigned to essentially all of the radiation-affected zone. Moreover, exploiting information obtained by Rutherford backscattering spectrometry, X-ray photoelectron spectrometry, and secondary ion mass spectrometry, the final configuration of the irradiated film can be depicted as shown in Fig. 6. The top layer is observed by X-ray photoelectron spectrometry to contain a small amount of iron and tin oxides.

3.2. Iron/Nickel

Iron–nickel alloys exhibit interesting magnetic and mechanical properties that make them industrially important materials. Of particular interest is the composition $Fe_{65}Ni_{35}$ about which the alloys exhibit Invar-like behavior, characterized by a close-to-zero thermal expansion coefficient over a wide temperature range and by anomalies in the magnetic moment and other properties.[15] It has been found that the structural and magnetic properties of thin films of iron–nickel alloys depend strongly on the preparation technique and can differ from that of bulk alloys.[16] In contrast to the bulk iron–nickel alloys, sputtered and unannealed evaporated films do not show any discontinuity in the saturation magnetization around the Invar composition. Ion beam mixing is an alternative process to prepare thin films of iron–nickel alloys, and can provide additional insight into the effect of the preparation conditions on the structural and magnetic characteristics of iron–nickel alloy films.

Gratton *et al.*[17] and Tosello *et al.*[18] have observed that 100-keV Ar^+ and 200-keV Kr^{2+} ion irradiation with a dose range from 1×10^{16} to 6×10^{16} ions/cm^2 and with a current density of 1–1.5 $\mu A/cm^2$ of iron/nickel bilayers, with either iron or nickel on top, is effective in mixing the interface and forming a heterogeneous iron–nickel alloy with a structure that is partly bcc and partly fcc. In order to evaluate the sputtering effect, quantitative information can be obtained from Rutherford backscattering spectrometry on the Ar^+ and Kr^{2+} retained dose and the film thickness before and after irradiation. The conversion electron Mössbauer spectra, shown in Fig. 7, are those of Kr^{2+}-irradiated samples. They can be analyzed by assuming two superimposed ranges of hyperfine field distributions. The first is fairly narrow, centered at about 33 T, and may be attributed to a bcc iron-rich phase. The second is rather broad, has lower field values, and may be assigned to fcc

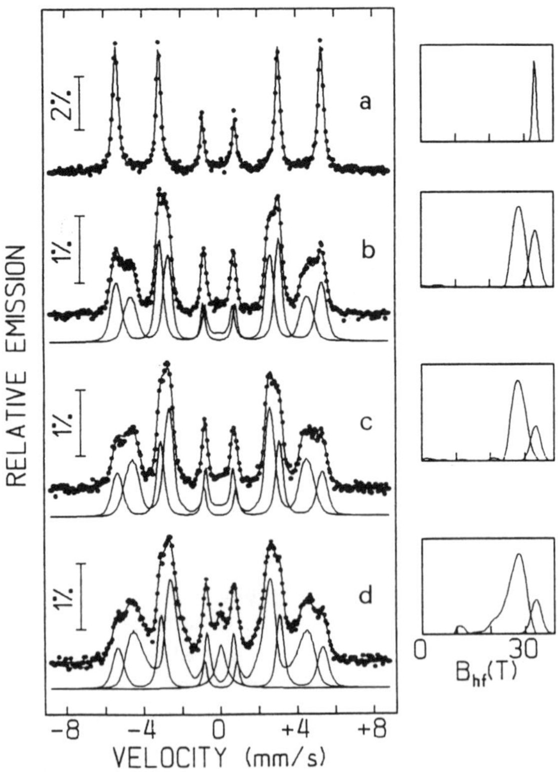

Figure 7. Conversion electron Mössbauer spectra and corresponding hyperfine field distributions of nickel/iron layers: (a) as evaporated, and Kr^{2+}-irradiated to the dose (b) 0.9×10^{16}, (c) 2.4×10^{16}, and (d) 3.5×10^{16} ions/cm^2 (Ref. 18).

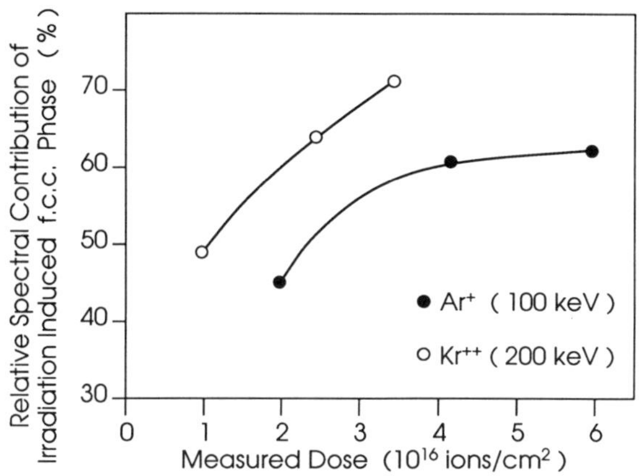

Figure 8. Relative amount of irradiation-induced Fe–Ni phases, evaluated from the Mössbauer spectra, versus the measured Ar^+ and Kr^{2+} dose (Ref. 18).

iron–nickel phases induced by ion beam mixing. The higher mixing efficiency of krypton relative to that of argon, as measured by the fcc relative spectral component, is depicted in Fig. 8. Nuclear reaction analysis based on the resonant reaction $^{58}Ni(p, \gamma)^{59}Cu$ ($E_p = 1424$ keV, $\Gamma = 45$ eV) and performed with a high-efficiency germanium detector,[19] confirms this higher mixing efficiency, otherwise expected being the weight of krypton ions twice that of argon ions. The amount, Q, of mixed nickel atoms in the case of the highest irradiation doses yielded 2.3×10^{17} Ni/cm^2 and 8.5×10^{16} Ni/cm^2 for krypton and argon, respectively.

Tosello *et al.*[20] have studied what happens with the ion beam mixing of $Fe_{65}Ni_{35}$ multilayers prepared by ultrahigh-vacuum evaporation at 77 K according to the scheme, $Si/SiO_2/^{56}Fe/Ni/^{56}Fe/Ni/^{57}Fe/Ni/^{56}Fe/Ni/^{56}Fe$, in which the individual layer thicknesses have been adjusted to give the overall desired composition. The position of the Mössbauer marker layer matches the projected range and range straggling of the 200-keV Kr^{2+} ions with a nominal dose range of 1×10^{15} to 5×10^{16} ions/cm^2 and a current density of 1.5 μA/cm^2. The glancing angle X-ray diffraction patterns, Fig. 9, show the evolution of the structure with the radiation dose. In the as-deposited sample the broad bcc Fe(110) peak is slightly asymmetric at lower reciprocal lattice parameter values because of an fcc Ni(111) contribution. The absence of an fcc Ni(220) component can be attributed either to the presence of preferential texture (111) normal to the growth plane, or to small crystallite size, or to lattice disorder. Progressive bcc-to-fcc phase transformation occurs after irradiation with up to 8.5×10^{15} Kr^{2+}/cm^2 with a variation in the lattice parameter of the fcc phase corresponding to a content of 47 at.% nickel. At higher dose the X-ray diffraction pattern indicates grain growth of both coexisting bcc and fcc structures and a distinct increase in the fcc lattice parameter. This increase can be understood as the contribution of interstitial inclusions of krypton.

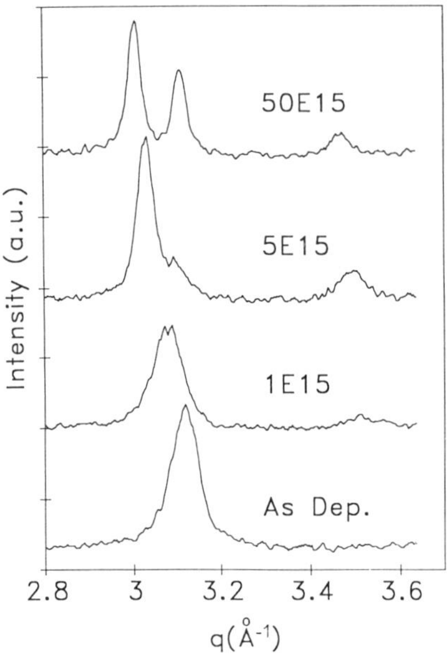

Figure 9. Partial X-ray diffraction patterns of iron–nickel multilayers: as deposited and Kr^{2+}-irradiated to the nominal dose 1.0×10^{15}, 5.0×10^{15}, and 5.0×10^{16} ions/cm^2 (Ref. 20).

The trend displayed by the conversion electron Mössbauer spectra of Fig. 10 strongly correlates with that of X-ray diffraction patterns. It is evident that the increase with dose of the lower magnetic field component occurs at the expense of the component attributed to the bcc phase. The contributions centered around 28 and 23 T are attributed to a nickel-rich alloy and to an Invar phase, respectively. The Invar component becomes important in the sample irradiated at a dose of 4×10^{16} Kr^{2+}/cm^2, in which the bcc and fcc phases are still present. The small amount of nonmagnetic iron-rich fcc phase, formed at intermediate dose and characterized by a small grain size, tends to be destabilized at higher dose. In spite of these promising results, a homogeneous iron–nickel Invar film has not been obtained to date by ion beam mixing, even by starting with multilayers having exactly the Invar composition.

3.3. Iron/Aluminum

The ion beam mixing of the iron/aluminum couple is one of the most extensively studied. Much work has been done on the evolution at the iron/aluminum interface and there are also several reports on mixed multilayers with a variety of compositions.

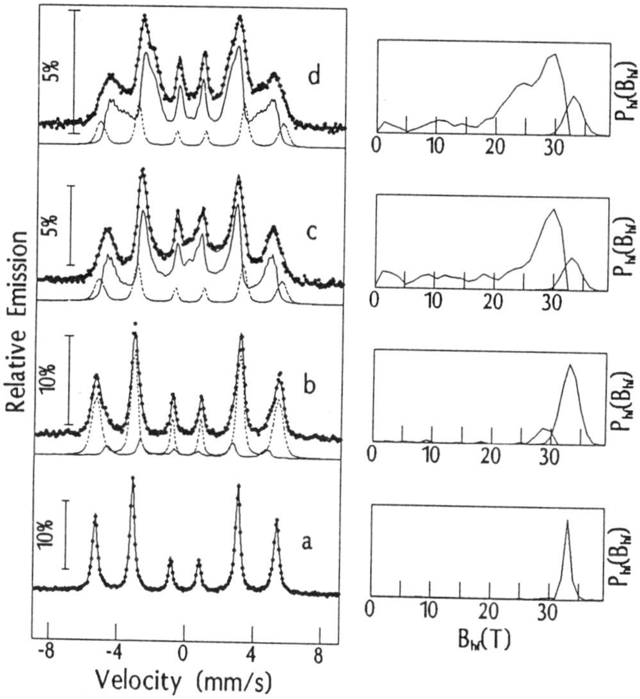

Figure 10. Conversion electron Mössbauer spectra and corresponding hyperfine field distributions of iron–nickel multilayers: (a) as deposited, and K^{2+}-irradiated to the dose (b) 1.3×10^{15}, (c) 8.5×10^{15}, and (d) 4×10^{16} ions/cm^2 (Ref. 20).

3.3.1. Single Iron/Aluminum Interface

Iron-coated aluminum has a poorly defined interface, because of the unavoidable oxidation and contamination of the aluminum substrate before iron deposition. This is shown in the conversion electron Mössbauer spectrum by Godbole *et al.*[21] of the as-deposited specimen, in which the α-iron sextet is a minor component. Auger profiling measurements by Rauschenbach *et al.*[22] have shown that intermixing of iron and aluminum at the interface decreases in the presence of oxygen. Moreover, the influence of oxygen on the structure at the interface is poorly understood.

The Rutherford backscattering spectrometry, transmission electron microscopy, and secondary ion mass spectrometry investigations by Gaboriaud *et al.*[23] have shown that (1) mixing takes place in iron/aluminum bilayers, with aluminum on top, starting at a dose level of 2.5×10^{15} Xe^{2+}/cm^2; (2) iron atoms migrate rapidly toward the surface which is reached at a dose level of 10^{16} ions/cm^2; (3) a bcc iron/aluminum solid solution along with crystalline precipitates of xenon are formed at a dose level of 3.6×10^{16} ions/cm^2. By Auger depth profiling of the mixed iron/aluminum bilayers[22] it has been shown that the mixing efficiency follows approximate logarithmic behavior with increasing dose. Further, the mixing is greater for samples with aluminum on top than for samples with iron on top.

Conversion electron Mössbauer spectroscopy, Rutherford backscattering spectrometry, and X-ray diffraction have been used by Vasconcellos *et al.*[24–26] to study a combined effect of the radiation dose, from 3×10^{15} to 1×10^{17} ions/cm^2, substrate temperature, from 77 to 500 K,and atomic number of the implanted ions, argon, krypton, or xenon. The radiation dose seems to affect only the efficiency of mixing, the phases induced remaining the same. It also seems that temperature is the only important parameter for phase formation and, in particular, that irradiation at 300 K with the three ions leads to an intermediate state in which intermixing occurs without the formation of well-organized intermetallic compounds. Stoichiometric Fe_2Al_5 or $FeAl_6$ phases are formed under ion bombardment at 500 K. Ion beam mixing induces interdiffusion and phase formation at temperatures lower than simple thermal annealing. It has also been found[27] that there is a general decrease in the temperatures required to reach specific interdiffusion or reaction stages for ion mixing compared with furnace annealing. This decrease is estimated to be about 120 K for Ar^+ and about 150 K for Kr^{2+} and Xe^{3+}.

3.3.2. Multiple Iron/Aluminum Interface

An important role is played by the overall composition of the multilayer.[28–30] Rauschenbach and Hohmuth[29] have shown by means of transmission electron microscopy on samples of Al_xFe_{1-x}, with $0.1 < x < 0.9$, that xenon irradiation with 1×10^{16} ions/cm^2 at a current density less than 0.5 μA/cm^2, leads to the formation of bcc disordered solid solutions of the B2 type,[28] when x is rather high at ca. 0.9 or low at ca. 0.2. However, irradiation of samples with $x = 0.8$ leads to an amorphous phase that is partially transformed to the icosahedral quasicrystalline state after annealing at ca. 470 K at a small heating rate of 5 to 20 K/min. For short annealing times at low temperatures the predominant product is a fine-grained icosahedral phase dispersed in a crystalline iron–aluminum matrix.[30] It should be noted that the formation of an amorphous film by ion beam mixing of iron/aluminum couple is in agreement with its ΔH_F value of -28 kJ/mol.

The possibility of forming a quasicrystalline phase by ion beam mixing without an additional postannealing has been investigated by Liu *et al.*[31] Quasicrystalline grains dispersed in an amorphous matrix have been observed by means of transmission electron microscopy after 100-keV argon ion bombardment of $Al_{65}Fe_{35}$ multilayers at a dose greater than 1×10^{16} ions/cm^2, and at a current density of less than 1 μA/cm^2. The size of these quasicrystalline grains was about 10 to 30 nm and their composition was found to be $Al_{80}Fe_{20}$. On this basis it seems that the quasicrystalline phase is precipitated from the amorphous matrix under irradiation.

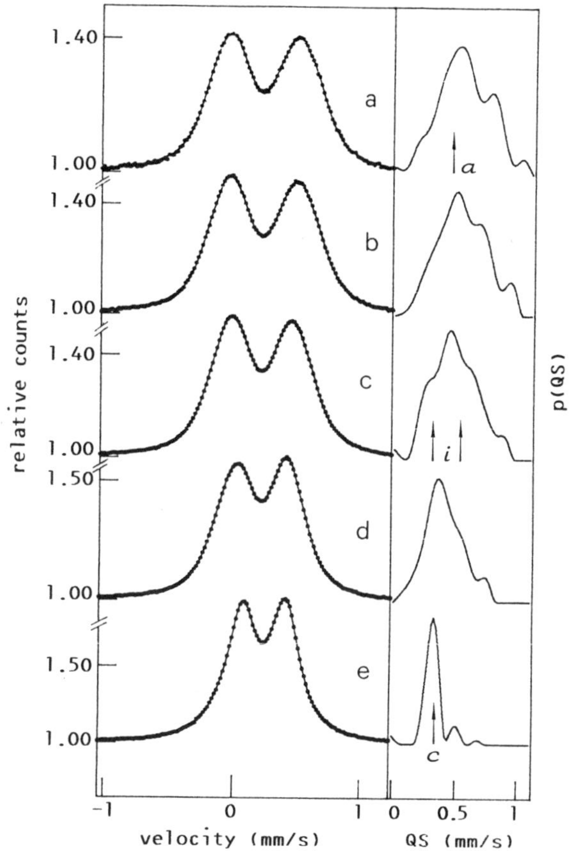

Figure 11. Conversion electron Mössbauer spectra and corresponding hyperfine field distributions of (a) as-irradiated iron–aluminum specimen, and after sequential vacuum annealing for 30 min at (b) 473, (c) 573, (d) 623, and (e) 673 K. The arrows indicate the values taken from the literature concerning the quadrupole splitting values for the amorphous a-$Al_{86}Fe_{14}$, the quasicrystalline i-$Al_{86}Fe_{14}$, and the crystalline c-Al_6 phases (Ref. 32).

Gratton *et al.*[32] have shown that only a nonmagnetic pattern appears in the conversion electron Mössbauer spectrum (see Fig. 11a) after krypton irradiation at a support temperature of 150 K of iron/aluminum multilayers having the average Al_6Fe composition made up to 50 nm of aluminum, 10 nm of iron-57, and 40 nm of aluminum sequentially deposited on an SiO_2 substrate. This means, in agreement with changes observed after irradiation in the AES depth profiles, that the sample is completely mixed. Figure 11 shows the Mössbauer, spectra, and the corresponding quadrupole splitting distributions, after irradiation and sequential annealing in the temperature range from 473 to 673 K. These spectra indicate that the initial irradiated specimen is structurally disordered and inhomogeneous. Upon annealing the structure gradually changes through an intermediate icosahedrally coordinated quasicrystalline $Al_{86}Fe_{12}$ phase to the crystalline Al_6Fe phase. Additional evidence of these structural transformations is obtained by differential scanning calorimetry, measured on similar iron/aluminum films deposited on KBr supports

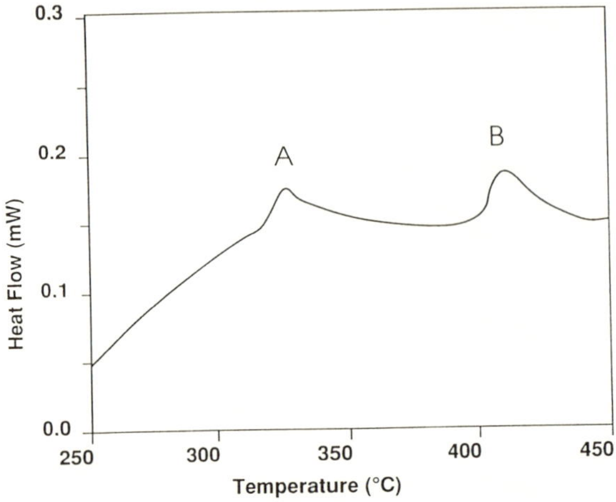

Figure 12. Differential scanning calorimetry scan, with a heating rate of 5 degrees/min, of krypton-irradiated iron–aluminum layers (Ref. 32).

and ion beam mixed. The differential scanning calorimetry diagram of the irradiated sample (Fig. 12) exhibits two peaks at ca. 500 and 680 K, which can be associated with the amorphous-to-icosahedral and with the icosahedral-to-crystalline transformations, respectively.

The correct interpretation of the Mössbauer spectra, shown in Fig. 11, is actually difficult and questionable, because the samples could be rather inhomogeneous. However, an unambiguous Mössbauer characterization of amorphous and quasicrystalline Fe–Al films, which can support the above interpretation, is possible. Homogeneous amorphous and quasicrystalline films, as confirmed by electron diffraction patterns, have been obtained by

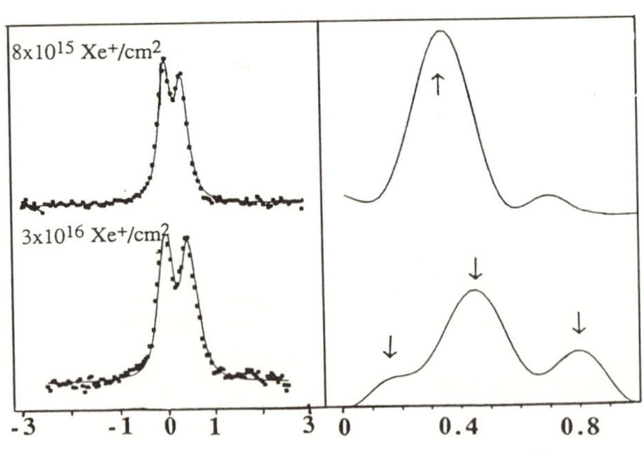

Figure 13. Conversion electron Mössbauer spectra and corresponding quadrupole splitting distributions of homogeneous $Al_{82}Fe_{18}$ quasicrystalline (top) and amorphous (bottom) films obtained by xenon ion irradiation of iron–aluminum multilayers (Ref. 33).

Plenet *et al.*[33] from $Al_{82}Fe_{18}$ multilayers irradiated with xenon ions at a dose of 3×10^{16} ions/cm^2 at 77 K and at a dose of 8×10^{15} ions/cm^2 at 493 K, respectively. Their conversion electron Mössbauer spectra and the corresponding quadrupole splitting distributions are shown in Fig. 13.

3.4. Iron/Copper

The possibility of obtaining nonequilibrium iron–copper alloys is even more attractive than for the iron–aluminum alloys. In fact, in each case the constituents are sparingly soluble in each other but, whereas the iron–aluminum system can give equilibrium alloys, such as Fe_3Al, $FeAl_6$, this is not possible for the iron–copper system. Moreover, the heat of mixing, ΔH_F, is -28 kJ/mol for iron–aluminum and $+19$ kJ/mol for iron–copper. There are reports of the formation of supersaturated fcc copper-rich and bcc iron-rich solid solutions prepared by vapor quenching[34] and sputter deposition.[35,36] Poor agreement has been found for the composition range in which the fcc copper-rich and the bcc iron-rich phase coexist.

Huang and Liu[37] have found by transmission electron microscopy that xenon ion mixing of $Fe_{20}Cu_{80}$ multilayers produces a mixture of fcc phases having a variable copper atomic percentage and an amorphous phase of composition $Fe_{70}Cu_{30}$. This was the first experimental proof of the possibility of obtaining amorphous iron–copper alloys by ion beam mixing, in spite of its large positive heat of formation. Moreover, it was shown[38] that the solid solubility of iron in copper can be extended to values as high as 45% and that it is possible to form amorphous iron–copper alloys by ion mixing within a specific composition range of 60–70% of iron. Subsequent ion beam mixing experiments on $Fe_{60}Cu_{40}$ multilayers[39] with 1×10^{16} Xe^{2+}/cm^2 revealed by selected area electron diffraction the formation of a quasicrystalline phase. The optical density of the electron diffraction pattern of the quasicrystalline iron–copper phase is shown in Fig. 14. It was thus proven that quasicrystals could be obtained apart from the restricted family of aluminum-based alloys. Because no amorphous phase was detected prior to the formation of the new quasicrystalline phase, this suggested that the formation of a quasicrystal could not be expected to arise from a preexisiting amorphous phase, as seems to occur in the iron–aluminum system. Another important experimental observation was that the formation of amorphous or quasicrystalline materials is composition dependent, because it could only be formed within a narrow composition range.

The above results are stimulating for Mössbauer spectral studies of the ion-beam-mixed iron–copper system. Conversion electron Mössbauer spectroscopy, X-ray diffraction, and Rutherford backscattering spectrometry investigations have been performed by Gupta *et al.*[40] on ion-beam-mixed and pulsed ruby laser-irradiated $Fe_{60}Cu_{40}$ and $Fe_{70}Cu_{30}$ multilayers. The X-ray diffraction patterns of Fig. 15 reveal the modifications of the Cu(111) and Fe(110) peaks

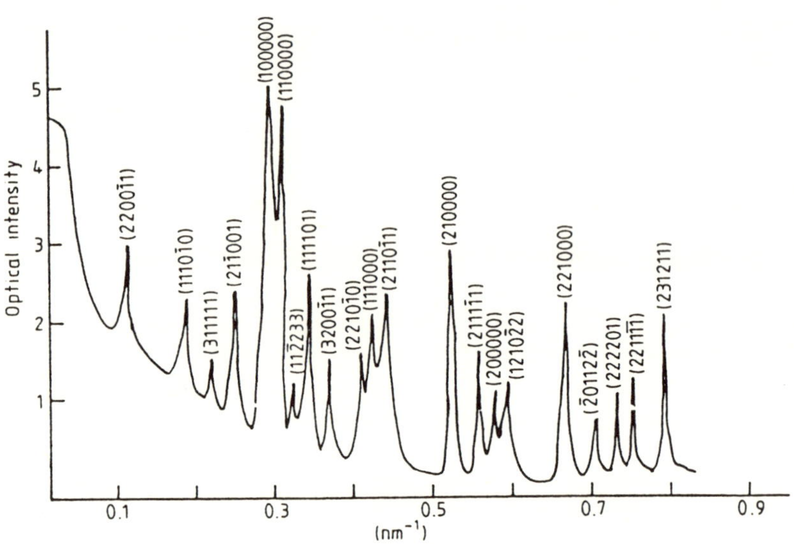

Figure 14. Optical density curve of the selected area electron diffraction pattern of $Fe_{60}Cu_{40}$ quasicrystalline phase with icosahedral indices (Ref. 39).

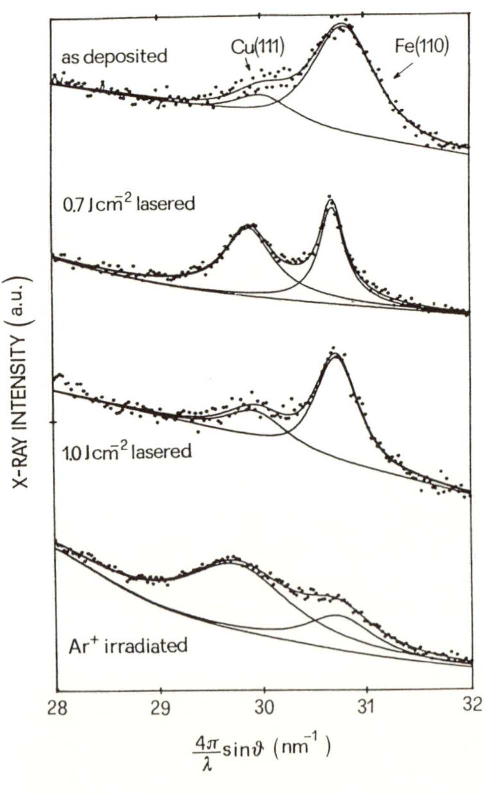

Figure 15. Partial X-ray diffraction patterns of $Fe_{70}Cu_{30}$ multilayers showing the evolution of Cu(111) and Fe(110) peaks under various treatments (Ref. 40).

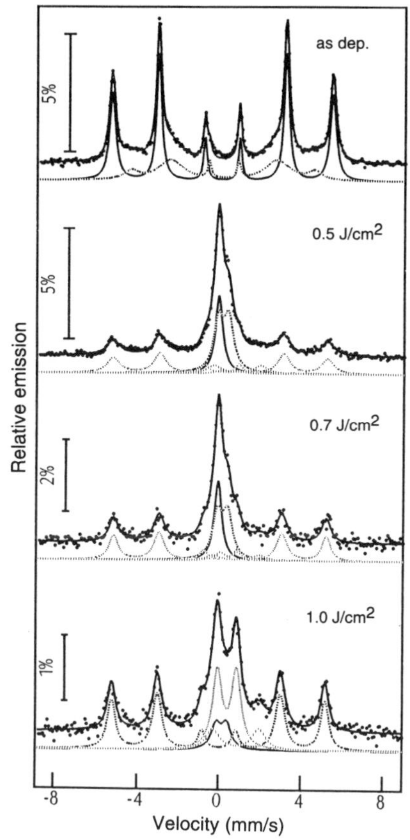

Figure 16. Conversion electron Mössbauer spectra of as-deposited and laser-irradiated, at different energy density, $Fe_{70}Cu_{30}$ multilayers (Ref. 40).

induced by the two types of treatments. Changes in peak positions, relative intensities, and linewidths indicate that laser quenching increases the grain size of both bcc and fcc structures, produces enrichment of copper atoms in the bcc solid solution, and diffusion of iron atoms into the copper matrix. Ion beam mixing leads to a remarkable reduction or disappearance of the Fe(110) peak, indicating that iron atoms have migrated to the fcc solid solution and/or have been converted into an amorphous phase.

Mössbauer spectroscopy clearly differentiates the two types of treatments, as shown in Figs. 16 and 17. Whereas laser-quenched samples are characterized by spectra indicating the coexistence of bcc and fcc solid solutions, the ion-beam-mixed samples yield very broad spectra, indicating the formation of a disordered structure. Compared with ion-mixed $Fe_{60}Cu_{40}$, the hyperfine field distribution in ion-mixed $Fe_{70}Cu_{30}$ has significantly more weight in the higher field region. This difference has been attributed to bcc and fcc phases mixed with an amorphous phase. Ion-beam-mixed $Fe_{70}Cu_{30}$ would consist of an amorphous phase and a bcc iron-rich solid solution. In contrast, ion-beam-

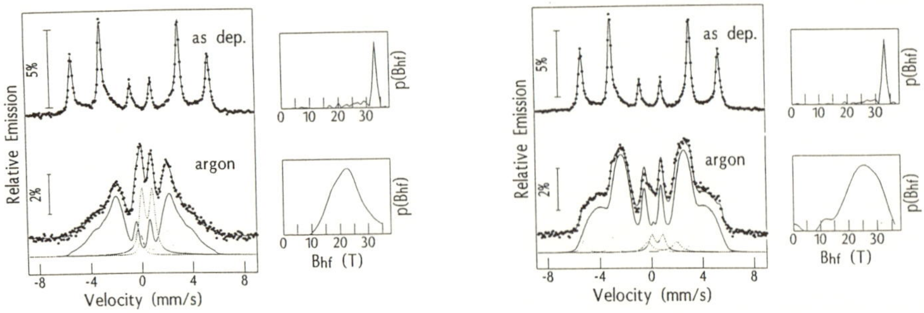

Figure 17. Conversion electron Mössbauer spectra and corresponding hyperfine field distributions of as-deposited and Ar^+-irradiated $Fe_{60}Cu_{40}$ (left) and $Fe_{70}Cu_{30}$ (right) samples (Ref. 40).

mixed $Fe_{60}Cu_{40}$ would consist of an amorphous phase and a copper-rich solid solution. This interpretation agrees with the above-mentioned studies[37] on the iron–copper system, in which the phenomenon of localized amorphization was observed within a rather narrow composition range. It seems that ion beam mixing leads to the formation of an amorphous iron–copper phase, which consists of a mixture of bcc and fcc short-range order with different copper proportions depending on the overall composition of the treated multilayers.

Laser-irradiated samples do not show any sign of amorphization over the beam energy density range of 0.5 to 1.0 J/cm^2. The formation of disordered structures was expected, because laser quenching gives rise to a melt–solid transition characterized by a very fast cooling rate of 10^9 to 10^{10} degrees per second. In contrast, the ion beam mixing method has several features that differ from the quenching method. Perhaps the most important difference, besides a faster cooling rate, is the highly defective regions created by the high amount of kinetic energy deposited by the incident ions. Moreover, the unavoidable although limited, carbon contamination in the implantation chamber may aid the stabilization of highly disordered structures.

Formation of a quasicrystalline phase is not observed in the iron–copper samples mixed with the argon ion beam. This can be attributed to the lighter argon projectiles used for mixing, which produce much less radiation damage in the samples than xenon ions.[39] Also, the sample irradiation temperature was perhaps too low to permit a reorganization of atoms in the damaged regions into the longer range rotational order required for the quasicrystalline structure.

3.5. Iron/Molybdenum

This is another interesting metal couple to be subjected to ion beam mixing, because it is characterized by maximum possible amorphization range and ΔH_F values in the range of possible glass-forming systems. Ion beam irradiation can be considered as a process that induces local composition variation; a variation that can be more critical for the possible glass-forming systems than for the easily glass-forming systems.

Liu *et al.*[6] have evidenced by transmission electron microscopy that iron–molybdenum multilayers can be amorphized under a xenon ion beam. They observed that, starting from a threshold dose of about 1×10^{16} ions/cm^2, the amorphization is easy for $Fe_{40}Mo_{60}$. In contrast, a dose twice this threshold value is required for $Fe_{50}Mo_{50}$. Further, energy dispersive analysis has revealed that the molybdenum concentration in $Fe_{50}Mo_{50}$ was about 70 at.% in the local amorphous region, and about 50 at.% in the neighboring crystalline phase. The completely amorphized $Fe_{40}Mo_{60}$ samples yielded an average value of 60 at.%. Thus, it has been considered likely that the most stable amorphous state, which corresponds to a free energy minimum in the iron–molybdenum system, has a molybdenum-rich composition of ca. 60–70 at.%.

Conversion electron Mössbauer measurements carried out by Li and Liu[41] on the same iron–molybdenum samples, in spite of the poor quality of the spectral data, could be used to draw some interesting conclusions. For both compositions the spectra after irradiation with 1.5×10^{16} Xe/cm^2 were a mixture of magnetic and nonmagnetic components, but with different hyperfine parameters and relative areas. The nonmagnetic contribution was attributed to a metastable solid solution as was also detected by transmission electron microscopy. A nonmagnetic subspectrum was also found by Joshee *et al.*[42] for krypton-irradiated iron layers deposited on molybdenum foils, but in this case it was likely caused by oxygen contamination of the iron–molybdenum interface. The magnetic subspectrum for samples $Fe_{40}Mo_{60}$ was broader than for samples $Fe_{50}Mo_{50}$, confirming the formation of an amorphous phase.

3.6. Iron/Titanium

The literature on the ion beam mixing of the iron–titanium system provides only two reports, both by Brenier *et al.*,[43,44] on multilayers. In the first work, 500-keV xenon irradiation in the dose range from 1×10^{15} to 3×10^{16} ions/cm^2 was unable to obtain complete mixing and formation of an amorphous structure; a result that would be expected on the basis of the criterion of Liu *et al.*[6] This failure was attributed to impurity content deriving from the sample preparation. In the second work, Fe_xTi_{100-x} (with x equal to 29, 33, 50, 58, and 67) multilayers were deposited under ultrahigh vacuum conditions and 800-keV xenon irradiation was conducted with the same dose range at 77 and 300 K. X-ray diffraction and conversion electron Mössbauer spectroscopy gave evidence in each case for the formation of amorphous films. It was argued that amorphization by ion beam mixing in the iron–titanium system can be possible over a wider range of compositions, because a dose of 8×10^{15} Xe$^+$/cm^2 was sufficient to amorphize the multilayer with $x = 50$, the case that seems to be the least favorable based on the phase diagram, apart from the extreme compositions. Iron–titanium amorphous alloys have also been obtained by rapid solidification methods for titanium concentration between 20 and 75 at.%.[45]

3.7. Iron/Zirconium

The iron–zirconium system easily yields amorphous alloy by means of rapid solidification and then is classified among the easy glass-forming materials. Even bilayers result in amorphized state in the krypton ion-beam-mixed region.[46] In contrast, there is evidence from transmission electron microscopy of a composition range over which amorphization can be carried out by melt spinning but not by ion beam mixing.[47] Conversion electron Mössbauer spectroscopy has been used by Kopcewicz *et al.*[48] to analyze proton beam-irradiated and solid state-reacted $Fe_{61}Zr_{39}$ multilayers. A strong similarity has been found for the two amorphization processes, and the results favor radiation-enhanced diffusion as the primary mixing mechanism.

3.8. Iron/Cobalt

Ion beam mixing can be used on thin metallic films to change their magnetic properties. Riviere *et al.*[49] have presented an interesting application for $Fe_{60}Co_{40}$ films evaporated from the bulk and then irradiated at 77 K with either light 15-keV helium ions or heavy 360-keV xenon ions up to a dose of 2×10^{16} ions/cm^2. The initially random spin orientation in the disordered film was completely modified by xenon ion irradiation, which led to a preferential magnetization parallel to the film plane. In Fig. 18 it is clearly evident that the

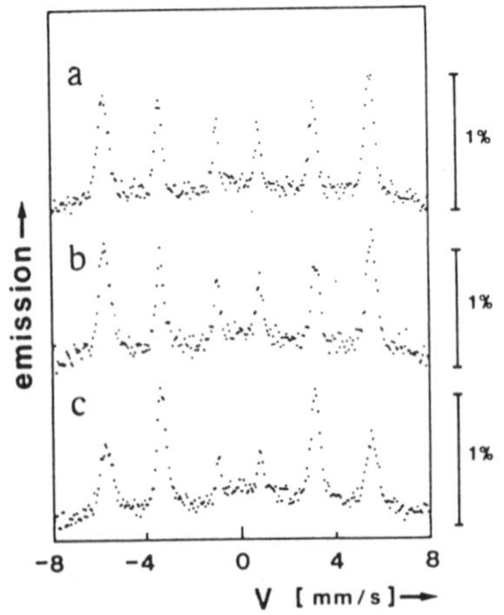

Figure 18. Conversion electron Mössbauer spectra of $Fe_{60}Co_{40}$ thin films: (a) as evaporated, (b) after 15-keV He^+ ion irradiation, and (c) after 360-keV Xe^+ ion irradiation (Ref. 49).

average angle made by the Fe spins with the direction perpendicular to the film plane, as measured from the spectral intensity line ratio $A_{2,5}/A_{3,4}$, changes from the as-evaporated film spectrum, (a), to the xenon-irradiated film spectrum, (c). Irradiation by helium ions, in contrast, does not produce substantial variations [see spectrum (b)]. Only heavy ions give rise to highly disordered and damaged regions in dense displacement cascades which produce strong compressive stresses. These stresses are probably responsible for the observed spin alignment after xenon irradiation of $Fe_{60}Co_{40}$ films.

3.9. Iron/Palladium

Ion beam mixing in the iron–palladium system has been studied to date only for bilayer configurations. Analyses with Rutherford backscattering and X-ray photoelectron spectrometries, X-ray diffraction, and transmission electron microscopy have been performed by Battaglin *et al.*[50,51] The samples had a palladium layer on top of an inner iron layer and were irradiated with 200-keV Kr^{2+} ions in the dose range 1.0×10^{16} to 2.5×10^{16} ions/cm^2. It could be inferred that a subsurface layer is formed in which palladium is completely mixed with iron, that iron is the dominant moving species responsible for the intermixing of the two layers, that the mixed region is dominantly formed as a disordered $FePd_3$ phase, and that the unmixed iron layer undergoes a possible transformation from α- to γ-iron. This final point agrees with an appreciable shift in the X-ray-diffracted Fe(110) peak, which could alternatively be attributed to the incorporation of krypton in the lattice.

Mössbauer measurements carried out by Gupta *et al.*[52] served to clarify this question and further to confirm the kind of phase change that occurs at the iron/palladium interface. By placing a 5-nm-thick layer of iron-57, the remaining iron being unenriched, at the iron/palladium interface, sample 1, and at the iron/substrate interface, sample 2, one obtains the spectra, before and after irradiation at a dose of 2.5×10^{16} Kr^{2+}/cm^2, shown in Fig. 19. It is evident that there is strong mixing in sample 1, because of the broad peak in the hyperfine field distribution centered around 27.3 T. This confirms the formation of a disordered $Fe_{0.28}Pd_{0.72}$ alloy. In sample 2 only a small change in the spectrum is detected after irradiation due, at least in part, to the natural iron layer mixed with palladium. No paramagnetic components are visible, which rules out the possibility of any transformation of the unmixed iron layer to the fcc phase.

3.10. Tin/Nickel and Tin/Aluminum

Mention should also be made of the ion beam mixing studies by Calliari *et al.*[53] and by Czako-Nagy *et al.*[54] on the tin–nickel and tin–aluminum systems. They reported scanning electron microscopy and Auger and Mössbauer spectrometry data of 100-keV xenon ion-irradiated tin coatings on nickel and

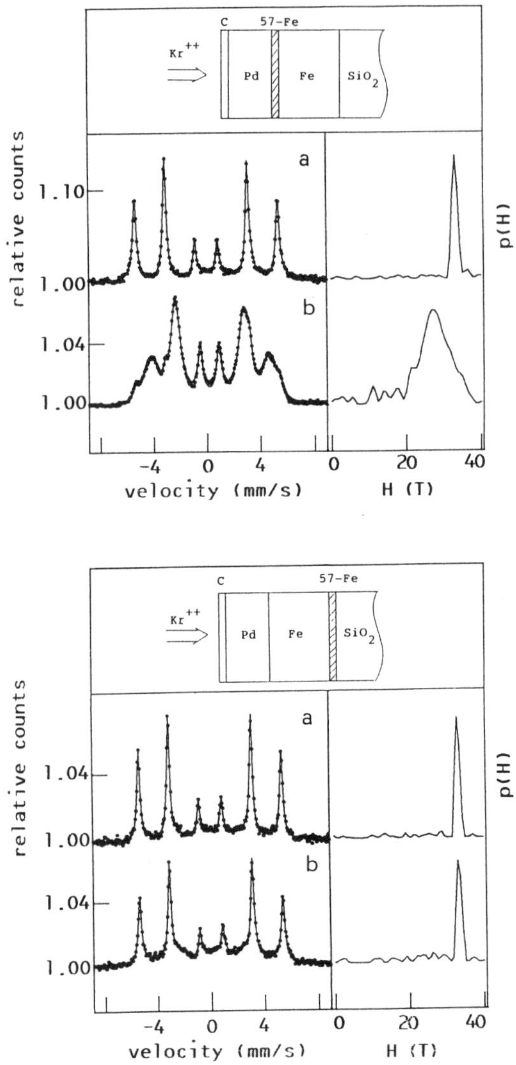

Figure 19. Conversion electron Mössbauer spectra and corresponding hyperfine field distributions of as-deposited (a) and krypton-irradiated (b) iron/palladium samples having the iron-57 maker interfaced with the palladium layer (top) and with the substrate (bottom) (Ref. 52).

aluminum sheets. It has been found that energetic ion bombardment is very effective in forming intermetallic compounds when the substrate is nickel. In contrast, only modifications of oxides at the interface occur when the substrate is aluminum. The result on the irradiated tin coatings of nickel is interesting from a technological point of view, because this treatment has been found to be very effective against corrosion in hydrogen sulfide environments.[53]

4. METAL/INSULATOR SYSTEMS

Ion beam mixing studies of metal/insulator combinations, such as Fe/SiO_2 and Fe/Al_2O_3, are important for several reasons. Such studies can provide a better understanding of the collisional, diffusional, and chemical processes involved in the ion beam mixing of metal/insulator systems. Iron is practically insoluble in SiO_2 and Al_2O_3 and thus it is of interest to examine ion beam induced metastable structures. Modifications of the interfacial region of metal/insulator couple can lead to improved film adhesion properties, improvements which are important for microelectronic device fabrication. Modifications of the surface optical properties of silica and sapphire are important for applications in optoelectronic devices.

4.1. Fe/SiO_2 Interface

4.1.1. Effects of 100-keV Ar^+ Irradiation

The effects of irradiation with 100-keV Ar^+ ions on the Fe/SiO_2 system have been investigated earlier with iron films of different thicknesses deposited onto oxidized silicon.[55–57] In order to avoid uncertainties in the analysis of the irradiated samples, Rutherford backscattering and Mössbauer measurements were performed before and after chemical etching to remove from the surface the iron that remained unmixed.

The partial Rutherford backscattering spectra of Battaglin *et al.*,[56] involving only the iron signal, for as-deposited and irradiated samples, before and after etching, are shown in Fig. 20. In the nonetched samples the amount of iron decreases with increasing argon dose because of sputtering effects. The areal density of iron atoms mixed with the SiO_2 substrate can be evaluated from the Rutherford backscattering spectra of the etched samples and clearly increases with the dose.

The conversion electron Mössbauer spectra, shown in Fig. 21, of the as-deposited and irradiated samples are completely different from those obtained from the irradiated and etched samples (see Fig. 22). The former spectra are combinations of magnetic sextets, which result essentially from iron atom reactions with contaminants in the irradiation chamber. In contrast, the latter, which are nonmagnetic, originate from iron atoms mixed with the substrate. This is further evidence that the etching procedure is necessary to avoid misinterpretation of both the Rutherford backscattering and Mössbauer data. Further, it has been proven by measurements with the same techniques that the deposition process by itself does not produce any combination between iron atoms and the substrate.

The spectra of the irradiated and etched samples, Fig. 22, can be well fit with a singlet and three doublets, characterized by unique values of average isomer shift, IS, and average quadrupole splitting, QS. The singlet, S, with an isomer shift of $\simeq 0$ mm/s is attributed to small clusters of metallic iron, whose superparamagnetic behavior at room temperature indicates an average particle size not larger than 10 nm. The doublet, D1, with an IS $\simeq 0.8$ mm/s and a QS $\simeq 0.8$ mm/s, can most likely be assigned to iron in an oxide phase

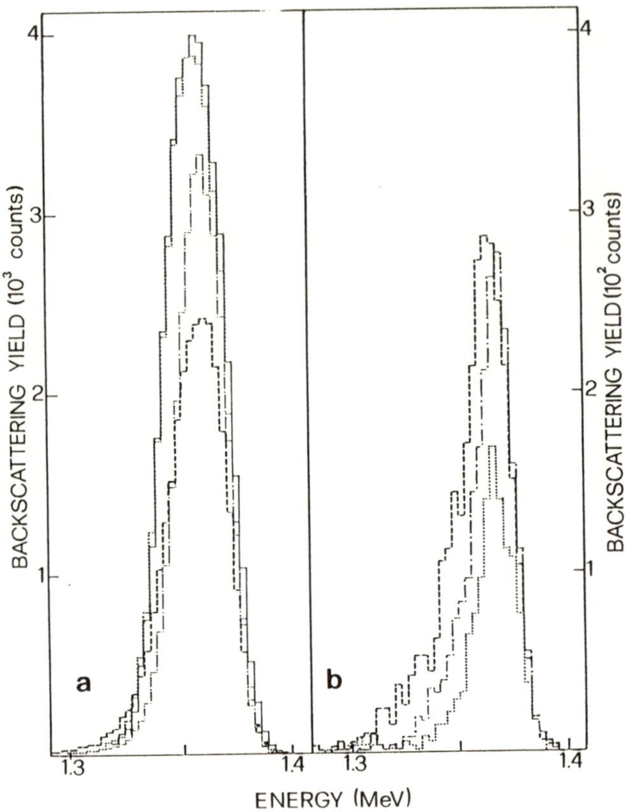

Figure 20. Partial Rutherford backscattering spectra (only the iron signal is shown) of as-deposited and irradiated Fe/SiO_2 samples before etching (a) and after etching (b): as deposited (——), 2.5×10^{16} Ar^+/cm^2 (···), 5×10^{16} Ar^+/cm^2 (·–·), and 1×10^{17} Ar^+/cm^2 (––––) (Ref. 56).

corresponding to defect-associated Fe^{2+}.[58] The doublet, D2, with an IS $\simeq 1.2$ mm/s and a QS $\simeq 1.5$ mm/s, and the doublet, D3, with an IS $\simeq 1.0$ mm/s and a QS $\simeq 2.4$ mm/s, are attributed to Fe^{2+} in octahedral sites of a silicate phase.[59] The iron areal density for each iron phase can be deduced by combining the absolute Rutherford backscattering values of the areal density of the mixed iron with the relative area of the Mössbauer spectral components. This, of course, assumes that the iron in each phase has the same recoil-free fraction, at least to a first approximation. The amount of iron in clusters increases noticeably and the amount of iron in the oxide phase increases slightly upon increasing the dose. In contrast, the iron silicate phase reaches its maximum value at an intermediate dose. These observed iron phases are substantially similar to those seen by Perez *et al.*[60–62] after the direct implantation of iron into synthetic fused silica.

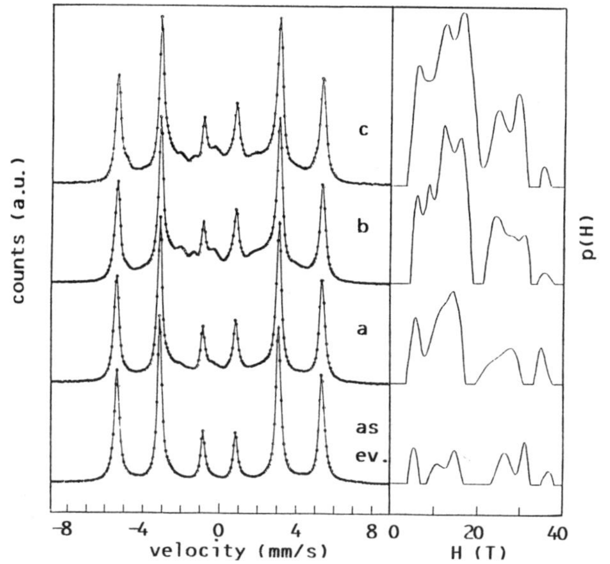

Figure 21. Conversion electron Mössbauer spectra (left) and hyperfine field distributions (right) of Fe/SiO_2 samples before etching: as evaporated and after Ar^+ ion irradiation at doses of (a) 2×10^{16}, (b) 5×10^{16}, and (c) 1×10^{17} ions/cm^2 (Ref. 57).

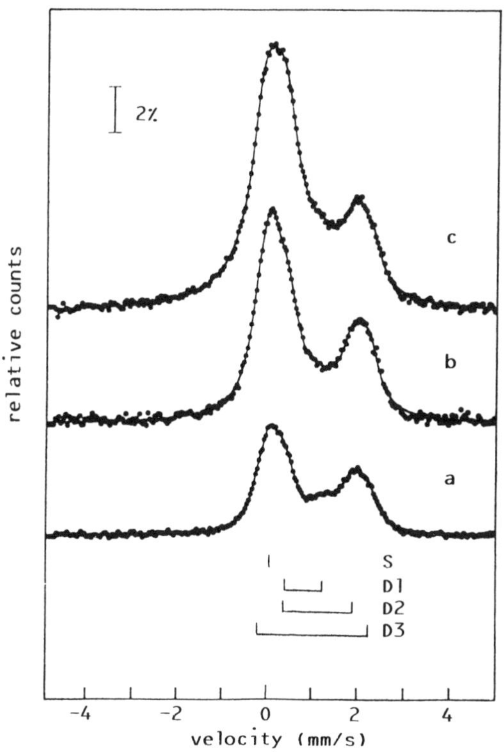

Figure 22. Conversion electron Mössbauer spectra of the Ar^+ ion-irradiated Fe/SiO_2 samples after etching: (a) 2.5×10^{16}, (b) 5×10^{16}, and (c) 1×10^{17} ions cm^{-2}. The 2% effect represented by the bar is the vertical scale reference for the three spectra. Stick diagrams represent spectral components described in the text (Ref. 57).

4.1.2. Effects of 200-keV Kr^{2+} Irradiation

Ion beam mixing in the Fe/SiO_2 system has also been studied by Battaglin and coworkers[63–66] with the 200-keV Kr^{2+} irradiation. High-purity SiO_2 slices were used as bulk material, rather than thermally oxidized silicon. This was done with the aim of studying modifications in the optical properties of the ion-mixed samples after removal of the unmixed iron film by means of the same chemical etching process as was used with the Ar^+-irradiated samples.

The amount of mixed iron, as determined from the area under the iron signal of the Rutherford backscattering spectra of the chemically etched samples, is shown as a function of the Kr^{2+} dose in Fig. 23. Even though the experimental points can be reasonably fitted in the log–log plot by a single straight line of slope 0.85, a better interpolation is achieved by using the solid line, which has a decreasing slope with increasing Kr^{2+} dose. Up to a dose of 3×10^{15} ions/cm^2, this curve lies very close to the dotted straight line, which has a unit slope and from which the experimental points deviate significantly only at higher doses. This behavior is similar to that observed by Banwell and Nicolet[67] for xenon ion-irradiated Ni/SiO_2, and can be explained on the basis of a demixing process. In Fig. 23, the dashed line represents the amount of the mixed iron obtained by irradiation with 100-keV argon ions. The Kr^{2+} mixing efficiency is about three times that obtained with Ar^+ and may be attributed to the higher mass of krypton.

The Mössbauer spectral profiles of samples etched after irradiation could be resolved with the same set of components observed in the argon-irradiated samples, but with different amount present. A straightforward comparison of samples irradiated with the two ions at doses giving the same amount of mixed iron is given in Fig. 24. On the left side are shown the spectra of the samples

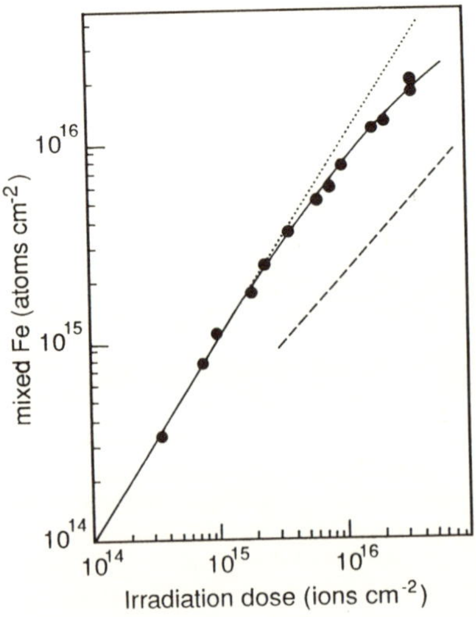

Figure 23. Measured iron mixed amount in Fe/SiO_2 samples as a function of 200-keV Kr^{2+} ion irradiation dose: the solid line is an eye guide connecting the experimental points and the dotted slope 1 straight line interpolates the low-dose points. The dashed line shows the iron mixing upon 100-keV Ar^+ ion irradiation (Ref. 63).

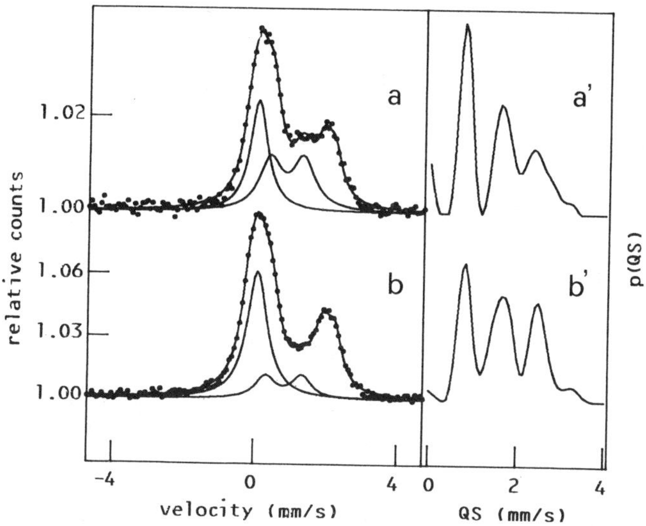

Figure 24. Comparison between the conversion electron Mössbauer spectra of (a) 200-keV, $1.8 \times 10^{16}\, Kr^{2+}/cm^2$ and (b) 100-keV, $5 \times 10^{16}\, Ar^{+}/cm^2$ irradiated and etched samples (for clarity, only the spectral components attributed to iron clusters and iron oxide are reported). The quadrupole splitting distributions for the doublets associated with iron oxide and iron silicate are also shown in (a′) and (b′) (Ref. 65).

irradiated with $1.8 \times 10^{16}\, Kr^{2+}/cm^2$ and with $5 \times 10^{16}\, Ar^{+}/cm^2$, the amount of mixed iron being 1.15×10^{16} and $1.03 \times 10^{16}\, Fe/cm^2$ for Kr^{2+} and Ar^{+} irradiation, respectively. On the right side of Fig. 24 are shown the distributions of quadrupole splitting, p(QS), in which the singlet, attributed to iron clusters, has been considered as a crystalline spectral component. The p(QS) plots are further evidence of the different phase distribution in the two cases. More iron oxide is formed during irradiation by krypton ions than by argon ions. This occurs at the expense of the small iron clusters and the iron silicate phase. It would appear that the radiation damage produced by the krypton ions is higher than that produced by the lighter argon ions, especially with regard to the number of atoms of the target displaced from their equilibrium sites. As a result, more oxygen than silicon is available for chemical reaction with the iron atoms. Further, it should be noted that the ratio of oxygen to silicon atoms displaced per incident ion is presumably higher than their stoichiometric ratio, in a fashion analogous to the mixing phenomenology observed in the Fe/Al_2O_3 system.[68]

Figure 25 summarizes the quantitative behavior of the areal density of the iron phases, as determined by Mössbauer and Rutherford backscattering measurements, versus the dose of the two different ions, and is further evidence of their different mixing efficiency.

Mössbauer spectra measured after thermal anneal (see Fig. 26) still show the same set of spectra components as observed before the 770 K annealing,

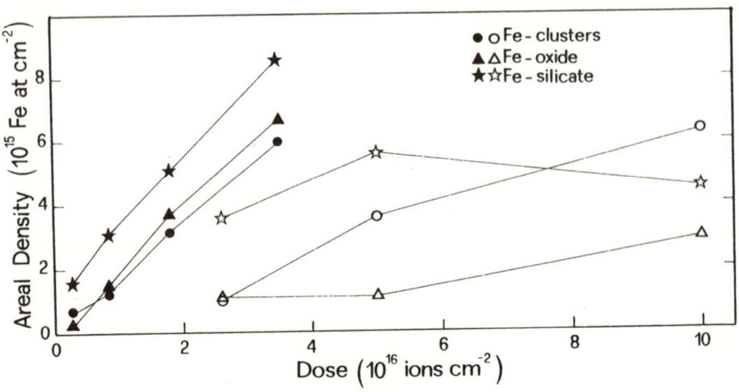

Figure 25. Dose dependence of the iron areal density of each iron phase at the Fe/SiO_2 interface after irradiation with krypton (solid symbols) and argon (open symbols) ions (Ref. 65).

plus a broad sextet after the 970 K annealing. As the temperature increases, the areal density of both iron clusters and iron oxide decreases. The iron silicate seems to be thermodynamically favored, because its relative amount, as determined by Mössbauer spectroscopy, increases considerably at the expense of other two phases upon annealing at 770 K. Even after annealing at 970 K,

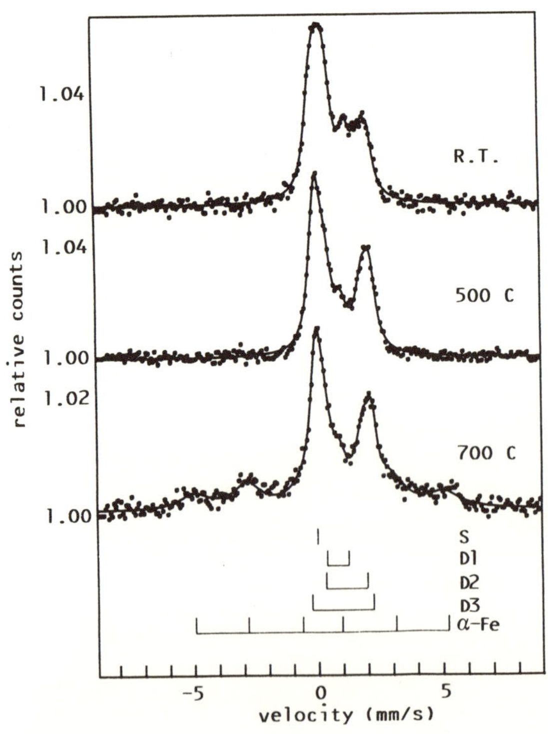

Figure 26. Conversion electron Mössbauer spectra of a $^{57}Fe/SiO_2$ sample Kr^{2+}-irradiated at a dose of 3.5×10^{16} ions/cm^2, etched (RT), and annealed at different temperatures. Stick diagrams represent spectral components described in the text (Ref. 66).

the relative amount of iron silicate is slightly higher than the initial value. The new spectral component that appears after heating to 970 K is characterized by an internal hyperfine field lower than that of crystalline α-Fe, and by a rather broad linewidth of 1.5 mm/s, for the external peaks. This indicates that, at this stage of precipitation, the metallic particles are characterized by a wide range of grain size distribution.

Bhagwat *et al.*[69] have reported a Mössbauer spectroscopy and X-ray diffraction study of ion beam mixing of the Fe/SiO_2 system without the etching procedure after irradiation. This study was carried out on a 30-nm-thick overlayer of iron, enriched to 30% in iron-57, deposited in ultrahigh vacuum onto SiO_2 single crystals and then irradiated with 100-keV Ar^+/cm^2 up to a dose of 5×10^{16} ions/cm^2. After irradiation, an Fe_7SiO_{10} phase as well as Fe–Si–O complexes were observed. Upon annealing at 720 K for 6 h, outmigration of α-iron occurred with an increase of the Fe^{3+}/Fe^{2+} ratio. These results agree qualitatively with those of the systematic study described above, although, because only unetched samples were analyzed, the result may not be well representative of the interfacial structure of the system after mixing.

4.2. Fe/Al_2O_3 Interface

Conversion electron Mössbauer spectroscopy was used by Ogale *et al.*,[68,70] in an early study of 23-nm-thick iron layers, 30% enriched with iron-57, deposited on polycrystalline alumina, which had been bombarded with 100-keV krypton ions at a dose ranging from 1×10^{15} to 5×10^{16} ions/cm^2. The spectrum of the as-deposited sample is shown in Fig. 27 and consists of a strong nonmagnetic contribution, in addition to the magnetic sextet of metallic iron. This nonmagnetic contribution may be attributed to Fe^{3+} substituted for the Al^{3+} sites in Al_2O_3 in the interface region. After irradiation, only two nonmagnetic components are present (see Fig. 27). The one with an isomer shift of 1.00 mm/s and a quadrupole splitting of 1.65 mm/s can be attributed to the $FeAl_2O_4$ phase. The other component, with an isomer shift of 0.64 mm/s

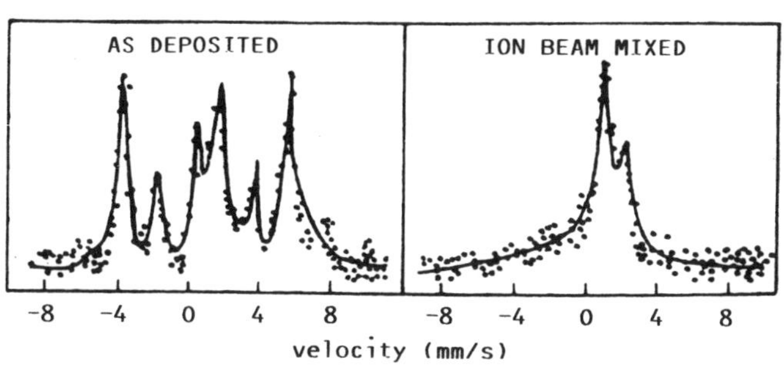

Figure 27. Conversion electron Mössbauer spectra of as-deposited and krypton ion-beam-mixed Fe/Al_2O_3 sample (Ref. 68).

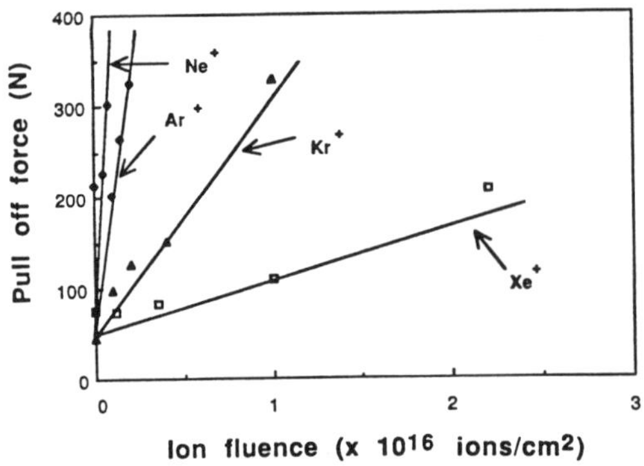

Figure 28. Evolution of the pull-off force versus ion dose for 60-nm Fe/Al_2O_3 samples mixed at room temperature with different ions (Ref. 73).

and a quadrupole splitting of 0.48 mm/s, can be assigned to a non-stoichiometric FeO-like oxide. Glancing angle X-ray diffraction measurements seemed to confirm these assignments. The as-deposited sample used in this experiment, however, did not present a sufficiently sharp interface, as evidenced by its Mössbauer spectrum. This fact probably influenced the measurements after ion beam irradiation.

The possibility of ion beam mixing enhancement of adhesion at the metal/ceramic interface has been explored by Abonneau *et al.*,[71,72] who studied, in particular, the ion-beam-mixed Fe/Al_2O_3 interface.[73] Adhesion tests carried out on iron films deposited on sapphire and irradiated with a variety of energetic ions give the results summarized in Fig. 28. It is remarkable that the adhesion ehancement in the case of neon-mixed samples is a factor of five for a dose as small as $2 \times 10^{15}\, Ne^+/cm^2$. Microstructural observations performed by high-resolution transmission electron microscopy reveal the existence of an interfacial roughness and a thin amorphous zone between the iron and sapphire in the ion-beam-mixed samples.

Conversion electron Mössbauer spectra of neon ion-irradiated samples are shown in Fig. 29. The spectrum of the as-deposited sample is predominantly the sextet of metallic iron, which ensures that the initial Fe/Al_2O_3 interface is rather abrupt. A nonmagnetic contribution is seen to increase with increasing dose and may be attributed to the presence of Fe^{2+} and Fe^{3+} or Fe^{4+} species. The Fe^{2+} may be associated with an $FeAl_2O_4$ spinel local environment, whereas the Fe^{3+} may be associated with Fe_2O_3 and/or Fe_3O_4 or with iron substitution in the Al sites in the near-interface region. The Fe^{2+} contribution seems to saturate for doses higher than $10^{16}\, Xe^+/cm^2$, whereas the signal associated with Fe^{3+}, or Fe^{4+}, increases linearly, as shown in Fig. 30. This result clearly indicates the dependence of the nature and composition of the interfacial zone on the energy deposited at the interface. Even if the interfacial zone is very thin, i.e., 3 to 5 nm, it is sufficient to improve the metal–ceramic adhesion in the present case.

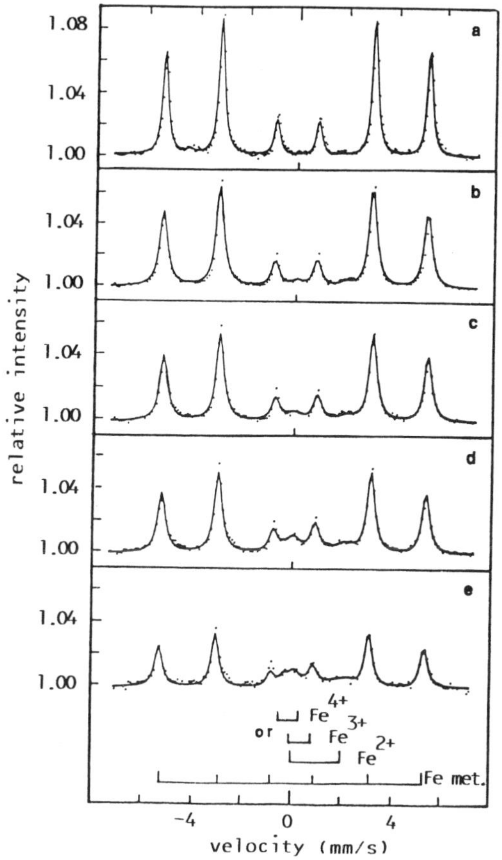

Figure 29. Conversion electron Mössbauer spectra of Fe/Al_2O_3 samples unmixed (a) and mixed at room temperature with 90-keV neon ions at doses of 1 (b), 2 (c), 6 (d), and 10×10^{16} ions/cm^2 (e) (Ref. 73).

4.3. Fe/Glass Interface

Another approach to the enhancement of the metal/insulator adhesion has been attempted by Carbucicchio *et al.*[74] with proton bombardment of iron films supported by glass. Scratch tests have shown that 100-keV proton irradiation of iron/glass samples induces an increase by a factor of four in the loads necessary for the onset of film removal. Depth-selective conversion electron Mössbauer spectroscopy has been used to analyze the irradiation-induced structural changes in the interfacial zone. The Mössbauer spectral profile of Fig. 31a is caused by iron atoms near the interface, observed by detecting conversion electrons in the 7.0 to 7.1 keV range, and reveals, in addition to the predominant sextet of bcc iron, a broad central contribution, which may be attributed to iron atoms in nonstoichiometric iron–carbon compounds. This is in agreement with the detection by X-ray photoelectron spectrometry of carbon contamination even after a very careful cleaning of the

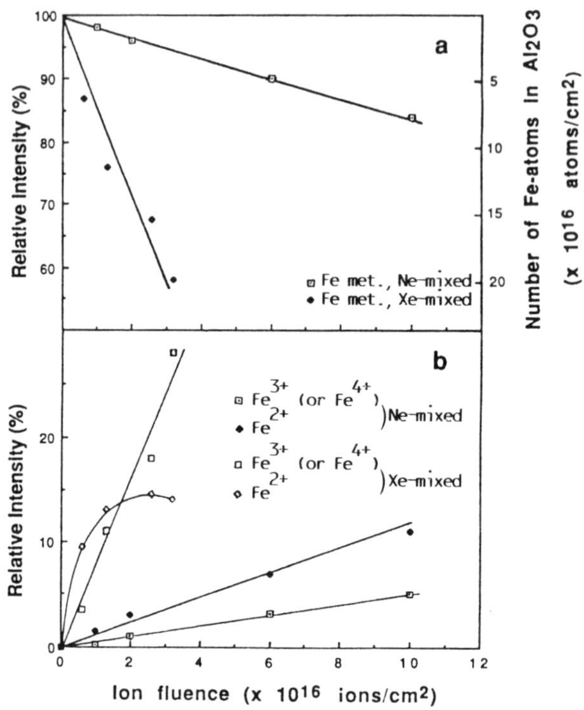

Figure 30. Dose dependence of relative intensity of various Mössbauer spectral components, metallic iron (a), and Fe^{2+} and Fe^{3+} (or Fe^{4+}) (b), for Fe/Al_2O_3 ion-beam-irradiated samples (Ref. 73).

surfaces prior to iron deposition. The broad central contribution is consistently lower in the spectrum of the sample surface (not shown), observed by detecting conversion electrons in the 7.2 to 7.3 keV range. After proton irradiation, the Mössbauer spectral profile at the interface changes noticeably, as shown in Fig. 31b. The decrease in the broad central component, and the significant change in the $A_{2,5}/A_{3,4}$ intensity ratio, leads to the conclusion that proton bombardment destabilizes the interfacial iron–carbon compounds and induces stress variations at the interface. The reduction of the interfacial compounds, which permits a more intimate contact between film and substrate, and the reduction of the magnetic anisotropy, i.e., the decrease in the mechanical stresses at the interface, have been proposed as responsible for the observed improved adhesion.

5. METAL/SEMICONDUCTOR SYSTEMS

Lau *et al.*[75] have shown that systems that can exist in a number of phases can be mixed over a wide range of compositions with energetic ion beams. This

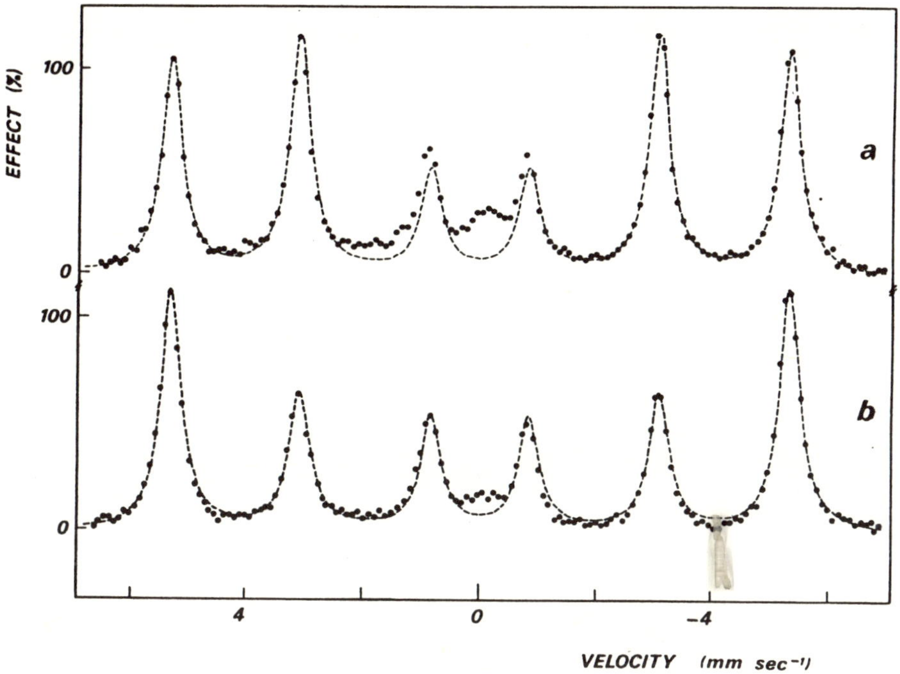

Figure 31. Depth-selective conversion electron (7.0–7.1 keV) Mössbauer spectra for 25-nm iron-57 on glass: (a) as deposited, and (b) bombarded with 100-keV protons at a dose of 10^{16} H^+/cm^2 (Ref. 74).

is the case of, for example, iron–germanium and iron–silicon couples, whose importance in fundamental and applied research is well known.

Early results on iron/germanium[76] and iron/silicon[77] ion-beam-mixed bilayers were probably invalidated by the sample preparation method, as indicated by the Mössbauer spectra of the as-deposited samples, very different from the simple sextet of metallic iron.

Conversion electron Mössbauer spectra of 400-keV krypton and xenon ion-bombarded bilayers of iron/silicon, with either iron or silicon on top, were measured by Hayashi *et al.*[78] and are shown in Fig. 32. Paramagnetic components, which are attributed to the iron silicides, $FeSi_2$ and FeSi, appear after irradiation. By properly selecting the energy window of the detected conversion electrons it is possible to obtain approximate depth-selective detection, and a spectrum more characteristic of the interface, with lower-energy electrons, or a spectrum more characteristic of the surface, with higher-energy electrons. The spectra of Fig. 32a,b were obtained in this way and show, as expected, that the concentration of the ion-beam-induced phases was higher near the interface than near the surface. Moreover, because the spectra with the elements in film and substrates exchanged appear quite

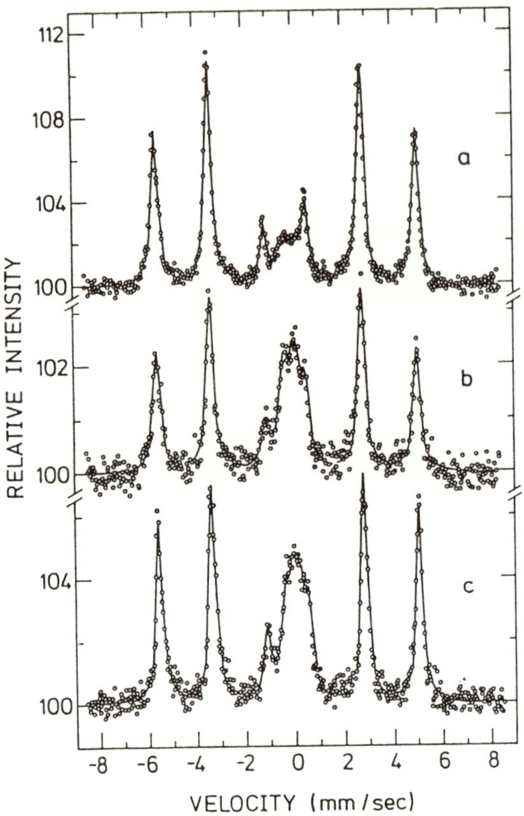

Figure 32. (a, b) Conversion electron Mössbauer spectra of iron-covered silicon irradiated to $9 \times 10^{16}\,Kr^+/cm^2$. (a) is measured with high-energy electrons and (b) with low-energy electron. (c) Integral conversion electron Mössbauer spectrum of silicon-covered iron irradiated to $5 \times 10^{16}\,Xe^+/cm^2$ (Ref. 78).

similar, a collisional mixing can be considered an important mechanism in ion beam mixing of the iron/silicon system. Ferromagnetic amorphous layers with a nonuniform distribution of iron and silicon atoms can be formed upon xenon ion irradiation of Fe_xSi_{1-x}, with x equal to 0.65 and 0.75.

Laser quenching has been used by Gupta *et al.*[79] as an additional technique, together with ion beam mixing, to induce nonequilibrium alloy phase formation in equiatomic iron/silicon multilayers. The samples consisted of six layers deposited on SiO_2/Si substrates and had an ~5-nm iron-57 layer deposited at one of the iron/silicon interfaces. Ion beam mixing was performed using 100-keV krypton ions at 150 K and laser irradiation was carried out under an argon protective atmosphere with a pulsed ruby laser.

Rutherford backscattering spectra of as-deposited and treated samples, shown in Fig. 33, indicate that the krypton ion beam is able to mix the outer part of the multilayer, while yielding little change in the inner iron and silicon layers, where the Mössbauer probe is located. In contrast, laser quenching produces more complete interdiffusion and even the possibility of interaction with the substrate. The corresponding Mössbauer spectra, shown in Fig. 34,

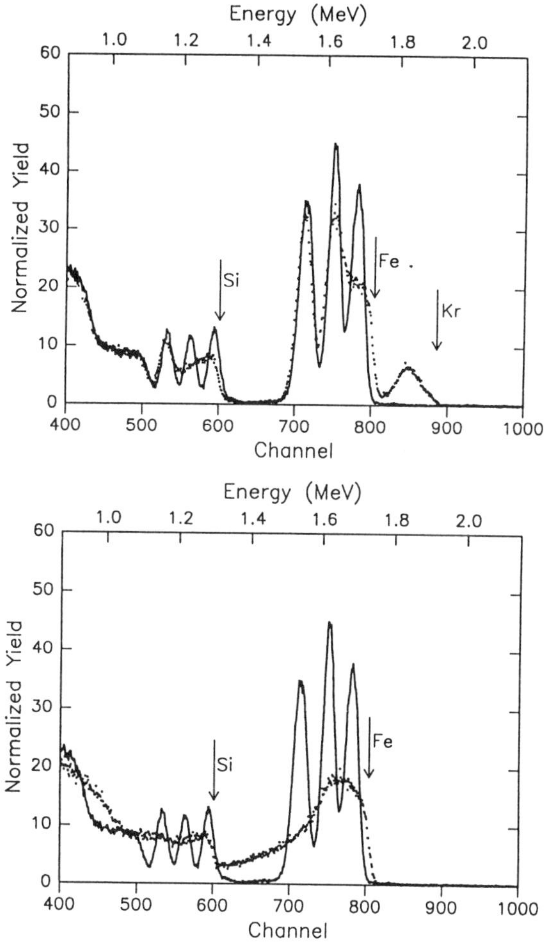

Figure 33. Partial Rutherford backscattering spectra of as-deposited (continuous line) and krypton ion (top)- and pulsed laser (bottom)-irradiated (points) iron–silicon multilayers (Ref. 79).

are consistent with the Rutherford backscattering results. Whereas the ion-beam-mixed sample gives rise to both a nonmagnetic and a broad magnetic component, the laser-treated sample shows only the nonmagnetic component. This nonmagnetic component is characterized in both cases by parameters intermediate between those of crystalline and amorphous FeSi alloys. The presence of the broad magnetic contribution is probably the result of a concentration gradient with regions richer in iron than would be expected in homogeneous $Fe_{50}Si_{50}$. This is, on the other hand, an iron concentration higher than that considered critical for the appearance of magnetism in amorphous Fe_xSi_{1-x} at room temperature.[80]

There are also literature reports of ion beam mixing in the tin–silicon system. Metallic 30- and 40-nm tin layers evaporated onto crystalline silicon

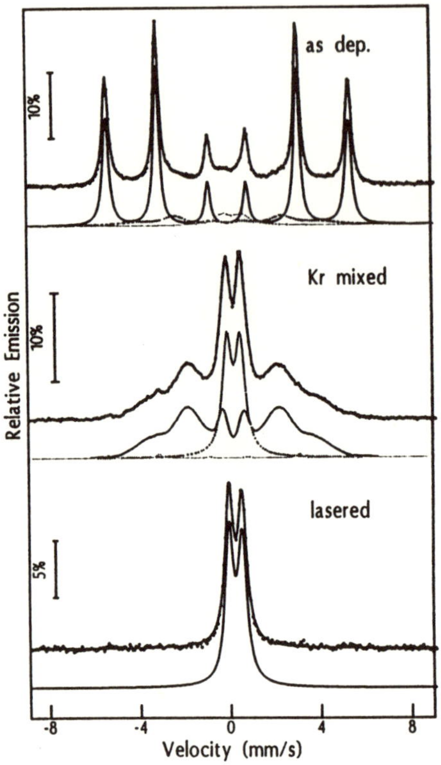

Figure 34. Conversion electron Mössbauer spectra of as-deposited and krypton ion- and pulsed laser-irradiated iron–silicon mutlilayers (Ref. 79).

and bombarded with argon ions have been analyzed by Prasad *et al.*[81] using Rutherford backscattering and conversion electron Mössbauer spectrometries. The Rutherford backscattering data suggest the occurrence of a recoil-type mixing, because the number of silicon atoms in the intermixed region is found to increase linearly with dose. The Mössbauer spectra (see Fig. 35) reveal an increase with dose of the number of tin atoms located substitutionally in silicon, resulting in the component at about 1.8 mm/s. β-tin is present even at the highest dose level, as indicated by the component at about 2.5 mm/s. The intensity of the component relating to SnO_2, at about zero velocity, decreases gradually with the dose, probably because of sputtering.

A study of 900-keV xenon ion-irradiated tin–silicon multilayers has also been performed by Massouras *et al.*[82] using Rutherford backscattering and Mössbauer spectrometries, transmission electron microscopy, and resistivity measurements. It shows that the product of irradiation is an amorphous semiconducting matrix with substitutional tin in silicon embedded with very small β-tin precipitates.

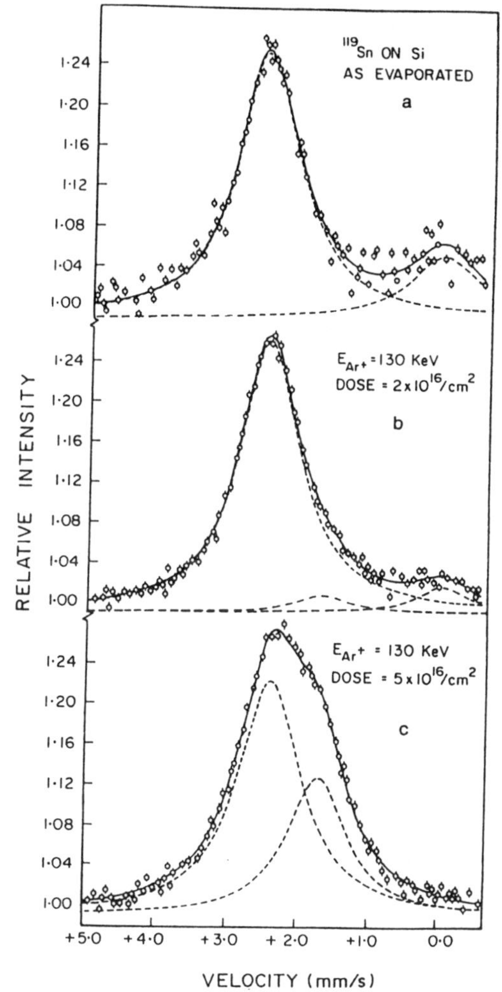

Figure 35. Conversion electron Mössbauer spectra of 45-nm tin-119 on silicon samples as evaporated (a) and 130-keV argon ion irradiated at a dose of 2 (b) and 5×10^{16} ions/cm^2 (c) (Ref. 81).

6. METAL/POLYMER SYSTEMS

An important technological problem is the improvement of the adhesion of thin metal films on polymeric substrates. The utility of ion beam irradiation to this problem has been explored and Mössbauer spectroscopy is of great benefit for the analysis when the metal is iron or tin.

Ion beam mixing of iron/polyethylene interfaces has been reported by Patil *et al.*[83] The as-deposited samples in this study give a Mössbauer spectrum in the form of a doublet with an isomer shift of 0.39 mm/s and a quadrupole splitting of 0.78 mm/s. This is probably the result of the formation of iron–carbon compounds via either vacuum contamination or excessive heating of the substrate. Unfortunately, this invalidates the subsequent process of ion beam mixing.

Griffith *et al.*[84] have demonstrated that ions in the megaelectron volt energy range have the ability to improve the adhesion of metals to substrates such as Teflon. It should be pointed out that the phenomena involved at the interfaces by ions in the megaelectron volt region are associated with the electronic stopping rather than the nuclear stopping, which is characteristic of the interactions produced by ions in the kiloelectron volt energy range. To exploit this difference, irradiation of iron films deposited onto Teflon or polyvinyl chloride with 16-MeV S^{3+} ions or 48-MeV Br^{8+} ions has been performed by Ingemarsson *et al.*[85,86] The deposited films consisted of a thermally evaporated layer of 2.5, 5, or 10 nm of iron-57 as well as a 15-nm overlayer of iron-56. The role of the iron-56 was to minimize atmospheric exposure of the Mössbauer probe layer.

Conversion electron Mössbauer spectroscopy shows that, even at doses as low as 10^{11} ions/cm^2, a considerable amount of the iron film is converted from α-iron into nonmagnetic compounds, which result from the chemical reactions with atoms from the underlying substrate. Typical results are shown in Fig. 36 for samples having polyvinyl chloride substrates. When the samples are bombarded with 16-MeV S^{3+} ions, an Fe^{2+} component, attributed to iron bonded with chloride ions, predominates. In contrast, irradiation with 48-

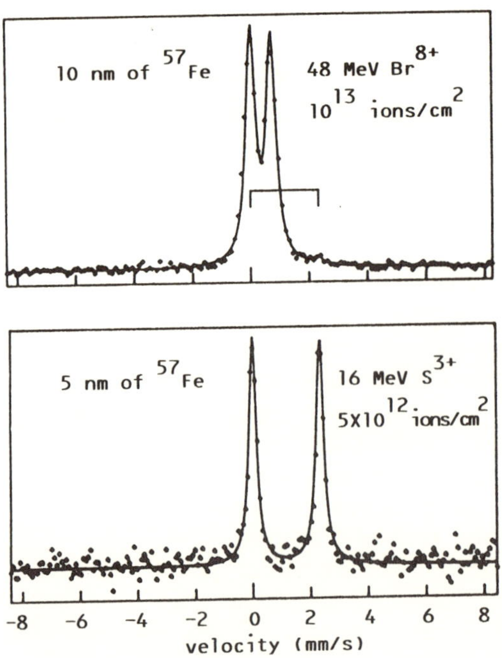

Figure 36. Conversion electron Mössbauer spectra of (top) 10-nm iron-57 on polyvinyl chloride bombarded with 48-MeV Br^{8+} to 10^{13} ions/cm^2 and (bottom) 5-nm iron-57 on polyvinyl chloride bombarded with 16-MeV S^{3+} to 5×10^{12} ions/cm^2 (Ref. 86).

MeV Br^{8+} ions seems to favor the formation of iron–carbon complexes. At ion doses lower than those used for the spectra shown in Fig. 36, the iron film remained partly unconverted as α-iron. It has been found that the relative amount of metallic iron, as measured by the area of the magnetic spectral component, decreases with decreasing film thickness, increasing stopping power of penetrating ions, and increasing dose.

7. CONCLUSIONS

The particular ability of Mössbauer spectroscopy to elucidate problems connected with irradiation-induced modifications on thin metallic layers has been demonstrated in this review. It has also been shown, by several examples, that the Mössbauer effect should be used in connection with other analytical methods so as to obtain a comprehensive understanding of the complex problems presented by these radiation-induced modifications.

ACKNOWLEDGMENTS. Financial contributions of CNR (91.02221.CT11) and MURST are gratefully acknowledged.

REFERENCES

1. H. De Waard and L. Niesen, in *Mössbauer Spectroscopy Applied to Inorganic Chemistry,* G. J. Long, ed., Vol. 2, p. 1., Plenum Press, New York, 1989.
2. J. W. Mayer and S. S. Lau, in *Surface Modification and Alloying,* J. M. Poate, G. Foti, and D. C. Jacobson, eds., p. 241, Plenum Press, New York, 1982.
3. S. Matteson and M.-A. Nicolet, *Annu. Rev. Mater. Sci.* 339 (1983).
4. L. C. Feldman and J. W. Mayer, *Fundamentals of Surface and Thin Film Analysis,* North Holland, Amsterdam, 1986.
5. J. A. Sawicki, *Mater. Sci. Eng.* **69**, 501 (1985).
6. B. X. Liu, E. Ma, J. Li, and L. J. Huang, *Nucl. Instrum. Methods* **B19/20**, 682 (1987).
7. A. R. Miedema, *Philips Tech. Rev.* **36**, 217 (1976).
8. J. A. Alonso and J. M. Lopez, *Mater. Sci.* **4**, 765 (1986).
9. L. M. Gratton, L. Guzman, and A. Molinari, *J. Microsc. Spectrosc. Electron.* **8**, 293 (1983).
10. L. M. Gratton, L. Guzman, A. Molinari, G. Principi, S. Lo Russo, and C. Tosello, in *Applications of the Mössbauer Effect,* Y. M. Kagan and I. S. Lyubutin, eds., p. 1415, Gordon & Breach, New York, 1985.
11. S. Moncher, G. Principi, L. M. Gratton, C. Tosello, I. Czako-Nagy, and A. Vertes, *Hyp. Interact.* **46**, 535 (1989).
12. I. Czako-Nagy, A. Vertes, L. M. Gratton, C. Tosello, and G. Principi, *J. Radioanal. Nucl. Chem. Lett.* **136**, 275 (1989).
13. J. H. Sanders, D. L. Edwards, J. R. Williams, and B. J. Tatarchuk, *J. Appl. Phys.* **67**, 3121 (1990).
14. B. Rodmaq, M. Piecuch, C. Janot, G., Marchal, and P. Mangin, *Phys. Rev.* B **21**, 1911 (1980).
15. K. Sumiyama, M. Kadono, and Y. Nakamura, *Trans. Jpn. Inst. Met.* **24**, 190 (1983).
16. G. Dumpich, E. F. Wassermann, V. Manns, W. Keune, S. Murayama, and Y. Miyako, *J. Magn. Magn. Mater.* **67**, 55 (1987).
17. L. M. Gratton, A. Gupta, W. Keune, S. Lo Russo, J. Parellada, G. Principi, and C. Tosello, *Mater. Sci. Eng.* **A115**, 161 (1989).

18. C. Tosello, L. M. Gratton, W. Keune, R. A. Brand, S. Lo Russo, J. Parellada, J. M. Fernandez, G. Marest, M. A. El Khakani, G. Principi, and A. Gupta, *Nucl. Instrum. Methods* **B59/60**, 541 (1991).
19. G. Marest and M. A. El Khakani, *Nucl. Instrum. Methods* **B59/69**, 833 (1991).
20. C. Tosello, F. Ferrari, R. A. Brand, W. Keune, G. Marest, M. A. El Khakani, J. Parellada, G. Principi, S. Lo Russo, V. Rigato, and S. Enzo, Proc. Int. Conf. IBMM, Heidelberg, Sept. 1992 (*Nucl. Instrum. Methods,* in press).
21. V. P. Godbole, M. S. Chaudari, S. V. Ghaisas, S. M. Kanetkar, and S. B. Ogale, *Phys. Rev. B* **31**, 5703 (1985).
22. B. Rauschenbach, K. Hohmuth, R. Kükler, and M. Posselt, *Nucl. Instrum. Methods* **B39**, 141 (1989).
23. R. J. Gaboriaud, J. J. Grob, and F. Abel, *Nucl. Instrum. Methods* **B19/20**, 648 (1987).
24. M. A. Z. Vasconcellos, S. R. Teixeira, F. L. Freire, M. C. S. Nobrega, P. H. Dionisio, W. H. Schreiner, and I. J. R. Baumvol, *Mater. Sci. Eng.* **A104**, 169 (1988).
25. M. A. Z. Vasconcellos, J. A. T. Borges da Costa, W. H. Schreiner, and I. J. R. Baumvol, *Phys. Status Solidi* **112**, 105 (1989).
26. M. A. Z. Vasconcellos, S. R. Teixeira, P. H. Dionisio, W. H. Schreiner, and I. J. R. Baumvol, *Nucl. Instrum. Methods* **A280**, 557 (1989).
27. M. A. Z. Vasconcellos and J. A. T. Borges da Costa, *Hyp. Interact.* **66**, 331 (1991).
28. C. Jaouen, J. P. Eymery, E. L. Mathe, and J. Delafond, *Mater. Sci. Eng.* **69**, 483 (1985).
29. B. Rauschenbach and K. Hohmuth, *Nucl. Instrum. Methods* **B23**, 323 (1987).
30. K. Hohmuth, V. Heera, and B. Rauschenbach, *Nucl. Instrum. Methods* **B39**, 136 (1989).
31. B. X. Liu, G. A. Cheng, and C. H. Shang, *Philos. Mag. Lett.* **55** 265 (1987).
32. L. M. Gratton, A. Miotello, C. Tosello, D. C. Kothari, G. Principi, and A. Tomasi, *Nucl. Instrum. Methods* **B59/60**, 541 (1991).
33. J. C. Plenet, A. Perez, J. Rivory, and O. Laborde, Proc. Int. Conf. IBMM, Heidelberg, Sept. 1992 (*Nucl. Instrum. Methods,* in press).
34. W. Keune, J. Lauer, and D. L. Williamson, *J. Phys.* **35**, 473 (1974).
35. F. Z. Cui, A. M. Vredenberg, R. de Reus, F. W. Saris, and H. J. G. Draaisma, *J. Less-Common Met.* **145**, 621 (1988).
36. C. L. Chien, S. H. Liou, D. Kofalt, W. Yu, T. Egami, and T. R. McGuire, *Phys. Rev. B* **33**, 3247 (1986).
37. L. J. Huang and B. X. Liu, *Nucl. Instrum. Methods* **B18**, 256 (187).
38. L. J. Huang, B. X. Liu, and H. D. Li, *Appl. Phys.* **A44**, 269 (1987).
39. L. J. Huang, M. Q. Cheng, Y. D. Fan, H. D. Li, and B. X. Liu, *J. Phys. F* **18**, L69 (1988).
40. A. Gupta, G. Principi, P. Trotta, E. Jannitti, C. Tosello, L. M. Gratton, S. Enzo, S. Lo Russo, and V. Rigato, *Surf. Coat. Technol.* **51**, 429 (1992).
41. J. Li and B. X. Liu, *Phys. Status Solidi A* **99**, K31 (1987).
42. R. Joshee, D. M. Phase, S. V. Ghaisas, S. M. Kanetkar, and S. B. Ogale, *J. Appl. Phys.* **59**, 388 (1986).
43. R. Brenier, A. Perez, P. Thevenard, M. Treilleux, and T. Capra, *Mater. Sci. Eng.* **69**, 83 (1985).
44. R. Brenier, P. Thevenard, T. Capra, A. Perez, M. Treilleux, L. Romana, J. Dupy, and M. Brunel, *Nucl. Instrum. Methods* **B19/20**, 691 (1987).
45. K. Sumiyama, Y. Hashimoto, and Y. Nakamura, *Trans. Jpn. Inst. Met.* **24**, 61 (1983).
46. F. R. Ding, P. R. Okamoto, and L. E. Rhen, *Nucl. Instrum. Methods* **B39**, 122 (1989).
47. J. Bottiger, K. Pampus, and B. Torp, *Nucl. Instrum. Methods* **B19/20**, 696 (1987).
48. M. Kopcewicz, D. L. Williamson, and F. E. Cecil, *Hyp. Interact.* **56**, 1599 (1990).
49. J. P. Riviere, P. Bouillaud, and J. P. Eymery, *Phys. Lett. A* **138**, 223 (1989).
50. G. Battaglin, S. Lo Russo, B. Bertoncello, G. Granozzi, G. Fagherazzi, and P. Mengucci, *J. Less-Common Met.* **145**, 513 (1988).
51. G. Battaglin, R. Bertoncello, G. Granozzi, G. Fagherazzi, S. Lo Russo, and P. Mengucci, *Mater. Sci. Eng.* **A115**, 165 (1989).

52. A. Gupta, G. Principi, G. Battaglin, S. Lo Russo, and Q. Zhang, *Hyp. Interact.* **56**, 1587 (1990).
53. L. Calliari, L. M. Gratton, L. Guzman, G. Principi, and C. Tosello, *M. R. S. Soc. Symp. Proc.* **27**, 85 (1984).
54. I. Czako-Nagy, L. Fedrizzi, L. M. Gratton, G. Principi, and A. Vertes, *Hyp. Interact.* **29**, 1187 (1986).
55. G. Battaglin, S. Lo Russo, A. Paccagnella, G. Principi, and P. Q. Zhang, *Nucl. Instrum. Methods* **B27**, 402 (1987).
56. P. Q. Zhang, G. Principi, A. Paccagnella, S. Lo Russo, and G. Battaglin, *Nucl. Instrum. Methods* **B28**, 561 (1987).
57. G. Principi, P. Q. Zhang, G. Battaglin, S. Lo Russo, and A. Paccagnella, *J. Mater. Sci. Lett.* **7**, 484 (1988).
58. C. Wilkinson, A. K. Cheetham, J. G. Long, P. D. Battle, and D. A. O. Hope, *Inorg. Chem.* **23**, 3136 (1984).
59. J. M. D. Coey, *J. Phys.* **35**, C6-89 (1974).
60. A. Perez, R. Maudre, P. Thevenard, and P. Sibut, in *Induced Defects in Insulators,* P. Mazzoldi, ed., p. 171, Les Editions de Physique, Paris, 1984.
61. A. Perez, M. Treilleux, T. Capra, and D. L. Griscom, *J. Mater. Res.* **2**, 910 (1987).
62. D. L. Griscom, J. J. Krebs, A. Perez, and M. Treilleux, *Nucl. Instrum. Methods* **B32**, 272 (1988).
63. G. Battaglin, S. Lo Russo, A. Paccagnella, P. Polato, and G. Principi, *Nucl. Instrum. Methods* **B39**, 126 (1989).
64. G. Battaglin, S. Lo Russo, A. Paccagnella, P. Polato, and G. Principi, in *Structure Property Relationships in Surface Modified Ceramics,* C. J. McHargue, R. Kossowsky, and W. O. Hofer, eds., p. 189, Kluver, Dordrecht, 1989.
65. G. Principi, P. Polato, A. Paccagnella, A. Maddalena, S. Lo Russo, and G. Battaglin, *Hyp. Interact.* **46**, 517 (1989).
66. G. Battaglin, S. Lo Russo, A. Paccagnella, P. Polato, and G. Principi, *J. Non-Cryst. Solids* **123**, 427 (1990).
67. T. Banwell and M.-A. Nicolet, *Nucl. Instrum. Methods* **B19/20**, 704 (1987).
68. S. B. Ogale, D. M. Phase, S. M. Chaudari, S. B. Ghaisas, S. M. Kanetkar, P. P. Patil, V. G. Bhide, and S. K. Date, *Phys. Rev. B* **35**, 1593 (1987).
69. S. Bhagwat, S. N. Yedave, D. M. Phase, S. M. Chaudari, S. M. Kanetkar, and S. B. Ogale, *Phys. Rev. B* **40**, 700 (1989).
70. S. B. Ogale, D. M. Phase, P. P. Patil, S. M. Kanetkar, S. V. Ghaisas, V. G. Bhide, and S. K. Date, *Hyp. Interact.* **29**, 1193 (1986).
71. E. Abonneau, G. Fuchs, M. Treilleux, and A. Perez, *Nucl. Instrum. Methods* **B46**, 111 (1990).
72. G. Fuchs, E. Abonneau, M. Treilleux, and A. Perez, *J. Mater. Sci. Lett.* **9**, 97 (1990).
73. A. Perez, E. Abonneau, G. Fuchs, M. Treilleux, C. J. McHargue, and D. L. Joslin, *Nucl. Instrum. Methods* **B65**, 129 (1992).
74. M. Carbucicchio, A. Valenti, G. Battaglin, P. Mazzoldi, and R. Dal Maschio, *Radiat. Eff.* **98**, 21 (1986).
75. S. S. Lau, B. Y. Tsaur, M. Allmen, J. W. Mayer, B. Stritkzer, C. W. White, and B. Appleton, *Nucl. Instrum. Methods* **182/183**, 97 (1981).
76. J. Patankar, Y. V. Bhandarkar, S. M. Kanetkar, S. B. Ogale, and V. G. Bhide, *Nucl. Instrum. Methods* **B7/8**, 720 (1985).
77. S. B. Ogale, R. Joshee, V. P. Godbole, S. M. Kanetkar, and V. G. Bhide, *J. Appl. Phys.* **57**, 291 (1985).
78. N. Hayashi, I. Sakmoto, and A. Tanque, *Hyp. Interact.* **29**, 1183 (1986).
79. A. Gupta, R. Gupta, G. Principi, E. Jannitti, C. Tosello, L. M. Gratton, S. Lo Russo, V. Rigato, and R. Frattini, *Surf. Coat. Technol.* **51**, 41 (1992).
80. G. Marchal, P. Mangin, M. Piecuch, and C. Janot, *J. Phys.* **37**, C6-763 (1976).

81. K. G. Prasad, M. B. Kurup, and A. Bhagawat, *Nucl. Instrum. Methods* **B15**, 698 (1986).
82. G. Massouras, J. A. Roger, A. Perez, G. Fuchs, and L. Romana, *Hyp. Interact.* **46**, 509 (1989).
83. P. P. Patil, N. Kulkarni, S. V. Ghaisas, S. B. Ogale, V. G. Bhide, and K. Date, *Hyp. Interact.* **35**, 919 (1987).
84. J. E. Griffith, Y. Qiu, M. H. Mendenhall, and T. A. Tombrello, *Nucl. Instrum. Methods* **198**, 607 (1982).
85. P. A. Ingermarsson, T. Ericsson, A. Gustavsson-Siedel, G. Possnert, B. U. R. Sundqvist, and R. Wäppling, in *Application of Ion Beams in Materials Science,* T. Sebe and Y. Yamamoto, eds., p. 323, Hosei University Press, Tokyo, 1987.
86. P. A. Ingemarsson, T. Ericsson, G. Possnert, and R. Wäppling, *Hyp. Interact.* **46**, 549 (1989).

3

Applied Field Mössbauer Spectroscopy of Magnetic Powders

Q. A. Pankhurst and R. J. Pollard

1. INTRODUCTION

The study of fine-particle magnetism is of importance because of the natural abundance of various fine-particle materials, and because the special properties of finely divided materials lead to their utilization in a wide range of manufactured goods. Naturally occurring fine particles include the chains of magnetite particles found in magnetotactic bacteria, the several grams of iron stored in the human body as the ferrihydrite cores of storage proteins, and the magnetic phases and weathering products found in rocks, soils, and coals. Manufactured magnetic fine particles include the magnetite particles used in fluid seals and bearings, and the particles of maghemite, barium ferrite, and others used in computer hard and floppy disks, credit cards, and audio and video recording tapes.

Fine particles exhibit special properties, different from those of the bulk material, which are of intrinsic interest in terms of the underlying physics governing the behavior, as well as being of direct utility in industry. For example, below a critical size, ferrimagnetic particles contain only a single magnetic domain, giving rise to a large magnetic moment per unit volume, and a large magnetic anisotropy. The large magnetization and anisotropy are advantageous in magnetic fluid and recording applications, while the single-

Q. A. Pankhurst • Department of Physics, University of Liverpool, Liverpool L69 3BX, England. *R. J. Pollard* • Department of Physics, Monash University, Clayton, Victoria 3168, Australia.

Mössbauer Spectroscopy Applied to Magnetism and Materials Science, Volume 1, edited by Gary J. Long and Fernande Grandjean. Plenum Press, New York, 1993.

domain structure allows relatively simple theoretical modeling of the physical processes involved in the response of such a system to an external field.

Other special properties of fine particles include the effects of the relatively high surface-to-volume ratio, and the related effect of particle coating, which in some cases can dramatically alter the magnetic properties of the base material. In the limit of very small particles, long-range magnetic order may break down, and the effects of lattice vacancies and other anomalies can become prominent. For example, the latter is observed in ultrafine particles of some antiferromagnets, where a net moment becomes apparent.

Superconducting magnets are now standard pieces of equipment in many Mössbauer laboratories. Most commonly, they are used in studies of relatively simple magnetic systems such as diamagnets, paramagnets, ferromagnetic and strongly ferrimagnetic powders, and antiferromagnetic single crystals. In these cases the Mössbauer spectra are relatively easy to understand. For example, oriented antiferromagnetic crystals display well-resolved subspectra corresponding to each magnetic sublattice. The direction of the sublattice magnetization may be determined directly from both the splittings and the intensities of the absorption lines,[1] thus enabling the study of magnetic phase transitions in antiferromagnets.

In comparison, the applied field Mössbauer spectra of powders with internal magnetic ordering are much more complex, often exhibiting a combination of narrow and broad line profiles. Frequently, such spectra are interpreted in purely qualitative terms, and the fundamental parameters relevant to the magnetic material are left undetermined and are not extracted from the data.

In this chapter we review recent advances in the modeling and analysis of the applied field Mössbauer spectra of magnetic powders. Such analysis is of basic scientific interest, since it removes the ambiguities present in qualitative interpretations of the spectra, and permits a rigorous testing of any models. It also enhances the potential of applied field Mössbauer spectroscopy as an analytical technique for studying fine-particle magnetism. The majority of the work presented refers to the authors' own studies of the past few years, but where appropriate the work and results of other research groups, using the same or different analysis methods, are included. Least-squares fitting results for ^{57}Fe Mössbauer spectra of all of the most common iron oxides and oxyhydroxides are presented. This is not an attempt to exclude other isotopes, but rather a reflection of the fact that iron is the world's most commonly encountered and most often used magnetic element. We conclude with some suggestions for possible future work in the field.

2. CONTEXT

Before embarking on a review of the physics and methodology of applied field Mössbauer spectroscopy of magnetic powders, it is important to establish a context for the work—to address questions such as: "Why is it relevant to investigate magnetic powders?" and "Why use Mössbauer spectroscopy?" We therefore provide, in this section, a brief review of selected contemporary applications involving fine-particle magnetic materials and their response to an applied field, a summary of some of the fundamental questions related to the physics of magnetism in small volumes, and some comments on the particular benefits offered by applied field Mössbauer spectroscopy to the study of fine-particle magnetism.

2.1. Applications

2.1.1. Magnetic Fluids

A wide range of manufacturing and engineering applications utilize magnetic powders suspended in a liquid or an oil to form a magnetic fluid.[2,3] The underlying principle of operation is that the fluid in essence behaves as if it were a "liquid magnet". The fluid is usually based on either magnetite (Fe_3O_4), Fe or Co particles, and the choice of carrier liquid is dependent on the fluid dynamical properties required.

The most common use of magnetic fluids is in O-ring seals and bearings, where a combination of permanent magnets and pole pieces are arranged so that intense magnetic field regions retain the fluid as distinct rings around a rotating shaft (Fig. 1). Seals capable of maintaining integrity under vacuums of 10^{-8} Torr and fluid–shaft interface speeds of up to 10 m/s are industry standards. Such seals are found in lasers, CAT scanners, sputtering systems, and Winchester-type computer disk drives.

A second important use of magnetic fluids is in the Bitter method of magnetic imaging,[4,5] particularly as applied to studies of permanent magnet materials[6,7] and geological specimens,[8,9] the quality control of magnetic recording media,[10,11] and the identification of structural defects in massive steel structures.[12] A fluid carrying fine particles of magnetite is spread on the object

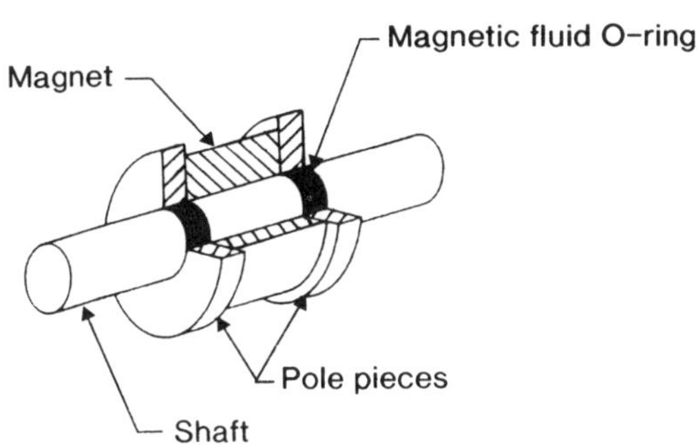

Figure 1. Schematic diagram of a magnetic fluid O-ring seal around a rotating shaft.

of interest. The particles are attracted to regions where the product of the field and the field gradient is largest, and thus congregate at domain walls.[13,14] The technique compares favorably with other methods of domain observation,[15] and continues to be developed, as in a recent application involving scanning tunneling microscopy.[16]

Magnetite particles of diameter ~70 Å are used in the quality control of recording media. This enables the resolution of the ~2-μm-wide uniformly magnetized regions on the recording surface that correspond to individual "bits" ("binary digits"). The same magnetic imaging principle applies in the oil and gas industries, albeit on a much larger scale, in the technique of "magnetic particle inspection." Here a suspension of magnetite particles is sprayed onto the transversely magnetized surface of, for example, a steel girder. Any cracks in the surface, provided they do not run parallel to the magnetizing direction, result in leakage fields above the surface, which in turn attract the magnetite particles and make the defects visible to an operator.

Among the many other applications of magnetic fluids are inertia dampers used in stepper motors and in audio loudspeakers, angular position sensors used in avionics, robotics, and medicine, and transducers used for aerodynamic, acceleration, and vibration measurements.

2.1.2. Magnetic Separation

Utilization of the effect of applied fields on fine particles to achieve filtering and phase separation is well established in the mineral processing industry.[17] From early days when simple magnets were used to extract strongly magnetic minerals from slurries, the technology of separation has developed to the stage where applied fields are now used on a commercial scale to extract micrometer-sized particles of feebly magnetic materials. This improvement has been largely the result of the development of magnet systems producing both high field intensity and high field gradient.[18,19] In a typical high-gradient magnetic filter, a slurry is passed through a stainless-steel wire mesh in a homogeneous background field of sufficient magnitude to saturate the magnetization in the wires (Fig. 2). Regions of high field and gradient attract any magnetic particles toward the wires. Filters comprising steel wool of diameter ~10 μm in an external field of ~10^7 A/m are capable of capturing particles of 1 to 100-μm diameter from a fluid flowing at a rate of up to ~10 cm/s. The captured particles can be collected by the simple expedient of switching off the external field and flushing the mesh out with distilled water.

Some separation techniques are based on particle deflection rather than particle capture, and can therefore be operated continuously. In "open gradient magnetic separation" the slurry falls under gravity either through the centre of a cylindrical Helmholtz split-pair solenoid, or on either side of the faces of a linear periodic array of magnets.[20] The applied field produces a lateral deflection force on the falling particles, so that the magnetic fractions are selected out.

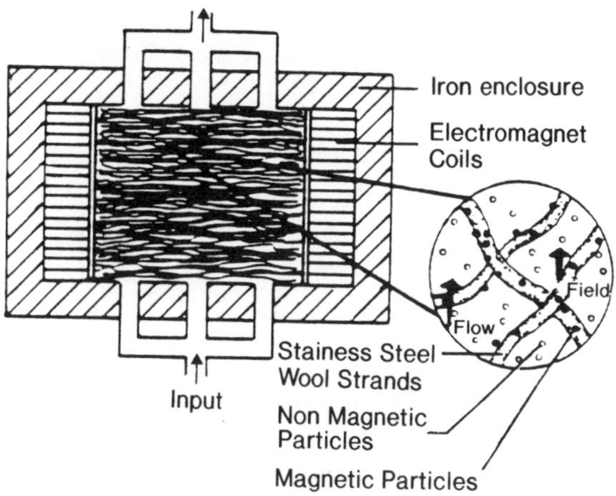

Figure 2. Cross section of a high-gradient magnetic filter. A slurry containing magnetic and nonmagnetic particles is forced through a stainless-steel wire mesh. The high magnetic field and gradient attracts the magnetic particles to the mesh, while the nonmagnetic particles are not affected and pass through the filter.

2.1.3. Magnetic Recording

The advent of high-density magnetic recording systems has contributed greatly to the prominence of both information technology and computer-based operating systems in our everyday lives. Today we are surrounded by magnetic records: audio and video tapes, identification and credit cards, computer floppy and hard disk memories, and so forth. There is continued research and development aimed at increasing storage capacity, density, and quality, much of which has been published and reviewed in the literature.[21–25]

The basic principles of the magnetic recording process are illustrated in Fig. 3. The signal is supplied as a current in a coil wound around a ring-shaped writing "head," and a fringe magnetic field emerges from the "gap" (a narrow cut in the ring) which magnetizes the particles on the tape or disk as it moves

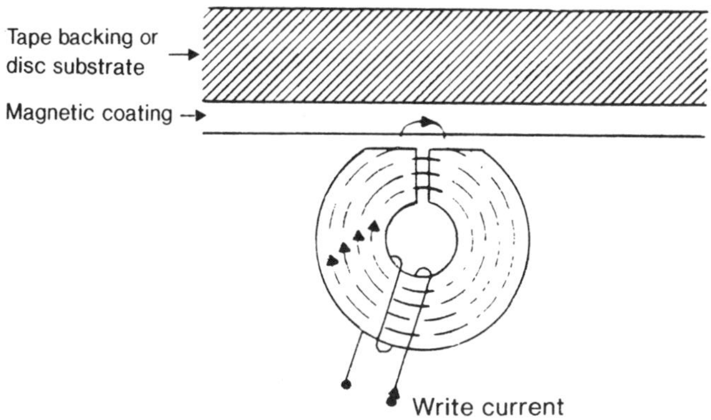

Figure 3. Schematic diagram of the writing process in magnetic recording. The "write current" produces a fringe field at a discontinuity in the ring-shaped "head," which magnetizes the particles in the magnetic coating.

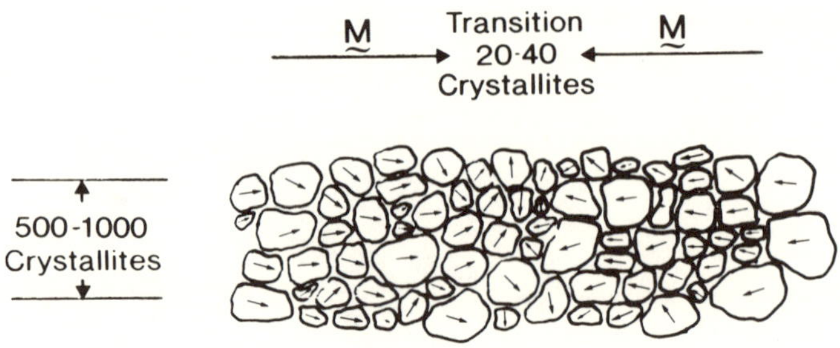

Figure 4. Schematic diagram, not to scale, of the transition region between two oppositely magnetized regions on the surface of a particulate recording medium track. The small arrows represent the magnetization direction within individual particles.

past the head. On replay, a flux is induced in the reading head by the spatially varying magnetization, which in turn produces a current in the output coil.

The recording medium should have properties similar to those of permanent magnets: large remanent magnetization and square hysteresis loops. It should also be easily written, read, and erased, but stable and hard to corrupt. Hence, the coercivity should be moderate. The elementary magnetic grains or particles should be physically small and magnetically independent to permit short-wavelength recording. To date, these prerequisites are best met by two types of media: particulate dispersions (γ-Fe_2O_3, Co-modified γ-Fe_2O_3, CrO_2, or barium ferrite) and thin metallic films (Fe or Co–Fe). During digital recording, the magnetization direction is switched between two stable preferred directions in the single-domain magnetic units. A schematic diagram of a transition region on the surface of a particulate recording medium is shown in Fig. 4. As bit densities and data transmission rates increase, a detailed understanding of the fundamental magnetization reversal process is becoming more important.

2.1.4. Biomagnetism and Medicine

Magnetite particles formed as biochemical precipitates have been identified in three of the five animal kingdoms,[26] although they are probably best known as the navigational "compasses" found in magnetotactic bacteria.[27] For orientational purposes, these magnetite particles are single-domain crystals in the form of a chain (a magnetosome). This chain acts as a bar magnet that, despite its small volume, is very sensitive to the geomagnetic field.

Since their discovery over a decade ago, magnetotactic bacteria are finding many applications. They have been coated onto transformer sheets for domain imaging[28] and give useful information for rock magnetism and paleomagnetism by acting as "biomagnetometers."[29] Their ability to ingest or precipitate ion species, and to be subsequently removed via a magnetic separator, provides a means of extracting heavy metals.[30,31] Their magnetic particles have been isolated for uses in enzyme immobilization.[32]

Magnetic particles have numerous medical uses. Magnetic particles can act as a contrast agent for magnetic resonance imaging[33,34] and can be used for cell separation[35,36] as in some cancer treatments.[37,38] Chromium dioxide particles are used as the solid support in enzyme immunoassay.[39] Magnetic powders have also been used to orient biological assemblies[40] and to isolate red blood cells.[41]

2.2. Fundamental Research

The impetus for fundamental research into fine-particle magnetism comes from two sources. First, such research is a natural component of the research and development programs of the manufacturers of products that rely on the properties of magnetic particles, especially magnetic fluids and magnetic recording media. Second, fundamental research is undertaken by independent scientists, usually working in universities or similar institutions, whose primary concern is to gain an understanding of the physical principles governing the behavior and properties of particulate matter. Not surprisingly, these two strands of research are closely related, as can be seen from the literature where one often finds papers that present "pure" research into materials and processes that are, in essence, application motivated.

The themes outlined below represent those questions that are currently the subject of concerted research efforts. These questions fall into two categories: those related to the macroscopic and dynamical behavior of magnetic powders, especially the problem of predicting the effects of interparticle interactions, and those related to the microscopic properties of individual particles, such as the effects of surface coatings, the nature of magnetization reversal in ferrimagnets, and the breakdown of magnetic order in particles of very small dimension.

2.2.1. Interparticle Interactions

In magnetic fluid applications it is important that the fluid be homogeneous and isotropic, and resistant to cluster formation and aggregation. In commercial fluids, adsorbed surfactants and surface electric charges are employed to prevent aggregation, but there are many unresolved issues that are the subject of research. For example, no dispersants have yet been found that can entirely offset the magnetic attractions between single-domain particles in the size range 15–500 nm.[42] In other fluids applications, the formation of aggregates is desirable, as in magnetic particle inspection, and in magnetic separation via flocculation. In high-density magnetic recording media it is found that as the packing density increases, interaction effects become apparent in the response of the medium, modifying the reversal mechanism and related parameters.[43–45]

Theoretical work on aggregations in fluids has largely been based on computer simulations and numerical modeling,[46,47] and scaling laws have been proposed.[48,49] Several models are used for magnetostatic interactions in particulate media, including the "Preisach" distribution function[50] and its modern "moving vector" variants,[51] mean field models,[52] analytical formulations,[53] and Monte Carlo simulations.[54,55] A wide range of experimental techniques are used to monitor interparticle interaction effects, including magnetic susceptibility[56,57] and optical microscopy[58] measurements, as well as other bulk magnetic measurements. Mössbauer effect studies have largely focused on the observable effects of interactions on the phenomenon of superparamagnetic relaxation.[59–62]

2.2.2. Surface Effects

As the physical dimensions of a particle get smaller, the proportion of those ions that are at or near the surface of the particle increases. Thus, small particles offer the potential of exhibiting special effects that are not observed in bulk materials. A great deal of research effort is devoted to this field, utilizing all manner of theoretical and experimental methods on a variety of particulate materials.

An illustration of the nature of research into surface effects comes from the magnetic recording industry. In the early 1970s a method was discovered whereby the uniaxial anisotropy, and therefore the coercivity, of iron oxide particles could be nearly doubled without adversely affecting the temperature dependence of the coercivity. The technique relies on the diffusion of cobalt ions into a surface layer on the particle.[63] This increased coercivity allowed the development of high-density recording media, and today Co-modified acicular particles of maghemite (γ-Fe_2O_3) are the preferred choice in a number of commercial applications, especially in video and audio tapes. Given the success of these surface-modified materials, it is surprising that as yet the mechanism by which the surface treatment increases the coercivity is not well understood.[22,64,65] It is clear that if such materials are to be able to meet future demands, a good deal more fundamental research work will be required. Along with other experimental methods, applied field Mössbauer spectroscopy has proved to be an invaluable tool in studying these surface-modified particles,[67–71] although there have recently been suggestions that some of the earliest interpretations of the Mössbauer results were not entirely correct.[72] This work will be discussed in more detail later.

2.2.3. Magnetization Reversal Mechanisms

Another area of basic research that is motivated by the requirements of the magnetic recording industry is in micromagnetic modeling of the process of magnetization reversal in single-domain ferro- or ferrimagnetic particles.[73,74]

The primary objective of the research is to develop an understanding of the writing process in particulate recording media that is sufficient to enable improvements in manufactured media.

In the writing process the magnetization direction is reversed between two energetically equivalent directions in the crystal lattice. The central question is whether or not the individual moments in the particle move in unison. Many years ago, analytical expressions were derived for the three physically possible modes of "cooperative" magnetization reversal (i.e., where the rotation is initiated and proceeds uniformly over the whole particle), namely "coherent rotation", "curling," and "buckling," for the simple case of an isolated particle shaped like an ellipsoid of revolution.[75–77] For deviations from the ellipsoidal shape, no analytic expressions have been derived, but for the case of acicular shaped particles, such as γ-Fe_2O_3, a model comprising a linear chain of touching spheres[76] proved to be successful. In this case the reversal mode is similar to buckling, and is called the "fanning" mode. More recently, experimental[78] and theoretical[79] work has suggested a "nucleation" model in which the magnetization reversal is initially localized at sites of lower energy because of pits on the surface and defects in the particles. Theoretical research continues, largely based on computer simulations, on three fronts: elongated particles treated as a single entity,[80–82] the chain of spheres model,[83,84] and particles with defects.[85] Experimentally, research continues to be based on bulk measurements of parameters such as the coercivity and the anisotropy field, although techniques such as applied field Mössbauer spectroscopy, which offer the potential of distinguishing the micromagnetic behavior of the individual moments, are being increasingly used.

There have recently been observations of macroscopic quantum tunneling of magnetization. This may occur by a variety of mechanisms,[86] including tunneling of the total magnetization between easy directions in single-domain particles,[87,88] the quantum nucleation of magnetization reversal, and the tunneling of domain walls. The materials include small ferromagnetic particles,[89,90] and antiferromagnets[91] including proteins.[92] Typically the particles are smaller than 20 nm, and the effects are observed at temperatures lower than 2 K. Implications of this research for applications of magnetic materials are an interesting area for further work.

2.2.4. Structural and Magnetic Disorder

There are many intriguing fundamental questions relating to the effect of small particle size on the nature of the magnetically ordered state of a material. We mention here two topics that have received recent attention in the literature: the appearance of a weak net magnetic moment in small antiferromagnetic particles, and the possibility of a unique type of magnetic ground state existing in disordered insulators. In both cases the work will be discussed in more detail later.

In 1962 Néel suggested that a size-dependent net magnetic moment might be observed in extremely small antiferromagnetic particles containing only a few magnetic atoms as the result of unequal sublattice occupancies.[93] Within a decade, weak magnetic moments were observed in naturally occurring goethite (α-FeOOH), with the strength of the moment increasing both with decreasing particle size and with increasing amounts of Al^{3+} ion impurities substituting for Fe^{3+} ions in the lattice.[94,95] Following Néel's suggestion,[94] the number of unbalanced moments in a particle containing N moments in total would be of order $N^{1/2}$, giving a net moment per magnetic atom of order $N^{-1/2}$. This was too small to account for the moments observed in goethite, and an alternative model was proposed, in which the diamagnetic Al^{3+} impurity ions preferentially occupied one of the two sublattices.[95] Recently, another explanation was put forward, in which structural defects give rise to significant numbers of unpaired surface moments.[96] However, this effect has been shown to be too small to account for the ferrimagnetic moment observed, and systematic applied field Mössbauer experiments have now been completed that support the notion that the net moment originates as a result of the nonrandom aluminum substitution.[97]

Another important problem in magnetism is the nature of the magnetic ground state in disordered materials. Twenty years ago, applied field Mössbauer effect measurements on an amorphous insulator, a ferric gel containing fine particles of $Fe(OH)_3 \cdot 0.9H_2O$ in an organic matrix, led to the proposed existence of a new type of magnetic order, termed "speromagnetism".[98,99] In speromagnets, the spin arrangement on an atomic nearest-neighbor level is disordered, despite the presence of strong antiferromagnetic exchange interactions. Very recently, it was suggested that since the identification of speromagnetism is largely a matter of scale, and that since it is not possible to use Mössbauer methods to distinguish between random spin structures on an atomic level and random structure at the level of domains, crystallites, and particles, the entire question of the existence or otherwise of speromagnetism is open.[100] However, given this proviso, it is clear that Mössbauer experiments will continue to be an important tool for research into disordered materials.

2.3. Applied Field Mössbauer Spectroscopy

Having established the motivation for conducting research into the properties of magnetic powders, we now consider the potential of the experimental technique of applied field Mössbauer spectroscopy. First, by far the majority of the magnetic materials mentioned in the preceding sections contained iron as their magnetic element. ^{57}Fe Mössbauer spectroscopy is ideally suited to the study of the magnetism of iron: it is a cheap, nondestructive, and reliable technique that offers high resolution of charge state, ordering temperature, magnetic moment direction, and other parameters. Second, the use of polycrystalline samples in Mössbauer spectroscopy offers very useful side benefits: absorbers of ideal thickness may be prepared, so as to maximize the spectral signal-to-noise ratio and reduce the required measurement time. Also, the homogeneity of the powder may be utilized to simplify subsequent analysis. This is an important consideration when the only available alternatives are poorly characterized single crystals for which mosaicity and alignment uncertainties are major problems.

There are several distinct regimes in which applied field Mössbauer spectroscopy is currently being used as a research tool. These include the study of superparamagnetic systems at temperatures near the blocking transition,[62,101,102] the determination of anisotropy fields in magnetic recording media using small applied fields,[103] and studies of the behavior of ferromagnetic iron and iron-containing particles supported in a carbon or alumina matrix.[104,105] However, a relatively new area of research is in the modeling and analysis of the low-temperature, high-applied-field Mössbauer spectra of antiferromagnetic and ferrimagnetic iron oxides and oxyhydroxides. The rich variety of magnetic behavior and phenomena that fall within this area are illustrated by the set of Mössbauer spectra shown in Fig. 5. The fact that the spectra are all recorded at low temperatures, generally at 4.2 K, means that the materials are in their magnetic ground states, which greatly simplifies the theoretical analysis. The use of large applied fields also simplifies analysis, since it accentuates any line splitting or broadening in the spectra. The theoretical modeling, and the experimental results obtained using applied field Mössbauer spectroscopy in this way, form the basis of the remainder of this chapter.

3. PHYSICAL BEHAVIOR AND MODELING

When faced with applied field Mössbauer spectra such as those in Fig. 5, one has many options as to how to analyse and interpret them. One option is to leave the data unfitted and draw purely qualitative conclusions from them. This is somewhat unreliable: in a number of cases the intuitive interpretation has subsequently been found to be the wrong one. A more sophisticated method is to fit numerous spectra with a nonphysical, but internally consistent, phenomenological model, and use the semiquantitative extracted data to compare the different materials. Such a method has recently been applied to a range of FeOOH polymorphs and ferrihydrite, which were fitted using arbitrary numbers of Voigtian profile spectral lines.[106] However, there is clearly scope for quantitative considerations of the physical interactions that occur between the applied field and the magnetic moments in the material. We now review three methods of analysis that, to varying extents, pursue this objective.

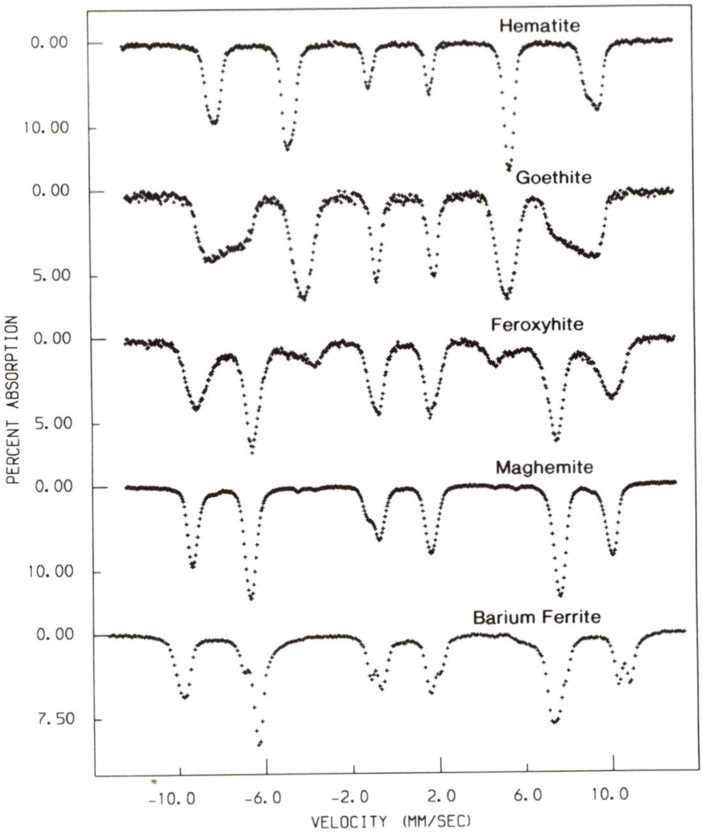

Figure 5. ^{57}Fe Mössbauer spectra, recorded at 4.2 K in an applied field of 9 T parallel to the γ-ray beam, of polycrystalline samples of hematite (α-Fe_2O_3), goethite (α-FeOOH), feroxyhite (δ'-FeOOH), maghemite (γ-Fe_2O_3), and barium ferrite ($BaFe_{12}O_{19}$).

3.1. Atomic Spin Hamiltonian

The formulation of an atomic spin Hamiltonian to describe the energy of simple magnetic systems in applied fields is a well-known technique that has been used for many years in studies of magnetic phase transitions, in particular in work on the spin-flop transition in single-crystal antiferromagnets.[107,108] It is therefore somewhat surprising that the extension of this methodology to studies of polycrystalline antiferromagnets and ferrimagnets is a comparatively recent event.[109,110]

The simplest model of the response of a two-sublattice magnetic system to an applied field is represented by a Hamiltonian including exchange interactions between nearest-neighbor moments and a uniaxial anisotropy establishing a preferred direction for the moments in the crystalline lattice. A generalized form of this Hamiltonian, which incorporates both ferrimagnetism and antiferromagnetism, is given by[110]

$$\mathcal{H} = -2J \sum_{\langle ij \rangle} \boldsymbol{S}_i \cdot \boldsymbol{S}'_j - K \sum_i (S_{zi})^2 - K' \sum_j (S'_{zj})^2 - g\mu_B \mathbf{B} \cdot \left(\sum_i S_i + \sum_j S'_j \right) \quad (1)$$

where J is an exchange constant, K and K' are anisotropy constants, and $\boldsymbol{S}$ and $\boldsymbol{S}'$ are sublattice spins corresponding to the sublattice atomic moments. $\mathbf{B}$ is the "local" field that is experienced by the sublattice spins as a result of the application of an external field $\mathbf{B}_{\text{app}}$. In the case of an antiferromagnet, $\mathbf{B} = \mathbf{B}_{\text{app}}$, but in a ferrimagnet the local field is modified by the net magnetization $\mathbf{M}$ in the material. For a crystal lattice with cubic symmetry, and in SI units,

$$\mathbf{B} = \mathbf{B}_{\text{app}} - \mu_0(N_{\parallel} - \tfrac{1}{3})\mathbf{M} \quad (2)$$

where $N_{\parallel}$ is the demagnetization factor parallel to the magnetization direction.

From the Hamiltonian, it is a straightforward matter to express the magnetic energy of the system in terms of the orientations of the two sublattice spins. One obtains[109,110]

$$E = NSg\mu_{\text{B}}[B_{\text{E}}\xi \cos(\theta - \theta') - \tfrac{1}{2}B_{\text{A}} \cos^2(\theta - t) - \tfrac{1}{2}B'_{\text{A}}\xi^2 \cos^2(\theta' - t) - B(\cos\theta + \xi \cos\theta')] \quad (3)$$

where N is the number of sublattice spins, $\xi = S'/S$ is the ratio of the spin magnitudes, and θ, θ', and t are the polar angles of the sublattice spins and of the easy anisotropy axis in a coordinate system defined by the applied field direction. The exchange and anisotropy terms are expressed as effective magnetic fields, with $B_{\text{E}} = -2JzS/g\mu_{\text{B}}$ where $J < 0$ and z is the number of exchange-coupled nearest-neighbors surrounding a given atom, $B_{\text{A}} = 2KS/g\mu_{\text{B}}$ and $B'_{\text{A}} = 2K'S/g\mu_{\text{B}}$.

The solutions of Eq. (3) are obtained by solving the simultaneous equations $\partial E/\partial\theta = \partial E/\partial\theta' = 0$, with the validity of the solution being ensured by requiring that the determinant of the 2×2 second-derivative matrix, $\nabla^2 E$, be greater than or equal to zero. These equilibrium spin configurations have been discussed in detail elsewhere[109,110] for a variety of different antiferromagnetic and ferrimagnetic systems, and a full range of applied field directions. The equilibrium spin directions define the direction of the hyperfine field $\mathbf{B}_{\text{hf}}$, which when added vectorially to the local field $\mathbf{B}$ gives rise to the effective field $\mathbf{B}_{\text{eff}}$, which determines the line positions in the corresponding Mössbauer spectra. The relative spectral line intensities are determined by the angle between $\mathbf{B}_{\text{eff}}$ and the Mössbauer γ rays.

In a homogeneous polycrystalline sample the crystallites are oriented in all directions relative to the applied field direction, so that the angle t is sampled continuously between 0 and 360°. The cumulative response of the powder to the applied field therefore amounts to a $\sin(t)$-weighted superposition of the responses of the individual crystallites. The resulting Mössbauer spectrum may be modeled by choosing a finite number of representative values of t, such as

$$t = \cos^{-1}[(2n - 1 - N_t)/N_t], \qquad n = 1, \ldots, N_t \tag{4}$$

and calculating and summing the resulting subspectra. In practice it is found that values of $N_t \sim 20$ are usually sufficient to give a good approximation (less than 1% error) to the full summation, while not exceeding readily available computer resources. Fewer subspectra are required in the case of antiferromagnets, since then the range $0 < t < 180°$ encompasses all unique solutions, and the representative values

$$t = \cos^{-1}[(2n - 1)/2N_t], \qquad n = 1, \ldots, N_t \tag{5}$$

may be used. Moderate deviations away from sample homogeneity, often referred to as "preferred orientation" effects, may be accounted for by modifying the $\sin(t)$ weighting function. Sampling functions based on both oblate and prolate ellipsoids of revolution have been successfully used in such cases.[72,111]

As a means of interpreting the applied field Mössbauer spectra of powders, the spin Hamiltonian model has both advantages and disadvantages. Perhaps its greatest advantage is that it is firmly grounded in a sensible, verifiable model of the atomic-scale magnetic interactions in the material. This allows the measurement of physically meaningful parameters, such as the exchange and anisotropy fields, and the ratio of the sublattice spin magnitudes, many of which are determinable from methods other than Mössbauer spectroscopy. There are relatively few variable parameters, so that unambiguous solutions are more likely to be obtained in any fitting procedures. The model predicts behavior that complements existing theoretical models of magnetic systems, as in the case of a ferrimagnet where the model predicts a coherent magnetization reversal mechanism.[112] The very simplicity of the model is, however, also a disadvantage. Macroscopic factors that affect the magnetic behavior of the system are included implicitly rather than explicitly, so that the anisotropy field is a quantity that incorporates both magnetocrystalline and shape anisotropy contributions. The assumption of uniaxial anisotropy may not always be a good approximation, and there are many cases where there are more than two magnetic sublattices in the material, so that the effectiveness of the two-sublattice model will depend largely on the uniformity of the exchange interactions between individual sublattices. On the other hand, there is ample scope for further theoretical work within the existing framework of the spin Hamiltonian formulation, so such difficulties may be seen as the impetus toward a more complete model.

3.2. Probability Distributions

Probability distributions are often used in the analysis of Mössbauer spectra to model the line broadening observed when a distribution of local atomic environments leads to site-by-site variations in quantities such as the hyperfine field, quadrupole splitting, and isomer shift. These are almost always one-dimensional distributions, although correlations are sometimes included between, for example, the hyperfine field and the isomer shift parameters. However, in 1980 a method of evaluating a two-parameter probability distribution was developed for the analysis of noncrystalline solids.[113] This method has recently been extended and used to fit the applied field Mössbauer spectra of magnetic small particles in terms of bidimensional probability distributions of the hyperfine field B_{hf} and the angle θ between $\mathbf{B}_{hf}$ and the applied field.[114] Within the fitting algorithm the $P(B_{hf}, \theta)$ distribution is constrained to adhere to user-supplied boundary conditions and is smoothed using two independent Lagrangian multipliers, one for each parameter. In addition, all of the probabilities are constrained to be ≥ 0 using a nonlinear programming technique.[115] In the last few years the method has been used to fit the spectra of several iron oxides and oxyhydroxides, including δ-FeOOH, aluminous goethite, and maghemite.[114,116,117]

The use of a hyperfine field distribution is well suited to describing the inherent disorder found, to varying extents, in many naturally occurring magnetic phases. In well-crystallized materials where the hyperfine field is single valued, the analysis reduces to a one-dimensional distribution of the orientation angles. Aside from the boundary conditions and the smoothing parameters that are selected by the user, the fitting process is relatively free of constraints, and the resultant calculated spectra are often very good fits to the data. Although it is clearly desirable that one achieves good-quality fits, this does not guarantee the physical validity of the fits. The $P(B_{hf}, \theta)$ distributions obtained amount to a transformation of the data contained in the spectra, and secondary analysis, either qualitative or quantitative, is required to interpret the distributions in physical terms. To date, such analysis has been qualitative only.

3.3. Superoperator Method

The ^{57}Fe Mössbauer spectra of many iron-containing magnetic materials are simple six-line spectra that may be adequately described by assuming that the electric quadrupole interaction is a small perturbation on a dominant magnetic dipole interaction. However, in some cases the electric field gradient (EFG) is comparatively large, and the spectra must be computed from a combined hyperfine Hamiltonian. The conventional method of doing this is to separately solve the secular equations of the matrices representing the nuclear ground and excited state Hamiltonians to obtain their eigenvalues and eigenvectors.[118] For polycrystalline samples where the EFG is randomly oriented with respect to the applied field direction, the resultant spectra are numerically averaged over a sphere to obtain the complete spectrum. The problem with this method is that one often has to compromise between the accuracy of the computed spectrum and the large amounts of computing time required for each calculation.

A faster and more precise method of calculating such spectra has been developed by Blaes *et al.* using a "superoperator" formalism.[119] A single 8 × 8 matrix, representing both the ground and excited state Hamiltonians, is used. For the case of an axially symmetric EFG, the average over the hyperfine field directions may be calculated analytically, while for asymmetric EFGs the averaging is reduced to one dimension. The method is suitable for diamagnets and paramagnets, extending earlier treatments,[120–122] and is also applicable to ferro- and ferrimagnetic materials in which the magnetic moments are completely aligned by the applied field.

4. FERRIMAGNETS

In this section, and the following section, we illustrate the modeling techniques described above with reference to a number of common ferrimagnets and antiferromagnets.

4.1. Maghemite

The applied field Mössbauer spectra of maghemite (γ-Fe_2O_3) powders have been studied extensively ever since it was reported by Coey, in 1971, that a field of 5 T was not sufficient to completely align the net ferrimagnetic moment in the applied field direction.[66] Evidence for this effect, which was coined the "spin-canting anomaly," came from the nonzero intensity of the $\Delta m_I = 0$ lines, the second and fifth of a magnetic sextet. Since in the experimental geometry that was used the Mössbauer γ rays were parallel to the applied field, the $\Delta m_I = 0$ lines would have had zero intensity had the moments and the applied field been collinear. Coey attributed the effect to a random canting of the surface spins resulting from a differing balance of exchange interactions at sites near the surface.[66] In ensuing years, similar effects were observed in a number of iron oxides, oxyhydroxides, and ferrites,[69,123,124] and the concept of surface spin canting became widely accepted.

However, recent work does not support the previous consensus that the surface spins are canted.[71,72,125] First, it was shown that unless high-quality spectra were recorded in large applied fields, the presence of surface layers of antiferromagnetic hematite (α-Fe_2O_3) formed during the oxidation of Fe_3O_4 into γ-Fe_2O_3 could easily be misinterpreted as anomalous spin canting.[125] This is illustrated in Figs. 6 and 7. The acicular magnetic recording particles shown in Fig. 6 were thought, on the basis of X-ray diffractograms, to be pure maghemite. Mössbauer spectra (Fig. 7) at room temperature and at 4.2 K in the absence of an applied field also showed the characteristic maghemite

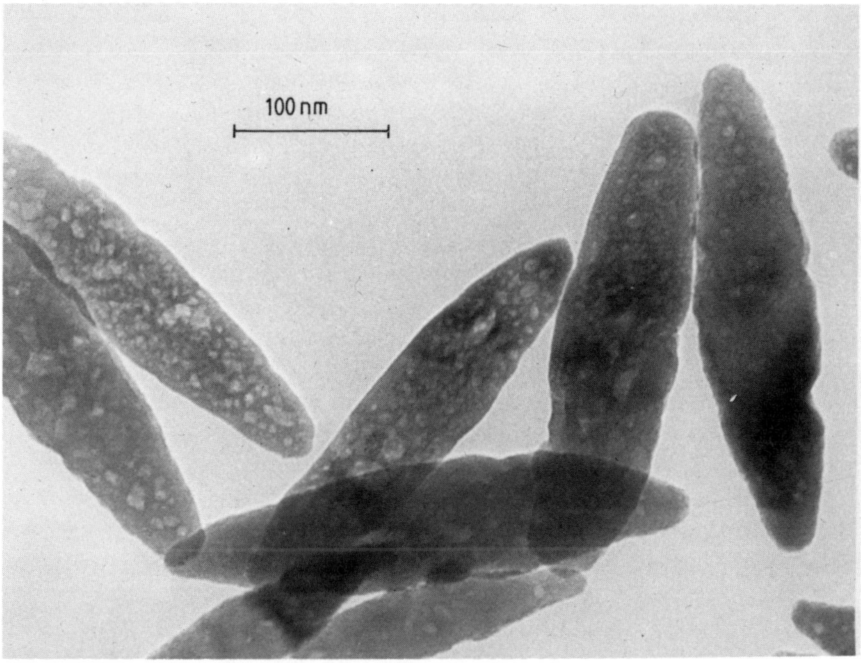

Figure 6. Transmission electron micrographs of spindle-shaped maghemite particles prepared for magnetic recording.

pattern. However, with a 6-T field applied parallel to the γ rays, an extra sextet is resolved, with relatively narrow $\Delta m_I = 0$ lines near -4.4 and $+5.6$ mm/s, which indicates the presence of α-Fe_2O_3. In retrospect, it seems likely that much of the earlier work on γ-Fe_2O_3 has suffered from the presence of unnoticed contributions from α-Fe_2O_3.

In a subsequent paper,[72] high-quality spectra of Co-adsorbed maghemite were recorded at 4.2 K in zero field and in applied fields of 6 and 7.6 T. These spectra are shown in Fig. 8. No α-Fe_2O_3 lines were observed. The small but significant $\Delta m_I = 0$ line intensities in the applied field spectra were found to be adequately described using the atomic spin Hamiltonian model of Section 3.1. The fit in the region of the $\Delta m_I = 0$ lines is shown in Fig. 9. Different anisotropy fields were fitted for the two sublattices, but they were rather ill-defined, and, being of order 1 T in magnitude, were larger than the 0.2- to 0.3-T anisotropy fields generally measured in Co-maghemite samples.[103] Such overestimates are not unusual in models that assume a coherent magnetization reversal mechanism. Even so, the results are conclusive in showing that it was not necessary to invoke noncollinear surface spins to explain the nonzero $\Delta m_I = 0$ line intensities.

In another study,[71] Parker and Berkowitz prepared samples by coating partially reduced γ-Fe_2O_3 particles with layers of either natural Fe or

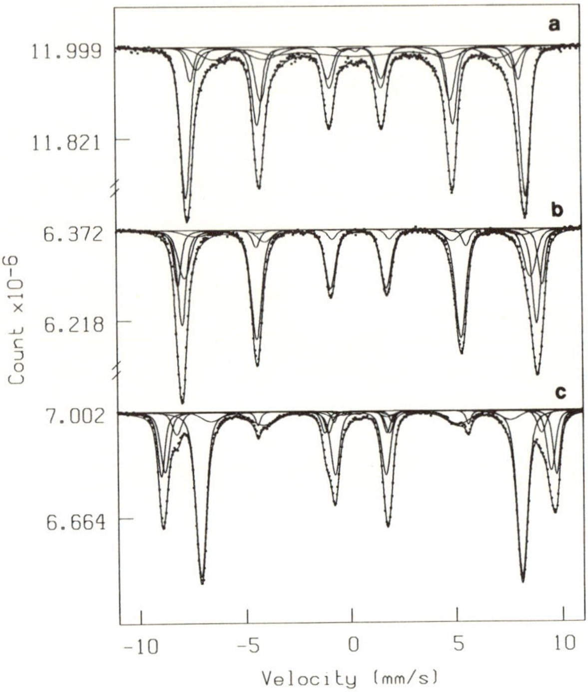

Figure 7. Mössbauer spectra of the maghemite particles in Fig. 6 at (a) room temperature, (b) 4.2 K, and (c) 4.2 K with a 6-T longitudinal field.[125]

isotopically enriched ^{57}Fe, followed by a layer of Co. Results obtained with both high fields (6 T at 4.2 K) and low fields (0.18 T at 296 K) were not consistent with the surface spin-canting model. They proposed a model in which, in applied fields, the high-anisotropy Co shell causes a parabolic dependence of sublattice spin angle as a function of lattice position. The predicted maximum deviation away from collinearity is then only a few degrees, which is consistent with the observed data.

There is continued discussion about the origin of the spin-canting anomaly. Although the random-spin surface-shell model has now been questioned, there is as yet no consensus as to the appropriate alternative model. One proposal is that the sublattice spins in a given particle remain collinear, and the incomplete alignment is a routine response for a high-anisotropy uniaxial ferrimagnetic powder in an applied field.[72] Another proposal sees the core spins aligned by the field and the surface spins canted, with the intermediate spins exhibiting a smooth progression between the two.[71] It is clear that further work, both theoretical and experimental, is required to fully resolve this issue.

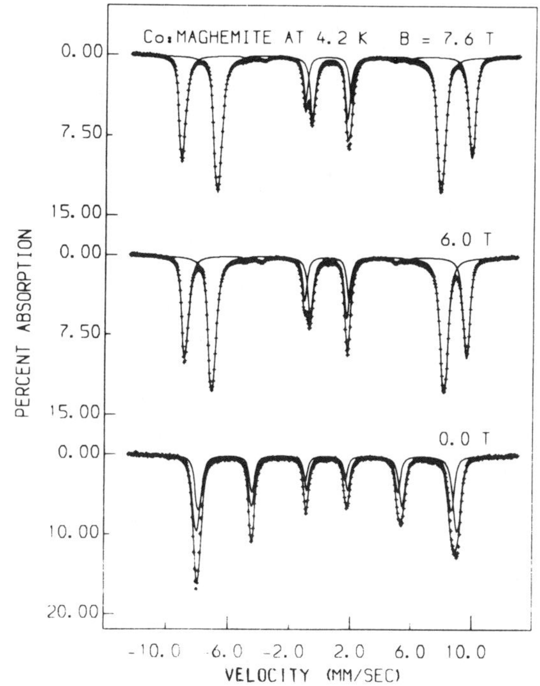

Figure 8. Mössbauer spectra of Co-adsorbed maghemite at 4.2 K in zero field and in longitudinal fields of 6 and 7.6 T.[72]

4.2. Barium Ferrite

The M-type hexagonal ferrite $BaFe_{12}O_{19}$ has five distinct magnetic sublattices. The 12k, 2a, and 2b sublattices are spin "up," while the $4f_{IV}$ and $4f_{VI}$ sublattices are spin "down," giving rise to a net moment in the "up" direction. The application of a field at 4.2 K separates the five different subspectra, as shown in Fig. 10, with the magnetic splitting at the spin "up" sites decreasing, and that at the spin "down" sites increasing, as the applied field increases.[110] This is logical, since the hyperfine field at the ^{57}Fe nucleus is antiparallel to the direction of the atomic Fe moment. The resolution of the five subspectra makes it possible to accurately measure the sublattice occupancy ratios, which are usually determined from series of Mössbauer spectra recorded both above and below the Curie temperature.[126] This is particularly useful when dealing with substituted barium ferrites, such as $BaFe_{12-2x}Co_xTi_xO_{19}$, where line broadening resulting from inhomogeneities in the local environments of the ferric ions otherwise complicates the analysis.[110]

The spectra in Fig. 10 were recorded in transverse drive geometry, with the applied field perpendicular to the γ-ray beam. Thus, the observed line intensity ratios in the subspectra, which are 3:4:1:1:4:3, correspond to a complete alignment of the ferrimagnetic moment with the applied field.

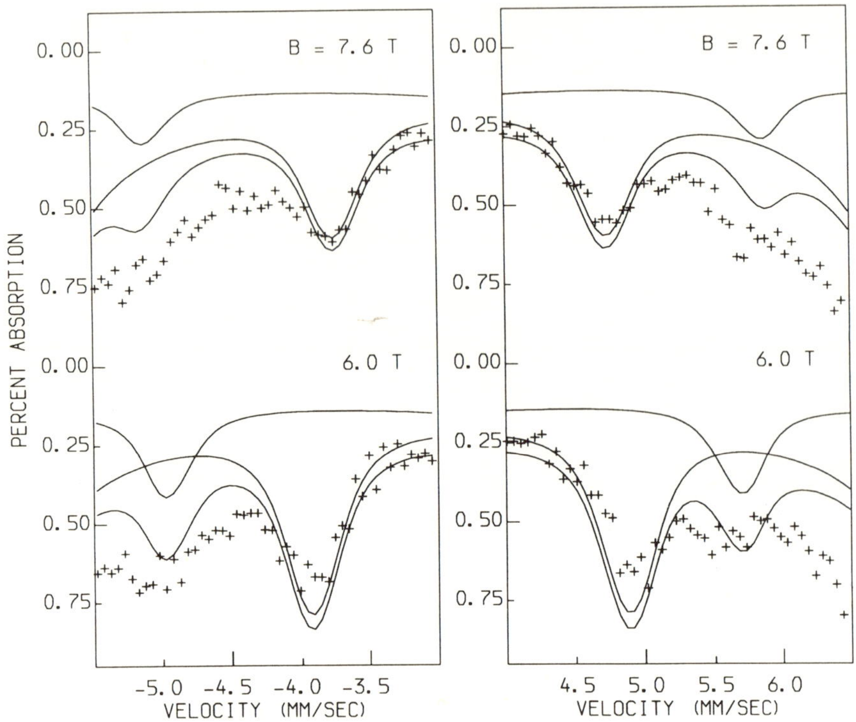

Figure 9. The applied field spectra of Fig. 8, showing the experimental data and curves calculated using the spin Hamiltonian model described in Section 3.1, in the vicinity of the $\Delta m_I = 0$ lines.[72]

4.3. Feroxyhite

Feroxyhite, δ'-FeOOH, is one of only two iron oxyhydroxides that are ferrimagnetic, the other being δ-FeOOH. The arrangement of oxygen and hydroxyl ions in feroxyhite is disordered. It is thought that the ferric ions randomly occupy octahedral sites,[127,128] although there have been suggestions that tetrahedral sites are also occupied.[129,130] Applied field Mössbauer experiments at 4.2 K show incomplete alignment of the ferrimagnetic moment in the direction of the field.[128,131] This effect is sample dependent. As illustrated in Fig. 11, the remanent $\Delta m_I = 0$ line intensities may be substantial, even in fields as high as 9 T.

The solid curves in Fig. 11 were fitted using the spin Hamiltonian model, assuming a simple two-sublattice ferrimagnetic system.[131] Three hyperfine fields were associated with each sublattice to model the effect of the crystallographic disorder, but the anisotropy and exchange fields were taken to be single valued. The reasonable agreement between the data and the calculated spectra indicates that anomalous effects resulting from surface spins are not significant. The isomer shifts and quadrupole terms for the two sublattices were consistent with those of ferric ions in octahedral sites.

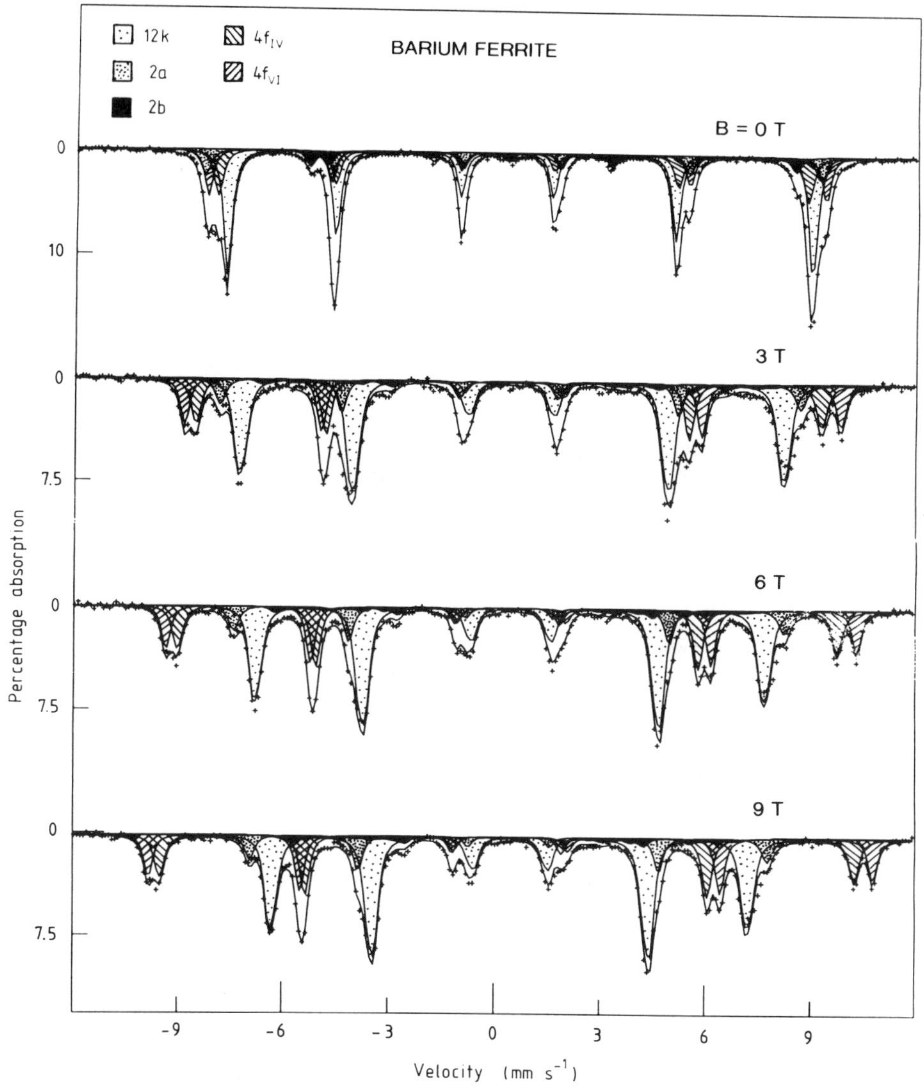

Figure 10. Mössbauer spectra of $BaFe_{12}O_{19}$ at 4.2 K in zero field and in transverse fields (perpendicular to the γ-ray beam) of 3, 6, and 9 T.[110]

4.4. Magnetite

Magnetite, Fe_3O_4, undergoes a metal–insulator transition at the Verwey temperature, $T_V = 121$ K. In the metallic state, above T_V, magnetite is readily saturated by applied fields (see Fig. 12). The applied field increases the magnetic splitting of ions on the 8 tetrahedral A sites, and reduces the splitting for ions on the 16 octahedral B sites. For all components the $\Delta m_I = 0$ lines vanish as the moments align parallel and antiparallel to the γ-ray beam.

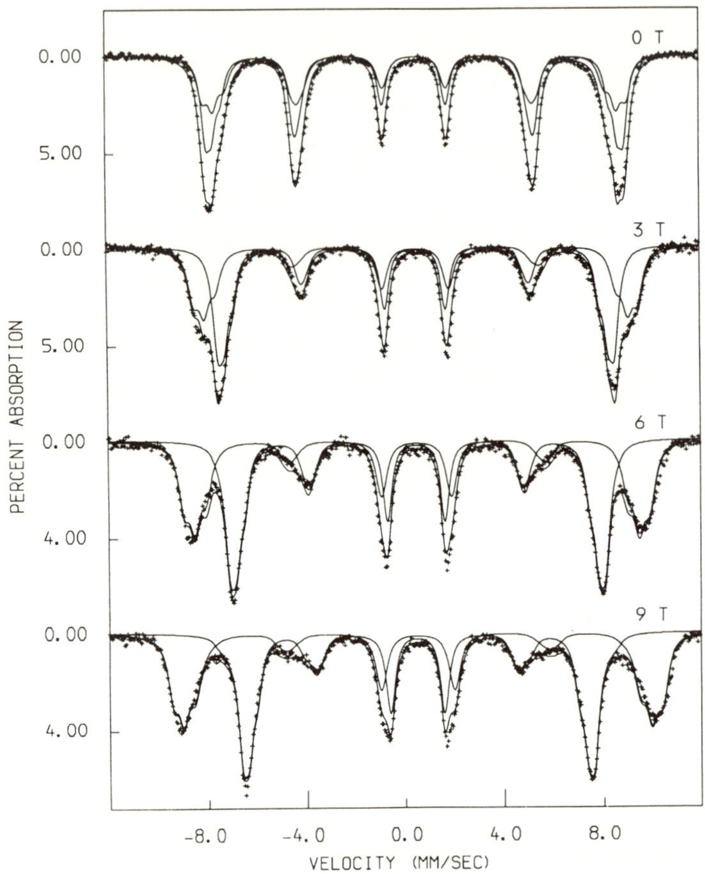

Figure 11. Mössbauer spectra of feroxyhite, δ'-FeOOH, at 4.2 K in zero field and in longitudinal fields of 3, 6, and 9 T.[131]

The low-temperature insulating state of magnetite has been investigated intensively, but the details remain uncertain. Early Mössbauer experiments showed at least six components: one ascribed to tetrahedral Fe^{3+}, two to octahedral Fe^{2+}, and three to octahedral Fe^{3+}.[132] Subsequently, a zero-field 4.2 K spectrum showed three or four octahedral Fe^{2+} sites.[133] NMR data show at least five Fe^{3+} octahedral sites.[134] Recently, Zuo *et al.* postulated ten different possible structures, and argued that electron diffraction data favour a structure with six Fe^{3+} octahedral sites and five Fe^{2+} octahedral sites.[135] Most recently, Mössbauer spectra with applied fields up to 14 T have been measured.[136] At least five Fe^{3+} octahedral sites and four octahedral Fe^{2+} sites are present, as shown by the solid lines in Fig. 13.

For Fe^{2+}, an appreciable quadrupole interaction leads to line broadening and nonzero $\Delta m_I = 0$ lines in applied fields, as illustrated in Fig. 14 for an

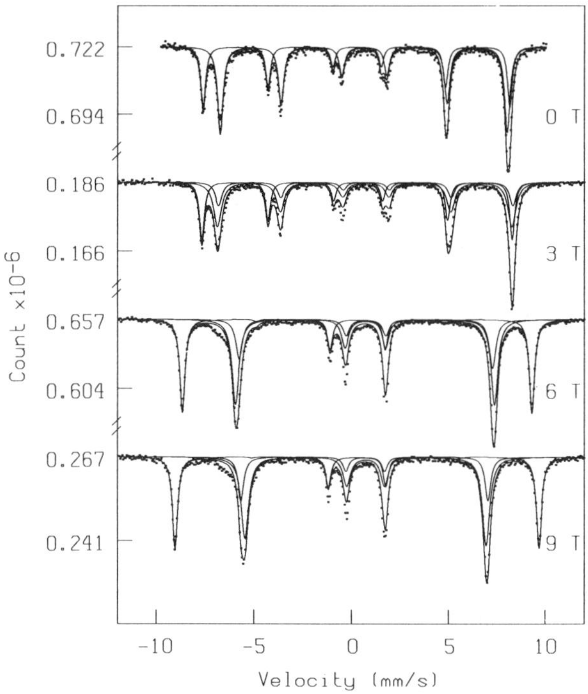

Figure 12. Mössbauer spectra of natural magnetite particles at 200 K (above the Verwey transition) in zero field and in longitudinal applied fields of 3, 6, and 9 T.

applied field of 5.8 T. The zero-field values of the quadrupole term, $\frac{1}{2}eQV_{zz}$, are between 0.3 and 2.6 mm/s, and increase markedly with applied field, up to 4.3 mm/s. Such large values are also responsible for the extreme asymmetry of the Fe^{2+} subpatterns when strong fields are applied (see Fig. 14), and presumably arise from the anisotropy of the 3*d* wave functions.

5. ANTIFERROMAGNETS

5.1. Hematite

Hematite, α-Fe_2O_3, undergoes a transition from a weakly ferromagnetic state to an antiferromagnetic state at the Morin temperature, $T_M = 260$ K. Above T_M the ferric spins lie in the basal (111) plane but are not precisely antiparallel. As a result of the presence of an antisymmetric "Dzyaloshinsky" exchange interaction between the spins, they are canted by a fraction of a degree, producing a small net moment in the basal plane.[137] Applied field Mössbauer spectra, recorded at 340 K, are shown in Fig. 15. The increase in

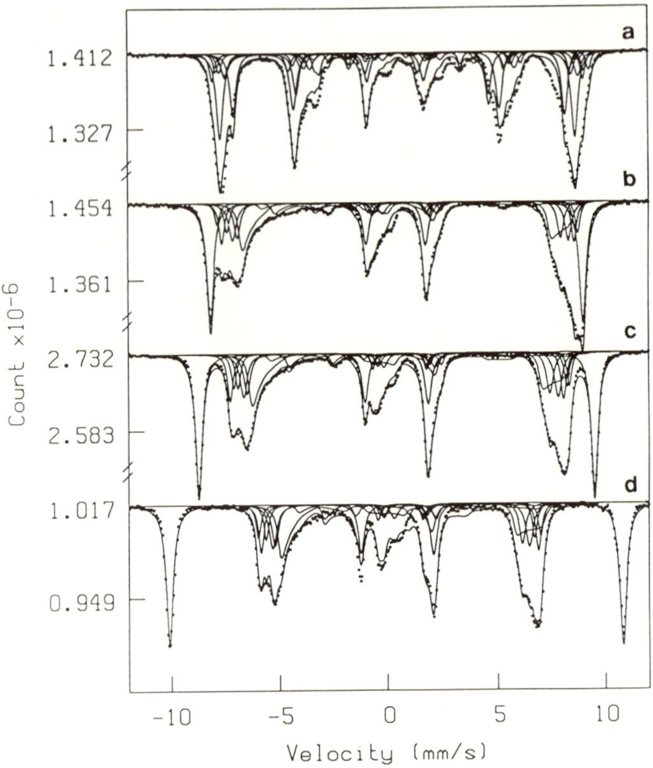

Figure 13. Mössbauer spectra of high-purity, crushed magnetite crystals at 4.2 K, below the Verwey transition. Longitudinal fields of (a) 0 T, (b) 2.8 T, (c) 5.8 T, and (d) 14.3 T were applied.[136]

$\Delta m_I = 0$ line intensity in response to the longitudinal field indicates that the weak net moment is aligned by the field, so that the atomic spins reorient perpendicular to the field.

Below the Morin temperature the Fe spins align antiferromagnetically along the trigonal [111] axis. Field-induced spin-flop transitions have been observed in single-crystal samples at a critical field, directed along the [111] axis, of $B_{sf} = 6.4$ T.[107,138,139] The spectra of a powdered sample of hematite are shown in Fig. 16. A reasonable fit was obtained using the simple spin Hamiltonian model, ignoring the effect of the Dzyaloshinsky interaction, and a fitted anisotropy field $B_A \simeq 0.02$ T was obtained.[109] This corresponds to a spin-flop field of about 6.2 T, which compares favorably with the value measured in single-crystal experiments.

5.2. Goethite

Goethite, α-FeOOH, is a simple antiferromagnet in which the Fe moments lie along the crystallographic c axis. Applied field spectra recorded at

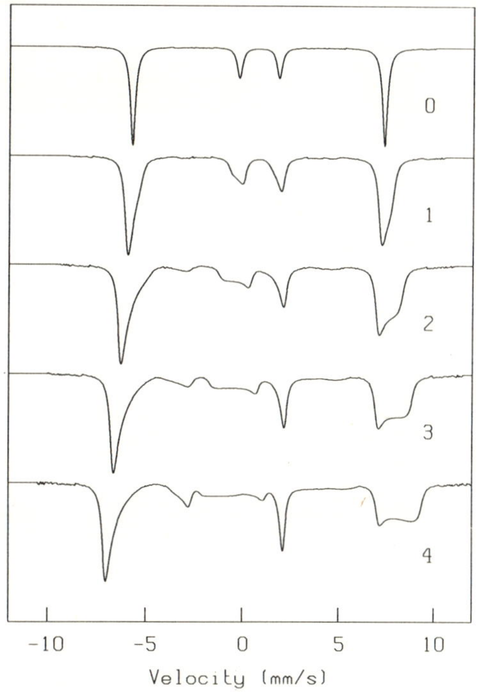

Figure 14. Subspectra computed for ferrous ions in magnetite at 4.2 K and in a longitudinal field of 5.8 T (refer to Fig. 13), using the super-operator method of Blaes *et al.*[119] The moments are assumed to be collinear with the applied field, and the numbers beside each trace refer to the assumed values of $\frac{1}{2}eQV_{zz}$ (mm/s). Data supplied courtesy of S. J. Harker.

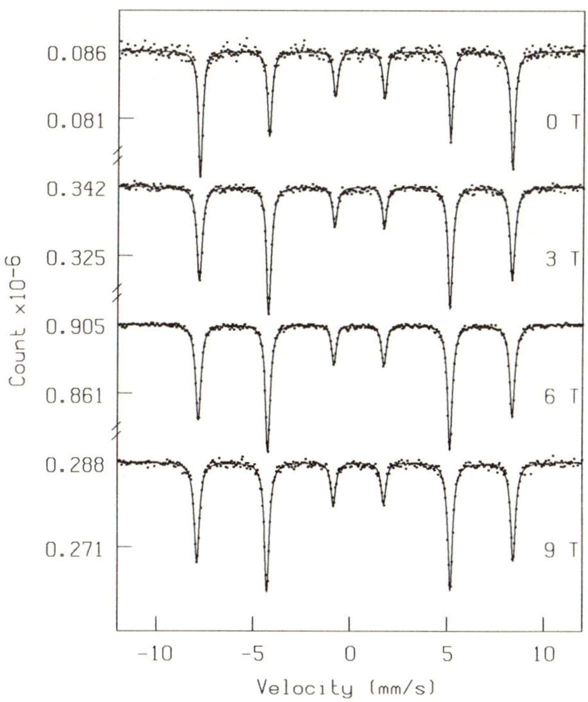

Figure 15. Mössbauer spectra of polycrystalline hematite taken at 340 K, above the Morin transition, in zero field and in longitudinal fields of 3, 6, and 9 T.[153]

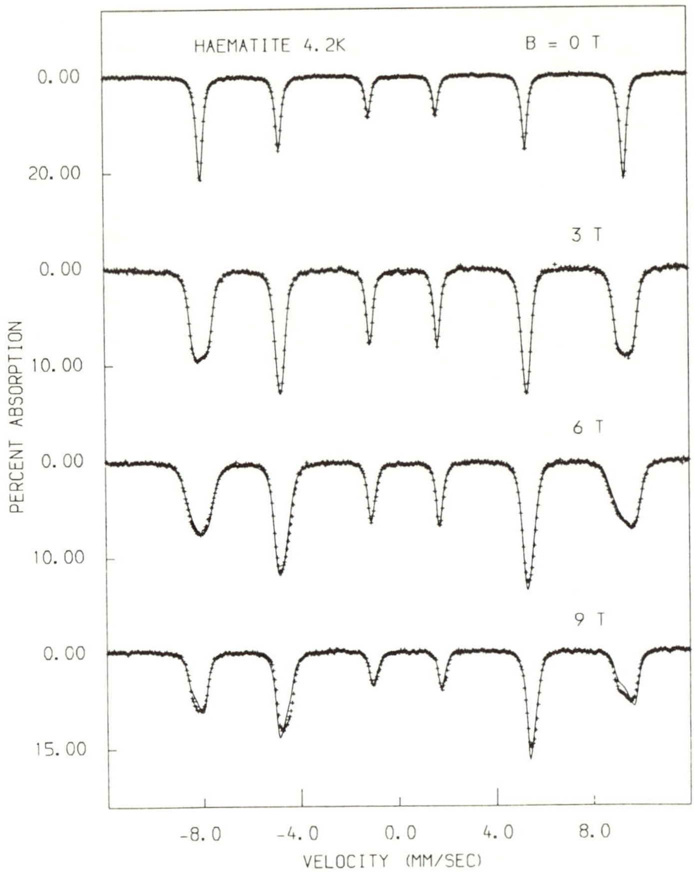

Figure 16. Mössbauer spectra of hematite at 4.2 K, below the Morin transition, in zero field and in longitudinal fields of 3, 6, and 9 T.[109]

4.2 K exhibit substantial line broadening, as in Fig. 17, which is very well described using the spin Hamiltonian method.[109] The fitted anisotropy field in his case is $B_A \simeq 0.4\,T$, which leads to a predicted spin-flop field of order 15 T. This is consistent with earlier single-crystal work where applied fields of up to 10 T were directed along the *c* axis, and no spin-flop was observed.[140]

5.3. Aluminous Goethite

In the past few years there have been a number of reports of applied field Mössbauer experiments on synthetically produced aluminous goethites.[97,114,141,142] Samples containing up to 31% atomic replacement of iron by aluminum have been studied. (The substitutional limit for aluminum in goethite is 32 mol%.[143]) Spectra of a 30 mol% sample are shown in Fig. 18. On

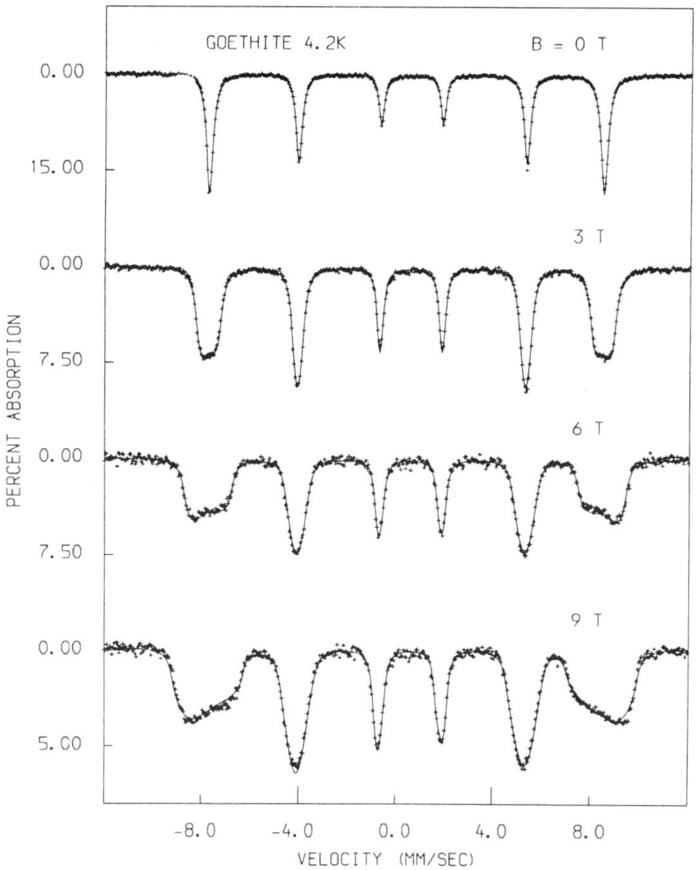

Figure 17. Mössbauer spectra of goethite at 4.2 K in zero field and in longitudinal fields of 3, 6, and 9 T.[109]

comparing these spectra with those of pure goethite, as in Fig. 17, it is clear that the aluminum substitution has a dramatic effect on the magnetic properties of the material.

Spectra similar to those in Fig. 18 have been analyzed with the bidimensional probability distribution method.[114,141] It was inferred that the aluminum substitution creates uncompensated Fe spins and weakens the magnetocrystalline anisotropy so that the net ferrimagnetic moment could be reoriented toward the applied field. In contrast, a quantitative spin Hamiltonian analysis of samples containing 5, 11, 18, and 30 mol% Al showed that the anisotropy experienced by the ferric ions increased with aluminum substitution.[97] The solid lines in Fig. 18 were fitted using the spin Hamiltonian method, with a wide distribution of anisotropy fields around a mean of order 15 T. Such a large anisotropy would, of itself, inhibit the reorientation of the

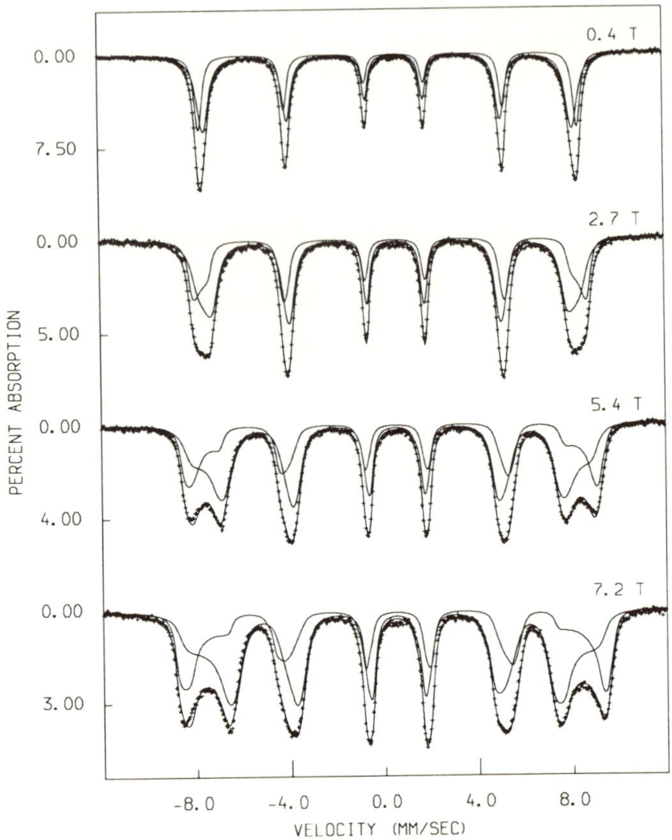

Figure 18. Mössbauer spectra of goethite substituted with 30 mol% aluminum at 4.2 K, in longitudinal fields of 0.4, 2.7, 5.4, and 7.2 T.[97]

moments in the applied field. However, this is overcome by a second consequence of the aluminum substitution, which is that the relative occupancies of Fe in the two sublattices are no longer equal. The tendency for the aluminum ions to occupy the same sublattice is strong: in the 30 mol% sample shown in Fig. 18, the fitted ratio of sublattice Fe occupancies is ~0.6. The resultant net ferrimagnetic moment accounts for the response of the aluminous goethite samples in applied fields.

5.4. Ferrihydrite

Ferrihydrite is the growth precursor to minerals such as hematite. It is a poorly crystalline material with the approximate composition $Fe_5HO_8 \cdot 4H_2O$. The poor crystallinity of ferrihydrite samples is evident in their X-ray diffraction patterns, which contain between two and six broad lines. Samples are described as, for example, two-line, four-line, or six-line ferrihydrites, as a means of indicating the degree of crystallinity.

Both the structural and the magnetic properties of ferrihydrite are currently the subject of debate. It is generally accepted that ferrihydrite has a hexagonal unit cell, with the iron in octahedral coordination.[144,145] However, a new structural model based on a trigonal unit cell has been proposed, in which some of the Fe appears in tetrahedral sites.[146] This proposal has subsequently been challenged.[147] The magnetic state of ferrihydrite is also controversial. It has been identified as being "speromagnetic,"[148] a state in which the atomic magnetic moments are individually oriented at random despite the presence of antiferromagnetic nearest-neighbor interactions. The prototype speromagnet was a natural ferric gel[98] that is now thought to have been ferrihydrite.[148,149] However, it has been demonstrated that the original applied field Mössbauer spectra of the prototype material can be explained most simply as resulting from disordered antiferromagnetic particles.[100] Consequently, the nature of the magnetic order in ferrihydrite is in doubt.

Spectra of synthetic samples of two-line and six-line ferrihydrite are shown in Figs. 19 and 20, along with fits obtained with the spin Hamiltonian model.[150] The two-line ferrihydrite sample was modeled as a simple two-sublattice ferrimagnet. It is anticipated that this ferrimagnetism arises from unequal sublattice occupancies: the fit gave a relative occupancy of 0.85, and a ratio of sublattice spins of $\xi \simeq 0.92$. The more crystalline six-line ferrihydrite sample was treated as a simple antiferromagnet. In neither case was it necessary to invoke speromagnetism in order to explain the data. The ferrimagnetism in two-line ferrihydrite has been analyzed in terms of random fluctuations in sublattice occupancy arising from the small numbers of ferric ions per particle.[150] (In the two-line ferrihydrite there were about 100 ions per particle, compared with about 1500 in the larger six-line ferrihydrite particles.) The fitted values of isomer shift and quadrupole splitting obtained from the data were consistent with octahedrally coordinated iron, although the presence of a small proportion of tetrahedral iron cannot be entirely ruled out.

6. CONCLUSIONS

We conclude this chapter with some considerations of a few possible directions and objectives for future work involving applied field Mössbauer spectroscopy of magnetic powders.

6.1. Industrial Applications

One area of commercial relevance where applied field Mössbauer spectroscopy is a strong contender is in the analysis of mixed-phase industrial and

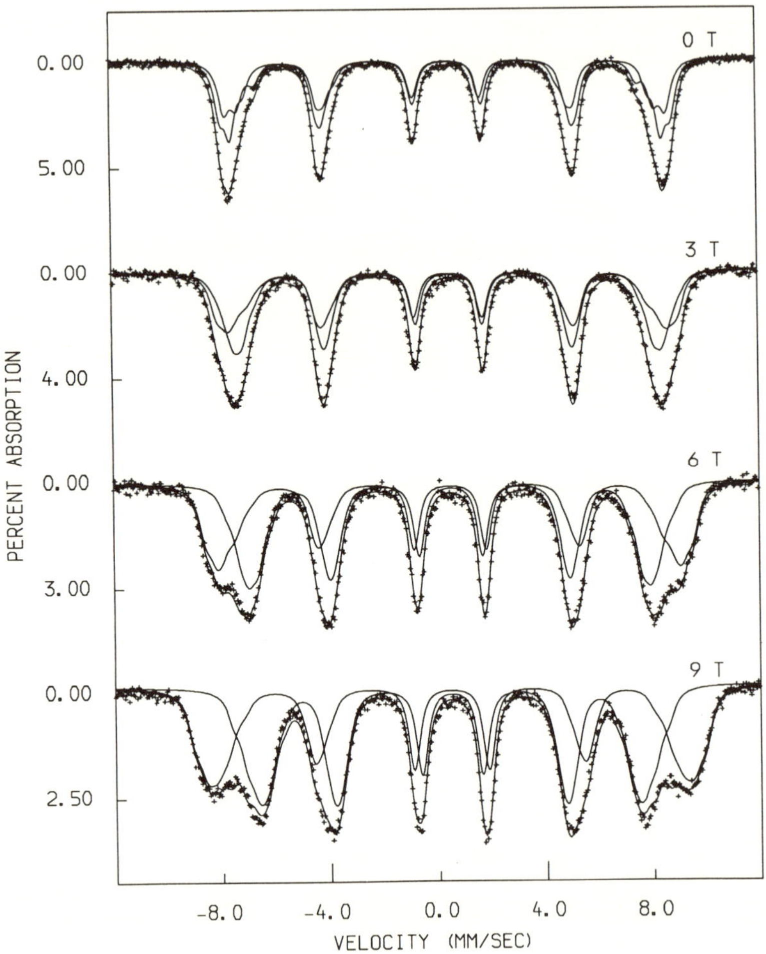

Figure 19. Mössbauer spectra of two-line ferrihydrite at 4.2 K, in zero field and in longitudinal fields of 3, 6, and 9 T.[150]

mineralogical samples. Zero-field Mössbauer analysis is already well established in this area, but there are occasions when it is nearly impossible to distinguish phases unless one uses an applied field. A common example of this is when line broadening caused by small particle sizes inhibits identification. Although the line broadening may be greatly reduced by cooling the sample to low temperatures, one then has the problem that the zero-field spectra of a number of common phases are very similar at low temperatures. It is therefore necessary to apply a magnetic field.

As an example of this type of mineralogical application, Fig. 21 shows Mössbauer spectra of a soil sample from Weipa, Queensland, Australia.[151] Zero-field spectra show that both goethite and hematite may be present. Upon

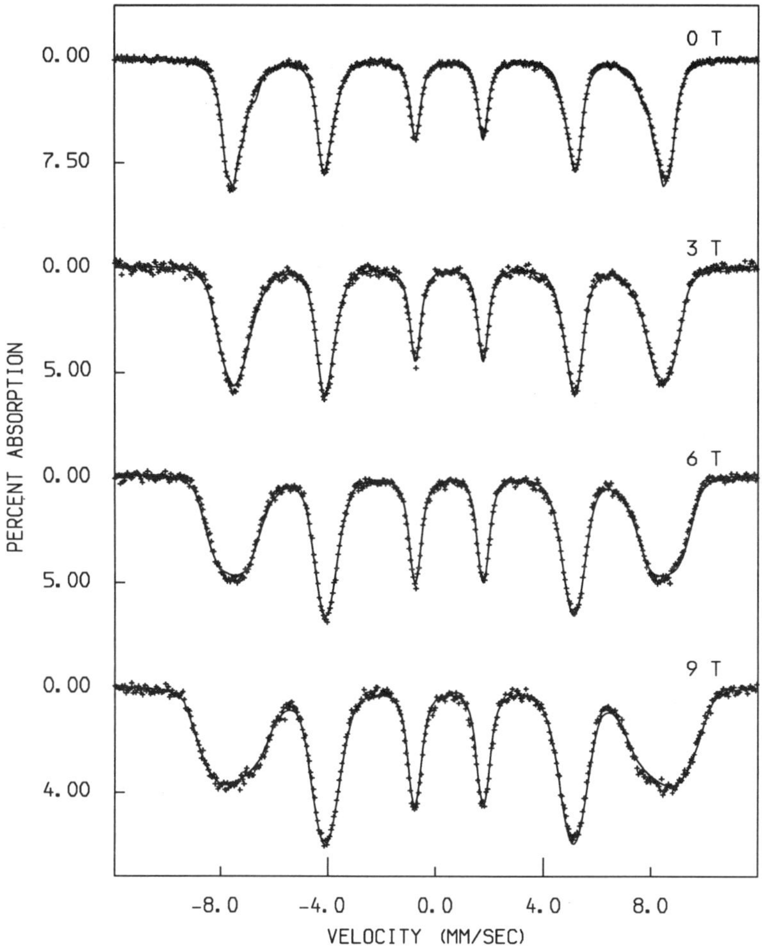

Figure 20. Mössbauer spectra of six-line ferrihydrite at 4.2 K, in zero field and in longitudinal fields of 3, 6, and 9 T.[150]

applying a magnetic field the $\Delta m_I = 0$ line intensities increase. This is characteristic of neither well-crystallized hematite nor well-crystallized goethite, since both are antiferromagnetic at 4.2 K, and an applied field slightly reduces the $\Delta m_I = 0$ lines. The increase demonstrates that the magnetic moments have reoriented toward the plane normal to the applied field direction, as is characteristic of hematite above the Morin transition. In order for the Morin transition to be suppressed, the hematite must either contain a sufficient amount of diagmagnetic ions, or be in the form of very fine particles.[152] By reference to the hyperfine field of the second component, it is identified as aluminous goethite. Using applied-field data,[97] the aluminum content is estimated to be 25 mol%. Thus, for this sample, the applied field

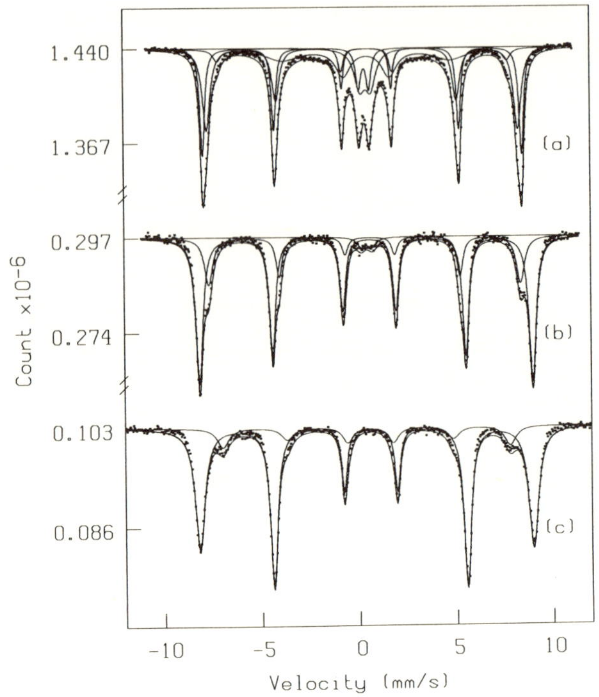

Figure 21. Mössbauer spectra of a soil sample from Weipa, Queensland, Australia, at (a) room temperature, (b) 4.2 K, and (c) 4.2 K with a 6-T longitudinal field applied.[151]

technique provides information that is not readily obtainable from the zero-field experiments, and enables partial analysis of the iron phases.

However, when considering the use of applied field Mössbauer analysis as a tool for phase identification within an industrial setting, one must weigh the monetary cost against the scientific benefit. The costs of applied field experiments are generally higher than those of zero-field experiments, because the solenoid is immersed in liquid helium and because the source–sample–detector geometry dictates that longer counting times are required. In addition, the results are often too complex to allow the convenient measurement of phase abundances and other desirable information. Therefore, usage of the applied field technique is likely to remain infrequent in direct industrial applications, although there will most likely be special cases where the technique will be found to be of net benefit.

6.2. Research Directions

The applied field experiments described in this chapter demonstrate the potential for Mössbauer spectroscopy to contribute significantly to fundamental studies of magnetic materials. In particular, the atomic spin Hamiltonian method of quantifying the behavior of magnetic powders in applied fields is a promising development that may form the basis of further work.

The measurement of anisotropy field strengths is a substantial advance. Experiments and analyses such as those described will lead to a better understanding of the origins of anisotropy. In general, the anisotropy of fine particles is much larger than expected from bulk materials, although whether this arises from surface effects or from greater densities of defects remains to be determined. This question is central to an unambiguous determination of the cause of the spin-canting anomaly in ferrimagnetic particles, and is obviously of importance in applications of fine particles.

A second advance concerns the relatively direct measurement of the magnetic ground states in disordered materials. This is especially evident in the work related to aluminous goethite and ferrihydrite, where a net moment appears on what are ostensibly antiferromagnetic particles. Quantitative analysis, based on applied field Mössbauer experiments, provides unique information on the origin of the net moments.[97,150] One therefore gains a new understanding of a phenomenon that has received considerable attention in theories of magnetic fluctuations at elevated temperatures, but for which the nature of the fluctuating magnetic entities has, until now, been the subject of conjecture only.

Looking ahead, we may anticipate many experimental and theoretical developments in applied field Mössbauer spectroscopy. Superconducting magnets delivering fields of up to 15 T are likely to become more widespread, and more research will be devoted to high-field effects. Taking advantage of ever-increasing computation capacities, more realistic (and therefore more complicated) analysis programs will be written, incorporating features such as nonuniaxial anisotropy and incoherent magnetization reversal mechanisms. These and similar developments should ensure the continued use of applied field Mössbauer spectroscopy in studies of the magnetic properties of particulate matter.

REFERENCES

1. C. E. Johnson, *Hyper. Interact.* **49**, 19 (1989).
2. S. W. Charles, *J. Magn. Magn. Mater.* **65**, 350 (1987).
3. K. Raj and R. Moskowitz, *J. Magn. Magn. Mater.* **85**, 233 (1990).
4. W. C. Elmore, *Phys. Rev.* **54**, 309 (1938).
5. K. Goto and T. Sakurai, *Appl. Phys. Lett.* **30**, 355 (1977).
6. T. Inoue, K. Goto, and T. Sakurai, *Jpn. J. Appl. Phys.* **22**, L695 (1983).
7. S. Chikazumi, *J. Magn. Magn. Mater.* **54–57**, 1551 (1986).
8. D. J. Dunlop, *Rep. Prog. Phys.* **53**, 707 (1990).
9. S. L. Halgedahl, *Phys. Earth Planet. Inter.* **46**, 127 (1987).
10. K. Goto, M. Ito, and T. Sakurai, *Jpn. J. Appl. Phys.* **19**, 1339 (1980).
11. J. Šimšová, R. Gemperle, and J. C. Lodder, *J. Magn. Magn. Mater.* **95**, 85 (1991).

12. C. E. Betz, *Principles of Magnetic Particle Testing,* Magnaflux Corp. Chicago, 1967.
13. U. Hartmann, *J. Magn. Magn. Mater.* **68**, 298 (1987).
14. J. Ye and R. T. Merrill, *Geophys. Res. Lett.* **18**, 593 (1991).
15. G. P. Cameron and J. H. Judy, *IEEE Trans. Magn.* **24**, 3063 (1988).
16. P. Rice and J. Moreland, *Rev. Sci. Instrum.* **62**, 844 (1991).
17. B. A. Wills, *Mineral Processing Technology,* Pergamon Press, Elmsford, N.Y., 1985.
18. R. R. Oder, *IEEE Trans. Magn.* **12**, 429 (1976).
19. R. Gerber, *IEEE Trans. Magn.* **18**, 812 (1982).
20. J. M. Turner, M. R. Parker, and D. Fletcher, *IEEE Trans. Magn.* **24**, 866 (1988).
21. J. C. Mallinson, *IEEE Proc.* **64**, 196 (1976).
22. A. E. Berkowitz, *IEEE Trans. Magn.* **22**, 466 (1986).
23. M. P. Sharrock, *IEEE Trans. Magn.* **25**, 4374 (1989).
24. R. W. Chantrell, *Textures Microstruct.* **11**, 107 (1989).
25. G. Bate, *J. Magn. Magn. Mater.* **100**, 413 (1991).
26. J. L. Kirschvink, *Bioelectromagnetism* **10**, 239 (1989).
27. R. B. Frankel, R. P. Blakemore, and R. S. Wolfe, *Science* **203**, 1355 (1979).
28. K. Futschik, H. Pfutzner, A. Doblander, P. Schonhuber, T. Dobeneck, N. Petersen, and H. Vali, *IEEE Eng. Med. Biol. Magn.* **8**, 46 (1989).
29. M. Funaki, H. Sakai, and T. Matsunaga, *J. Geomagn. Geoelectr.* **41**, 77 (1989).
30. A. S. Bahaj, J. H. P. Watson, and D. C. Ellwood, *IEEE Trans. Magn.* **25**, 3809 (1989).
31. A. S. Bahaj, D. C. Ellwood, and J. H. P. Watson, *IEEE Trans. Magn.* **27**, 5371 (1991).
32. T. Matsunaga and S. Kamiya, *Appl. Microbiol. Biotech.* **26**, 328 (1987).
33. S. Saini, J. T. Ferrucci, D. D. Stark, and J. Wittenberg, *Gastrointestinal Radiol.* **11**, 296 (1986).
34. S. Cerdan, H. R. Lotscher, B. Kunnecke, and J. Seelig, *Magn. Reson. Med.* **12**, 151 (1989).
35. J. Ugelstad, A. Berge, and T. Ellingsen, *Bone Marrow Transplant* **2**, 74 (1987).
36. S. B. Sato, S. Ohnishi, Y. Sako, and S. Yamashina, *J. Biochem.* **100**, 1481 (1986).
37. G. Kvalheim, O. Fodstad, S. Funderud, K. Nustad, A. Pharo, A. Pihl, and A. Ugelstad, *J. Cancer Res.* **47**, 846 (1987).
38. A. S. Mavrichev and V. Y. Fertman, *Vopr. Onkol.* **37**, 11 (1991).
39. R. C. Birkmeyer, R. Diaco, D. K. Hutson, H. P. Lau, W. K. Miller, N. V. Neelkantan, T. J. Pankratz, S. Y. Tseng, D. K. Vickery, and E. K. Yang, *Clin. Chem.* **33**, 1543 (1987).
40. S. W. Charles, *J. Magn. Magn. Mater.* **85**, 277 (1990).
41. S. Roath, A. R. Smith, and J. H. P. Watson, *J. Magn. Magn. Mater.* **85**, 285 (1990).
42. P. C. Scholten and J. A. P. Felius, *J. Magn. Magn. Mater.* **85**, 107 (1990).
43. G. Bottoni, *J. Appl. Phys.* **69**, 4499 (1991).
44. G. W. D. Spratt, N. Kodama, H. Inoue, Y. Uesaka, and M. Katsumoto, *IEEE Trans. Magn.* **27**, 4661 (1991).
45. M. El Hilo, K. O'Grady, and R. W. Chantrell, *IEEE Trans. Magn.* **27**, 4666 (1991).
46. A. B. Eriksson and M. Jonson, *Phys. Rev. B* **40**, 884 (1990).
47. J. J. M. Janssen, J. J. M. Baltussen, A. P. van Gelder, and J. A. A. J. Perenboom, *J. Phys. D* **23**, 1447 (1990).
48. G. Helgesen, A. T. Skjeltorp, P. M. Mors, R. Botet, and R. Jullien, *Phys. Rev. Lett.* **61**, 1736 (1988).
49. A. B. Eriksson and M. Jonson, *Phys. Rev. Lett.* **62**, 1698 (1989).
50. F. Preisach, *Z. Phys.* **94**, 273 (1935).
51. E. Della Torre, *J. Appl. Phys.* **61**, 4016 (1987).
52. D. C. Jiles and D. L. Atherton, *J. Magn. Magn. Mater.* **61**, 48 (1986).
53. V. Christoph and K. Elk, *J. Magn. Magn. Mater.* **74**, 143 (1988).
54. A. Lyberatos and E. P. Wohlfarth, *J. Magn. Magn. Mater.* **59**, L1 (1986).
55. M. T. Rekveldt and R. Rosman, *J. Magn. Magn. Mater.* **95**, 221 (1991).
56. J. Popplewell, B. Abu Aisheh, and N. Y. Ayoub, *J. Appl. Phys.* **64**, 5852 (1988).
57. M. Hanson and C. Johansson, *J. Magn. Magn. Mater.* **101**, 45 (1991).

58. G. Helgesen and A. T. Skjeltorp, *J. Appl. Phys.* **69**, 8277 (1991).
59. S. Mørup, *J. Magn. Magn. Mater.* **37**, 39 (1983).
60. J. L. Dormann, L. Bessais, and D. Fiorani, *J. Phys. C* **21**, 2015 (1988).
61. D. H. Jones, *Hyper. Interact.* **47**, 289 (1989).
62. S. Mørup, *Hyper. Interact.* **60**, 959 (1990).
63. S. Umeki, S. Saitoh, and Y. Imaoka, *IEEE Trans. Magn.* **10**, 655 (1974).
64. W. M. Mularie and M. P. Sharrock, *J. Appl. Phys.* **69**, 4938 (1991).
65. A. E. Berkowitz, in *Science and Technology of Nanostructured Magnetic Materials,* G. C. Hadjipanayis and G. A. Prinz, eds., p. 533. Plenum Press, New York, 1991.
66. J. M. D. Coey, *Phys. Rev. Lett.* **27**, 1140 (1971).
67. A. Ochi, K. Watanabe, M. Kiyama, T. Shinjo, Y. Bando, and T. Takada, in *Ferrites: Proc. 3rd Int. Conf.* (*Sept.–Oct. 1980, Japan*), H. Watanabe, S. Iida, and M. Sugimoto, eds., p. 618, Center for Acad. Publ., Japan, 1981.
68. M. P. Sharrock, P. J. Picone, and A. H. Morrish, *IEEE Trans. Magn.* **19**, 1466 (1983).
69. A. H. Morrish and K. Haneda, *J. Magn. Magn. Mater.* **35**, 105 (1983).
70. A. E. Berkowitz, F. T. Parker, E. L. Hall, and G. Podolsky, *IEEE Trans. Magn.* **24**, 2871 (1988).
71. F. T. Parker and A. E. Berkowitz, *Phys. Rev. B* **44**, 7437 (1991).
72. Q. A. Pankhurst and R. J. Pollard, *Phys. Rev. Lett.* **67**, 248 (1991).
73. A. Aharoni, *IEEE Trans. Magn.* **22**, 478 (1986).
74. E. Della Torre, *IEEE Trans. Magn.* **22**, 484 (1986).
75. E. C. Stoner and E. P. Wohlfarth, *Philos. Trans. R. Soc. London Sec.* A **240**, 599 (1948).
76. I. S. Jacobs and C. P. Bean, *Phys. Rev.* **100**, 1060 (1955).
77. A. Aharoni, *Phys. Status Solidi* **16**, 3 (1966).
78. J. E. Knowles, *IEEE Trans. Magn.* **20**, 84 (1984).
79. J. E. Knowles, *J. Magn. Magn. Mater.* **61**, 121 (1986).
80. A. Aharoni, *J. Appl. Phys.* **63**, 4605 (1988).
81. M. E. Schabes and H. N. Bertram, *J. Appl. Phys.* **64**, 5832 (1988).
82. M. E. Schabes and H. N. Bertram, *J. Appl. Phys.* **67**, 5149 (1990).
83. Y. Ishii and M. Sato, *J. Appl. Phys.* **59**, 880 (1986).
84. P. C. Kuo, *J. Appl. Phys.* **64**, 5071 (1988).
85. Y. Uesaka, Y. Nakatani, and N. Hayashi, *J. Appl. Phys.* **69**, 4847 (1991).
86. P. C. E. Stamp, E. M. Chudnovsky, and B. Barbara, *Int. J. Mod. Phys. B* **6**, 1355 (1992).
87. A. DeFranzo, I. Klik, L. Gunther, A. G. Swanson, and J. S. Brooks, *J. Appl. Phys.* **63**, 4234 (1988).
88. A. Garg and G. H. Kim, *Phys. Rev. B* **45**, 12921 (1992).
89. E. M. Chudnovsky and L. Gunther, *Phys. Rev. Lett.* **60**, 661 (1988).
90. C. Paulsen, L. C. Sampaio, B. Barbara, D. Fruchard, A. Marchand, J. L. Tholence, and M. Uehara, *Phys. Lett. A* **161**, 319 (1991).
91. I. V. Krive and O. B. Zaslavskii, *J. Phys. Condens. Matter* **2**, 9457 (1990).
92. D. D. Awschalom, J. F. Smyth, G. Grinstein, D. P. Divincenzo, and D. Loss, *Phys. Rev. Lett.* **68**, 3092 (1992).
93. L. Néel, *J. Phys. Soc. Jpn. Suppl. B-1* **17**, 676 (1962).
94. D. W. Strangway, R. M. Honea, B. E. McMahon, and E. E. Larson, *Geophys. J. R. Astron. Soc.* **15**, 345 (1968).
95. I. G. Hedley, *Z. Geophys.* **37**, 409 (1971).
96. D. Brož, A. Šolcová, J. Šubrt, B. Sedlák, F. Zounová, and S. I. Reiman, *Acta Phys. Slov.* **39**, 235 (1989).
97. R. J. Pollard, Q. A. Pankhurst, and P. Zientek, *Phys. Chem. Miner.* **18**, 259 (1991).
98. J. M. D. Coey and P. W. Readman, *Nature* **246**, 476 (1973).
99. J. M. D. Coey and P. W. Readman, *Earth Planet. Sci. Lett.* **21**, 45 (1973).
100. R. J. Pollard and Q. A. Pankhurst, *J. Phys. Condens. Matter* **4**, L317 (1992).
101. T. G. St. Pierre, D. H. Jones, and D. P. E. Dickson, *J. Magn. Magn. Mater.* **69**, 276 (1987).

102. S. Bocquet, R. J. Pollard and J. D. Cashion, *Phys. Rev. B* **46**, 11657 (1993).
103. F. T. Parker and A. E. Berkowitz, *IEEE Trans. Magn.* **25**, 3647 (1989).
104. J. L. Dormann, C. Djega-Mariadassou, and J. Jove, *J. Magn. Magn. Mater.* **104–107**, 1567 (1992).
105. F. Bødker, S. Mørup, C. A. Oxborrow, S. Linderoth, M. B. Madsen, and J. W. Niemantsverdriet, *J. Phys. Condens. Matter* **4**, 6555 (1992).
106. R. J. Pollard, C. M. Cardile, D. G. Lewis, and L. J. Brown, *Clay Miner.* **27**, 57 (1992).
107. Q. A. Pankhurst, C. E. Johnson, and M. F. Thomas, *J. Phys. C* **19**, 7081 (1986).
108. D. H. Jones and Q. A. Pankhurst, *J. Phys. C* **20**, 2453 (1987).
109. Q. A. Pankhurst and R. J. Pollard, *J. Phys. Condens. Matter.* **2**, 7329 (1990).
110. Q. A. Pankhurst, *J. Phys. Condens. Matter* **3**, 1323 (1991).
111. R. J. Pollard, Q. A. Pankhurst, and P. Zientek, *J. Magn. Magn. Mater.* **104**, 1557 (1992).
112. Q. A. Pankhurst, *J. Magn. Magn. Mater.* **101**, 291 (1991).
113. P. Levitz, D. Bonnin, G. Calas, and A. P. Legrand, *J. Phys. E* **13**, 427 (1980).
114. P. M. A. de Bakker, E. De Grave, R. M. Persoons, L. H. Bowen, and R. E. Vandenberghe, *Meas. Sci. Technol.* **1**, 954 (1990).
115. G. Le Caer and J. M. Dubois, *J. Phys. E* **12**, 1083 (1979).
116. P. M. A. de Bakker, E. De Grave, R. E. Vandenberghe, and L. H. Bowen, *Hyper. Interact.* **54**, 493 (1990).
117. P. M. A. De Bakker, E. De Grave, R. E. Vandenberghe, L. H. Bowen, R. J. Pollard, and R. M. Persoons, *Phys. Chem. Miner.* **18**, 131 (1991).
118. W. Kündig, *Nucl. Instrum. Methods* **48**, 219 (1967).
119. N. Blaes, H. Fischer, and U. Gonser, *Nucl. Instrum. Methods Phys. Res.* **B9**, 201 (1985).
120. J. R. Gabriel and S. L. Ruby, *Nucl. Instrum. Methods* **36**, 23 (1965).
121. J. R. Gabriel and D. Olson, *Nucl. Instrum. Methods* **70**, 209 (1969).
122. G. Lang and B. W. Dale, *Nucl. Instrum. Methods* **116**, 567 (1974).
123. J. M. D. Coey, *Can. J. Phys.* **65**, 1210 (1987).
124. K. Haneda, *Can. J. Phys.* **65**, 1233 (1987).
125. R. J. Pollard, *J. Phys. Condens. Matter* **2**, 983 (1990).
126. Q. A. Pankhurst, D. H. Jones, A. H. Morrish, X. Z. Zhou, and A. R. Corradi, *Crystal Properties and Preparation* **27–30,** 323 (1989).
127. E. Murad and J. H. Johnston, in *Mössbauer Spectroscopy Applied to Inorganic Chemistry,* G. J. Long, ed., Vol. 2, p. 507, Plenum Press, New York, 1987.
128. R. M. Persoons, D. G. Chambaere, and E. De Grave, *Hyper. Interact.* **28**, 647 (1986).
129. I. Dészi, L. Keszthelyi, D. Kulgawczuk, B. Molnár, and N. A. Eissa, *Phys. Status Solidi* **22**, 617 (1967.
130. A. Vértes, T. C. Huang, J. W. Vanderhoff, K. Lázár, M. Bennetch, and W. M. Reiff, *Radiochem. Radioanal. Lett.* **59**, 307 (1983).
131. R. J. Pollard and Q. A. Pankhurst, *J. Magn. Magn. Mater.* **99**, L39 (1991).
132. M. Rubinstein and D. W. Forester, *Solid State Commun.* **9**, 1675 (1971).
133. J. L. Dormann, C. Djega-Mariadassou, and V. A. M. Brabers, in *International Conference on the Applications of the Mössbauer Effect, Jaipur, India,* V. G. Bhide, ed., p. 196, Indian National Science Academy, India, 1981.
134. M. Mizoguchi, *J. Phys. Soc. Jpn.* **44**, 1501 (1978).
135. J. M. Zuo, J. C. H. Spence, and W. Petusky, *Phys. Rev. B* **42**, 8451 (1990).
136. S. J. Harker and R. J. Pollard, *Nucl. Instrum. Methods Phys. Res.* **B76,** 61 (1993).
137. I. Dzyaloshinsky, *J. Phys. Chem. Solids* **4**, 241 (1958).
138. P. J. Besser and A. H. Morrish, *Phys. Lett.* **13**, 289 (1964).
139. S. Foner, in *Proc. Int. Conf. Magnetism (Nottingham) 1964, p.* 438, Physical Society, London, 1965.
140. A. Meagher, Q. A. Pankhurst, and D. P. E. Dickson, *Hyper. Interact.* **28**, 533 (1986).
141. L. H. Bowen, E. De Grave, P. M. A. de Bakker, and R. E. Vandenberghe, *Hyper. Interact.* **54**, 467 (1990).

142. E. De Grave, R. E. Vandenberghe, and L. H. Bowen, in *Condensed Matter Studies by Nuclear Methods,* J. Stanek and A. T. Pedziwiatr, eds., Vol. 2, p. 186, World Scientific, Singapore, 1990.
143. E. Murad and L. H. Bowen, *Am. Mineral.* **72**, 194 (1987).
144. K. M. Towe and W. F. Bradley, *J. Colloid Interface Sci.* **24**, 384 (1967).
145. F. V. Chukrov, B. B. Zvyagin, L. P. Ermilova, and A. I. Gorshkov, in *Proc. Int. Clay. Conf. Madrid, 1970,* J. M. Serratosa, ed., p. 333, Div. de Ciencias, CSIC, Madrid, 1973.
146. R. A. Eggleton and R. W. Fitzpatrick, *Clays Clay Miner.* **36**, 111 (1988).
147. A. Manceau, J. M. Combes, and G. Calas, *Clays Clay Miner.* **38**, 331 (1990).
148. M. B. Madsen, S. Mørup, and C. J. W. Koch, *Hyper. Interact.* **27**, 329 (1986).
149. E. Murad, L. H. Bowen, G. J. Long, and T. G. Quin, *Clay Miner.* **23**, 161 (1988).
150. Q. A. Pankhurst and R. J. Pollard, *Clays Clay Miner.* **40**, 268 (1992).
151. R. J. Pollard, *Nucl. Instrum. Methods Phys. Res.* **B76,** 188 (1993).
152. E. Murad, in *Iron in Soils and Clay Minerals,* J. W. Stucki, B. A. Goodman, and U. Schwertmann, eds., p. 309, Reidel, Dordrecht, 1988.
153. S. J. Harker, *1989 Physics Honours Report,* Monash University (unpublished).

4

Mössbauer Effect Studies of Magnetic Soils and Sediments

Lawrence H. Bowen, Eddy De Grave, and Robert E. Vandenberghe

1. INTRODUCTION

In 1984, Bowen and Weed reviewed the applications of Mössbauer spectroscopy to soils and sediments.[1] Since that time there has been a steady increase in such applications and a number of reviews of various aspects of these have appeared,[2–9] including two by the present authors.[2,3] It is not the intention to cover the same material in the present review, but to emphasize two newer areas that appear to be growing rapidly: the study by Mössbauer spectroscopy of highly magnetic soils and sediments (i.e., those that are attracted to a hand magnet) and the application of external field Mössbauer spectroscopy to elucidate the magnetic properties of soil components. In the first case the spectra often are very complex, as is the case with soils and sediments in general, and thus have naturally resulted in studies of synthetic materials that relate to the environmental samples, but are simpler and more thoroughly characterized. In the latter case, the experimental technique has only been available in a few laboratories having a liquid helium cryostat with superconducting magnet. Of the laboratories with suitable equipment only a small percentage have worked with soil-related materials and in most cases the

Lawrence H. Bowen • Department of Chemistry, North Carolina State University, Raleigh, North Carolina 27695-8204. *Eddy De Grave and Robert E. Vandenberghe* • Laboratory of Magnetism, University of Gent, B-9000 Gent, Belgium. E.D.G. is now Research Director of the National Fund for Scientific Research, Belgium.

Mössbauer Spectroscopy Applied to Magnetism and Materials Science, Volume 1, edited by Gary J. Long and Fernande Grandjean. Plenum Press, New York, 1993.

studies have been limited to synthetic samples or particularly simple environmental ones.

The use of external field Mössbauer spectroscopy, although becoming available in more laboratories, must be considered a nonroutine analysis method because of its expense and because of the difficulties involved in fitting and interpreting the spectra. The magnetic characteristics seen in such spectra are thus of a more fundamental rather than analytical nature. As these results become better understood it is to be expected that the analytical applications to environmental samples will become increasingly important. The review of De Grave *et al.*[2] concentrates on this area, with particular emphasis on results from our collaborative research.

From a practical standpoint the most useful applications of Mössbauer spectroscopy to soils and soil-related materials are limited to room temperature (RT) and liquid nitrogen temperature (77 K), which is relatively easy to obtain in the laboratory. A review of the systematic studies on soil-related iron oxides and hydroxides with emphasis on analytical application has recently been made by Vandenberghe *et al.*[3]

In the context of the present review we use the term *magnetic soil* in a broad sense, to include not only highly magnetic samples that generally contain the ferrimagnetic oxides maghemite or magnetite, but also soils with an appreciable component that undergoes magnetic ordering at RT or below, producing a sextet in the Mössbauer spectrum. Thus, soils with any of the common iron oxides (hematite) or oxyhydroxides (goethite, lepidocrocite) are included. In order to provide a useful overview of research activity related to soils, we include a summary of work on related synthetic samples, organized according to compound. However, we have avoided discussion of work appearing before 1984, as these papers were reviewed earlier.[1,4]

It is generally understood, but should be repeated in this introduction, that Mössbauer spectroscopy will always be a complementary technique in soil science to be used in conjunction especially with X-ray diffraction, thermal analysis, infrared spectroscopy, and chemical methods such as selective dissolution. It will probably never be routine, as the obtaining and analysis of the spectra are too time-consuming to apply to large numbers of samples. Nevertheless, it has made important contributions to the understanding of a number of natural processes occurring in the environment, and the work reviewed here indicates that applications will become more common in the future.

By far the most complete survey of Mössbauer spectroscopy applied to iron oxides and oxyhydroxides is that of Murad and Johnston.[4] Their 1987 review has over 300 references and is divided like the present chapter into two major sections: the first on spectra of the specific oxides and the second on applications to soils, sediments, and meteorites. The reader of this present chapter should consult that reference especially for background work prior to that covered here and for more information relating the Mössbauer spectra to

Table 1. Magnetic Properties of Iron Oxides Found in Soils[7]

Mineral	Formula	T_0 (K)[a]	Character[b]	Hyperfine field (T)		
				RT	77 K	4 K
Goethite	α-FeOOH	400	af	38.2	50.0	50.6
Lepidocrocite	γ-FeOOH	77	af	—	—	45.8
Akaganeite	β-FeOOH	295	af	—	47.3	48.9
				—	46.3	47.8
				—	43.7	47.3
Feroxyhite	δ'-FeOOH	400	fi	42.0	52.0	52.5
Ferrihydrite	$Fe_2O_3 \cdot nH_2O$	*c*	sp	—	<46	46–50
Hematite	α-Fe_2O_3	955	wf[d]	51.8	53.5	53.5
			af		54.2	54.2
Magnetite	Fe_3O_4	850	fi	49.0	*e*	*e*
				46.0	*e*	*e*
Maghemite	γ-Fe_2O_3	>750	fi	50.0	52.8	52.8
				50.0	52.4	52.4

[a] Magnetic ordering temperature and hyperfine field for chemically pure, well-crystallized compounds. The T_0 is a Néel temperature for antiferromagnetic order and a Curie temperature for ferro- or ferrimagnetic order.
[b] Magnetic character: fi: ferri-, af: antiferro-, wf: weakly ferro-, sp: speromagnetic.
[c] Magnetic ordering occurs over a range of temperatures usually below 77 K.
[d] Weakly ferromagnetic above the Morin temperature at about 260 K.
[e] Below the Verwey temperature at about 120 K, the spectrum splits into additional discrete sextets, typically five.

the structure of the oxides and to their formation processes. Murad has also written several more specialized reviews on the general subject of Mössbauer spectroscopy applied to soil-related problems: a review for soil scientists and clay mineralogists of the information gained about the iron oxides from Mössbauer spectroscopy,[5] a summary of Mössbauer spectroscopy of poorly crystalline minerals including, but not limited to, iron oxides,[6] and a discussion of applications to clay mineralogy and soil science with emphasis on what can be done and on limitations of the technique.[7] From this latter reference we list in Table 1 the important iron oxides and oxyhydroxides to be found in soils, together with their magnetic properties. According to Table 1, these compounds have widely varying hyperfine fields resulting from their different magnetic ordering, and the field B observed in the Mössbauer spectrum in conjunction with isomer shift δ and quadrupole interaction 2ε (not given in the table) should be sufficient to distinguish these compounds by RT and 77 K spectra for most, and for all if 4 K liquid helium spectra are included.

However, of course the situation with environmental samples is more complicated. Both the poor crystallinity of these and the common substitution for iron(III) by other metal ions, especially aluminum, will have drastic effects on the magnetic properties. This was well known in 1984 and is discussed in all of the earlier reviews mentioned above. However, a more recent review by Mitra *et al.*[8] summarizes in particular the results on aluminum substitution for goethite and hematite; basically, a reduction in the observed hyperfine and a broadening of linewidths, both of which properties are also affected by crystallinity if the particles size is small. Detailed studies on the effect of substitutional metal ions are not available for the other oxides in Table 1, although some recent results on this subject will be reviewed in the present chapter.

One additional recent review should be mentioned in this introduction to a chapter on Mössbauer spectroscopy of soils, that of Parfitt and Childs,[9] which is concerned with the evaluation of chemical extraction techniques to remove iron and aluminum from soils: pyrophosphate, acid–oxalate, and dithionite–citrate. Their review covers the literature on these techniques and the different forms of the metals that may be extracted. In addition, an extensive comparison of soils before and after extraction, using Mössbauer spectroscopy and X-ray diffraction, is included. The major iron oxides studied were ferrihydrite (oxalate extracted, although sometimes incompletely by a single extraction), goethite, and hematite (dithionite extracted but not by oxalate). However, some soils contained maghemite or magnetite and the different response of these to dithionite is mentioned (maghemite preferentially removed). It seems that the use of these dissolution techniques in conjunction with Mössbauer spectroscopy deserves much further study, in particular with either synthetic materials or soils with high concentrations of the less common oxides: lepidocrocite, maghemite, and magnetite. The effect of aluminum substitution and/or particle size on extractibility is not currently known for these.

2. HYDRATED IRON OXIDES

The hydrated iron oxides include the oxyhydroxides of stoichiometry FeOOH (although these may contain additional absorbed water) and the poorly crystalline ferrihydrite, which has somewhat more water in the structure and a stoichiometry close to $5Fe_2O_3{\cdot}9H_2O$.[4] The oxyhydroxide structures are all based on octahedral iron surrounded by oxygen with different spatial arrangements of the octahedra, and hydrogen bonds between some of the oxygens. Of these, by far the most commonly occurring is goethite, and that species has also been most thoroughly studied by Mössbauer spectroscopy.

2.1. Goethite

Goethite, α-FeOOH, has a Néel temperature T_N of 400 K and is antiferromagnetic below that temperature, with a hyperfine field of about 38 T at RT, saturating at 50.6 T at 4 K (Table 1). The sextet has a quadrupole shift $2\varepsilon \cong -0.25$ mm/s while the paramagnetic doublet above T_N has quadrupole splitting $\Delta \cong 0.5$ mm/s. Thus, the principal axis of the electric field gradient is perpendicular to the spin direction. Small particles exhibit superparamagnetism, with a doublet appearing in the Mössbauer spectrum at temperatures well below the Néel point.[3] There will be a range of temperatures over which the relaxation is obvious with various proportions of extremely broadened sextet and doublet, usually specified by a blocking temperature T_B, which roughly corresponds to the point where the doublet is 50% of the total peak area. Often for soil goethites the T_B is well below RT, which then produces a simple doublet in the RT Mössbauer spectrum with $\Delta \cong$ 0.55 mm/s, typical of iron(III) in octahedral oxygen sites. Substitutional ions such as aluminum are commonly found in soil goethites and have effects on the Mössbauer parameters that resemble those of particle size. There is a reduction of the Néel temperature T_N with aluminum, first of all, such that only a doublet appears at RT for aluminum greater than 12 at.%, and also a reduction of the observed hyperfine field by about 0.14 T at 80 K and 0.05 T at 4 K per atomic percent.[3]

The general behavior outlined above was already established by the early 1980s. Since that time there has been further investigation of synthetic systems. Schwertmann *et al.* studied a series of goethites synthesized at varying temperatures and thus having a range of crystallinities.[10] Many different properties were measured and correlated in the paper: surface area, mean crystal dimensions, chemical extraction behavior, and Mössbauer spectra at RT and 4 K. The samples had surface areas of 153 to 9 m^2/g and mean crystal diameter along [100] of 9 to 52 nm, with the smallest particles synthesized at lower temperature. All RT Mössbauer spectra were sextets, but the sample 39/4 prepared at the lowest temperature had very broad and asymmetric lines, requiring a distribution of hyperfine fields to fit the data. Some type of distribution program has become common for fitting sextet spectra of poorly crystalline materials.[6] Such a distribution is characterized by a field of maximum probability, B_{max}, and some measure of the spread of values such as the half-width of the distribution or the average value, B_{ave}, and its standard deviation σ. In the case of 39/4 the lower half-width was 8.2 T and the upper half-width was 3.2 T, indicating a very asymmetric distribution with broad tail at lower fields. The peak B_{max} was 31.8 T (compared with 38.2 T for well-crystallized goethite, Table 1). In contrast, samples prepared at 60 to 90°C had B_{max} of 38.0 T and much narrower distributions. The distribution effect was minimized at 4 K but a small reduction in field was observed, from 50.6 T to 50.3 T, for the sample 39/4.

Murad and Bowen report a study of crystallinity effects in goethites with a high degree of aluminum substitution.[11] Two synthetic goethites, one with 31 at.% aluminum and mean crystal diameter along [111] of 74 nm and the other with 26 at.% aluminum but diameter of only 9 nm, were compared. Mössbauer spectra were taken for both samples over a range of temperatures from 4 K where both gave fairly broad sextets with almost the same hyperfine fields, 49.0 ± 0.2 T, to RT where both were doublets. However, the intermediate temperatures showed dramatic differences between the two: there was a broad range of temperatures in which both doublet and sextet contributed to the spectra, but this range and the position of the 50% doublet temperature differed by about 50°, with the sample having less aluminum but smaller particles (26% 9 nm) being lower (T_B = 110 K versus 165 K for the other). At 80 K the sample with smaller particles had an appreciable amount of doublet, but the field distribution of the sextet component was similar to that of the other sample which had no doublet. These results indicate that, although both particle size and aluminum substitution affect the magnetic ordering, in the temperature region where sextet and doublet coexist the sextet is due to the better-crystallized portion of the sample.

Two papers by Amarasiriwardena *et al.* concern the quantitative determination of mixtures of synthetic aluminum goethite and hematite.[12,13] In the first,[12] the pair had about 18 at.% aluminum each and were mixed in varying proportions down to 3 wt.% hematite. Mössbauer spectra were taken at various temperatures from 12 K to 130 K. A field distribution was used for fitting. It was shown that the two components had almost the same recoilless fraction, and that any temperature from 80 K down gave about the same result. The hematite fraction tended to be underestimated by the thin absorber approximation. This underestimation was minimized by using distribution fits and was less pronounced for samples with goethite as major component. It was shown that 3 wt.% hematite in goethite could be readily detected. In the second paper,[13] various other substitutions were used for the mixtures and a comparison was made between Mössbauer spectroscopy and two other techniques for determining the mixtures quantitatively: X-ray diffraction and infrared spectroscopy. It was shown that both Mössbauer spectroscopy and X-ray diffraction were more sensitive than infrared spectroscopy for detection of small amounts of hematite in goethite, and that the sensitivity of X-ray diffraction decreased with increasing substitution while that of Mössbauer spectroscopy did not. The decreased sensitivity of X-ray diffraction is caused by broadening of the lines by decreasing the particle size. The observed peak ratios in Mössbauer spectroscopy for hematite to goethite were in agreement with the actual proportions within 10% for all substitutions.

The reaction kinetics for the hydrothermal decomposition of goethite into hematite have been studied by Barb *et al.*,[14] using Mössbauer spectroscopy to monitor the production of hematite at temperatures of 180–200°C. First-order rate constants were obtained, which gave an activation energy of 56 kJ/mol. Meagher and Pankhurst[15] studied a single oriented crystal of goethite above the Néel temperature and related the asymmetric doublet intensity ratio to the orientation of the electric field gradient axes. However, there is ambiguity in the interpretation, as the ratio depends on two angles and on the asymmetry parameter. They used point charge calculations for a theoretical estimate of the electric field gradient and, by comparison with experiment, concluded that the electric field gradient axes did not coincide with the crystallographic axes and further, that the asymmetry parameter was nonzero. Several aspects of the temperature variation in Mössbauer parameters for microcrystalline goethite have been explored recently. These particles exhibit what has been called superferromagnetism,[16] in which the particles interact magnetically with each other, and this interaction results in a reduction of hyperfine field at low temperatures and collapse to a doublet over a narrow temperature range, unlike the usual superparamagnetic behavior where the doublet and sextet coexist over a broad temperature range. Koch *et al.*[17] have shown that goethites made from a solution containing SiO_3^{2-} exhibit the more usual latter behavior and thus the magnetic interactions were decoupled by the amorphous SiO_2 coprecipitated with the goethite. Madsen *et al.*[18] consider the doublet spectra obtained from microcrystalline goethite above its transition temperature and in their analysis explain the asymmetry and line broadening as resulting from magnetic interactions between particles, in agreement with the low-temperature behavior mentioned above. It appears that aluminum-substituted goethites do not have this superferromagnetic behavior.

In contrast to the effects of aluminum substitution on the Mössbauer parameters of goethite, those of manganese have been much less studied. However, Vandenberghe *et al.*[19] report Mössbauer spectra at 80 K and RT for a series of manganese-substituted goethites with manganese up to 8 at.%. The average and most probable fields decreased with substitution, but the decrease was much less than for aluminum substitution because manganese is not diamagnetic. The effects of silicon and phosphorus on goethite properties are discussed by Quin *et al.*[20] Unlike manganese or aluminum, these species most likely adsorb on crystal growth sites rather than become a true part of the goethite lattice. Thus, very complex Mössbauer behavior was observed relating to loss of crystallinity and production of mixtures of microcrystalline goethite, ferrihydrite, and amorphous phases. This effect is possibly related to the results of Koch *et al.*[17] discussed earlier.

The difference in magnetic properties between unsubstituted and aluminum-substituted goethite has been utilized by Dorman *et al.* to obtain a separation of these two species and also to monitor the separation by Mössbauer spectroscopy.[21] The technique of high-gradient magnetic separation was used, in which a dilute suspension of the mixture is passed through a steel wool filter placed in an external magnetic field. This technique has been used for many years to concentrate iron-containing particles from mixtures with diamagnetic species. However, the new application separates two very similar materials: pure goethite and a goethite with 30 at.% aluminum. As the pure α-FeOOH gave a sextet and the aluminum sample a doublet at RT, the degree of separation under different conditions was measured by the relative peak areas in the Mössbauer spectra of both the magnetic fraction retained on the filter and the tailings that passed through. From a 50–50 wt.% mixture over a range of applied fields from 0.1 to 1.1 T the aluminum goethite was concentrated in the magnetic fraction by a factor of about 2.4. This effect was explained as resulting from the aluminum goethite being paramagnetic and having a higher magnetic susceptibility than the antiferromagnetic pure goethite. However, particle size effects were shown to be important by using another goethite with similar aluminum content but higher surface area: it was concentrated to an even greater extent even though its susceptibility was lower.

In 1986 two independent studies of goethite Mössbauer spectra in an external field appeared. Meagher *et al.*[22] used a single crystal oriented so that the γ beam was parallel to the *b* axis and perpendicular to the spin direction, giving a 3:4:1 intensity pattern for the sextet at 4 K. The external field (6 T, 10 T) was applied to the spin and split the observed sextet into two, with fields approximately $B_{hf} \pm B_{ext}$. No reorientation of the spins was observed, placing a lower limit of about 60 kJ/m^3 for the anisotropy constant K. They also estimated K theoretically to be even larger. The high value of anisotropy prevents the external field from reorienting the spins. De Grave and Vandenberghe[23] studied a powder sample made from well-crystallized natural goethite. The temperature variation of the Mössbauer spectra in zero applied field was studied up to and above the Néel temperature of 400 K. At 19 K an external field of 6 T produced no reorientation of the spins, in agreement with the single-crystal experiment discussed above. The spectrum was in good agreement with a simulated one calculated assuming random orientation of particles and a spin direction perpendicular to the electric field gradient axis. An external field applied to the doublet spectrum at 405 K showed that the electric field gradient was positive.

More recently, Mössbauer spectra in external field have been reported for aluminum-substituted goethites. Bowen *et al.*[24] report spectra at 4 K and 6 T for hydrothermally prepared samples with aluminum up to 31 at.%. Two methods were used to fit the spectra, both using a distribution of hyperfine fields. In the simplified method the spectra were decomposed into three distributions: parallel, perpendicular, and antiparallel orientations of internal and external field. In the second, model-independent fit, a bidimensional distribution of both the internal hyperfine field B_{hf} and the angle θ it makes with the external field was used. This latter method has been described in more detail by de Bakker *et al.*[25] The effect of aluminum substitution is readily seen as a drastic change in the shape of the outermost lines (Fig. 1) and in the three distribution fits shown corresponds to an increase with aluminum in the relative proportion of the antiparallel orientation (the one where B_{ext} subtracts from B_{hf}) at the expense of the other orientations. The bidimensional distribution fits (Fig. 2) show an increase with aluminum in both the average θ and the θ_{max} corresponding to maximum probability. Both fitting techniques gave the same conclusion, that the goethites with high aluminum have a nonrandom canting of the spins so that B_{hf} orients away from B_{ext}, and thus these latter samples appear to have uncompensated spins that give rise to a ferrimagnetic-like behavior.

A quite different procedure to fit the applied field Mössbauer spectra of antiferromagnetic materials is presented by Pankhurst and Pollard.[26] This

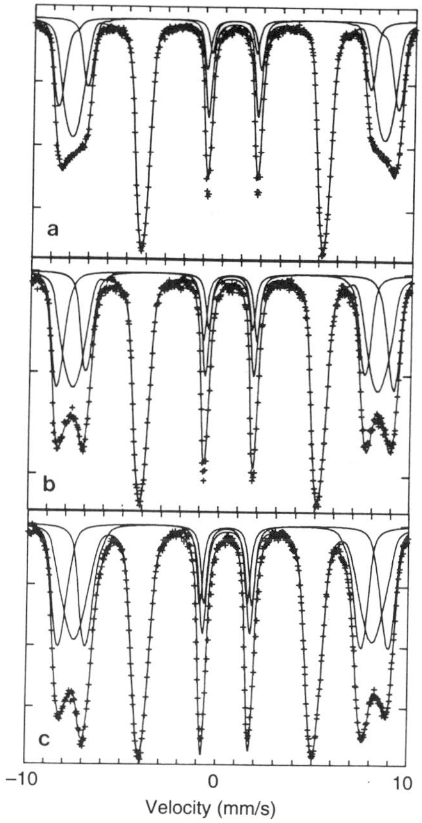

Figure 1. Spectra of goethite at 4 K in external field of 6 T, showing the effect of Al substitution.[24] (a) Pure; (b) 17 at.% Al; (c) 31 at.% Al. The spectra have been fitted by three distributions.

method uses a mean-field model with uniaxial anisotropy and in the form presented assumes a random distribution of particle orientations, but with spin orientation determined by the relative strengths of the exchange and anisotropy terms. It was applied with success to external field spectra of goethite and hematite. Extensions of the model to more complicated cases were mentioned as feasible in principle. This model has been applied by Pollard *et al.* to aluminum-substituted goethite spectra[27] of similar samples to the ones studied by Bowen *et al.* The fits are good, but a number of assumptions had to be made, most important of which were that there were four equally probable anisotropy fields, that the applied field decayed during measurement, and that there was a nonrandom orientation parameter. Simulation studies were performed and interpreted to show that neither random aluminum substitution nor surface effects could account for the observed ferrimagnetism of aluminum goethites. Instead, these authors propose that there is a preferential crystallization of aluminum on one of the magnetic sublattices. Surprisingly, their model indicates that this preferential effect is at a maximum at 10–20 at.%

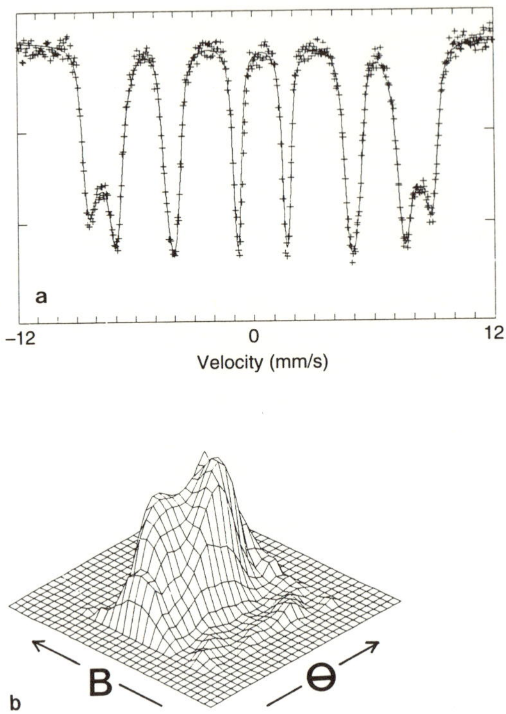

Figure 2. Spectrum of 31 at.% Al goethite at 4 K in external field of 6 T, fitted by a bidimensional distribution. (a) Fitted spectrum; (b) probability distribution of B_{hf} and θ.

aluminum and less for 30 at.% aluminum. Likewise, the mean anisotropy field has a maximum at about 11 at.% aluminum. Both of these results disagree with the analysis of Bowen *et al.*[24] It should be pointed out that no independent evidence of nonrandom aluminum substitution has been presented (by X-ray diffraction for example). Clearly, more samples of aluminum goethite need to be investigated and more tests of this model need to be carried out. Pollard *et al.*[28] have specifically investigated the use of a nonrandom orientation parameter to fit goethite spectra in an external field by measuring two absorbers prepared in different ways and showing that the spectral shape differed in the applied field. The goethite was not aluminum-substituted and thus consisted of nonspherical particles. Although an orientation effect may well be the cause of the difference observed, it is noted that a simple method to eliminate texture effects is to place the absorber at about 55° from the γ-ray beam (the magic angle). We have made such a test of the 31 at.% aluminum goethite whose 90° spectrum is shown in Fig. 2. There was no change in the spectral shape, indicating that texture effects were negligible for that absorber.

In a recent paper, Pollard *et al.*[29] have reported external field spectra of goethite and other unsubstituted FeOOH species arbitrarily fitted with three sextets for characterization purposes. The line shape of the sextets was assumed to be Voigtian, thus accounting for a degree of variation in the hyperfine fields. Of course the Voigtian profile is, like the Lorentzian, symmetric, and thus cannot be used for those microcrystalline and substituted materials that have asymmetric distributions.

2.2. Lepidocrocite

The basic structure of lepidocrocite, γ-FeOOH, differs from goethite in being based on cubic rather than hexagonal close packing of the oxygens. It has quite different magnetic properties (Table 1): a Néel temperature well below RT and very close to the boiling point of liquid nitrogen, such that even at 80 K the Mössbauer spectrum will be a doublet. The low-temperature saturation field is 45.8 T for well-crystallized lepidocrocite, much lower than those for the other iron oxides and oxyhydroxides except for ferrihydrite. Although much less common than goethite, lepidocrocite is found in many soils. Murad and Schwertmann[30] studied two synthetic samples with rather poor, but different crystallinities, mean crystal diameter along [010] of 66 and 42 nm (the 19 nm value reported in this paper was subsequently corrected to 42 nm) with surface area 18 and 65 m^2/g. Both samples ordered over a range of temperatures with the 66-nm sample showing onset of ordering at 77 K, but some doublet persisting at least to 55 K. The 42-nm sample had a lower ordering temperature beginning only at about 50–60 K. At 4 K the samples were completely ordered, but required hyperfine field distributions to fit the spectra. These distributions had about the same maximum, $B_{max} \cong 45–46$ T, but differed in width, that for the smaller particles naturally being broader and more asymmetric. In a later paper, De Grave *et al.*[31] extended the study of lepidocrocites to include samples of even poorer crystallinity, with surface areas from 86 to 147 m^2/g. Surprisingly, the field distributions were very similar at 13 K, $B_{max} \cong 44–45$ T with little difference in the widths of the distribution. Variation in particle size caused slight differences in the shape of the doublet at RT due to changes in the relative contribution of surface ions (those with a higher average quadrupole splitting). There were some differences in the relative doublet-to-sextet proportions in the transition region 30–60 K, but these did not clearly relate to particle size. Thermal scans at zero velocity, with and without application of external field of 4 T, were made for several samples and showed no shift of the transition to higher temperature, which should occur if the particles were superparamagnetic. These results and the behavior of the doublet spectrum at 55 K in external field were interpreted to indicate that the samples had a range of Néel temperatures due to different water contents, but no superparamagnetic relaxation.

Pollard *et al.*[29] report 4 K Mössbauer spectra for lepidocrocite (with dimensions about 10–33 nm) both with and without applied field. The spectra were fitted with Voigtian line shape: two sextets for zero field with $B = 46$ and 44 T, the latter being very broad, and the spectrum in 9 T applied field with

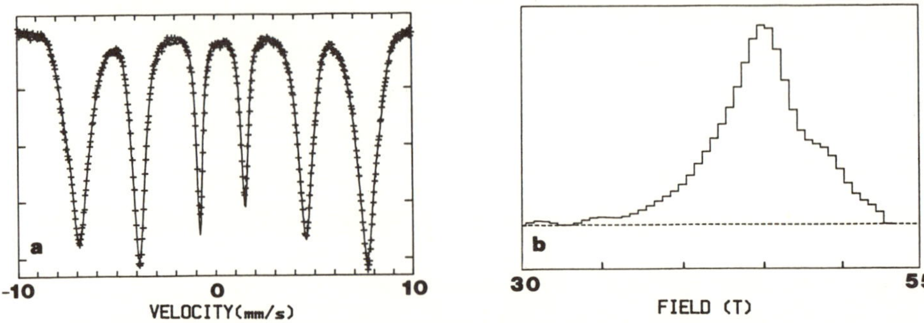

Figure 3. Spectrum of poorly crystalline lepidocrocite at 4 K fitted with a distribution of hyperfine fields. (a) Fitted spectrum; (b) Probability distribution of B_{hf}.

three sextets. A spectrum obtained in our laboratory at 4 K for a poorly crystalline lepidocrocite (mean crystal diameter along [020] $\cong$ 3 nm) is shown in Fig. 3. The fit is to a broad distribution of hyperfine fields with $B_{max} \cong 45$ T and a correlation of quadrupole shift with field. Unlike the microcrystalline goethites, the sample has a much more symmetric probability distribution with contributions from both higher and lower fields than B_{max}. This shape could be approximated by a superposition of one broad and one narrow sextet, as used by Pollard *et al.* for their fit.[29]

Lepidocrocite and goethite often occur together in soils[7] and on the basis of their Mössbauer parameters are readily distinguished in mixtures. Substitution of aluminum for iron in lepidocrocite, unlike the case for goethite, has not been clearly shown in soils, although synthetic samples can be made. In Fig. 4 we show spectra of lepidocrocites in external field, both a pure one similar to that in Fig. 3 and one with 18 at.% aluminum. Although the shape and the detailed parameters are quite different from goethites (Figs. 1 and 2), the differences between the aluminum-substituted and nonsubstituted samples are small. Clearly this area needs further investigation.

2.3. Akaganeite

Akaganeite, β-FeOOH, requires a small amount of chloride or fluoride ion to stabilize the structure. It has a Néel temperature just below RT and at low temperature requires three sextets to fit its Mössbauer spectrum (Table 1, Fig. 5). Chambaere and De Grave have made a detailed study of the temperature variation of the hyperfine field,[32] with the objective in particular to clarify the nature of the third sextet when the RT spectrum is clearly composed of two, not three, doublets. The components in Fig. 5 are labeled in order of decreasing field: at 30 K, B_1 for S_1 was 49.0 T, $B_2 = 47.5$, and $B_3 = 46.9$, with the S_3 peak being the least intense (~12%).[32] What is

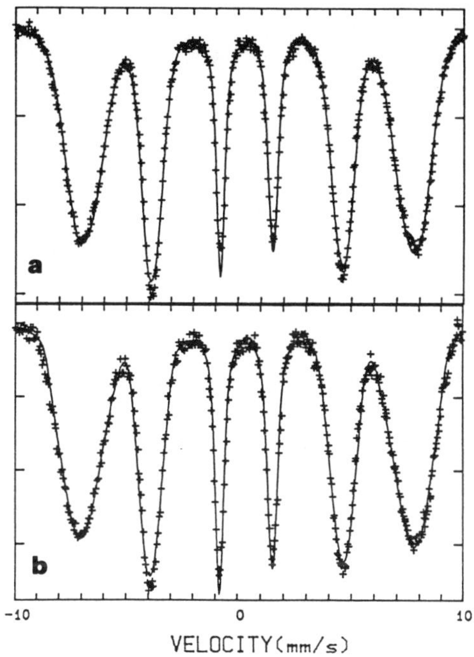

Figure 4. Spectra of poorly crystalline lepidocrocites at 4 K in external field of 6 T fitted by a bidimensional distribution. (a) pure; (b) 18 at.% Al.

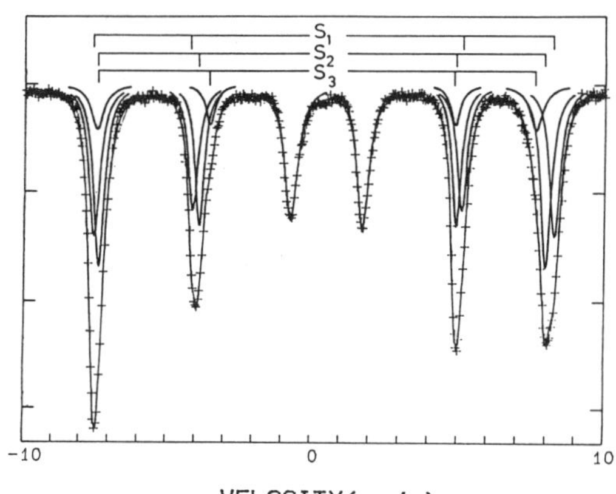

Figure 5. Spectrum of akaganeite at 30 K fitted with three discrete sextets.[32]

particularly interesting is the observation that the intensity of the third component was barely significant at 4 K but increased markedly as the temperature was raised. Thus, these authors attribute S_1 and S_2 to the two sites seen in the doublet, S_1 to an iron(III) in $O_2(OH)_3$ coordination and S_2 to $O_2(OH)_4$, while the S_3 sextet is caused by a spin hopping between the two sites which have different canting angles. Above about 60 K the sextet lines became very broad and could not be fitted to distinct components, an effect presumably caused by interstitial water.[2] The nature of the two sites, in particular their electric field gradient, was studied in the doublet region by detailed measurement of spectra as a function of temperature and also by applying an external field.[33] These latter spectra were obtained at 302 K and showed that the fields, up to 6 T, transform the doublet to a triplet. The shape of the external field spectra required that the two sites have opposite signs of the electric field gradient and distribution of values for both quadrupole splittings. Thermal behavior of the two sites resulted in an assignment of the inner doublet with $\Delta \cong 0.55$ mm/s to $O_2(OH)_4$ coordination and $\Delta \cong 0.95$ mm/s to $O_3(OH)_3$. In a subsequent paper, Chambaere and De Grave report a study of the changes in magnetic structure as akaganeite decomposes thermally to hematite.[34] Interstitial water had a drastic effect on the thermal transformation.

The spectrum of an oriented sample of akaganeite in an external field of 6 T at 4 K, with the crystallographic *c* axis perpendicular to the applied field,[2] showed reduced effective fields and a change in intensity ratios indicating the spins of both sites rotated toward the external field. Only two sextets were required for the fit, just as only two were found essential for the 4 K spectrum without external field.[32] Pollard *et al.*[29] have fitted their spectrum of akaganeite at 4 K and 9 T to three Voigtian sextets. Presumably their absorber was randomly oriented, however.

2.4. Feroxyhite

The synthetic δ-FeOOH and feroxyhite, δ′-FeOOH, are similar in structure but only the latter is found naturally.[4] At RT the latter gives a superparamagnetic doublet because of its poor crystallinity, while the synthetic materials have varied behavior from broadened sextets to doublet. The values in Table 1 are for a well-crystallized δ-FeOOH. The magnetic structure of synthetic δ-FeOOH was studied by Persoons *et al.*[35] By external field Mössbauer spectra at 4 K they clearly showed that their two samples were ferrimagnetic with about the same value for B_{hf} for both sites, almost the same isomer shift, and quadrupole shift near zero, but unequal occupancy. The two samples differed in particle size and this difference was reflected in a different degree of canting, the sample with smaller particles showing greater canting. Mössbauer spectra in zero field were fitted with a single hyperfine field distribution having an average B_{hf} that decreased much more rapidly with temperature for the sample with smaller particles. These 4 K spectra in external field have been more recently fitted using a bidimensional distribution of B_{hf} and θ.[25] In this latter procedure the integrated B_{hf} distributions of the two sites were shown to be essentially the same and equal to the distribution for the spectrum in zero field. In addition the integrated distribution of canting angles could be quantified for the two sites.

Madsen *et al.*[36] report Mössbauer spectra at RT and 80 K for synthetic feroxyhites of varying particle size, with mean crystal diameter along [100] of 24 nm down to 2.4 nm. Spectra were also obtained in external field of 1 T. The larger particle samples showed ferrimagnetic behavior, while the smaller ones had the weak response to an external field of an antiferromagnet or spin glass at 80 K. These latter samples had pure doublet spectra at RT even in external field.

Pollard and Pankhurst[37] also studied several samples of synthetic feroxyhite at 4 K in external field. They used the mean-field model[26] to fit the spectra. However, just as with the goethite spectra described earlier,[27] a number of somewhat arbitrary assumptions had to be made: three hyperfine fields (rather than the broad distribution of the earlier work[35,25]) and two field-dependent values of ξ, the ratio of spin magnitudes on the two sites. The fits to the spectra were good, although clearly exhibiting fine structure in the theoretical curves not seen in the experimental data. The most important result of the analysis seems to be the quantification of the anisotropy field, much larger for the sample with larger particle size, although the authors preferred to ascribe the difference to morphology and/or microscopic disorder, rather than size.

2.5. Ferrihydrite

Ferrihydrite is the least crystalline of the hydrous iron oxides, and even its stoichiometry is somewhat uncertain. When iron(II) is oxidized in aqueous solution, various forms of ferrihydrite may be precipitated. The X-ray diffraction pattern of ferrihydrite is always broad and can vary from the poorly crystalline samples with only two X-ray diffraction peaks to better crystalline samples with six peaks. Environmental samples differ appreciably, but all ferrihydrites have a doublet in the RT Mössbauer spectrum with quadrupole splitting $\Delta \cong 0.7\text{–}0.9$ mm/s and isomer shift characteristic of iron(III) in octahedral oxygen coordination. These RT doublets require either a two doublet fit or preferably fitting with a distribution of quadrupole splittings.[3] Ferrihydrite generally requires temperatures below liquid nitrogen to order completely, and its 4 K spectrum is a broad sextet with almost zero quadrupole interaction and a field of about $B_{hf} \cong 46\text{–}50$ T (Table 1). There is often a small asymmetry in the sextet line intensities, which can be accounted for by correlations between the field and the quadrupole and isomer shift interaction if the field distribution method is used.

Murad *et al.*[38] have studied three naturally occurring ferrihydrites by

Mössbauer spectroscopy as a function of temperature. The drastic change in magnetic ordering at low temperature caused by difference in crystallinity was particularly noted. Examples of two and six X-ray diffraction-line ferrihydrites were measured. Distribution fits were required. At RT the mean quadrupole interaction Δ was distinctly larger, as was the spread of the distribution, for the sample with smallest particle size. The low-temperature spectra showed the transition from sextet to doublet occurring over a range of temperatures, as expected from superparamagnetic interaction. However, the temperature range and 50% ordering temperature, T_B, were both much smaller for the less crystalline sample, $T_B \cong 28$ K, than for the better crystalline one. The latter showed a broad sextet at 80 K and had a blocking temperature $T_B \cong 115$ K, well above the usual values for ferrihydrite. However, the temperature range of the transition was large and some small proportion of doublet remained down to about 35 K.

Ferrihydrites synthesized with varying proportions of aluminum incorporation have been studied by Chadwick *et al.*[39] These samples were made by rapid evaporation of a solution at high temperature, 250°C to 500°C, and thus are not directly comparable to the natural samples discussed above, although they gave the characteristic two-line X-ray diffraction pattern of poorly crystalline ferrihydrite. The sample without aluminum, for example, had a sextet-to-doublet transition at a much higher temperature range, 110 to 240 K. The samples prepared with aluminum had greater thermal stability, i.e., they resisted crystallization to hematite. There were observed various changes in isomer shift, quadrupole splitting, and linewidth for the doublet spectra at RT and at 77 K for samples with aluminum >30 at.%, the spectra being fitted with two discrete components, and these changes were qualitatively related to the aluminum incorporation and differences in crystallinity. The temperature range of the transition region decreased drastically as aluminum content increased and was ascribed to relaxation effects important particularly for low aluminum content. Above about 50% aluminum, relaxation effects were not noticeable, although doublet and sextet still coexisted over a range due presumably to superparamagnetism. These authors mention that the sextet at 4 K for most of their samples had a small intensity asymmetry with the lowest velocity line appearing more intense and narrower than the highest velocity. In Fig. 6 is shown the 4 K spectrum of a synthetic ferrihydrite obtained in our laboratory that exhibits a similar asymmetry. This spectrum was satisfactorily fitted with a distribution of magnetic fields including correlations between B_{hf} and δ, 2ε, which might be expected in a disordered structure.

Murad[40] has examined the Mössbauer properties of a sample initially considered to be well-crystallized ferrihydrite synthesized by bacterially induced oxidation of $FeSO_4$ solution. The RT doublet had an asymmetry that was best fitted by two distributions of quadrupole splittings with different isomer shifts. Similarly, the 4 K spectrum was fitted with two distributions of hyperfine field differing in both δ and 2ε. Although Murad noted that a

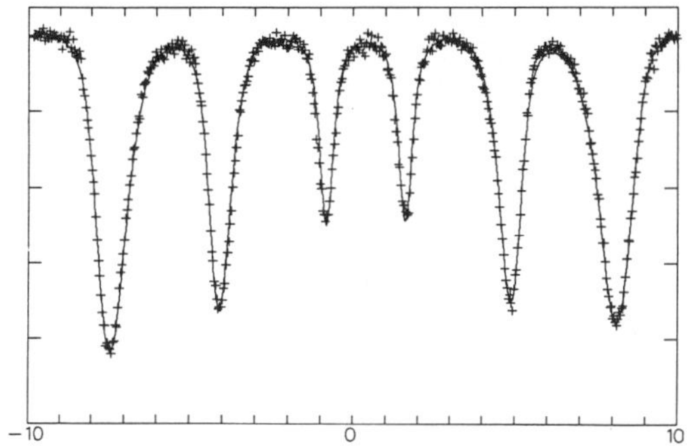

Figure 6. Spectrum of ferrihydrite at 4 K fitted with a distribution of fields. $B_{max} = 48.3$ T.

correlation of 2ε with B_{hf} also gave a reasonable fit, he suggested that the two distributions might correspond to the proposed two sites for iron in ferrihydrite, tetrahedral and octahedral, for which there has been no clear Mössbauer evidence previously. In a further study of this and other similar samples, including natural ones precipitated in acid mine water, Murad *et al.*[41] noted that the samples contained appreciable sulfate which might affect the magnetic properties. In particular, these materials had a narrower distribution of fields at low temperature, more negative quadrupole interaction $2\varepsilon \cong -0.2$ mm/s, and lower values of $B_{max} \cong 45$ T at 4 K than most ferrihydrites.

Cianchi *et al.*[42] have examined the relaxation behavior of a synthetic ferrihydrite in order to identify the mechanism of relaxation. They noted that vacuum removal of water at 350 K changed the low-temperature spectra drastically and gave more reproducible results. The temperature variation of the mean field in the low-temperature region and the line broadening of the doublet above the point where the sextet disappears (about 55 K) were in agreement with a strong interparticle interaction that could be described by superferromagnetism and a correlation of spin orientation between clusters.

Mössbauer spectra of ferrihydrite in external field have been reported and discussed by two groups in recent years. Madsen *et al.*[43] used fields up to 4 T at 5 K and at 80 K to assess the nature of the magnetic order. The fact that the sextet lines were broadened at 5 K but neither their general shape nor their relative intensities were changed by the external field was attributed to the amorphous nature of the magnetic order, i.e., speromagnetism. However, a small decrease in the intensity of the middle lines of the sextet could be assigned to uncompensated spins because of the small particle size. Application of the field at 80 K caused the doublet spectrum to decompose into a broad sextet, indicative of superparamagnetism. The average magnetic moment could be determined from the field variation at 80 K. Pollard and co-workers[26,44] also analyzed the external field behavior of ferrihydrites. Two samples were studied at 4 K and applied fields up to 9 T.[26] Careful measurement of the intensity ratios of the middle lines to the inner + outer showed a small but significant decrease for one sample (the less crystalline one) and an increase for the other, perhaps because of differing anisotropy. Most important, Pollard and Pankhurst[44] show that the ferrihydrite behavior in applied field and those of previously reported speromagnets could equally be described as antiferromagnetic structures with a large anisotropy preventing reorientation of the spins in response to an external field. Thus, the designation speromagnetism seems to be a questionable one.

3. NONHYDRATED IRON OXIDES

Ferrous and ferric iron are widespread in soils and sediments in various layered silicate minerals such as biotite and chlorite and many more. However, these species never have magnetic ordering at RT or even 80 K, and only a few order even at lower temperature. Thus, they are not discussed further in this review. The pure ferrous oxide wustite, FeO, is nonstoichiometric and has many interesting properties. However, it is not found in soils and is thus also excluded. There are three forms of magnetically ordered iron oxides found in soils (Table 1). Hematite, α-Fe_2O_3, is by far the most common and has quite interesting magnetic properties which will be discussed first in this section. Magnetite, Fe_3O_4, and its oxidized form maghemite, γ-Fe_2O_3, have quite different structure from hematite, and two distinct iron sites, octahedral and tetrahedral. Because the sites are unequally populated, these compounds are ferrimagnetic at RT and impart large magnetic susceptibility to any soil or sediment in which they are found. The distinction between the two becomes unclear in the common occurrence of oxidized magnetite, which could also be considered a mixture with maghemite. Substitution of iron by other metals, in particular aluminum, manganese, and titanium, has been observed in environmental samples, and in fact appears to be common.

3.1. Hematite

The hematite, α-Fe_2O_3, structure is based on hexagonal close packing of oxygens with iron in two-thirds of the octahedral vacancies. At low temperature, hematite is antiferromagnetic with spins oriented along the electric field gradient axis. As the temperature is raised, it undergoes a spin-flop transition known as the Morin transition at about 260 K[2] where the spins shift by about 90° and become slightly canted to each other, resulting in a weak ferromagnetic moment along the electric field gradient axis. The Mössbauer spectra reflect this transition, having different values of B_{hf} and 2ε above and below, with an increase in B_{hf} of about 0.8 T and a change in 2ε from -0.2 to about $+0.4$ mm/s going from high-temperature weakly ferromagnetic to antiferromagnetic state.[2] The Morin transition is sensitive to particle morphology. In natural samples it usually occurs over a range of temperatures, even though it is presumably a first-order transition.

Although most authors have considered the Mössbauer spectra in the transition region to consist of two distinct sextets corresponding to the weakly ferromagnetic and antiferromagnetic states, Gangas *et al.*[45] report a third sextet in the transition region, this component associated with a second-order transition. Two commercial powders were investigated. Chadwick *et al.*[46] studied the superparamagnetic behavior of synthetic hematites prepared under different thermal conditions. In some samples ferrihydrite was present. The 77 K spectra showed the effect of thermal treatment on the Morin transition: heating at 350°C gave a mixed sextet while heating at 400°C gave a hematite that was completely antiferromagnetic at 77 K. Pankhurst *et al.*[47] examined single crystals of hematite prepared in different fluxes. For spectra in the transition region of oriented crystals, two distinct components, one for each magnetic state, were observed. The line intensities confirmed the shift of spins by 90°, i.e., a change in intensities of the sextets from 3:4:1 to about 3:0:1. The effect of applied field was also studied,[47,48] and some differences were noted between samples prepared in different ways. The external field induced a spin-flop transition over a transition region that was narrow or broad, depending on the sample. Some evidence of a gradual rotation of the spins, corresponding to a second-order transition, was observed near the Morin transition when the external field was perpendicular to the γ-ray beam and the [111] axis.

Nikolov *et al.*[49] noted that conversion electron Mössbauer spectra and transmission spectra on the same oriented sample of hematite gave different values for the Morin transition temperature, with the transition occurring at significantly higher temperatures in the conversion electron spectra, thus indicating a difference between the surface ions, sampled by the conversion electrons, and the bulk. In contrast, Ji-Sen *et al.*[50] studied the Morin transition of iron-57-coated hematites and concluded that surface effects were unimportant. Both uniformly sized and nonuniform hematites were investigated and the latter had a broader range of temperatures for the Morin transition, although the average temperature was almost the same as that for the uniform ones, these latter having spherical shape and diameters of about 100 nm.

Verbeeck *et al.*[51] have studied hematites of varied morphology to determine its effect on the Morin transition. The morphology had little effect on the properties of the low-temperature antiferromagnetic phase, but changed the Mössbauer parameters of the weakly ferromagnetic phase in a systematic way indicating a shift in canting angle. Also the T_M temperature was decreased and its transition range increased with lower crystallinity. De Grave and Vandenberghe[52] looked specifically at natural hematite samples, three of these with markedly different crystallinity. The average Morin transition varied somewhat between the samples, and all had a component (8–16%) that remained in the weakly ferromagnetic phase at 80 K. Surprisingly, two samples had an antiferromagnetic contribution up to RT. The quadrupole interaction 2ε varied in the transition region in a way that suggested a gradual reorientation of the spins of both phases as the transition temperature is approached. Da Costa *et al.*[53] studied synthetic hematites of varied morphology prepared by heating three different goethites, each at varying temperatures from 440 to 1000°C. Many of these hematites had a small superparamagnetic doublet at RT. The major changes observed were at 85 K where the Morin transition would be complete for bulk hematite. No antiferromagnetic phase was found for samples prepared at 530°C or below. The proportion of antiferromagnetic phase for samples prepared at higher temperature depended on the original goethite and in a complicated way on the particle dimensions, although simple correlations with mean crystal diameter values were not successful.

Three studies of very small hematite particles deserve mention. Rancourt *et al.*[54] investigated the properties of α-Fe_2O_3 precipitated in zeolites. The mean diameter was 2.5 nm, calculated from the average magnetic moment. They indicated that the magnetic transition in their sample was characterized by two temperatures, the blocking temperature $T_B \sim 65$ K and the superparamagnetic transition temperature $T_{sp} \sim 23$ K. The particles had a very large magnetization, about 100 times greater than the bulk, which caused the low-temperature Mössbauer spectra to be quite different from other reported small-particle hematites. Dormann *et al.*[55] studied small-particle α-Fe_2O_3 prepared by co-sputtering with Al_2O_3. The particle diameters were 2.4 to 4.1 nm and blocking temperatures ranged from 35 K for the smallest particles to 145 K for the largest. The anisotropy constants calculated from the spectral variation with temperature were $\sim 10^3$ kJ/m^3, about a factor of 10 larger than that determined by Rancourt *et al.*[54] Dormann *et al.*[55] emphasize that the relaxation behavior of their particles depends on the uncompensated spins. Bentzon *et al.*[56] studied the behavior of 6.9-nm iron oxide particles of very uniform size prepared by oxidation of a ferrofluid. These were actually mixtures, however, of α-Fe_2O_3 and γ-Fe_2O_3 with RT Mössbauer spectra consisting mostly of sextets. The conclusion from electron microscopy and electron diffraction was that ordered aggregates of the particles, or "supercrystals," were formed.

The influence of aluminum substitution on the Mössbauer spectra of hematite was much studied in the earlier literature.[1,4] A number of additions to this field have appeared, most from either our group or the lab of Murad and Schwertmann. De Grave *et al.*[57] made a detailed analysis of the temperature variation of isomer shift for hematites containing up to 8 at.% aluminum. These were fitted by the Debye model to obtain two parameters, the intrinsic isomer shift δ_I and the characteristic Mössbauer temperature θ_M. Both of these quantities showed a distinct maximum at 4 at.% aluminum, which was related to the lattice strain relief obtained by substituting small amounts of aluminum for iron. Murad[58] studied the Mössbauer absorption for two series of aluminum hematites with substitutions up to about 10 at.%. Iron metal was used as an internal standard for peak areas. The low-temperature series prepared at 70°C showed an absorption maximum at ~ 4 at.% aluminum, but the series prepared at 1000°C did not. The data were interpreted in terms of a multiple regression with aluminum content and particle size represented by a reciprocal mean crystalline diameter.

Murad and Schwertmann[59] have published a complete Mössbauer study of these two hematite series at RT, with aluminum substitution up to 10 at.%. The 1000°C series showed good linear correlations of B_{hf}, linewidth, and quadrupole interaction 2ε with aluminum. The 70°C series showed nonlinear variations in all of these parameters, with high field and low widths for 4 at.% aluminum. Both widths and fields, however, gave a good multiple correlation when size was included in a similar way as above.

A hematite sample with 6.5 at.% aluminum was used by De Grave *et al.* to study the low-temperature behavior and in particular the Morin transition.[60] This sample was annealed at various temperatures up to 900°C in order to distinguish the effect of crystallinity from that of the aluminum. Although there was some evidence of surface enrichment by aluminum for the samples annealed at high temperature, the effect was minor and did not influence the bulk properties. This sample did not undergo a complete Morin transition at any temperature, but about 40% converted from the weakly ferromagnetic to the antiferromagnetic phase over a range centered at 170 K for the unannealed sample, and both the amount converted and transition temperature increased for annealed samples. One interesting feature was the variation of the quadrupole interaction for the two phases, which indicated a canting of the spins of both phases toward a more uniform spin orientation. In addition to this sample, a number of other low-aluminum samples were studied in less detail. Many magnetic properties showed changes in slope or maxima at about 4 at.% aluminum, including the Morin transition temperature.

De Grave *et al.* also studied hematites with a high degree of aluminum substitution up to 28 at.%.[61] In particular, the objectives were to quantify the hyperfine field distributions and the dependence of the average and maximum probability fields on aluminum content at RT and 80 K. Both of these of course decrease with increasing aluminum, but because of the asymmetry of the lines, the average B_{hf} decreases more rapidly than B_{max}. Those samples with aluminum > 10 at.% remain in the weakly ferromagnetic phase at all temperatures.

Vandenberghe *et al.* report studies on the Morin transition in manganese-substituted hematites,[62] with manganese up to 7 at.%. The substitution of manganese has a more drastic effect than aluminum and in fact no transition occurs for manganese > 4 at.%. This effect was proposed to result from the manganese contribution to the magnetocrystalline anisotropy. Rather than a substitutional effect, cobalt has been adsorbed on the surface of hematite, which markedly increases its magnetic coercivity.[63] Mössbauer spectra of samples with various amounts of adsorbed cobalt were taken as a function of temperature, and from these data an anisotropy constant could be calculated as a function of cobalt. This became constant at about 1.3×10^3 kJ/m^3, more than twice the value for hematite without cobalt.

3.2. Magnetite

Magnetite, Fe_3O_4, is the only pure iron oxide of mixed valence (Table 1). At RT it has a cubic spinel structure with iron in both tetrahedral and octahedral sites: $(Fe^{3+})_{tet}[Fe^{3+}Fe^{2+}]_{oct}O_4$. The two valence states on the octahedral sites are not distinct, however, as electron delocalization takes place. Thus, the stoichiometric magnetite has at RT only two distinct sextets[4] in 1:2 intensity ratio, corresponding to the tetrahedral iron (A sites) with $B_{hf} = 49$ T, isomer shift $\delta = 0.26$ mm/s, and octahedral (B sites) with $B_{hf} = 46$ T and $\delta = 0.67$ mm/s. The latter value is intermediate between Fe(III) and Fe(II). Because the two sites are unequally occupied and are coupled antiferromagnetically, magnetite is ferrimagnetic at RT. In external field (Fig. 7), the two sextets are distinctly separated[64] with the external field subtracting from the B_{hf} of the B sites and adding to that of the A sites. Because the spins orient parallel to the external field, the middle lines 2 and 5 of the sextets are absent. Below the Verwey transition of about 120 K the crystal structure is no longer cubic and the Mössbauer spectrum is complex. A spectrum of pure magnetite at 4 K fitted with five sextets is shown in Fig. 8. The fit is not perfect and the components cannot readily be related to the structural sites. It is shown to illustrate the difficulties that may be encountered for natural samples containing magnetite—for these it is preferred to stay above the Verwey transition if the Mössbauer spectra are to be fitted precisely.

Stoichiometry effects in magnetite have been studied recently by two

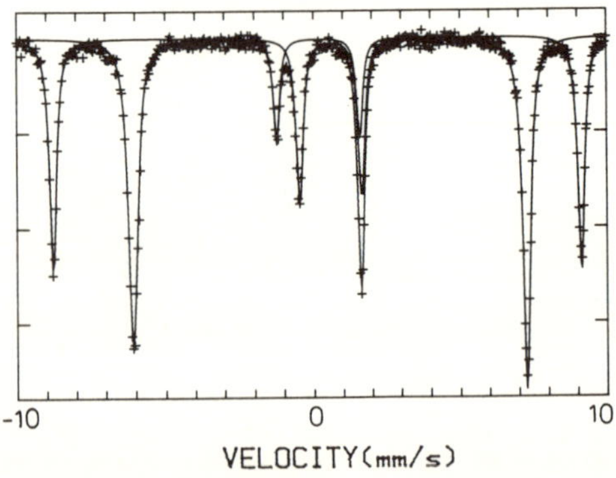

Figure 7. Spectrum of magnetite in external field of 6 T and 240 K, above the Verwey transition.[64] This sample has a small (0.3 at.%) Co substitution.

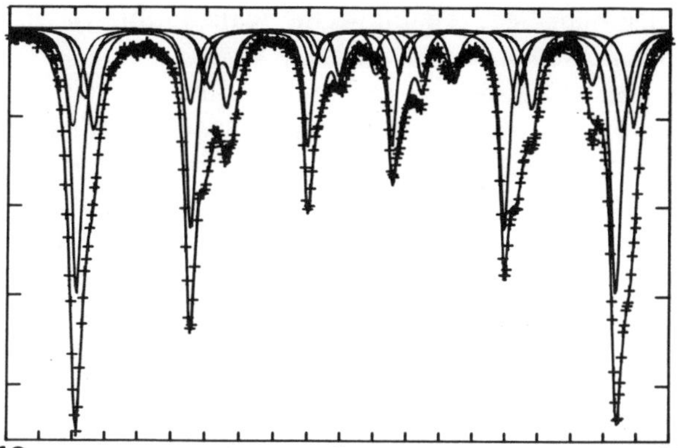

Figure 8. Spectrum of magnetite at 4 K fitted with five sextets.

groups. Ramdani *et al.*[65] showed that the Verwey transition disappears below 4 K for $x = 0.09$ in $Fe_{3-x}O_4$ (note that $x = 0.333$ corresponds to fully oxidized Fe_2O_3). Mössbauer spectra in external field of 5 T at 150 K resolved three components: Fe^{3+} on A, Fe^{3+} on B, and $Fe^{2.5+}$ on B. Since the Fe^{3+} on A corresponded to 33% of the intensity, it was clear that vacancies developed on B sites only. Even at 4 K in external field of 8 T, about 10% of the B sites for $x = 0.09$ showed a mixed valence signal, evidence of an electron tunneling effect. The transition behaviors of stoichiometric magnetite and one with $x = 0.03$ were compared by Haley *et al.*[66] At the Verwey temperature of about 121 K the Mössbauer pattern of stoichiometric magnetite undergoes sudden alteration of the hyperfine field parameters typical for a first-order transition. The nonstoichiometric sample in contrast shows gradual changes over a broad range of temperature, indicative of a higher-order transition at a lower temperature.

Substitution of other metal ions in Fe_3O_4 has been the subject of a number of Mössbauer studies. Rosenberg *et al.*[67] studied aluminum and gallium substitution. Mössbauer spectra in external field prove convenient in all of these substitution studies, as the substitutional site can be readily determined. In the case of gallium there was no preference and both A and B sites were substituted, while aluminum showed strong preference for B sites, displacing the iron(II) to the tetrahedral A sites. It should be noted that these samples were prepared at high temperature, $T \geq 1300°C$, with controlled oxygen pressure, and thus may not be comparable to substituted magnetites prepared at lower temperature.

Persoons *et al.*[64] studied cobalt-substituted magnetites, also using external fields. As cobalt(II) substitutes for iron(II) in the B sites only, it affects the Verwey transition. In some samples with low cobalt substitution the two B sites could be distinguished, enabling the calculation of the dipolar field contribution as a function of temperature.

Schwertmann and Murad[68] prepared aluminum magnetites at RT by slow oxidation of $FeCl_2$–$AlCl_3$ alkaline solutions. However, their procedure generally gave mixed products of magnetite and goethite, with goethite favored for higher aluminum content. The magnetite formed was partially oxidized and RT Mössbauer spectra were used to determine the degree of oxidation and the relative contribution of paramagnetic goethite. The hyperfine fields of both sites decreased with increasing aluminum as might be expected, although that for the tetrahedral A sites appeared to be more affected at low aluminum substitution, which was interpreted as indicating a preferential substitution at A. However, subsequent measurements in external field[69] showed that there was no preferential substitution of aluminum below an Al/Fe ratio of 0.05 and a preference for octahedral substitution at higher ratios.

Single crystals of synthetic titanomagnetites $Fe_{3-c}Ti_cO_4$ with $c = 0.24$, 0.50, and 0.72 were studied by Melzer *et al.*[70] The titanium was all Ti^{4+} and for the first two samples there was clear evidence in the Mössbauer spectra of electron spin transfer on the B sites; that is, a sextet due to $Fe^{2.5+}$, although both samples also had a separate iron(II) component. The spectra for $c = 0.72$ were complex at RT and at 4 K. This sample was also measured above its ordering temperature. Unlike magnetite, which has a single line in the paramagnetic state, the titanomagnetite had a doublet plus a singlet in approximately the correct proportions for iron(II) and iron(III). No evidence was seen in this spectrum for electron delocalization, which was presumably related to the fact that the high-titanium compound had *p*-type electrical conductivity whereas the lower-titanium species had *n*-type. The spinel Fe_2TiO_4, magnetically ordered below 120 K, has only a broad doublet above that temperature due to iron(II) on both A and B sites. By applying external magnetic fields, Vanleerberghe and Vandenberghe[71] were able to resolve the two sites and to evaluate the quadrupole splitting distributions for both.

3.3. Maghemite

Maghemite, γ-Fe_2O_3, is the completely oxidized form of magnetite. It also contains iron in tetrahedral A and octahedral B sites, but there are vacancies (usually in the octahedral positions) to compensate for the increased positive charge. Thus, in analogy to Fe_3O_4, the stoichiometry can be indicated as $Fe[Fe_{5/3}(Vac)_{1/3}]O_4$. Ordering of the vacancies leads to superstructure lines in the X-ray diffraction pattern and reduces the symmetry to tetragonal.[4] The two sites in maghemite have almost the same hyperfine field (Table 1) and are difficult to distinguish because of overlap between the sextets. Maghemite is also ferrimagnetic at RT and thus imparts a high susceptibility to any sample containing it as a component. Maghemite is used for magnetic recording tapes, and this industrial utility means there have been many studies of various synthesis techniques and resulting properties. Surprisingly few characterizations of industrial materials have emphasized Mössbauer spectroscopy, even though, as illustrated by the recent literature surveyed below, it can be a powerful technique for studying magnetic properties, in particular for ferrimagnetic species.

A natural sample of titanomaghemite has been studied by Collyer *et al.*[72] using Mössbauer spectroscopy and X-ray diffraction. This sample with approximate stoichiometry $Fe_{1.64}Ti_{0.32}O_3$ was compared with several pure synthetic maghemites. Mössbauer spectra were obtained at 300 and 4 K, the latter with an applied longitudinal field of 4 T. As for pure maghemite, the asymmetry in the sextet at RT showed it to be composed of at least two components, but the relative intensities were difficult to quantify. In addition, this sample had two doublets, assigned to a small amount of "hemo-ilmenite," $FeTiO_3 \cdot xFe_2O_3$. The magnetic components chosen were three, with about 90% corresponding to iron(III), but ~10% iron(II). The external field spectrum at 4 K was also fitted by three sextets, and the iron(III) parameters agreed almost exactly with these for the A and B sites in pure maghemite. The isomer shift for the presumed iron(II) sextet was, however, lower than that for one of the iron(III) components. They pointed out that evidence for iron(II) was inconclusive. It is certainly unlikely for iron(II) to have such a low isomer shift in an oxide.

A comprehensive summary of the magnetic properties of fine particles and especially the use of Mössbauer spectroscopy for studying them has been given by Morrish and Haneda.[73] Work with maghemite and other materials was included. In particular, for the maghemite samples it was noted that in longitudinal external field, small particles of γ-Fe_2O_3 do not align perfectly, but exhibit canting as shown by the nonzero intensity in lines 2 and 5 of the sextets. This canting depends on the particle size and diminishes as size increases. When the surface was enriched in ^{57}Fe, the middle lines were broader, indicating that the canting angles were distributed rather than fixed at certain value for A and B sites. Both particle morphology and chemical composition were shown to have important effects.

Tronc and Bonnin[74] consider the Mössbauer spectra of very fine spinel iron oxide particles and give a theoretical interpretation based on the Weiss local field. Magnetic coupling between particles was found to be important. Mössbauer studies of closely packed maghemite microcrystals by Tamura and Hayashi[75] are also interpreted by a theoretical model assuming interparticle magnetic interactions.

Several papers by Pollard and Pankhurst discuss the interpretation of applied field Mössbauer spectra of maghemite. Pollard[76] compared two morphologies of γ-Fe_2O_3, one rod-shaped and one spindle-shaped, both of about the same length, ~300 nm. The former had no intensities for the 2,5 lines in a longitudinal external field, the latter a small contribution. It was shown that quadrupole interaction effects did not account for the presence of these lines even though a large value for the interaction was obtained from the spectral fits. This seems to be in disagreement with other work, as the quadrupole interaction reported previously by most investigators has been close to zero for both sites. In a later paper, Pollard[77] presented convincing evidence that part of the 2,5 line intensity in γ-Fe_2O_3 was due in his samples to a hematite impurity. Although B sites contributed to the intensity because of canting, the A sites were collinear with the external field. Pankhurst[78] discussed the mean-field model for such external field spectra and Pankhurst and Pollard[79] applied it to Co-adsorbed γ-Fe_2O_3. Their interpretation is that surface effects are not needed to explain spin canting—that it is a natural response of ferrimagnetics that have opposing sublattice spins and large anisotropy.

The spectra of small-particle γ-Fe_2O_3 in external field have been interpreted in terms of spin canting in the recent work of de Bakker *et al.*[80] Two important additions have been made by these authors to the traditional fitting procedure. Both hyperfine field B_{hf} and canting angle θ were assumed to vary, and the fitting procedure gave probability distributions of each. This bidimensional distribution has been discussed earlier[25] in connection with goethite spectra. In the present case the procedure gave a good fit to the experimental data and rather large average canting angles for both sites, in keeping with the large intensity of the 2,5 lines in the spectrum. An example of such a fit is shown in Fig. 9. For contrast, in Fig. 10 is shown an external field spectrum of a much better crystallized maghemite with reduced 2,5 intensity and thus close to zero canting angle. The fact that the latter sample contains aluminum is another complication that is under study in our laboratory. However, it seems difficult to ignore particle size (and thus surface) effects in any interpretation of such spectra.

The γ-Fe_2O_3 thermal transformation to α-Fe_2O_3 has been studied by Tronc *et al.*,[81] specifically for small particles of about 10-nm diameter.

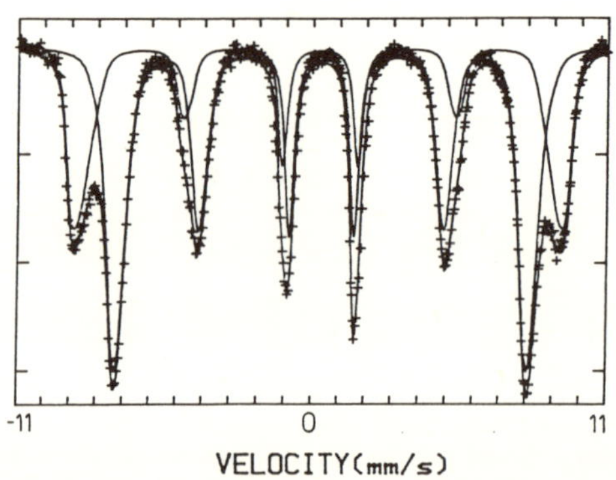

Figure 9. Spectrum of a poorly crystalline maghemite at 10 K in external field of 6 T fitted with two bidimensional distributions of B_{hf} and θ.[80]

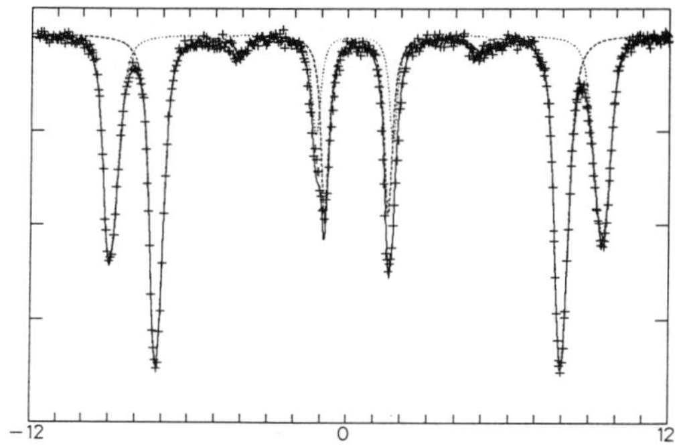

Figure 10. Spectrum of a well-crystallized maghemite at 4 K in external field of 6 T. This sample contained 11 at.% Al and the spectrum has been fitted with two distributions of B_{hf} only.

Mössbauer spectroscopy was used to characterized the product for various temperatures and times. Some evidence was seen that chemisorbed water stabilized the maghemites.

Two additional papers on cobalt-adsorbed γ-Fe_2O_3 should be mentioned. In an earlier work, Okada *et al.*[82] proposed that a distribution of canting angles should be used. They assumed a Gaussian distribution, and showed that surface treatment to remove defects narrowed the distribution. In a recent paper, Tsuji *et al.*[83] used emission Mössbauer spectroscopy to study the surface magnetism of Co-adsorbed maghemite. These materials are important commercially because they have much increased coercivity over maghemite itself. The high coercivity was directly associated with a broad magnetic sextet in the emission spectrum not seen in samples with lower coercivity. This was interpreted as evidence of a strong interaction between surface and bulk, although the average field of the surface particles was smaller. The effect of surface pretreatment with other cations was also studied. Their results seem to contradict the mean field interpretation of cobalt–maghemite discussed earlier.[79]

4. ENVIRONMENTAL SAMPLES

The application of Mössbauer spectroscopy in the study of minerals other than the iron oxides has been and continues to be important for geology. The subject of this present review excludes this more general application and concentrates on soils and sediments. The nature of the weathering process that produces these almost always results in iron oxides as components, sometimes major and sometimes minor, but crucial in either case to the magnetic behavior of the soil or sediment. In this review the distinction between soil and sediment is made by the authors of the papers reviewed, but can generally be understood as based on whether the process of formation occurred on land or under water. Even terrestrial sediments such as in caves have usually been produced by action of water. Another distinction might be made on the basis of the biological activity involved in formation of soils, which is generally lacking for sediments.

4.1. Soils

It seems of interest that in our survey of pertinent recent literature we found studies from soils of most of the continents and authors from all of these locations, although a strong evidence of international cooperation is evident. This field has, in addition to its scientific contribution, promoted a healthy cooperative effort between many countries to share projects, samples, methods, and results. The present section is generally arranged according to continent, with a few studies of clays at the end because they differ from the other examples by having only a minor magnetic component, at least before heating.

Graham *et al.* have studied weathering processes occurring on the Blue Ridge Front in North Carolina, using a variety of techniques.[84,85] Mössbauer spectroscopy was used to characterize primary iron-bearing minerals in the soil and saprolite, specifically chlorite, almandine, and magnetite.[84] In spite of visual evidence of surface oxidation, the Mössbauer spectra of sand-sized magnetite grains showed no indication of bulk hematite, only a partial oxidation of the magnetite. The clay fraction of these soils, in contrast, contained measurable hematite,[85] as well as goethite. Mössbauer spectroscopy at 77 K and down to 16 K for selected samples was used to quantify the relative goethite–hematite proportions as well as to estimate aluminum substitution. The iron oxide species in the clay were related to the minerals in the parent materials. In particular, hematite formation was correlated with almandine weathering and goethite with slow release of iron(II) from biotite. Furet *et al.*[86] studied the iron concretions in weathered crusts of nickeliferous deposits in Cuba. Mössbauer spectra showed goethite, hematite, and maghemite in the concretions. The Mössbauer spectra of goethite-containing samples at 77 K were fitted with an asymmetric distribution function showing magnetic field reduction that was ascribed to relaxation rather than aluminum or other substitutional effects.

A number of papers have appeared in recent years dealing with Brazilian soils. Fabris *et al.*[87] report studies on Oxisols using Mössbauer spectroscopy in conjunction with X-ray diffraction and chemical methods. Three sets of samples were examined: a soil profile with varying red and yellow bands, a toposequence from a shallow hill slope, and soils from adjacent sites on the Coastal Plain which differed in color. The observed hyperfine fields indicated changes in aluminum substitution and a decrease in crystallinity of the oxides as the depth of the profile decreased. The authors suggest that both goethite and hematite undergo re-formation reactions in response to the environmental conditions.

Resende *et al.*[88] concentrate on the magnetic nature of Brazilian Latosols. Many of these have such a high spontaneous magnetization that they are attracted to a hand magnet. The magnetic component in one sample at least was shown by Mössbauer spectroscopy to be maghemite (elemental analysis indicated titanium substitution: $Fe_{1.9}Ti_{0.6}O_4$). The saturation magnetization of the clay fraction approached 50 J/T per kg iron oxide, higher than that for synthetic titanomaghemite. The soils with the highest magnetization were those formed on mafic or itabiritic rocks. Experiments on a toposequence from top to bottom of a shallow hill slope showed the effect of xanthization (yellowing) on the magnetic properties as well as the color: the maghemite was sharply reduced by the xanthization even more so than hematite.

In a continuation of this work, Allan *et al.*[89] discuss the properties of four sets of magnetic soils. These include one formed on mafic rock from the earlier paper, but with 4 K spectra reported, and three other sets from very iron-rich rocks. Of the latter, one set was also studied by Mössbauer spectroscopy. Two distinct processes were observed, using primarily magnetization measurements in conjunction with X-ray diffraction. In one process, the soil inherits maghemite and/or magnetite from the rock which weathers in varying degrees to other iron oxides and thus the soil is less magnetic than the rock. In the other process, hematite in the rock is transformed to maghemite, making the soil more magnetic than the rock. The mafic rock samples and their magnetic extracts, as shown earlier, had no evidence of magnetite in the Mössbauer spectra. The maghemite could not be clearly resolved from hematite when present in a mixture. One of the magnetic extracts was studied in an external field of 5 T at 4 K. In that case no hematite was observed in the X-ray diffraction and the spectrum without external field appeared to be mostly a broad sextet characteristic of maghemite (the reported percent contribution of this sextet of 36% is apparently a typographical error). The external field spectrum showed clearly the A and B sites of maghemite with some spin canting indicated by the nonzero intensities of lines 2 and 5 (similar to Fig. 10). The spin canting was ascribed to substitution of nonmagnetic ions (Ti^{4+}, Al^{3+}), primarily on the A sites. The other set for which Mössbauer spectra were reported consists of an itabiritic rock of high hematite concentration and its surrounding magnetic soil. The rock had a sharp hematite spectrum at RT and this hematite underwent the Morin transition, evidenced by the different sign of 2ε at 4 K. The soil, in contrast, gave a spectrum that was mostly doublet at RT, a broad sextet at 80 K and even at 4 K. These sextets were fitted by two components, ascribed to goethite and hematite, although the Mössbauer parameters did not agree very well with these compounds. It was noted that according to magnetization measurements, there should have been about 10% maghemite, which was not resolved in the Mössbauer spectra. Coey *et al.*[90] summarize their results on Mössbauer spectroscopy of Brazilian Oxisols in a review paper emphasizing the varied problems of soil science that may be profitably addressed by this technique when used in conjunction with other techniques. One non-Brazilian example was included, a hematite-rich Latosol from Madagascar. In that case, the magnetization of the soil, 1.1 J/T per kg Fe_2O_3, was ascribed to superparamagnetic hematite on the basis of a qualitative examination of Mössbauer spectra as a function of temperature. Another example was that of the magnetic soil formed from dolomitic limestone, discussed in more detail in a separate paper.[91] This latter paper indicates the dolomite first produces "poorly crystalline ferric hydroxide," which then transforms to crystalline iron oxides, including substituted maghemite. The substitution, in this case by aluminum, was inferred mainly because of the low magnetization of the maghemite. As maghemite is generally unstable with respect to hematite, the detailed path of transformation was left open for further study. However, it was pointed out that limestone produces aluminum–maghemite, while basalt yields titanium–maghemite.

Monazite beach sand from Espirito Santo in Brazil was studied by Rodrigues *et al.*[92] The Mössbauer spectra showed hematite, magnetite, and ilmenite, $FeTiO_3$. However, the outer sextet of magnetite was not resolved from that of hematite so that the degree of oxidation of the former could not be reported. A Frantz magnetic separator was used and X-ray diffraction analysis of the subfractions indicated a difference in mineralogy, although these fractions were not analyzed by Mössbauer spectroscopy.

Fontes *et al.*[93] report Mössbauer spectroscopy data for a large number of Brazilian Oxisols from Minas Gerais. These were derived from four different parent materials: high-clay sediments, sandstone, basalt, and schist. The iron oxide mineralogy varied considerably between the soil clays. Field distribution methods were necessary to fit the low-temperature sextets which had both hematite and goethite components. The soils from mafic rocks had some maghemite, which was not clearly resolved in the Mössbauer spectra. The hyperfine field distributions as a function of temperature showed evidence of both particle size effects and aluminum substitution, although the Mössbauer results indicated some discrepancies with calibration curves for B_{hf} based on synthetic materials, generally overestimating aluminum. The goethite/hematite ratios could be calculated with reasonable certainty except for the mafic soils which contained maghemite. In a recent paper, Fontes *et al.*[94] have studied the association of humic acid with goethite in the Oxisols of Bahia, Brazil. These soils have large amounts of organic matter to depths greater than 2 m, and alkali extraction with NaOH removes humic acid plus goethite. Neither hematite nor ilmenite from the original soils was extracted, as shown clearly by the Mössbauer spectra. The organic matter appears to have complexed the goethite particles, which were $\sim$10 nm in diameter and almost spherical. Mössbauer estimates of aluminum substitution from the hyperfine field distributions at 80 K were high compared with chemical analysis and X-ray diffraction values, indicating the influence of the small particle size.

The nature of iron concretions formed on a sloping landscape in Taiwan by weathering of the underlying shale was determined by King *et al.*[95] A variety of techniques were used, with low-temperature Mössbauer spectra providing the goethite-to-hematite ratios. The results were interpreted in terms of a dynamic process of formation, disintegration, and re-formation of the concretions, which appear to be distinctive from other iron-rich concretions previously reported. The concretions consist of three layers: a gray core depleted in iron, a hard inner layer enhanced in both goethite and hematite, and a soft outer layer that is mostly goethite. The surrounding soil contains no hematite, but a large amount of aluminous goethite. It seems clear that these concretions are temporary, and a life cycle was proposed based on erosion, solubilization, and reprecipitation.

In north-central China, deposits of silt-sized particles from the deserts are common. This loess and the soil formed from it are highly magnetic. Vandenberghe *et al.*[96] have used Mössbauer spectroscopy to study samples of these materials. Both loess and soil were magnetically separated with a Frantz separator into strong and weak magnetic fractions. The weak fractions contained hematite and goethite as well as paramagnetic components. The strong magnetic fractions were considered in more detail. In order to test for low-temperature ordering, the Mössbauer spectra were obtained at 130 K as well as RT. The temperature 130 K was chosen rather than liquid nitrogen at 80 K in order to avoid the complication of the Verwey transition in magnetite. Sample spectra are shown in Fig. 11. These are of the strong magnetic

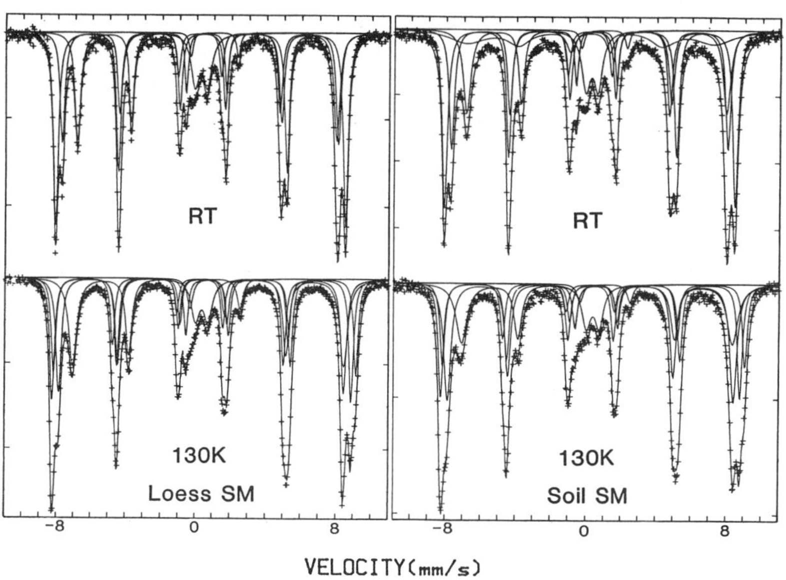

Figure 11. Spectra of Chinese loess (left) and associated soil (right) at RT and 130 K. These samples are the strongly magnetic fraction from a Frantz separator.[96]

fractions and have been fitted to the following components: loess—hematite, magnetite (~50%), and silicate doublets; soil—hematite, magnetite (~40%), maghemite (~16%), and silicate doublets. The magnetite was somewhat oxidized in both cases (about 1:1 rather than 1:2 ratio A:B sites). However, the hematite component of both (30–40%) had a partial Morin transition at 130 K indicating it was not substituted by titanium to any extent and the particles are larger than 20 nm in size. Maghemite was found only in the soil and was likely a weathering product from magnetite.

Several other Mössbauer studies of environmental samples from China have appeared recently. Zhang *et al.*[97] report spectra of sand samples from Flaming Mountain, named for its reddish color. The spectrum at 80 K contained two sextets, one due to hematite and the other with broad linewidths and smaller value of B_{hf} (49 T versus 53 T for hematite). These authors used thermal treatment to show that the broad sextet transformed into hematite after the sample was heated to 573 K. Conversion electron Mössbauer spectroscopy showed a decreased iron(II) doublet intensity consistent with surface oxidation processes. The relation between water-soaked paddy soils and their surrounding red soil was studied by Wang *et al.*[98] They conclude that the aqueous environment caused transformation of hematite to goethite by way of intermediate dissolution and recrystallization. These latter processes were inferred from the significant decrease in aluminum content of the goethites found in paddy soils. However, the reported goethite hyperfine fields at 77 K did not show the large variation in values expected from a reduction in aluminum substitution from 23 at.% to 8 at.%, and the method used to determine aluminum substitution was unclear.

A green shale from a Precambrian banded iron formation was studied by Das *et al.*[99] The sample was from the base of an oxide facies overlying the volcanogenic lower shale formation, and exhibited a marked variation in magnetization, becoming highly magnetic toward the younger side. Mössbauer spectra were used to characterize the nonmagnetic (chlorite, siderite) and magnetic (magnetite) components. The magnetite in the shale was essentially stoichiometric Fe_3O_4 according to its RT Mössbauer spectrum and underwent the Verwey transition between RT and 80 K as expected for pure magnetite.

A depth profile in the lateritic regolith of Western Australia was examined by St. Pierre *et al.*[100] with the primary objective of testing earlier suggestions that Mössbauer spectra could be related to the laterite aging. Both hematite nodules and the rock matrix were measured. The nodule showed partial superparamagnetism at RT but had only hematite in the 15 K spectrum. Lack of a Morin transition indicated poor crystallinity. The matrix had in some cases a second sextet due to goethite. However, there appeared to be no relation between degree of crystallization and depth.

Iron oxide pairs formed in soils around Mount Egmont in Taranaki, New Zealand, have been examined using many techniques by Childs *et al.*[101] These hard, dark deposits are 55–70% iron oxides and enriched especially in vanadium. Mössbauer spectra at 77 K were used to estimate the relative proportions of goethite and ferrihydrite, the two major components in X-ray diffraction patterns. Oxalate and dithionite extractions were also used. In general, the Mössbauer spectra gave high estimates of ferrihydrite and low estimates of goethite compared with the extraction results, indicating that some goethite was superparamagnetic at 77 K. Because some ferrihydrite may not have been extracted by oxalate, the authors indicate that the best estimate of relative proportions is likely in between the extraction and the Mössbauer values. Aldridge and Churchman[102] discuss the role of iron in weathering of a climosequence of soils formed from schist in South Island, New Zealand. The RT Mössbauer spectra were used to characterize the forms of ferrous and ferric iron in the soils and in their clay fractions. The main weathering process was shown to be oxidation of iron(II) in mica to iron(III) in mica and in oxyhydroxides. The latter appeared only in the iron(III) doublet component and was estimated by dithionite extraction of the whole soils.

Two environmental papers on North African soils deserve mention. Vandenberghe *et al.*[103] used Mössbauer spectroscopy to characterize the goethite and hematite in a red soil profile from Tunisia. Both aluminum and manganese were substitutional ions in the oxides, but the manganese had a particularly noticeable effect on the suppression of the Morin transition in hematite. It was shown that fitting the broad asymmetric sextets with a distribution of hyperfine fields allowed the determination of quantitative proportions of the two oxides from 80 K spectra and even, in most of these soils, from RT spectra. Eissa *et al.*[104] studied a series of Nile clays by Mössbauer spectroscopy on the original samples and after thermal treatment at varying temperatures up to 1000°C. Relative to desert clays the Nile samples had a larger fraction of magnetic component in their RT spectra as well as more ferrous doublets. Dehydroxylation of the clay minerals was seen at about 400°C and caused increase in the ferric quadrupole splitting. However, the clay minerals were stable against oxidation of iron(II) up to 600°C or higher.

Several other studies of clays seem pertinent. In these, the iron oxide phases are minor but important constituents. Diagenetic illitic clays from several locations were studied by Goodman and Nadeau.[105] In three out of seven samples, varying proportions of poorly crystalline goethite were seen in the 77 K spectra, while one sample had hematite which remained a sextet at RT. Murad and Wagner[106] made a detailed study of industrial clays by Mössbauer spectroscopy, including their thermal transformations and also the nature of the iron oxide contamination found in many clays. This species contributes to the RT ferric doublet and can be seen by its thermal transformation to hematite. This latter behavior is an alternative to the more traditional methods of oxide analysis by going to low temperature to produce magnetic ordering or, of course, removal by chemical extraction. In a later paper,[107] these authors review the Mössbauer studies on kaolinite and halloysite and report new results on a large number of samples. Mössbauer spectra were determined at RT and down to 4 K. These new results were on well-characterized kaolinites, measured with high-precision spectra. Thermal transformations were studied in detail for one kaolinite. Although the most predominant state of iron in kaolinite is iron(III), the occurrence of iron(II) in the structure was shown to be common. Goethite in kaolins could be detected at concentrations as low as 0.1% from 4 K Mössbauer spectra.

4.2. Lake and Ocean Sediments

This section is divided in two, first on fresh water and then marine sediments. In addition to sediments taken directly from underwater, several studies of sediments formed by aqueous deposition in the past are included.

The first group includes two studies of cave sediments of particular interest because of their magnetic properties. In an early study, Papamarinopoulos *et al.* used a magnetic concentration procedure to enhance the ferrimagnetic components in Greek lake sediments.[108] Mössbauer spectroscopy was essential in showing that the magnetism of the sediments was caused by nonstoichiometric magnetite (about $Fe_{2.95}O_4$). Although a small amount of hematite (10–15% of the iron) was present in the three studied sediments, it was not considered to be important for determining the magnetic properties. The possibility of impurity substitution in the magnetite was not excluded by the measurements. Later work by this group on cave sediments has emphasized the paleomagnetic information preserved in such sediments over long periods of time. The Greek cave of Petralona[109] was of interest because of dating problems associated with a human skull found in it. Again, magnetic concentration in conjunction with Mössbauer spectroscopy determined the importance of nonstoichiometric magnetite, in this case $Fe_{2.82}O_4$, mixed with or coated by a small amount of hematite. A more recent study concerns sediments in Ball's Cavern, USA,[110] presumably deposited during flood episodes, but over a very extended period of about 10,000 years per 2 m core. The two cores, however, dated from two separate periods, before and after the last ice age. Mössbauer spectra of the bulk sediments showed a distinct difference in degree of oxidation, with the younger sediment having more iron(II). Although magnetite, hematite, and possibly other iron oxides were present, their contribution to the Mössbauer spectra was so small as to prevent detailed analysis. The magnetic concentrates were clearly defined, consisting primarily of magnetite, with about 10% hematite and a ferrous doublet that was attributed possibly to wustite (FeO) even though that species is not expected in natural samples.

Hilton *et al.*[111] used sequential chemical extraction in combination with Mössbauer spectroscopy to interpret the complex doublet spectra observed at 78 K from sediments of a eutrophic lake in the Lake District of England. The idea promoted by their study is that Mössbauer spectroscopy in conjunction with extraction can be much more useful than previously supposed in studying redox chemistry in sediments. Schwertmann *et al.*[112] related the properties of iron oxides in two Finnish lakes to their formation environment. Both goethite and ferrihydrite were important iron minerals, but the proportions of the two oxides and their aluminum substitution differed markedly. These differences in mineralogy of the sediments reflected the differences in the aqueous environments of the two lakes.

König *et al.*[113] obtained Mössbauer spectra of sediments from the Elbe river in Germany. The importance of keeping the anoxic samples wet and excluded from oxygen was shown by the drastic oxidation of iron(II) when the samples were dried. Low-temperature spectra were required to distinguish between structural and oxidic iron.

Two papers by Mayer and Manning[114,115] discuss various problems related to sediments in the Great Lakes area, using Mössbauer spectroscopy in conjunction with other techniques. The relationship of phosphates to iron in sediments of the Lake Erie–Grand River confluence was explored in one paper,[114] while a general study of inorganic contaminants in the sediments of Hamilton Harbor, Lake Ontario, was the subject of the second paper.[115] Mössbauer spectroscopy was used in both to quantify the iron mineralogy. In the latter study, both wustite and hematite were identified in the Mössbauer spectra and related directly to industrial pollution.

Salt lakes in China were the site of sediments studied by Qi *et al.*[116] Chlorite, siderite, hematite, and ferric gels (ferrihydrite) were found in various proportions. In one lake, siderite was important and the ferrous/ferric ratio increased with sediment depth. In the other, ferric gels were more important and the ferrous fraction was much smaller. Chemical extraction with oxalate or with dilute HCl was used to clarify the mineralogy of some samples, while low-temperature (5 K) Mössbauer spectra were obtained for some.

The ferrous carbonates siderite, $FeCO_3$, and ankerite, $Ca(Mg, Fe)(CO_3)_2$, are important constituents of the sediments found in oil wells both off shore and on land in Denmark. Mørup *et al.*[117] have studied the complex iron mineralogy as a function of depth. It should be pointed out that dolomite, $CaMg(CO_3)_2$, allows a wide range of iron(II) for magnesium substitution ranging from small amounts in ferroan dolomite to over 20 at.% in ankerite. Both ankerite and siderite have distinctive quadrupole splittings, about 1.4–1.6 mm/s for the former and 1.8 mm/s for the latter at RT. The major component of pyrite, FeS_2, could be seen by its thermal decomposition to iron oxides at about 700 K, a temperature lower than that for breakdown of most clay minerals. Drastic changes in iron mineralogy were seen for the offshore wells at depths of about 3 km, below which ankerite appeared in the spectra at the expense of siderite. This depth also corresponded to a source rock maturity capable of formation of oil and gas. The formation of ankerite by release of magnesium may then be related to the temperature and pressure needed to transform organic matter to oil.

Torrance *et al.*[118] studied the iron mineralogy of postglacial marine clays found in eastern Canada and Japan in order to investigate the connection between iron oxides in these clays and their unusual rheological properties, characterized by the tendency to remold into an almost fluid state when disturbed (so-called "quick clays"). Mössbauer spectroscopy and X-ray diffraction were used. In addition to chlorite, mica, and amphibole seen clearly in both measurements, the Mössbauer spectra showed hematite to be a minor constituent in most of the samples, while magnetite was present in one. The iron oxide in a Japanese bay sediment in contrast was mostly ferrihydrite. Extraction methods were used to remove the iron components and appreciably lowered the geotechnical yield stress at given water content.

A number of studies have been made on ferromanganese nodules and crusts. Kraczka *et al.*[119] measured nodules from the Carpathian mountains including both a low- and high-manganese nodule. Small-particle goethite was seen in both, while the nodule with higher iron content had hematite as well. Estimates of particle size from the Mössbauer spectra ranged from 5 nm to about 20 nm. Deep-sea ferromanganese crusts from the Pacific Ocean were investigated by Murad and Schwertmann.[120] Poor crystallinity of the iron-containing minerals made X-ray diffraction identification difficult. Mössbauer spectra at 4 K showed most of the iron to be present in a magnetically ordered ferrihydrite-like phase which was closely associated with manganese oxides. The use of hydroxylamine chloride to extract manganese also affected the iron oxides, and thus cannot be recommended for the study of the iron mineralogy in this case.

Marine sediments and a deep-sea manganese nodule were the subject of a recent investigation by Chen *et al.*[121] Chemical leaching with HCl and H_2O_2 was used, but only RT Mössbauer spectra were obtained. In this case the primary concern was enhancement of the ferrous iron component, which was attributed to entrapment of terrigenous aluminosilicates during growth of the nodule. Deep-sea nodules from the Pacific and Indian oceans were studied recently by Rusanov *et al.*[122] Their nodules also contained a ferrous component which they attributed to siderite, although its quadrupole splitting of $\Delta \cong$ 2.2 mm/s is high for that species. Mössbauer spectra were obtained at RT and 77 K, but not below. They discuss the possibility that a ferrihydrite-like species was the major iron-bearing phase, although the evidence is indirect without low-temperature magnetic spectra.

Two interesting papers by Ellwood *et al.* relate the ferrous carbonates to paleomagnetism.[123,124] In both papers, Mössbauer spectroscopy had an important, although secondary part. As indicated in a number of the papers discussed above, siderite and ankerite are often found in deep-sea sediments. The earlier paper[123] considers especially siderite and its relation to the formation process of pyrite, possibly by bacterial reduction of iron(III). Siderite is easily oxidized to magnetic oxides such as magnetite or maghemite. If this occurs, the secondary minerals may mask completely the original magnetic properties. Thus, the presence of siderite is considered good evidence that the primary magnetic record is intact. In the second paper,[124] Mössbauer studies, magnetic and thermal properties of ankerite are presented to explore the possibility that ankerite also might be important in paleomagnetism for similar reasons. Their conclusion is affirmative. At temperatures as low as 250°C, thermal demagnetization produces iron oxides from both ankerite and siderite. Obviously the amount of secondary iron oxides formed will depend on the iron content: siderite > ankerite > ferroan dolomite.

4.3. Volcanic Samples

Soils developed on volcanic rock reflect the distinctive characteristics of their source. The lava and ash from volcanoes often are strongly magnetic because of the presence of magnetite and/or maghemite which may be titanium-substituted. The importance of iron-bearing minerals in volcanic samples invites the use of Mössbauer spectroscopy and indeed it has been applied to a variety of such materials.

Domingues *et al.*[125] report Mössbauer studies of lava samples from both outside and inside the crater of the extinct volcano of Quimsachata in Peru. Most of the iron was divalent, in olivine and pyroxenes. One sample from outside the crater contained maghemite, which was partially superparamagnetic at RT, and one sample of extruded rock from the inner rim contained partially oxidized magnetite. Hong *et al.*[126] measured four lava ash samples from ancient eruptions on Jeju Island south of the Korean coast, finding olivine, pyroxene, and ilmenite, in addition to varying proportions of hematite with very broad lines resulting presumably from aluminum substitution. These authors note the similarity between the ferrous mineral composition and lunar basalts, relating this to the collision ejection model for the formation of the moon.

Marion Island is an active volcanic island in the south Indian Ocean and because of high precipitation, the basalts have been strongly weathered. Murad and Schwertmann[127] used Mössbauer spectroscopy to characterize the poorly crystalline iron oxides. These, like ferrihydrite, were almost completely removed by oxalate extraction, had ill-defined X-ray diffraction patterns, and RT ferric doublets that became mostly ordered at 4 K with broad sextets. Oxalate extraction also removed considerable aluminum and silicon, and infrared spectroscopy showed Si–O–Fe bonds. The authors conclude that a close association of silicon and aluminum with the ferrihydrite is caused by the rapid weathering rate, and that these elements stabilize the ferrihydrite and affect the Mössbauer spectra in particular by causing a broad distribution of hyperfine fields. Childs *et al.*[128] have studied ferrihydrite in Japanese volcanic ash soils from a variety of sites. In their investigation, oxalate extraction was combined with X-ray diffraction and Mössbauer spectroscopy at RT and 77 K. Most of the samples had a minor sextet component relating to maghemite and/or magnetite, which was unaffected by the extraction. By using the

Mössbauer peak area ratios in conjunction with the extraction, they were able to quantify the ferrihydrite to other forms of iron in the ash. This technique was more sensitive to ferrihydrite than X-ray diffraction. They also noted the importance of silicon in the formation of natural ferrihydrites.

Schlinger *et al.*[129] made extensive and varied measurements in order to determine the nature of magnetic domains in volcanic glasses from the East African Rift. These are microcrystalline, 2–10 nm in size, mostly superparamagnetic at RT and were determined to be nonstoichiometric magnetite, primarily from the average isomer shift which was intermediate between maghemite and magnetite. The variation of the magnetic fraction with temperature was used to estimate a particle size distribution and also an anisotropy constant $K = 13\,\text{kJ/m}^3$, larger than that for well-crystallized magnetite, although an order of magnitude smaller than that for other small particles of magnetite.

Volcanic titanomagnetites have been studied by several groups. Julian *et al.*[130] compare titanomagnetites extracted by magnetic separation from basaltic lavas and tephra found in Iceland. All of the samples had nonmagnetic doublets due to pyroxene among others, but also varying proportions of a broad, ill-defined sextet at RT, which sharpened and increased in intensity at low temperatures, indicative of superparamagnetism. There was considerable difference in the sextet peaks between samples indicating the possible use of such spectra as a fingerprint of the tephra. Lipka *et al.*[131] studied titanomaghemites from basalts in Nigeria. Magnetic separation was used to concentrate the magnetic species. Mössbauer spectra were obtained as a function of temperature and the sextet increased in intensity with lower temperature. However, in this case, application of an external field of 0.6 T showed that this was not due to superparamagnetism, but to a distribution of Curie temperatures, presumably caused by inhomogeneous distribution of iron. The titanium substitution estimated by Mössbauer spectroscopy agreed well with that from X-ray diffraction. Heating at 675°C for 3 h caused the titanomagnetite to segregate into pure magnetite and various paramagnetic titanium-rich oxides, with some oxidation to hematite.

The basalt lavas of Iceland have been the subject of a number of interesting investigations by Helgason and co-workers.[132–135] The strong magnetic anomaly near Stardalur was probed by core sampling to a depth of 200 m.[132] Mössbauer spectroscopy of bulk samples at various depths all showed magnetite, but the ratio of B to A sites changed sharply at about 40 m depth from a highly oxidized value <1 [in which "A" sites include iron(III) in both tetrahedral and octahedral positions] to about 2, characteristic of stoichiometric magnetite. In contrast, other basalts showed low B-to-A ratios and in fact could often be fitted to three sextets, including a separate one for maghemite in addition to the magnetites. It was suggested that both the purity and small grain size of the magnetite in the deep core were due to hydrothermal alteration which separates out impurities and might also amplify the local magnetic field during the process.

Some of the basalts containing maghemite in addition to magnetite were subjected to thermal magnetization measurements.[133] The magnetization curves typically showed Curie temperatures of about 580°C and were reversible, indicating the thermal stability of the maghemite. Heating experiments at 600°C showed that the maghemite was stable in both oxidizing and inert atmospheres for at least several hours, while magnetite was less stable toward oxidation to hematite. In a reducing atmosphere at 600°C, both hematite and maghemite readily transformed to magnetite. Mössbauer determination of iron(II)/iron(III) ratios was used to study redox rates in a basalt melt at 1300°C.[134] The ratios changed essentially reversibly depending on the nature of the atmosphere (oxidizing or reducing) and reaction times were on the order of 100 min. Thus, it was concluded that the ratios in volcanic glasses directly reflect the oxygen fugacity in the magma. These results were applied to interpret oxygen stratification in a magma chamber responsible for the Askja eruption of 1875.[135]

5. SPECIAL ENVIRONMENTS

Soils and sediments are found over all of the surface of the earth. Even volcanic samples, the third subject of the previous section, are widespread. Is it possible that our planet is not unique, and in fact extraterrestrial application of Mössbauer spectroscopy to the surface composition of other planets might be feasible? There were, of course, the moon rocks. Meteorites provide a continuing and varied access to extraterrestrial samples. One planet stands out for its surface—Mars, the red planet, and also a highly magnetic one, according to the Viking mission. Both properties of color and magnetism point to iron oxides. There has been no mineralogical study of Mars and the only evidence is indirect. However, comparisons of Mössbauer spectra of terrestrial samples with similar color and magnetism to Mars regolith suggest an exciting possibility that is actively being pursued by groups in both the United States and Europe—send a Mössbauer spectrometer to Mars to find out what its surface really is! Considering the distance, it seems much more realistic to send the spectrometer there than to wait for a mission to bring samples back.

5.1. Meteorites

Two recent papers are included here because of their emphasis on Mössbauer spectroscopy. Earlier papers on meteorites have been reviewed by Murad and Johnston.[4] Madsen *et al.*[136] studied the Mössbauer properties of a carbonaceous chondrite, Murchison, which fell in 1969. The low-temperature spectra of the nonmagnetic matrix showed an amorphous ferric oxide component exhibiting superparamagnetism at low temperature, with a broad sextet first appearing at about 40 K. This component was related to that in the meteorite Orgueil, studied earlier. De Grave *et al.*[137] determined the temperature dependence of the Mössbauer parameters for the iron–nickel phases in a sample of the Santa Catherina meteorite. The low-temperature spectral fit was improved using a distribution of fields with correlated quadrupole splitting. Magnetic ordering of the 28% nickel phase was observed below 25 K.

5.2. Martian Soil Analogues

No studies of Martian samples by Mössbauer spectroscopy can be reported at this time. Several authors have discussed the similarities between Martian regolith and earth analogues. Coey *et al.*[138] emphasize titanomaghemite, i.e., the fully oxidized form of titanomagnetite, as the likely species responsible for the magnetic surface of Mars. Morris *et al.*[139] present evidence, including Mössbauer spectra, for the possible importance of superparamagnetic hematite on Mars (because of its small particle size, often designated nanocrystalline). A palagonitic soil from Mauna Kea in Hawaii has been the subject of Mössbauer and other studies by Morris *et al.*[140] The magnetic character of the soil, similar to that of Mars, was shown to be due to titanomagnetite. Magnetic separation with a Frantz apparatus was used to concentrate this fraction. Nanophase hematite, which was likely present, was not important for the magnetic properties of this soil.

Agresti *et al.* describe the backscatter Mössbauer spectrometer proposed for the Mars mission of NASA.[141] This spectrometer also has the capability for X-ray fluorescence, so that elemental abundance as well as iron mineralogy may be determined. Knudsen *et al.*[142] discuss the reasons why such measurements are of importance in the context of the proposed Russian mission. Jackiewicz *et al.*[143] have made Mössbauer studies of basalt rocks of Poland considered to be analogues for the Martian regolith. Weathering was determined by the iron(II)/iron(III) ratio. Only RT Mössbauer results were obtained, but evidence of magnetically ordered oxide phases were seen.

6. CONCLUSIONS

There has been much progress over the past 5–10 years in the application of Mössbauer spectroscopy to study the magnetic properties of soils and sediments. Major developments occurred both in the study of related synthetic systems and the more complex environmental samples. One improvement is the use of stronger sources and thus spectra of much higher precision, which allow more detailed interpretations. Another important development has been in methods for fitting complex spectra. The model-independent distribution method has been applied in a number of cases discussed in this review, and it often gives useful information relating, for example, to degree of crystallinity, foreign ion substitution, and relative proportions of oxides in a mixture. Of course, a model must be developed separately to interpret the results. However, this seems superior to an arbitrary assumption of two or three sextets, which might be falsely interpreted as indicating two or three distinct sites. There has been progress also in developing relatively simple models such as those based on the molecular field, which have a limited number of adjustable parameters to be determined from the fit.

In some cases, such as for maghemite, there are indeed distinct sites that are not clearly separated in the Mössbauer spectrum. Often the application of an external field is useful in unraveling complex magnetic spectra, and a number of papers showing this utility have been discussed. External field spectra will hopefully become more common in the future.

The combination of chemical or physical treatment with Mössbauer spectra before and after has been explored to a much greater extent in recent years, but there is certainly considerable opportunity for further study. Among the treatment methods showing great promise are chemical extraction (e.g., dithionite and oxalate extractions), controlled heating to produce thermal transformation, and magnetic separation, especially high-gradient magnetic separation.

The magnetic behavior of small particles of iron oxides needs further investigation in relation to poorly crystalline environmental samples. Such effects as speromagnetism, superparamagnetism, and superferromagnetism which depend critically on particle size need more study. The importance of this is shown by the recent work of Bocquet and Kennedy,[144] which clearly indicates a lack of any supermagnetic behavior in fine particles of goethite, but rather a reduction in Néel temperature due to particle size. Their results are in contradiction to many of the previous interpretations of spectra from similar materials.

The range of environmental applications of Mössbauer spectroscopy has considerably expanded. A bright future seems certain for those fortunate collaborators who combine expertise in both the technique and the environmental aspects. Although the possibility of an extraterrestrial Mössbauer spectrometer to study the regolith of Mars is an extremely exciting prospect, the application to earth science will surely be at least as rewarding.

ACKNOWLEDGMENTS. The authors are grateful to colleagues who provided samples and assisted in various aspects of the work reviewed. These include especially S. B. Weed, B. W. Ayers, M. A. Porterfield, D. D. Amarasiriwardena, S. C. Dorman, and A. M. Bryan of North Carolina State University, and D. Chambaere, R. M. Persoons, and P. M. A. de Bakker of the University of Gent. The Fund for Joint Basic Research of Belgium and NATO (Collaborative Grants 0370/86, 91/0116) have supported our research over a number of years. E. Murad provided a helpful review of this manuscript. The expert secretarial assistance of J. A. Weatherspoon is gratefully acknowledged.

REFERENCES

1. L. H. Bowen and S. B. Weed, in *Chemical Mössbauer Spectroscopy,* R. H. Herber, ed., p. 217, Plenum Press, New York, 1984.
2. E. De Grave, R. E. Vandenberghe, and L. H. Bowen, in *Proc. XXIV Zakopane School on Physics,* J. Stanek and A. T. Pedziwiatr, eds., Vol. 2, p. 186, World Scientific, Singapore, 1990.
3. R. E. Vandenberghe, E. De Grave, C. Landuydt, and L. H. Bowen, *Hyp. Interact.* **53**, 175 (1990).
4. E. Murad and J. H. Johnston, in *Mössbauer Spectroscopy Applied to Inorganic Chemistry,* G. J. Long, ed., Vol. 2, p. 507, Plenum Press, New York, 1987.
5. E. Murad, in *Iron in Soils and Clay Minerals,* J. W. Stucki, B. A. Goodman, and U. Schwertmann, eds., p. 309, Reidel, Dordrecht, 1988.
6. E. Murad, *Hyp. Interact.* **47**, 33 (1989).
7. E. Murad, in *Advances in Soil Science,* B. A. Stewart, ed., Vol. 12, p. 125, Springer-Verlag, Berlin, 1990.
8. S. Mitra, T. Pal, and T. Pal, *Indian J. Pure Appl. Phys.* **29**, 313 (1991).
9. R. L. Parfitt and C. W. Childs, *Aust. J. Soil Res.* **26**, 121 (1988).
10. U. Schwertmann, P. Cambier, and E. Murad, *Clays Clay Miner.* **33**, 369 (1985).
11. E. Murad and L. H. Bowen, *Am. Mineral.* **72**, 194 (1987).
12. D. D. Amarasiriwardena, E. De Grave, L. H. Bowen, and S. B. Weed, *Clays Clay Miner.* **34**, 250 (1986).
13. D. D. Amarasiriwardena, L. H. Bowen, and S. B. Weed, *Soil. Sci. Soc. Am. J.* **52**, 1179 (1988).
14. D. Barb, L. Diamandescu, D. Mihaila-Tarabasanu, A. Rusi, and M. Morariu, *Hyp. Interact.* **53**, 285 (1990).
15. A. Meagher and Q. A. Pankhurst, *Phys. Status Solidi A* **101**, K73 (1987).
16. S. Mørup, M. B. Madsen, J. Franck, J. Villadsen, and C. J. W. Koch, *J. Magn. Magn. Mater.* **40**, 163 (1983).
17. C. J. W. Koch, M. B. Madsen, and S. Mørup, *Hyp. Interact.* **28**, 549 (1986).
18. M. B. Madsen, S. Mørup, and C. J. W. Koch, *Hyp. Interact.* **42**, 1059 (1988).
19. R. E. Vandenberghe, A. E. Verbeeck, E. De Grave, and W. Stiers, *Hyp. Interact.* **29**, 1157 (1986).
20. T. G. Quin, G. J. Long, C. G. Benson, S. Mann, and R. J. P. Williams, *Clays Clay Miner.* **36**, 165 (1988).
21. S. C. Dorman, L. H. Bowen, and S. B. Weed, *J. Magn. Magn. Mater.* **98**, 28 (1992).
22. A. Meagher, Q. A. Pankhurst, and D. P. E. Dickson, *Hyp. Interact.* **28**, 533 (1986).
23. E. De Grave and R. E. Vandenberghe, *Hyp. Interact.* **28**, 643 (1986).
24. L. H. Bowen, E. De Grave, P. M. A. de Bakker, and R. E. Vandenberghe, *Hyp. Interact.* **54**, 467 (1990).
25. P. M. A. de Bakker, E. De Grave, R. M. Persoons, L. H. Bowen, and R. E. Vandenberghe, *Meas. Sci. Technol.* **1**, 954 (1990).
26. Q. A. Pankhurst and R. J. Pollard, *J. Phys. Condens. Matter* **2**, 7329 (1990).
27. R. J. Pollard, Q. A. Pankhurst, and P. Zientek, *Phys. Chem. Miner.* **18** 259 (1991).
28. R. J. Pollard, Q. A. Pankhurst, and P. Zientek, *J. Magn. Magn. Mater.* **104–107**, 1557 (1992).

29. R. J. Pollard, C. M. Cardile, D. G. Lewis, and L. J. Brown, *Clay Miner.* **27**, 57 (1992).
30. E. Murad and U. Schwertmann, *Mineral. Mag.* **48**, 507 (1984).
31. E. De Grave, R. M. Persoons, D. G. Chambaere, R. E. Vandenberghe, and L. H. Bowen, *Phys. Chem. Miner.* **13**, 61 (1986).
32. D. G. Chambaere and E. De Grave, *J. Magn. Magn. Mater.* **44**, 349 (1984).
33. D. G. Chambaere, E. De Grave, R. L. Vanleerberghe, and R. E. Vandenberghe, *Hyp. Interact.* **20**, 249 (1984).
34. D. G. Chambaere and E. De Grave, *Phys. Chem. Miner.* **12**, 176 (1985).
35. R. M. Persoons, D. G. Chambaere, and E. De Grave, *Hyp. Interact.* **28**, 647 (1986).
36. M. B. Madsen, S. Mørup, C. J. W. Koch, and O. K. Borggaard, *Surf. Sci.* **156**, 328 (1985).
37. R. J. Pollard and Q. A. Pankhurst, *J. Magn. Magn. Mater.* **99**, L39 (1991).
38. E. Murad, L. H. Bowen, G. J. Long, and T. G. Quin, *Clay Miner.* **23**, 161 (1988).
39. J. Chadwick, D. H. Jones, M. F. Thomas, G. J. Tatlock, and R. W. Devenish, *J. Magn. Magn. Mater.* **61**, 88 (1986).
40. E. Murad, *J. Magn. Magn. Mater.* **74**, 153 (1988).
41. E. Murad, J. M. Bigham, L. H. Bowen, and U. Schwertmann, *Hyp. Interact.* **58**, 2373 (1990).
42. L. Cianchi, M. Mancini, G. Spina, and H. Tang, *J. Phys. Condens. Matter* **4**, 2073 (1992).
43. M. B. Madsen, S. Mørup, and C. J. W. Koch, *Hyp. Interact.* **27**, 329 (1986).
44. R. J. Pollard and Q. A. Pankhurst, *J. Phys. Condens. Matter* **4**, L317 (1992).
45. N. H. J. Gangas, T. Bakas, and A. Moukarika, *Hyp. Interact.* **23**, 245 (1985).
46. J. C. Chadwick, D. H. Jones, M. F. Thomas, C. J. Tatlock, and M. Devenish, *Hyp. Interact.* **28**, 541 (1986).
47. Q. A. Pankhurst, C. E. Johnson, and M. F. Thomas, *J. Phys. C* **19**, 7081 (1986).
48. Q. A. Pankhurst, C. E. Johnson, and M. F. Thomas, *J. Magn. Magn. Mater.* **54–57**, 1163 (1986).
49. O. Nikolov, T. Ruskov, T. Tomov, and A. Toshev, *Hyp. Interact.* **39**, 409 (1988).
50. J. Ji-Sen, Y. Xie-Long, C. Long-Wu, and Z. Nai-Fu, *Hyp. Interact.* **41**, 483 (1988).
51. A. E. Verbeeck, E. De Grave, and R. E. Vandenberghe, *Hyp. Interact.* **28**, 639 (1986).
52. E. De Grave and R. E. Vandenberghe, *Phys. Chem. Miner.* **17**, 344 (1990).
53. G. M. da Costa, E. Galvao da Silva, and M. F. de Jesus Filho, *Hyp. Interact.* **67**, 501 (1991).
54. D. G. Rancourt, S. R. Julian, and J. M. Daniels, *J. Magn. Magn. Mater.* **49**, 305 (1985).
55. J. L. Dormann, J. R. Cui, and C. Sella, *J. Appl. Phys.* **57**, 4283 (1985).
56. M. D. Bentzon, J. van Wonterghem, S. Mørup, A. Thölen, and C. J. W. Koch, *Philos. Mag. B* **60**, 169 (1989).
57. E. De Grave, A. E. Verbeeck, and D. G. Chambaere, *Phys. Lett.* **107A**, 181 (1985).
58. E. Murad, *Phys. Lett.* **IIIA**, 79 (1985).
59. E. Murad and U. Schwertmann, *Clays Clay Miner.* **34**, 1 (1986).
60. E. De Grave, L. H. Bowen, R. Vochten, and R. E. Vandenberghe, *J. Magn. Magn. Mater.* **72**, 141 (1988).
61. E. De Grave, L. H. Bowen, D. D. Amarasiriwardena, and R. E. Vandenberghe, *J. Magn. Magn. Mater.* **72**, 129 (1988).
62. R. E. Vandenberghe, A. E. Verbeeck, and E. De Grave, *J. Magn. Magn. Mater.* **54–57**, 898 (1986).
63. S. Ke, L. Ming-xi, and L. He-Lie, *Physica B* **175**, 401 (1991).
64. R. M. Persoons, E. De Grave, and R. E. Vandenberghe, *Hyp. Interact.* **54**, 655 (1990).
65. A. Ramdani, J. Steinmetz, C. Gleitzer, J. M. D. Coey, and J. M. Friedt, *J. Phys. Chem. Solids* **48**, 217 (1987).
66. G. Haley, J. G. Mullen, and J. M. Honig, *Solid State Commun.* **69**, 285 (1989).
67. M. Rosenberg, P. Deppe, H. U. Janssen, V. A. M. Brabers, F. S. Li, and S. Dey, *J. Appl. Phys.* **57**, 3740 (1985).
68. U. Schwertmann and E. Murad, *Clays Clay Miner.* **38**, 196 (1990).
69. E. Murad, U. Schwertmann, and J. D. Cashion, *Hyp. Interact.* **54**, 619 (1990).

70. K. Melzer, Z. Simsa, M. Lukasiak, and J. Suwalski, *Cryst. Res. Technol.* **22**, K132 (1987).
71. R. Vanleerberghe and R. E. Vandenberghe, *Hyp. Interact.* **23**, 75 (1985).
72. S. Collyer, N. W. Grimes, D. J. Vaughan, and G. Longworth, *Am. Mineral.* **73**, 153 (1988).
73. A. H. Morrish and K. Haneda, *J. Magn. Magn. Mater.* **35**, 105 (1983).
74. E. Tronc and D. Bonnin, *J. Phys. (Paris) Lett.* **46**, L437 (1985).
75. I. Tamura and M. Hayashi, *J. Magn. Magn. Mater.* **72**, 285 (1988).
76. R. J. Pollard, *Hyp. Interact.* **41**, 509 (1988).
77. R. J. Pollard, *J. Phys. Condens. Matter* **2**, 983 (1990).
78. Q. A. Pankhurst, *J. Magn. Magn. Mater.* **101**, 291 (1991),
79. Q. A. Pankhurst and R. J. Pollard, *Phys. Rev. Lett.* **67**, 248 (1991).
80. P. M. A. de Bakker, E. De Grave, R. E. Vandenberghe, and L. H. Bowen, *Hyp. Interact.* **54**, 493 (1990).
81. E. Tronc, J. P. Jolivet, and J. Livage, *Hyp. Interact.* **54**, 737 (1990).
82. T. Okada, H. Sekizawa, F. Ambe, S. Ambe, and T. Yamadaya, *J. Magn. Magn. Mater.* **31–34**, 903 (1983).
83. T. Tsuji, K. Ando, K. Naito, and Y. Matsui, *J. Appl. Phys.* **69**, 4472 (1991).
84. R. C. Graham, S. B. Weed, L. H. Bowen, and S. W. Buol, *Clays Clay Miner.* **37**, 19 (1989).
85. R. C. Graham, S. B. Weed, L. H. Bowen, D. D. Amarasiriwardena, and S. W. Buol, *Clays Clay Miner.* **37**, 29 (1989).
86. N. R. Furet, C. Diaz, V. I. Nikolaev, and V. S. Rusakov, *Hyp. Interact.* **57**, 2195 (1990).
87. J. D. Fabris, M. Resende, J. Allan, and J. M. D. Coey, *Hyp. Interact.* **29**, 1093 (1986).
88. M. Resende, J. Allan, and J. M. D. Coey, *Earth Planet. Sci. Lett.* **78**, 322 (1986).
89. J. E. M. Allan, J. M. D. Coey, M. Resende, and J. D. Fabris, *Phys. Chem. Miner.* **15**, 470 (1988).
90. J. M. D. Coey, J. D. Fabris, and M. Resende, *Hyp. Interact.* **66**, 51 (1991).
91. A. Moukarika, F. O'Brien, J. M. D. Coey, and M. Resende, *Geophys. Res. Lett.* **18**, 2043 (1991).
92. O. D. Rodrigues, R. Garg, E. Galvao da Silva, J. C. Machado, and V. K. Garg, *Hyp. Interact.* **35**, 913 (1987).
93. M. P. F. Fontes, L. H. Bowen, and S. B. Weed, *Soil Sci. Soc. Am. J.* **55**, 1150 (1991).
94. M. R. Fontes, S. B. Weed, and L. H. Bowen, *Soil Sci. Soc. Am. J.* **56**, 982 (1992).
95. H. B. King, J. K. Torrance, L. H. Bowen, and C. Wang, *Soil Sci. Soc. Am. J.* **54**, 462 (1990).
96. R. E. Vandenberghe, E. De Grave, J. J. Hus, and J. Han, *Hyp. Interact.* **70**, 977 (1992).
97. G. L. Zhang, S. Y. Chen, S. Ambe, K. Maeda, K. Suzuki, Y. Sasa, E. Yagi, H. Yabuki, Z. W. Huang, and F. Ambe, *Hyp. Interact.* **70**, 973 (1992).
98. Y. Wang, X. Zhou, and J. Wu, *Hyp. Interact.* **70**, 1037 (1992).
99. D. Das, M. B. Chakrabarti, C. K. Majumdar, and M. K. Bose, *Pramana J. Phys.* **27**, 331 (1986).
100. T. G. St. Pierre, J. Webb, and C. R. M. Butt, *Hyp. Interact.* **57**, 2279 (1990).
101. C. W. Childs, R. W. P. Palmer, and C. W. Ross, *Aust. J. Soil Res.* **28**, 245 (1990).
102. L. P. Aldridge and G. J. Churchman, *Aust. J. Soil Res.* **29**, 387 (1991).
103. R. E. Vandenberghe, E. De Grave, G. De Geyter, and C. Landuydt, *Clays Clay Miner.* **34**, 275 (1986).
104. N. A. Eissa, S. S. Gomaa, M. Y. Hassaan, and H. A. Sallam, *Hyp. Interact.* **41**, 775 (1988).
105. B. A. Goodman and P. H. Nadeau, *Clay Miner.* **23**, 301 (1988).
106. E. Murad and U. Wagner, *Hyp. Interact.* **45**, 161 (1989).
107. E. Murad and U. Wagner, *Neues Jahrb. Mineral. Abh.* **162**, 281 (1991).
108. S. Papamarinopoulos, P. W. Readman, Y. Maniatis, and A. Simopoulos, *Earth Planet. Sci. Lett.* **57**, 173 (1982).
109. S. Papamarinopoulos, P. W. Readman, Y. Maniatis, and A. Simopoulos, *Archaeometry* **29**, 50 (1987).
110. S. Papamarinopoulos, P. W. Readman, Y. Maniatis, and A. Simopoulos, *Earth Planet. Sci. Lett.* **102**, 198 (1991).

111. J. Hilton, G. J. Long, J. S. Chapman, and J. P. Lishman, *Geochim. Cosmochim. Acta* **50**, 2147 (1986).
112. U. Schwertmann, L. Carlson, and E. Murad, *Clays Clay Miner.* **35**, 297 (1987).
113. I. König, H.-D. Knauth, C. Koopmann, F. E. Wagner, and U. Wagner, *Hyp. Interact.* **41**, 811 (1988).
114. T. Mayer and P. G. Manning, *J. Great Lakes Res* **15**, 687 (1989).
115. T. Mayer and P. G. Manning, *J. Great Lakes Res.* **16**, 299 (1990).
116. Y. Qi, C. Xu, J. Ensling, and P. Gütlich, *Hyp. Interact.* **70**, 1021 (1992).
117. S. Mørup, J. Franck, J. van Wonterghem, H. Roy-Poulsen, and L. Larsen, *Fuel* **64**, 528 (1985).
118. J. K. Torrance, S. W. Hedges, and L. H. Bowen, *Clays Clay Miner.* **34**, 314 (1986).
119. J. Kraczka, A. Z. Hrynkiewicz, and T. Wieser, *Hyp. Interact.* **29**, 1125 (1986).
120. E. Murad and U. Schwertmann, *Am. Mineral.* **73**, 1395 (1988).
121. S. Y. Chen, S. Ambe, G. L. Zhang, N. Takematsu, and F. Ambe, *Hyp. Interact.* **70**, 969 (1922).
122. V. Rusanov, V. Angelov, V. Jordanov, and S. Ormandjiev, *Phys. Chem. Miner.* **18**, 517 (1992).
123. B. B. Ellwood, T. H. Chrzanowski, F. Hrouda, G. J. Long, and M. L. Buhl, *Geology* **16**, 980 (1988).
124. B. B. Ellwood, B. Burkart, K. Rajeshwar, R. L. Darwin, R. A. Neeley, A. B. McCall, G. J. Long, M. L. Buhl, and C. W. Hickcox, *J. Geophys. Res.* **94**, 7321 (1989).
125. A. B. Domingues, R. B. Scorzelli, and E. Mattievich, *Hyp. Interact.* **41**, 763 (1988).
126. S. R. Hong, G. Haley, and J. G. Mullen, *Hyp. Interact.* **57**, 2221 (1990).
127. E. Murad and U. Schwertmann, *Hyp. Interact.* **41**, 835 (1988).
128. C. W. Childs, N. Matsue, and N. Yoshinaga, *Soil Sci. Plant Nutr.* **37**, 299 (1991).
129. C. M. Schlinger, D. Griscom, G. C. Papaefthymiou, and D. R. Veblen, *J. Geophys. Res.* **93**, 9137 (1988).
130. S. R. Julian, J. A. Westgate, J. M. Daniels, D. G. Rancourt, and P. Sullivan, *Hyp. Interact.* **41**, 807 (1988).
131. J. Lipka, M. B. Madsen, D. Orlicky, C. J. W. Koch, and S. Mørup, *Phys. Scr.* **38**, 508 (1988).
132. Ö. Helgason, S. Steinthorsson, M. B. Madsen, and S. Mørup, *Hyp. Interact.* **57**, 2209 (1990).
133. Ö. Helgason, H. P. Gunnlaugsson, S. Steinthorsson, and S. Mørup, *Hyp. Interact.* **70**, 981 (1992).
134. Ö Helgason, S. Steinthorsson, and S. Mørup, *Hyp. Interact.* **70**, 985 (1992).
135. Ö. Helgason, N. Oskarsson, and G. E. Sigvaldason, *Hyp. Interact.* **70**, 989 (1992).
136. M. B. Madsen, S. Mørup, T. V. V. Costa, J. M. Knudsen, and M. Olsen, *Hyp. Interact.* **41**, 827 (1988).
137. E. De Grave, R. E. Vandenberghe, P. M. A. de Bakker, A. Van Alboom, R. Vochten, and R. Van Tassel, *Hyp. Interact.* **70**, 1009 (1992).
138. J. M. D. Coey, S. Mørup, M. B. Madsen, and J. M. Knudsen, *J. Geophys. Res.* **95**, 14423 (1990).
139. R. V. Morris, D. G. Agresti, H. V. Lauer, Jr., J. A. Newcomb, T. D. Shelfer, and A. V. Murali, *J. Geophys. Res.* **94**, 2760 (1989).
140. R. V. Morris, J. L. Gooding, H. V. Lauer, Jr., and R. B. Singer, *J. Geophys. Res.* **95**, 14427 (1990).
141. D. G. Agresti, R. V. Morris, E. L. Wills, T. D. Shelfer, M. M. Pimperl, M.-H. Shen, B. C. Clark, and B. D. Ramsey, *Hyp. Interact.* **72**, 285 (1992).
142. J. M. Knudsen, M. B. Madsen, M. Olsen, L. Vistisen, C. B. Koch, S. Mørup, E. Kankeleit, G. Klingelhöfer, E. N. Evlanov, V. N. Khromov, L. M. Mukhin, O. F. Prilutski, B. Zubkov, G. V. Smirnov, and J. Juchniewicz, *Hyp. Interact.* **68**, 83 (1991).
143. E. Jackiewicz, M. Lukasiak, M. Kopcewicz, K. Szpila, and N. Bakun-Czubarow, *Hyp. Interact.* **70**, 993 (1992).
144. S. Bocquet and S. J. Kennedy, *J. Magn. Magn. Mater.* **109**, 260 (1992).

5

Mössbauer Effect Studies of Iron Fluorides

J. M. Grenèche and F. Varret

1. INTRODUCTION

Fluorine is the most electronegative element of the periodic chart of Mendeleev, and as a result it has some uncommon properties. In addition, the small size of this ion favors anionic conductivity, and its low polarizability strongly influences the optical properties of fluorides. In the frame of this review, we mention finally that (1) the π-donor character and the absence of any π-acceptor character of the fluorine ligand essentially determine its low crystalline field (high-spin) character and (2) the fluorine ion produces more effective superexchange interactions between magnetic cations than does oxygen.

A major advantage of fluorides results from the well-defined coordination of metallic ions; coordination that avoids departures from stoichiometry, provided sufficient care against hydrolysis is taken during synthesis and further investigations, and that may lead to polycrystalline samples of high structural quality; however, large single crystals are rarely obtained. The interest in ferric fluorides has been stimulated by the ease of their Mössbauer investigation. As a result, transition metal fluorides have been actively investigated because of their incredible wealth of structural forms (see, e.g., two reviews[1,2]) associated with a high ionic character of the chemical bond. This character permits a

J. M. Grenèche • Equipe de Physique de l'Etat Condensé, URA CNRS No. 807, Université du Maine, F-72017 Le Mans Cedex, France *F. Varret* • Département de Recherches Physiques, URA CNRS, No. 71, Université Pierre et Marie Curie, F-75252 Paris Cedex 05, France.

Mössbauer Spectroscopy Applied to Magnetism and Materials Science, Volume 1, edited by Gary J. Long and Fernande Grandjean. Plenum Press, New York, 1993.

quantitative analysis of the properties of fluorides within the framework of well-established models. For instance, lattice summations of point charges and dipoles lead to calculated values of the electric field gradients that are in excellent agreement with the experimental values determined from the iron-57 Mössbauer quadrupolar spectra (see Section 2).

Numerous crystalline fluorides exhibit[3] ferromagnetic or ferrimagnetic properties. Valuable information concerning the magnetic properties can be derived from Mössbauer spectroscopy as recently discussed.[4] Herein, we focus attention on fluorides that exhibit noncollinear magnetic structures.

The huge variety of fluoride structural forms is nicely illustrated by the simple example of ferric fluoride, FeF_3, which has rhombohedral, hexagonal tungsten bronze, pyrochlore, and amorphous forms. The discovery of these new forms resulted from different powerful synthesis methods, such as hydrothermal synthesis, chemical vapor deposition synthesis, and "chimie douce" synthesis. Such structural variety offers several examples of magnetic frustration which may be associated with structural disorder, leading to "pathologic" behavior, such as idle-spin or spin-glass like behavior, which will be discussed in Section 3 (see also Refs. 5 and 6).

Ferric fluorides, which are discussed in detail here, have the peculiarity that the iron environment is exclusively octahedral. This results in a well-defined value of the intrinsic magnetic moment of the ferric ion, and in turn leads to a well-defined value of the saturation hyperfine field. With a good accuracy, μ_{intrins} is 5.0 μB, H_{hyp} is -25 T per spin unit, and $H_{\text{hyp}}^{\text{sat}}$ is -62.5 T. The hyperfine field is antiparallel to the Fe^{3+} magnetization. Such well-established values permit the determination of the various contributions, i.e., supertransferred, and dipolar to the saturation internal field, as well as their zero-point reductions in low-dimensional systems.[7] We restrict ourselves to the structural consequences of octahedral coordination, which are the basic assumptions of various models for amorphous structures. This is developed in Section 4.

2. ELECTRIC FIELD GRADIENTS IN FLUORIDES

2.1. The Quadrupole Interaction

Mössbauer spectroscopy is an excellent method to measure the electric hyperfine interactions, through the electric field gradient (EFG) tensor, with the nucleus; these interactions reflect the electric charge distribution of the probe atom and its surroundings. A complete theoretical description of the quadrupole interaction has been given by Spiering.[8]

In the case of iron ions, whose nuclear spin value $I = \frac{3}{2}$, the quadrupolar splitting, Δ, is expressed as

$$\Delta = \frac{eQ\,|V_{zz}|}{2}\left(1 + \frac{\eta^2}{3}\right)^{1/2}$$

where V_{zz} is the main component of the electric field gradient tensor which is diagonal in its principal axes system OXYZ, η is the asymmetry parameter, given by $(V_{xx} - V_{yy})/V_{zz}$ with $0 \leq \eta \leq 1$ for $|V_{xx}| \leq |V_{yy}| \leq |V_{zz}|$, and Q is the nuclear quadrupole moment.

One can generally distinguish two contributions to the electric field gradient tensor, a lattice contribution created by the charge distribution of the surroundings of the iron sites in the lattice and a valence contribution reflecting the electronic density distribution in the valence shells of the probe atom. Consequently, the electric field gradient tensor at a nuclear site is given by

$$[V_{ij}] = (1 - \gamma_\infty)[V_{ij}^{\mathrm{Latt}}] + (1 - R)[V_{ij}^{\mathrm{Val}}]$$

where γ_∞ and R are the antishielding and shielding factors, respectively.

Because of the spherical symmetry of the electronic density, no valence contribution to the electric field gradient is expected from Fe^{3+} in the high-spin state. Consequently, the quadrupole splitting value arises from the lattice contribution alone.

In the case of highly ionic compounds, such as fluorides, one can estimate the EFG tensor from a summation of polarizable point charges. To do this, an ionic system is assumed to be composed of discrete deformable cations and anions located on a three-dimensional crystalline or amorphous network. Only monopolar and dipolar contributions are accounted for. The main techniques used are based on direct[9,10] and reciprocal[11,12] lattice summations. It should be mentioned that the direct lattice method requires that attention be paid to convergence problems.[9] In the next subsections, the EFG calculation method is described and quadrupole data obtained on some crystalline fluorides are discussed.

2.2. The Summation Methods

Theoretical EFG calculations for systems lacking spontaneous polarization are based on the following points. The local electric field, **E**, acting on a given site distorts the electronic shells of the ion and induces a dipole moment, $\mathbf{p} = [\alpha_d]\mathbf{E}$, where $[\alpha_d]$ is the dipolar polarizability tensor which is assumed to be isotropic. In turn, the induced dipole moments contribute to the total electric field, **E**. The induced quadrupole moments and their contributions to the electric field are neglected. Analytic techniques then lead to a set of coupled equations which must be solved self-consistently to determine the electric field. Then, the EFG components are easily obtained analytically through similar summations.

In the case of a crystal, the approach consists of $3n$ self-consistent equations, where n represents the number of nonequivalent iron sites. Some criteria for absolute convergence, in either direct or reciprocal lattice summations, were established by Coogan.[13] In the electric field calculation, the elementary cell is required to have neither monopole nor dipole moment.

The lattice summations may be performed either in reciprocal or in direct space. In direct space, iterative volumes are chosen as multiples of the elementary cell with fractional charges in order to satisfy Coogan's rules. The successive volumes consist of $(2L - 1)^3$ cells symmetrically located around the origin, where L represents the iteration indice. In the calculations of Teillet and co-workers,[9] the convergence of the lattice summations occurs by the fourth iteration and these authors have systematically performed up to seven iterations. In order to calculate the monopole electric field, the sums, corresponding to the infinite crystal, were extrapolated to $L = \infty$, by using a Neville plot[14] in which the computed results are plotted versus $(2L - 1)^{-2}$. To summarize, the iterative procedure consists of first calculating the monopole field. Then the dipole contribution to the total field is obtained by performing a double iteration involving the resolution of the self-consistent system in which the V_{ij} components at the iron sites are finally obtained. A further diagonalization provides the principal values V_{xx}, V_{yy}, and V_{zz} and the set of principal axes. In all cases, accurate structural data are required.

In the case of amorphous systems, successive spherical volumes are considered with a "smooth" limiting shell.[15] Unfortunately, the dipolar contribution had to be dropped because of computational time limitations.

2.3. Computed EFG and Discussion

Extensive attention has been given to the ferric $AFeF_4$ compounds where A is K, Rb, Cs, and NH_4, because of the unusually large values of their quadrupole splittings; splitting values estimated at 1.5 mm/s at room temperature[16–22] (see Fig. 1). The crystallographic structure of these two-dimensional compounds consists of puckered layers of tetragonally distorted fluorine octahedral units separated by A^+ cations, as shown in (Fig. 2). These

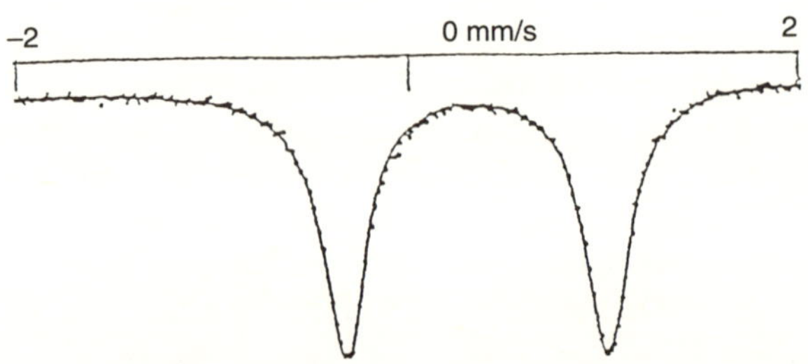

Figure 1. Room-temperature Mössbauer spectrum of crystalline $KFeF_4$.

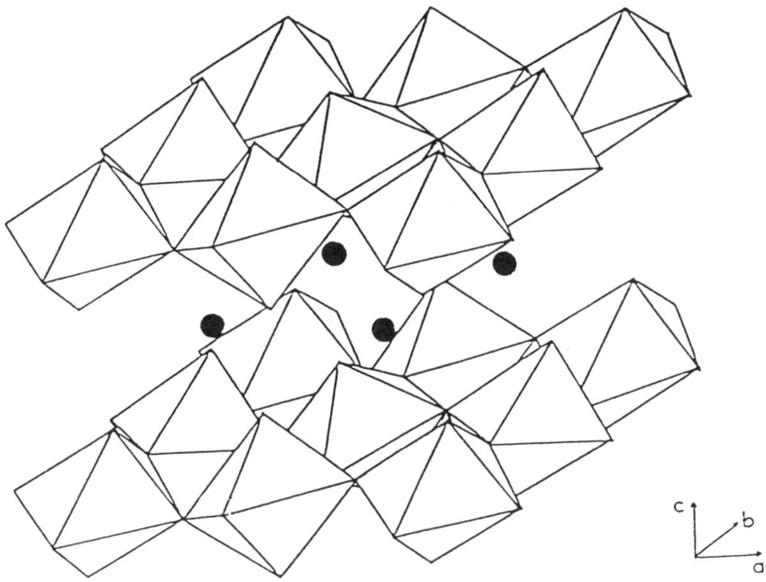

Figure 2. Crystallographic view of crystalline $AFeF_4$. Circles represent A cations.

octahedra are tilted by an angle of between 8° and 18°, depending on the nature of A^+ and the temperature. Structural transitions can occur with changing temperature.

Except in NH_4FeF_4, the very large values of the quadrupole splitting result from a dipolar contribution arising essentially from the polarization of the "terminal" fluorine ions, the nonbridging fluorines in a layer of FeF_6 octahedra. These terminal fluorine ions are subjected to a strong electric field whereas the in-plane fluorines are subjected to a zero electric field and do not contribute to the dipolar EFG. Similar computations have not been undertaken for NH_4FeF_4 because of the nonsphericity of the ammonium cation.[9]

The point charge model was also applied to the binary fluoride, $LiMnFeF_6$, in order to explain the EFG at the iron sites.[23] This compound exhibits a dimorphism; the transition temperature between the two trigonal phases, α and β, the low- and high-temperature phases, respectively, occurs at 560°C.[24] The structure of α-$LiMnFeF_6$, an isotype of Na_2SiF_6, results from a slightly distorted hexagonally close-packed framework of fluorines with cations located in one-half of the octahedral sites. The three-dimensional lattice is built up from corner and/or edge-sharing octahedra, as shown in Fig. 3a. The structure of β-$LiMnFeF_6$, an isotype of $LiMnGaF_6$,[25] differs from that of the α form by the inversion of the positions of the alkaline and trivalent ions, an inversion that leads to a fourfold increase in the unit cell volume (see Fig. 3b). The particular interest in the $AM^{2+}M^{3+}F_6$ series lies in the different crystallographic positions available for the cations.

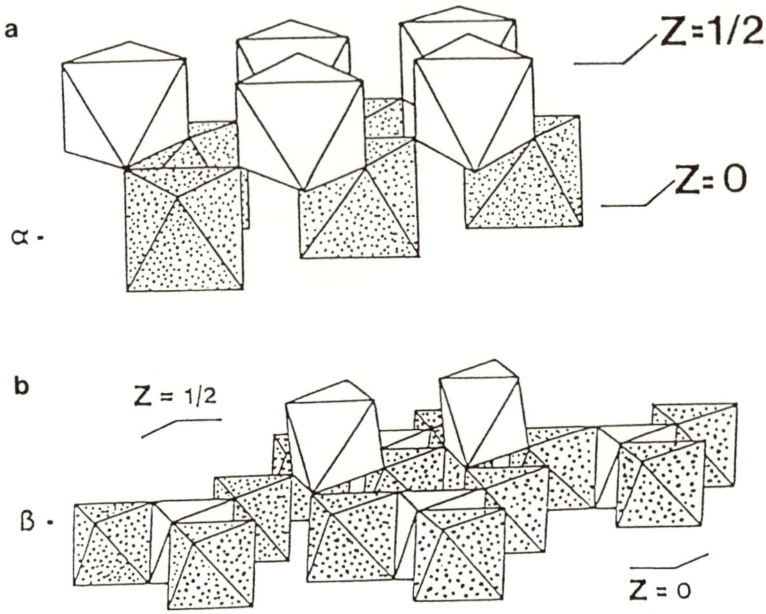

Figure 3. Perspective view of cationic ordering in α-$LiMnFeF_6$ (a) and β-$LiMnFeF_6$ (b). The manganese-containing octahedra are shaded. Reproduced from Ref. 60.

The EFG calculations, including its monopolar and dipolar contributions, have been performed at the different iron sites for both phases of $LiMnFeF_6$.[23] The fluorine polarizability is found to be in reasonable agreement with the expected value. The calculated quadrupole splitting value agrees well with the experimental data in the case of α-$LiMnFeF_6$ and reveals an important dipolar contribution. For β-$LiMnFeF_6$, the situation is less clear because of the large uncertainty in the analysis of paramagnetic spectra; however, the prediction of the sign of the quadrupole splitting, the asymmetry parameter, and the orientation of the EFG axes are in qualitative agreement with Mössbauer data in the magnetic state.

3. CRYSTALLINE FLUORIDES WITH NONCOLLINEAR MAGNETIC STRUCTURES

3.1. The Polymorphic Crystalline Phases of FeF_3

Among the ferric fluorides, the basic FeF_3 composition has received much attention because of its simple chemical nature and because of the various magnetic properties exhibited by its different crystalline and amorphous polymorphic phases. It should be noted that all of their structures result from corner-sharing octahedral packing. The amorphous forms of FeF_3 will be considered in the next section.

In the crystalline state, the three crystallographic structures are presented in Fig. 4. The more stable rhombohedral form, α-FeF_3, consists of weakly tilted regular octahedral units centered on a cubic lattice.[26,27] The hexagonal tungsten bronze phase of FeF_3 results from the superposition of hexagonal tungsten bronze planes in which the cationic lattice generates both hexagonal and triangular rings.[28] The latter form can be obtained by dehydration of the hydrate, $(H_2O)_{0.33}FeF_3$, which also exhibits the hexagonal tungsten bronze structure with water molecules located in the hexagonal cavities.[28] In the third pyrochlore-like phase, the cationic skeleton of FeF_3 is built up from corner-sharing tetrahedra.[29]

The magnetic structures, illustrated in Fig. 5, were determined by neutron diffraction, and are strongly different, depending on their respective cationic topology. The rhombohedral form of FeF_3 is a weak ferromagnet below 363 K,[30–32] a canted antiferromagnet, the canting of which was attributed to the weak antisymmetric exchange interaction represented by the Dzialozhinsky–Moriya Hamiltonian, $DS_i \times S_j$.[33,34] In the hexagonal tungsten bronze form of FeF_3, below $T_N \simeq 97$ K, the magnetic structure contains three ferromagnetic sublattices in the HTB planes which are antiferromagnetically coupled.[35] The pyrochlore form of FeF_3 orders below 20 K with four ferromagnetic sublattices oriented at 109° from each other.[36] The reduction of the magnetic ordering temperatures and the noncollinear magnetic structures were attributed to the frustration of the antiferromagnetic interactions implied by the cationic topology.

Mössbauer spectroscopy has provided a large contribution to the understanding of the magnetic and structural properties of these forms of FeF_3. For the rhombohedral form of FeF_3, the Mössbauer spectrum consists of a single resonance line above $T_N = 363$ K and of a Zeeman spectrum with very sharp lines in the magnetic ordered phase. This agrees well with the unique iron site in a regular fluorine octahedron. In addition, the hyperfine parameters are consistent with the high-spin state of ferric ions. A nice Mössbauer study was reported by Wertheim and co-workers[37] on the sublattice magnetization in the critical range just below the Néel temperature. They analyzed the thermal evolution of the hyperfine field by using the critical exponent description, $H(T) \propto D(1 - T/T_N)^{\beta}$. They were careful to determine accurately and independently both the critical exponent β, and T_N as well as D. They found $\beta = 0.352 \pm 0.005$, $T_N = 363.12 \pm 0.02$ K, and $D = 1.207 \pm 0.015$ within the temperature range $0.94 < T/T_N < 0.9995$. The critical exponent is closer to the Heisenberg than the Ising value expected in the case of a three-dimensional magnetic order, a result in good agreement with the isotropic properties of the Fe^{3+} ion.

The analysis by Mössbauer spectroscopy of the hexagonal tungsten bronze form of FeF_3 proved to be rather difficult, because of the unavoidable presence

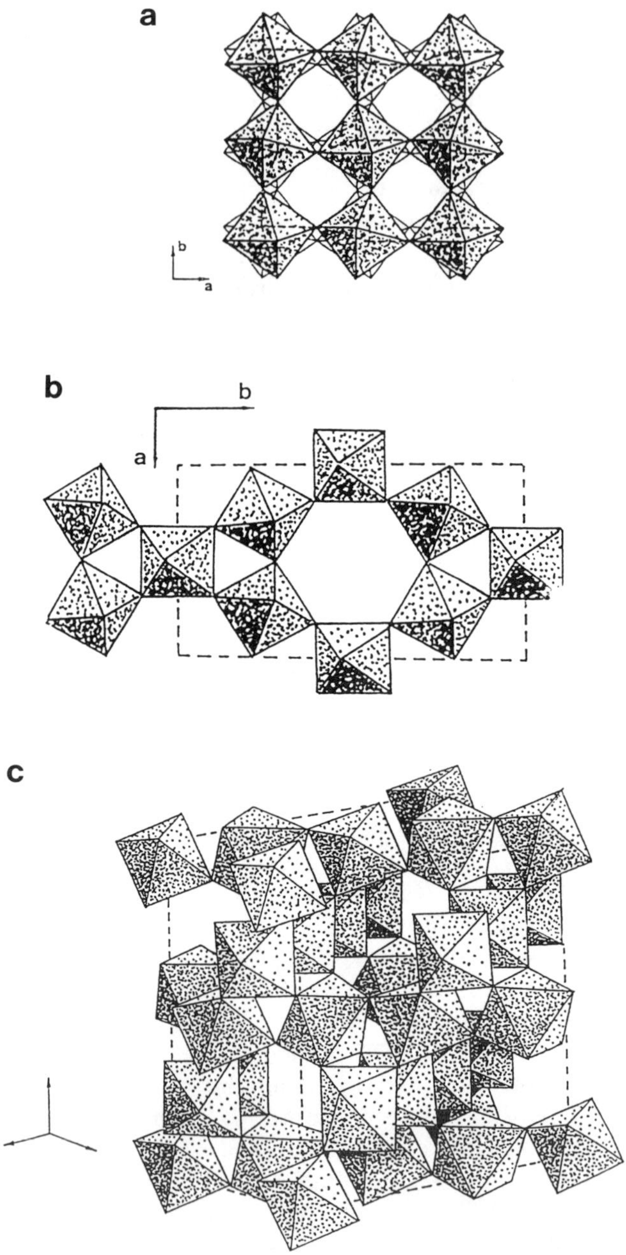

Figure 4. Crystal structures of the polymorphic crystalline forms of FeF_3: (a) rhombohedral FeF_3, (b) hexagonal tungsten bronze FeF_3, and (c) pyrochlore FeF_3. Reproduced from Ref. 76.

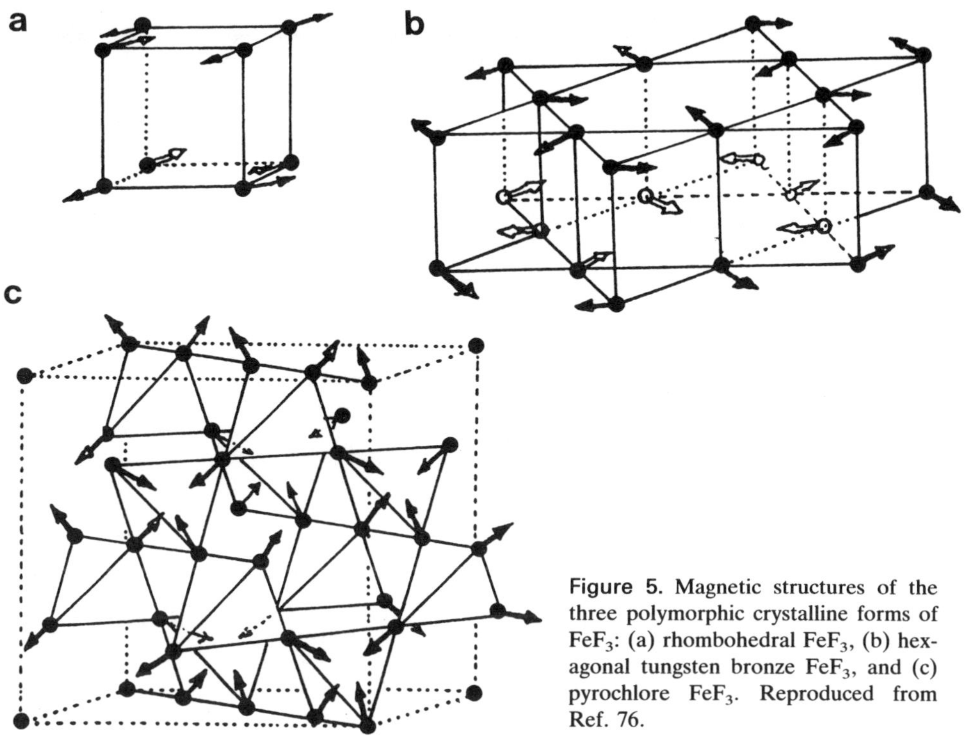

Figure 5. Magnetic structures of the three polymorphic crystalline forms of FeF_3: (a) rhombohedral FeF_3, (b) hexagonal tungsten bronze FeF_3, and (c) pyrochlore FeF_3. Reproduced from Ref. 76.

of the isostructural hydrate. We analyzed the role of zeolitic water, in the hexagonal tungsten bronze structure compounds, on the mechanism of dehydration, hydration, and hydrolysis, by studying a hydrate sample grown by hydrothermal synthesis. This study used a "cryofurnace" for an "*in situ*" Mössbauer analysis.[38] The quadrupole lines exhibited by the hydrated form were attributed to the existence of two slightly different crystallographic iron sites in the rhombohedral structure.[39] The quadrupole splitting value varies slightly with temperature, probably because of the motion of the hydrogen atoms in the H_2O molecules, as has been observed in other hydrated ferric salts. The presence of zeolitic water serves to increase the quadrupole splitting value, whereas the isomer shift is weakly reduced. In contrast, a decrease in water content leads to an increase in the magnetic ordering temperatures and to a reduction of the transition width established by thermal scanning method. This method consists in measuring the depth of the paramagnetic Mössbauer lines as a function of temperature, using a constant-velocity driving mode. In addition, the similar values of the saturation hyperfine field in the anhydrous and hydrated states (see Table 1) confirm that the water molecules are not coordinated to iron ions in the hexagonal tungsten bronze structure.

In the case of the hydrated form, the magnetization dependence just below T_N, as previously expressed, led to a β value of 0.35 ± 0.02, a value consistent with three-dimensional magnetic order. However, in contrast to the rhombohedral form of FeF_3, the magnetization curve strongly deviates from

Table 1. Magnetic and Mössbauer Spectral Parameters for the Three Crystalline Phases of FeF_3 and for $FeF_3(H_2O)_{0.33}$

	r-FeF_3	HTB-FeF_3	Pyr-FeF_3	$FeF_3(H_2O)_{0.33}$
Space group	$R\bar{3}c$	$P6/mmm$	$Fd3m$	$Cmcm$
T_N (K)	363.12	~97	~20	129
Θ_P (K)	~ −600	~ −600	—	—
μ (μ_B)	4.45	4.07	3.32	—
H_{Hyp} (T)	61.8	56.0	52.3	56.5
IS[a] (mm/s)	0.49	0.47	0.48	0.44
Δ[b] (mm/s)	0.05	−0.20	−0.22	0.64
β	0.352	0.35	—	—
Spin structure	Antiferromagnet Weak ferromagnet	Triangular 120°	Tetrahedral 109°	Triangular 120°

[a] Isomer shift values relative to metallic iron at 300 K.
[b] Quadrupolar splitting values in the paramagnetic range.

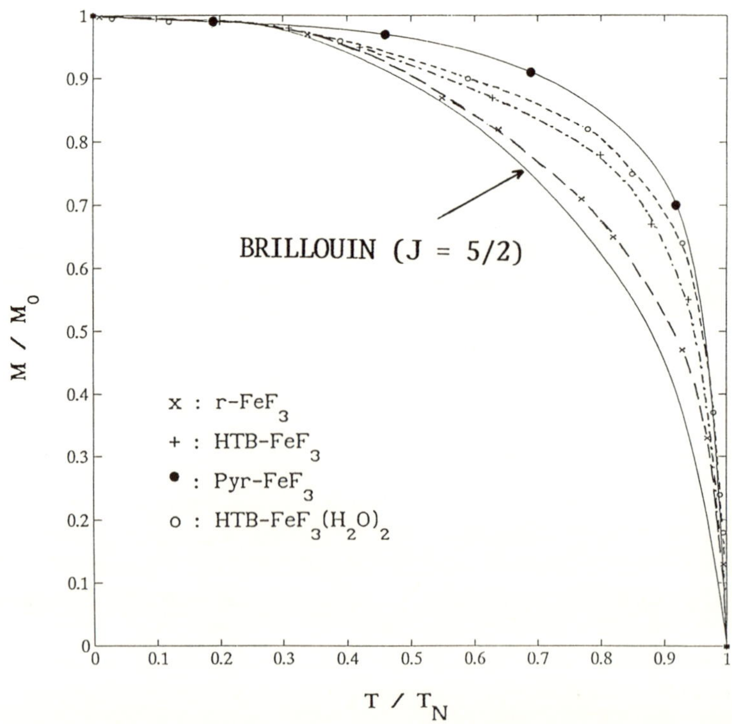

Figure 6. Evolution of the reduced hyperfine field values as a function of the reduced temperature for the different phases of FeF_3.

the Brillouin function, B $(J = \frac{5}{2})$, for the hexagonal tungsten bronze forms of FeF_3, as illustrated in Fig. 6.[35]

A Mössbauer study of the pyrochlore form synthesized by topotactic oxidation was more difficult because of the presence of impurities such as the hexagonal tungsten bronze form of FeF_3 and divalent FeF_2.[40] Nevertheless, the hyperfine data could be accurately analyzed and revealed a small negative value of V_{zz} and an inhomogeneous broadening of the magnetic lines, a broadening that may be attributed to a molecular field distribution induced by a lack of structural crystalline perfection.

By comparing all three crystalline forms, the hyperfine field values at 4.2 K could be linearly correlated with $\cos^2 \vartheta$, where ϑ represents the superexchange Fe–F–Fe angle.[36] A similar dependence was previously observed[41] for a series of orthoferrites, $MFeO_3$, where the superexchange angle is the Fe–O–Fe angle.

High-field Mössbauer spectroscopy on powder samples below the magnetic ordering temperature provides useful information on the magnetic structure, information that complements that given by magnetic measurements and neutron diffraction.[42,43]

For the rhombohedral form of FeF_3, the presence of sharp lines (see Fig. 7A) is indicative of a well-defined orientation of the ferric moments with respect to the external field applied parallel to the γ ray. This is in contrast to the other crystalline phases in which a lack of alignment of the ferric moments clearly results in a broadening of the outer lines. The intensities of the intermediate lines of the Mössbauer spectrum, which are sensitive to the orientation of the effective field with respect to the Mössbauer γ-ray propagation direction, increase for the rhombohedral form of FeF_3 as a function of the external field. However, the ideal intensity ratio of 3:4:1:1:4:3, expected for an antiferromagnet, is not reached; the discrepancy is attributed to the canting associated with the antisymmetric exchange interaction proposed by Dzialhozinsky and Moriya.[33,34] A more quantitative estimation of the canting angle would require high-quality crystals.

In spite of the absence of topological disorder in the two other crystalline phases, a "speromagnetic" model was used to fit their high-field spectra[44] (see Fig. 7B,C). A "speromagnet" is a frozen random magnetic structure,[45,46] and the "speromagnetic" line shape results from the vectorial addition of the external field and the randomly oriented internal fields. This line shape essentially reflects the complete lack of orientation of the magnetic moments under the influence of the applied field. Such a lack of orientation occurs in the case of speromagnets, and can be expected for many-sublattice antiferromagnets, the total magnetic anisotropy of which is reduced or canceled by the complex magnetic structure. The "speromagnetic" response of the hexagonal and pyrochlore forms of FeF_3 obviously originates from their rather complex magnetic structure which results from the frustration of the antiferromagnetic interactions in the triangular cationic rings of the atomic structure.

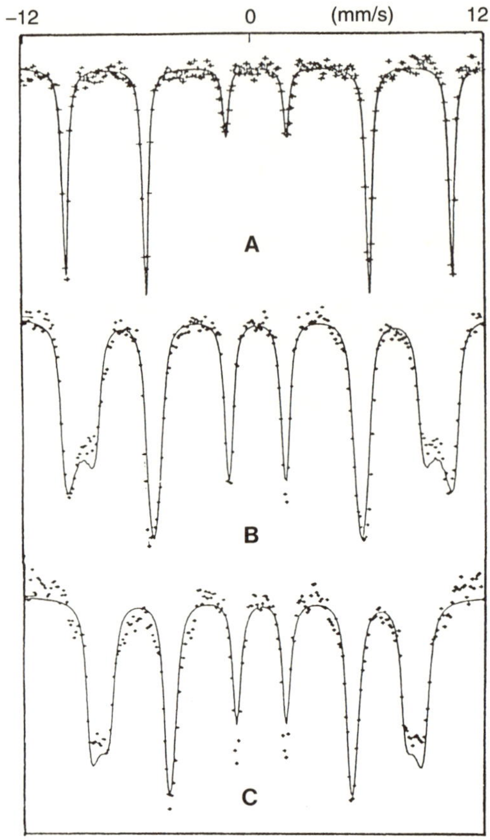

Figure 7. Powder Mössbauer spectra recorded at 4.2 K under an external magnetic field parallel to the γ ray on (A) rhombohedral FeF_3 (70 kG), (B) hexagonal tungsten bronze FeF_3 (60 kG), and (C) pyrochlore FeF_3 (40 kG). (B) and (C) are fitted using a "speromagnetic model". Reproduced from Ref. 44.

3.2. Frustrated Ferrimagnetic Inverse Weberites

Among the various crystalline fluorides that exhibit noncollinear magnetism due to their frustrated cationic topology, we report in this section a Mössbauer effect study performed on the series of compounds $AFeF_5(H_2O)_2$, where A is Mn, Fe, Co, Ni, or Zn.[47–49] These isotypes belong to the inverse weberite structural type, whose structure is shown in Fig. 8. In this structure, the cationic network forms triangles in the hexagonal tungsten bronze plane as previously defined. Frustration effects inherent to the triangular cationic rings are expected to be seriously influenced by the magnetic properties of the different cations.

Three-dimensional frustrated ferrimagnetic behavior was observed in neutron diffraction and magnetic measurements on all of these isotypic compounds, except for the Zn compound which exhibits one-dimensional magnetism because of the diamagnetic character of Zn^{2+}.[47] No peculiar information was deduced from the electric hyperfine data, and, as shown in

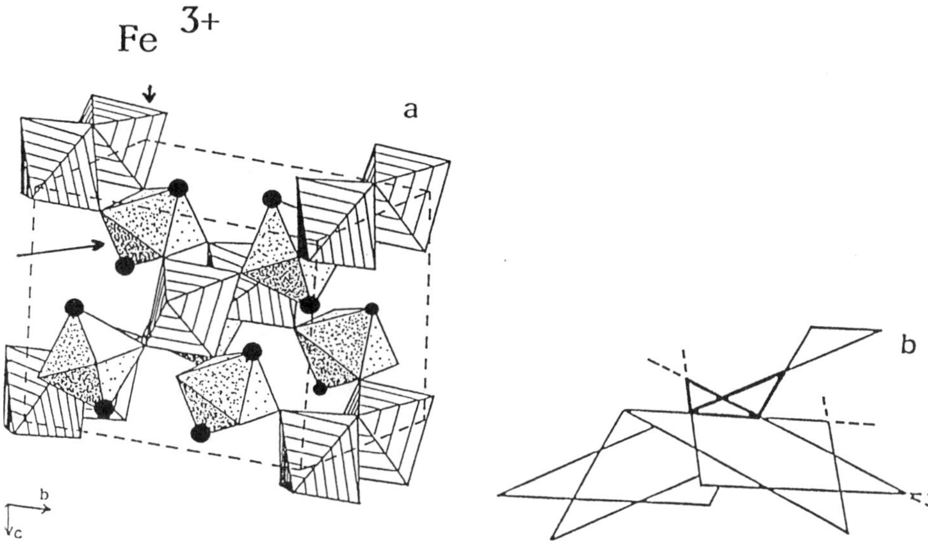

Figure 8. (a) Perspective view of the $AFeF_5(H_2O)_2$ structure. The Fe^{3+} and A^{2+} octahedra are respectively weakly and strongly hatched and the water molecules represented by black circles are linked to A^{2+}. (b) Partial view of the cationic network surrounding the cationic triangle. Reproduced from Ref. 49.

Table 2, the quadrupole splitting values are very similar, for all of the compounds. Concerning the thermal evolution of the hyperfine field exhibited by the ferric component, only the mixed valence iron weberite, $Fe_2F_5(H_2O)_2$, exhibits a singularity,[48] as shown in Fig. 9. Imbert *et al.* have reported both the onset of the magnetic order below $T_N = 48.6 \pm 0.5$ K, and a magnetic transition at $T_\pm = 26.5 \pm 0.5$ K. According to these authors, this magnetic

Table 2. Mössbauer Spectral Parameters for the Inverse Weberite Series $AFeF_5(H_2O)_2$, Adapted from Refs. 47–49

	Zn^{2+}	Mn^{2+}	Ni^{2+}	Co^{2+}	Fe^{2+}
T_N (K)	9(2)	39.5(5)	31.0(1.0)	34.5(1.0)	48.6(5)
IS^a (mm/s)	0.44(1)	0.45(1)	0.45(1)	0.45(1)	0.48(2)
Δ^b (mm/s)	0.60(1)	0.69(1)	0.57(2)	0.63(2)	0.59(2)
$H_{hyp}{}^c$ (T)	46.3(2)^d	53.4(2)	50.8(3)	53.6(2)	53.8(1)
H^e_{eff} (T)	—	57.7(5)	52.7(5)	56.6(5)	51.5(5)
$\langle\theta\rangle^f$ (deg)	—	—	127(5)	84(5)	88(5)

^{a}The isomer shift values are relative to metallic iron at 300 K.
^{b}Quadrupolar splitting values obtained in the paramagnetic temperature range.
^{c}Hyperfine field at 4.2 K.
^{d}Value obtained at 1.5 K.
^{e}Average angle between the hyperfine field and the applied field.
^{f}Average effective field: fitted values obtained from Mössbauer spectra recorded at 4.2 K in an 80-kG external magnetic field applied parallel to the γ ray.

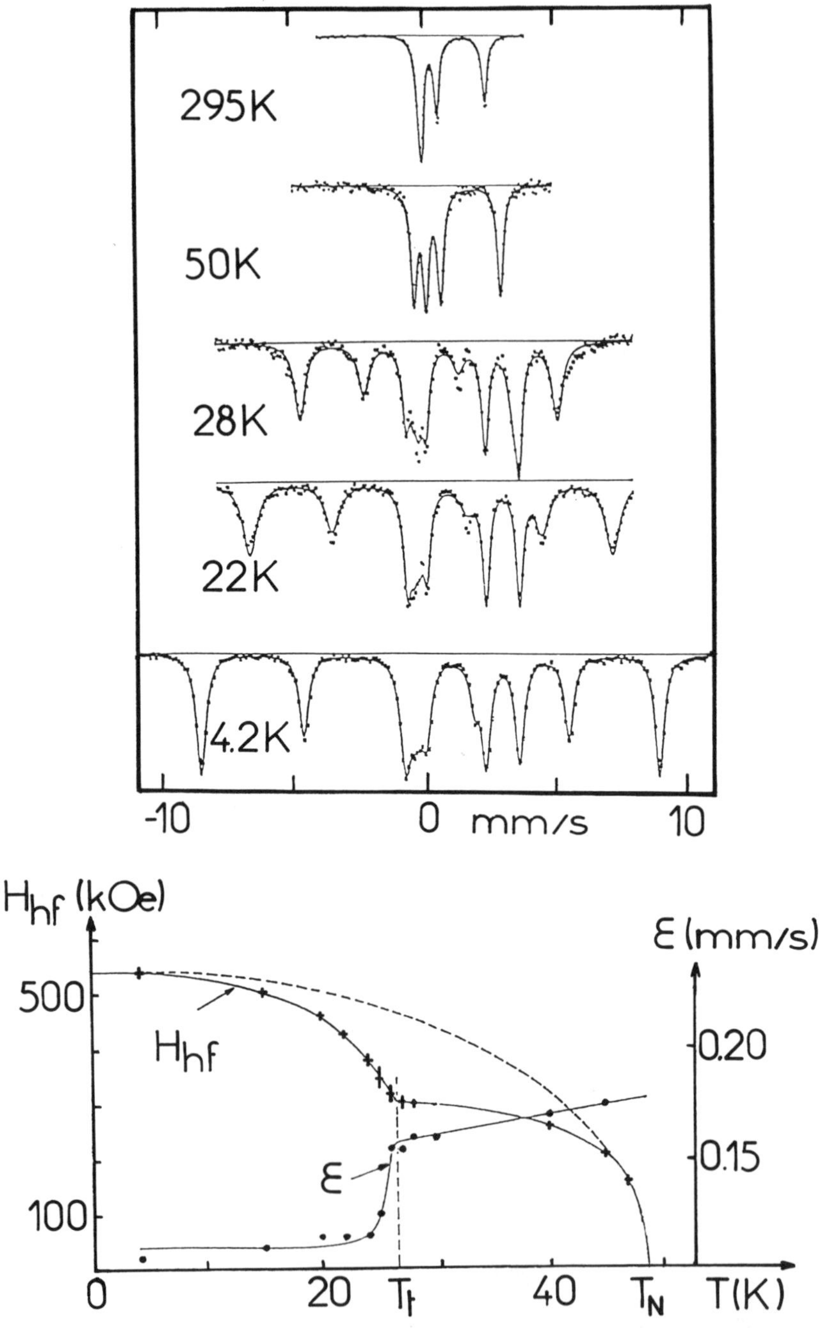

Figure 9. Mössbauer spectra, recorded at several temperatures, of $Fe_2F_5(H_2O)_2$ and thermal evolution of the quadrupole shift and the hyperfine field of the Fe^{3+} sites. The latter is compared to the Brillouin function with $S = 5/2$ given by the dashed line. Reproduced from Ref. 48.

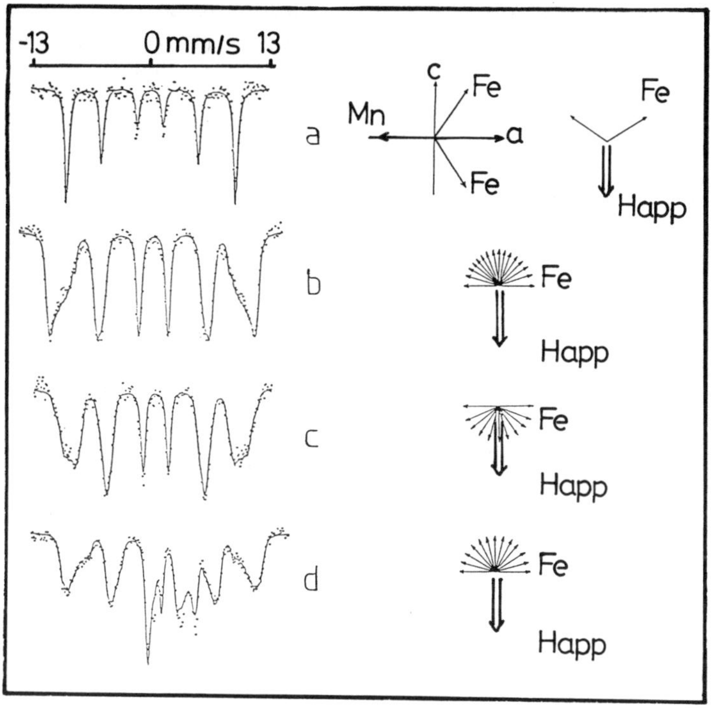

Figure 10. Powder Mössbauer spectra recorded at 4.2 K in an 80-kG external magnetic field applied parallel to the γ ray and schematic representations of the orientations of the iron magnetic moments and the applied field in the case of (a) $MnFeF_5(H_2O)_2$, (b) $CoFeF_5(H_2O)_2$, (c) $NiFeF_5(H_2O)_2$, and (d) $Fe_2F_5(H_2O)_2$. Reproduced from Ref. 49.

transition is associated with a reorientation of the ferric moments while the ferrous moments remain fixed along their local anisotropy axis.

We applied external magnetic fields to powder samples of all of these compounds below the magnetic ordering temperature.[49] As shown in Fig. 10, which displays the high-field Mössbauer spectra recorded at 80 kG applied field, the shape of the outer lines is rather different from each other. In $MnFeF_5(H_2O)_2$, the sharp lines reveal that the ferric moments are pointing along discrete orientations; the intensity of the intermediate lines and the effective field vary as a function of the applied field. It was concluded that the manganese magnetic lattice imposes the magnetization direction, which is opposite to that of the ferric lattice. Further, the line intensities permit the determination of the canting of the moments, as illustrated Fig. 10. The canting is in qualitative agreement with the structure observed by neutron diffraction[50] and with that refined from simulation techniques.[51]

In the case of the weberites where A is Co, Ni, and Fe, we observed a large broadening of the outer lines, a broadening that differs according to the nature of the divalent cation. This broadening was attributed to a lack of orientation of the ferric moments which results from the magnetocrystalline anisotropy of the Co^{2+}, Ni^{2+}, and Fe^{2+} ions. Schematically, in the case of a collinear magnetic structure, the effect of the external field on a given crystallite is to flip the total magnetization, which remains parallel to its anisotropy axis, into the half-space containing the external field. In a polycrystalline material, the ferric moments occupy the half-sphere centered either around the direction of the external field, if the Fe^{3+} moment and the local magnetization are parallel, as is the case of $NiFeF_5(H_2O)_2$, or are centered about the opposite direction if the Fe^{3+} moments and the local magnetization are antiparallel, as is the case in $CoFeF_5(H_2O)_2$ or $NiFeF_5(H_2O)_2$ (see Fig. 10). By referring to the taxonomy of magnetism previously proposed by Hurd,[52] these spectra can be described as "asperomagnetic-like" and "sperimagnetic-like," respectively. The noncollinear character of the magnetic structure obviously results in a departure from the random distribution of the moments in half space, a distribution that has been observed.[49] A complete quantitative analysis of these magnetic structures under the influence of an external field has not been undertaken.

3.3. Idle Spin Behavior

In some magnetic materials, such as rare-earth orthoferrites, some of the magnetic moments, i.e., the rare-earth moment, behave quasiparamagnetically, at temperatures far below the magnetic ordering temperature. This is due to the RE–RE and RE–TM magnetic interactions, which are much weaker than the TM–TM interactions, the interactions responsible for the high temperature of the magnetic ordering. The magnetic frustration in some fluorides described in this section leads to the extreme situation of a total cancellation of the molecular field at a given site; this situation strictly defines an "idle spin", which was also found in $Fe^{2+}Fe_2^{3+}Si_2O_8$.[53]

The first such examples are $Fe_3F_8(H_2O)_2$ and $MnFe_2F_8(H_2O)_2$. As shown in Fig. 11, the structure of these two isotypes prepared by hydrothermal synthesis[54,55] results from a packing of shifted hexagonal bronze-type layers, the hexagonal tungsten bronze planes being identical to those existing in the inverse weberite-type structure.

The paramagnetic Mössbauer spectrum of $Fe_3F_8(H_2O)_2$ consists of two well-resolved quadrupolar doublets which may be attributed to Fe^{2+} and Fe^{3+} sites, respectively. The experimental area ratio of these doublets agrees with the stoichiometry ratio of 1:2. Cationic order between the Fe^{2+} and Fe^{3+} ions is evidenced through the sharp lines observed at all temperatures. By lowering the temperature, Leblanc *et al.*[55] have observed the presence of a ferric sextet below 157 K, as shown in Fig. 12. In contrast, the ferrous component remains paramagnetic down to 35 K, below which it splits magnetically. This indicates that the ferrous lattice behaves as "idle spins" in the temperature range

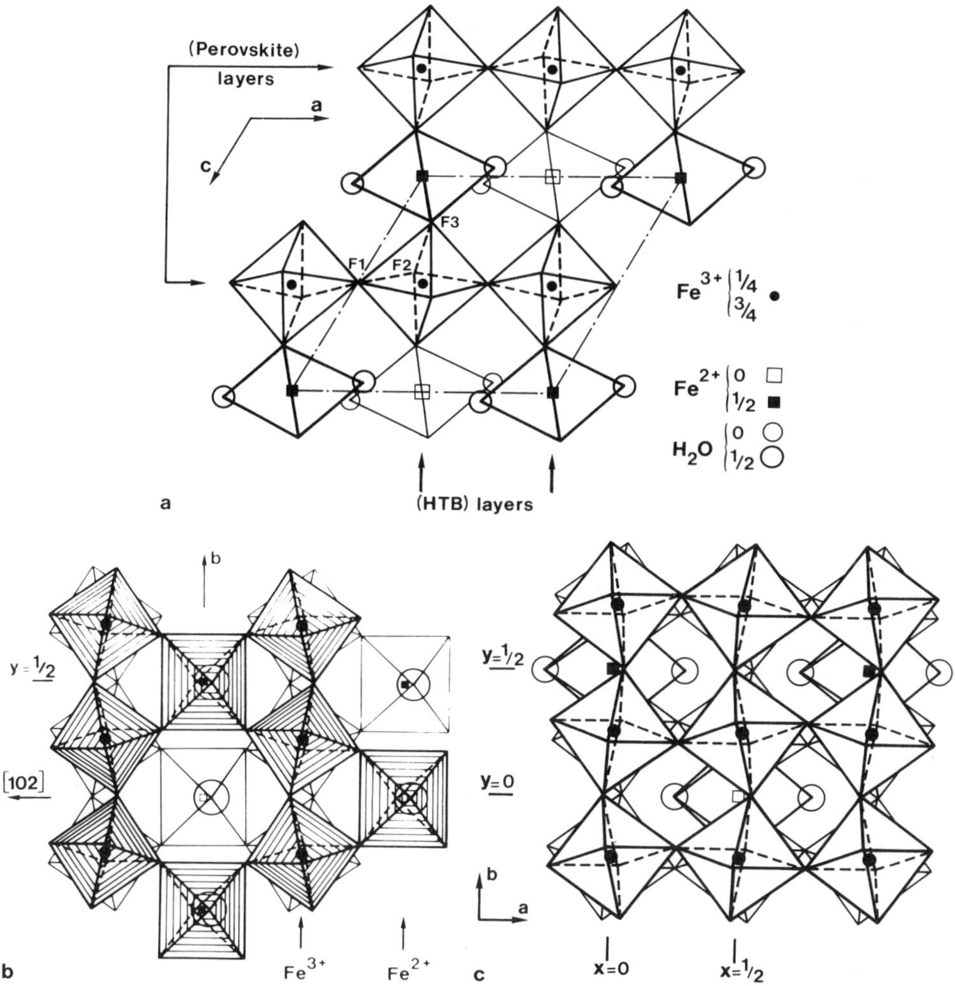

Figure 11. The (010) projection (a), the (201) projection showing the hexagonal tungsten bronze layers (b), and the (002) projection exhibiting perovskite layers (c) in $Fe_3F_8(H_2O)_2$. Reproduced from Ref. 55.

between 35 and 157 K. As illustrated in Fig. 13, the thermal evolution of the ferric and ferrous hyperfine fields clearly reveals that the ferrous magnetic ordering influences the ferric sublattice. From magnetic measurements, shown in Fig. 14, it is apparent that the linear thermal evolution above 157 K of the inverse susceptibility curve, which obeys the Curie–Weiss law, reveals the presence of strong antiferromagnetic interactions. Between 157 and 35 K, the magnetic measurements are consistent with the superimposition of a parasitic ferromagnetic behavior upon the antiferromagnetic ordering; the remanent magnetization reflects a canted antiferromagnetic phase at low temperature.[56]

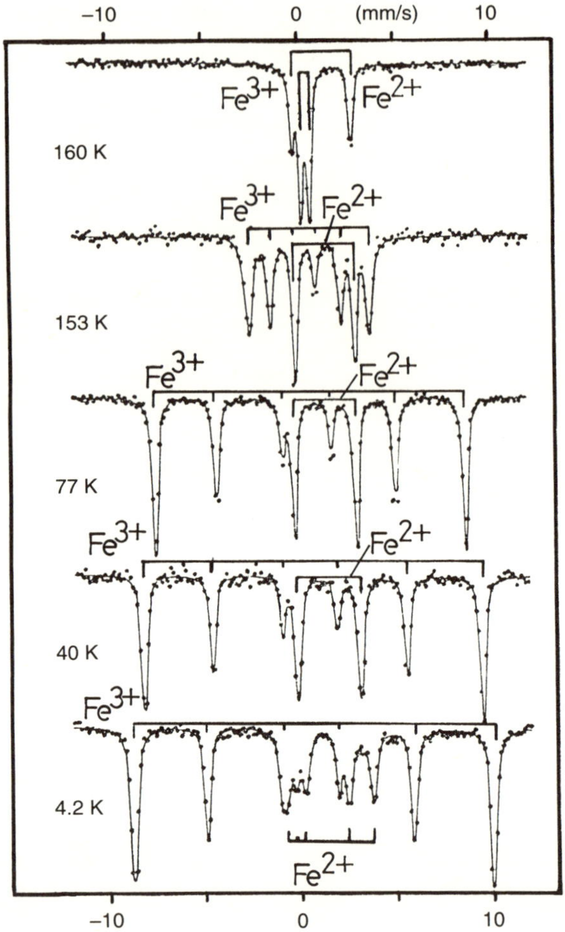

Figure 12. Thermal evolution of the Mössbauer spectra of $Fe_3F_8(H_2O)_2$. Reproduced from Ref. 55.

The $Fe_3F_8(H_2O)_2$ structure can also be described as perovskite planes of corner-sharing $Fe^{3+}F_6$ octahedra, linked by isolated $Fe^{2+}F_4(H_2O)_2$ octahedra which share equatorial fluorines, and serve to connect adjacent planes. Below $T_N = 157$ K, the magnetic ordering is due to the Fe^{3+} subnetwork; below $T_N = 35$ K, the magnetic ordering of the ferrous subnetwork induces a canting of the ferric structure. This behavior was confirmed by neutron diffraction studies.[57]

In the case of the manganese–iron-based fluoride, $MnFe_2F_8(H_2O)_2$, only one quadrupolar doublet and one magnetic sextet, respectively, are observed above and below $T_N = 156$ K. No sharp variation of the magnetic hyperfine field is observed at low temperatures. In addition, the inverse susceptibility

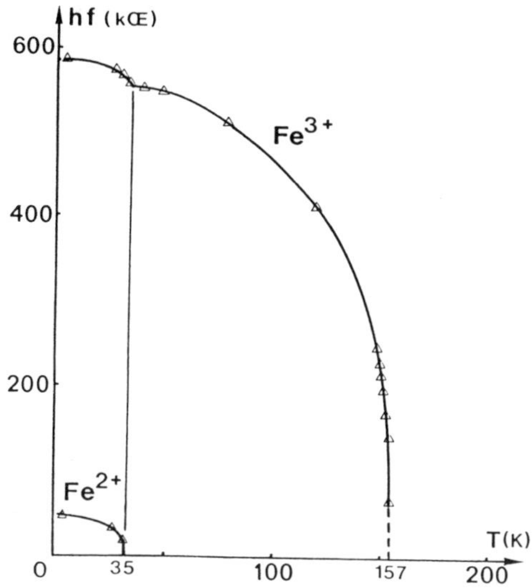

Figure 13. Temperature dependence of the hyperfine field values for the Fe^{3+} and Fe^{2+} sites in $Fe_3F_8(H_2O)_2$. Reproduced from Ref. 55.

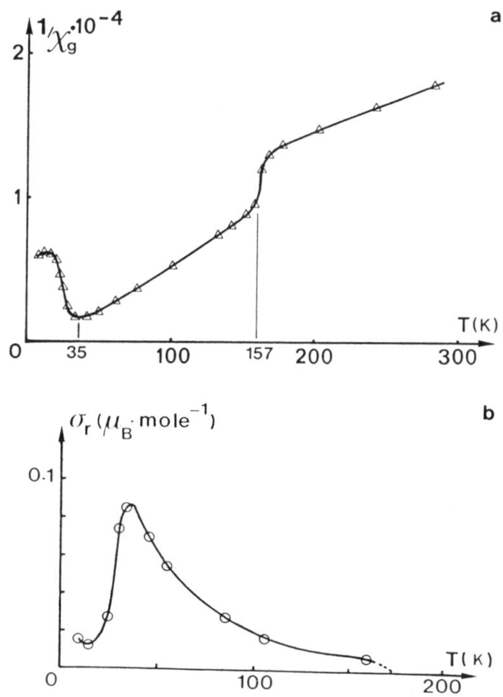

Figure 14. Temperature dependence of the inverse susceptibility (a) and the remanent magnetization (b) of $Fe_3F_8(H_2O)_2$. Reproduced from Ref. 55.

reveals strong antiferromagnetic interactions and the presence of a weak ferromagnetic component at low temperature. No unusual effect due to the presence of Mn^{2+} is observed in $MnFe_2F_8(H_2O)_2$ because only identical d^5–d^5 interactions occur due to the substitution of Fe^{2+} by Mn^{2+}.

The origin of the "idle" spin behavior resides in the cationic topology. It appears that the Fe^{2+} moments have only four magnetic neighbors, all of which belong to frustrated rings. In contrast, the ferric moments interact with six magnetic neighbors, only two of which belong to the frustrated rings. So, it is clear that only the Fe^{2+} ions are entirely subjected to magnetic frustration. The reason why such frustration leads to a complete cancellation of the molecular field is only quantitative, and a recent computer simulation, based on a Monte–Carlo algorithm[51,58] introduced by Metropolis,[59] has reproduced this idle spin behavior. The orientation of the magnetic moments located in a triangular platelet depends crucially on the local anisotropy and spin of the cations as well as the nature and magnitude of the superexchange interactions.

As a second example, we have considered the dimorphic ferrimagnetic fluoride, $LiMnFeF_6$, the structure of which was discussed briefly in Section 2. Classical hyperfine interactions were observed by Courbion *et al.*[60] in α-$LiMnFeF_6$. Further, applied field Mössbauer experiments revealed an increase of the intermediate line intensities, an increase that was attributed to a spin reorientation and is typical of an antiferromagnet with low anisotropy. These results, which are consistent with those obtained by neutron diffraction and magnetic measurements, led to a magnetic structure that consists of two ferromagnetic sublattices, which are antiferromagnetically coupled, as shown in Fig. 15a.

In the case of β-$LiMnFeF_6$,[60] a large uncertainty in the hyperfine parameters prevents a detailed interpretation of the rather complex paramagnetic spectra. By lowering the temperature below 112.7 K, the component attributed to the 2*d* sublattice, is split into a magnetic sextet, whereas the second component attributed to the 1*a* sublattice remains unsplit down to 4.2 K.[60] The authors have noted an unusual feature of the two linewidths of this component: one linewidth is constant whereas the second is strongly temperature dependent. It was suggested that the 1*a* iron moments are weakly coupled to the 2*d* magnetic sublattice which includes both Fe^{3+} and Mn^{2+} magnetic moments. In contrast, the large negative value of Θ_p is consistent with predominant antiferromagnetic interactions and a ferrimagnetic behavior which is deduced from magnetic studies. The magnetic structure suggested by the neutron diffraction study is illustrated in Fig. 15b.

Low-temperature high-field Mössbauer experiments have provided decisive information on these magnetic interactions.[60] The intermediate Zeeman lines of the 2*d* component have disappeared, as shown in Fig. 16. A complete spin rotation toward the applied field direction has occurred, as in the case of a ferromagnet. Because of the opposite directions of the hyperfine field and the magnetization and because of the increase in the effective field with the applied field, the 2*d* ferric moments are antiparallel to the applied field, and,

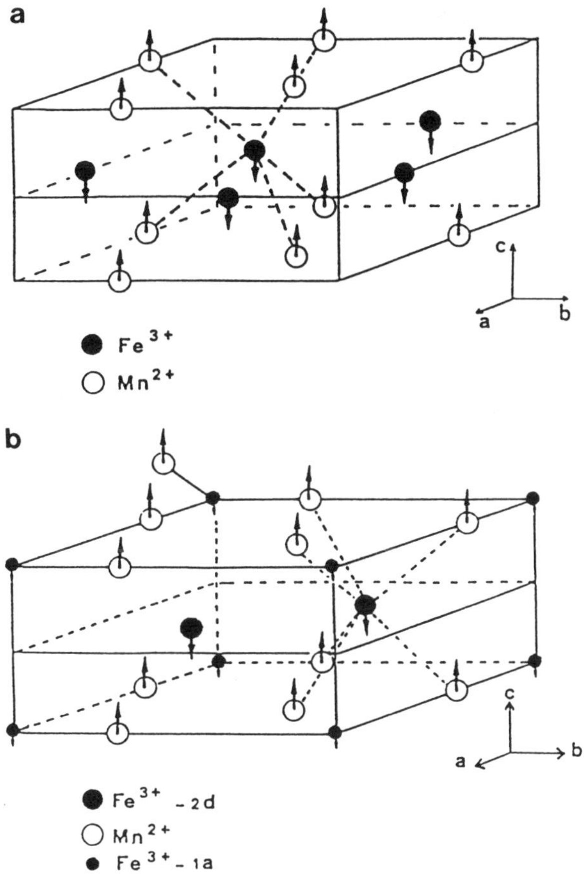

Figure 15. Magnetic structures of (a) α-$LiMnFeF_6$ and (b) β-$LiMnFeF_6$. Reproduced from Ref. 60.

consequently, the Mn^{2+} moments are parallel to the applied field. Such behavior confirms that β-$LiMnFeF_6$ is a ferrimagnet. The 1*a* iron magnetization is parallel to the applied field but the rather small value of the hyperfine field suggests that the sublattice is not saturated, even at 4.2 K.

These results indicate the presence of a weak antiferromagnetic coupling of the Fe(1*a*) sublattice to the net magnetization. The "idle spin" behavior results from the cancellation of the superexchange interactions, interactions that probably originate from the variety of connecting modes of the MF_6 octahedra. According to the Kanamori–Goodenough rules,[61,62] corner-sharing octahedra with angles greater than 130° induce strong antiferromagnetic interactions. In contrast, edge-sharing octahedra with angles close to 100° induce ferro- or weak antiferromagnetic exchange. This "idle spin" behavior was also recently observed in the mixed fluoride, β-$LiMnFe_{1-x}V_xF_6$, with

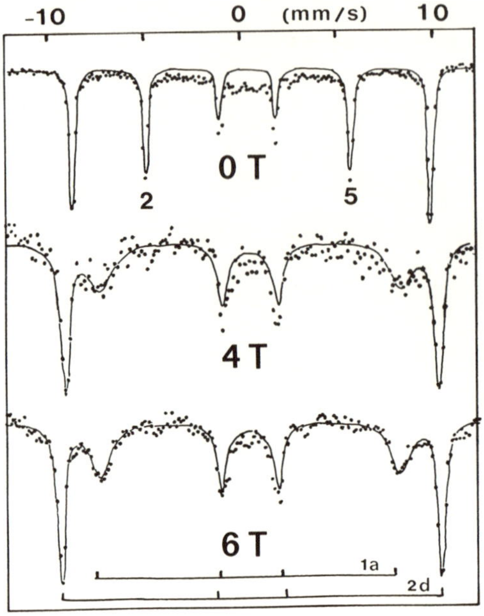

Figure 16. Powder Mössbauer spectra of β-$LiMnFeF_6$ recorded at 4.2 K under external magnetic fields applied parallel to the γ rays. Reproduced from Ref. 60.

$x \leq 0.3$ and the Mössbauer study revealed the nonnegligible role of the second-nearest neighbors exchange interactions.[63]

Another magnetic behavior which received the designation of "soft-spin" was displayed by $NaMnFeF_6$.[64] Its structure, derived from the Na_2SiF_6 structural type, exhibits a different cationic ordering. Figure 17 illustrates the crystallographic structure and the magnetic surroundings of iron atoms (a), the Mössbauer spectra (b), and the temperature dependence of the hyperfine fields on the 1*a* and 2*d* iron sites (c). In agreement with magnetic measurements, the magnetic transition which occurs at 32 K suggests a weak magnetic coupling of the Fe(2*d*) sublattice to the net magnetization. It is clear from the cationic topology that this behavior is not due to the presence of magnetic frustration. Rather it is attributed to a fortuitous compensation of the Fe^{3+}–Mn^{2+} magnetic interactions with ferromagnetic and antiferromagnetic superexchange coupling, due to the d^5–d^5 superexchange angles of 90° and 180°, respectively. The "soft-spin" behavior was also well reproduced by a Monte-Carlo simulation, by assuming only an incomplete balance of these Fe–Mn superexchange interactions.[58]

4. AMORPHOUS FERRIC FLUORIDES

In this section, we will discuss the properties of the amorphous insulating ferric fluorides, FeF_3, $KFeF_4$, $NaFeF_4$, $CsFeF_4$, and $RbFeF_4$.

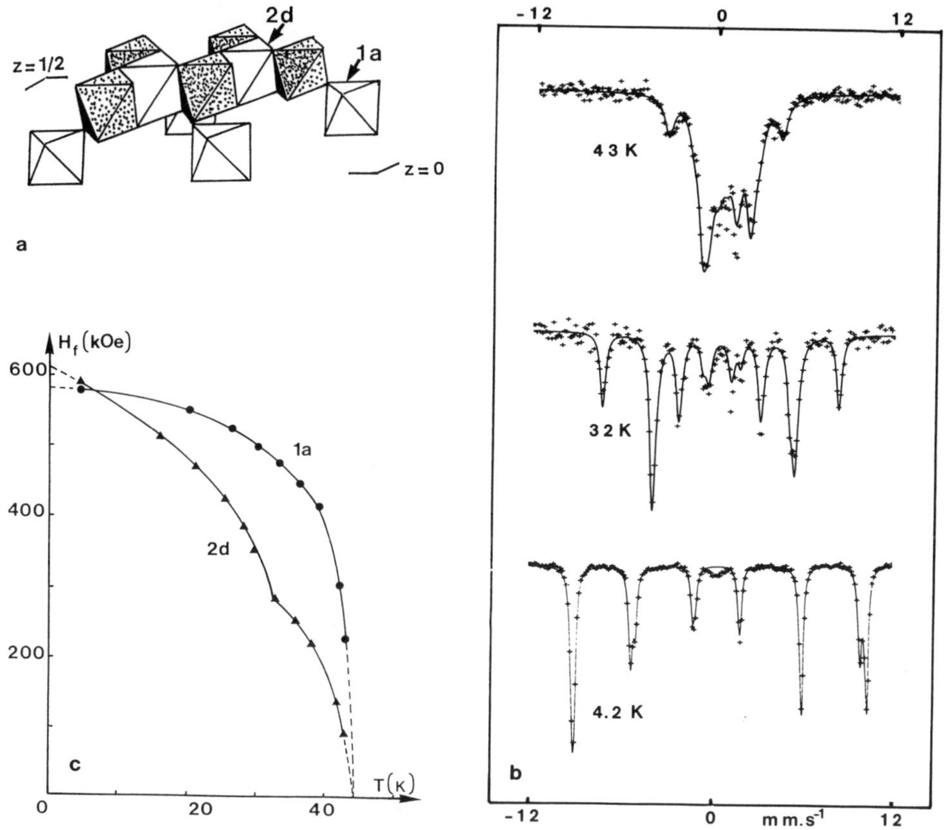

Figure 17. (a) Perspective view of cationic ordering in $NaMnFeF_6$. The manganese-containing octahedra are shaded. (b) Mössbauer spectra of $NaMnFeF_6$. (c) Temperature dependence of the hyperfine fields at Fe(1*a*) and Fe(2*d*). Reproduced from Ref. 64.

4.1. Preparation of the Samples

These amorphous ferric compounds were first synthesized by Ferey and co-workers[65] using the vapor deposition technique on a cold substrate, starting with powdered crystalline material. More recently, another amorphous form of FeF_3 was prepared by the low-temperature chemistry reaction of gaseous HF with crystalline $FeCl_3$.[66] The composition of this amorphous fluorinated material is in fact FeF_3, xHF with $0 \leq x \leq 1$. Structural investigations, including extended X-ray absorption fine structure spectroscopy (EXAFS) and neutron and X-ray diffraction, were carried out on both the deposited and the amorphous fluorinated forms of FeF_3.[44] It has been shown that the Fe–Fe distances are compatible with exclusively corner-sharing octahedra for the fluorinated material, and with a mixture of corner-sharing and edge-sharing octahedra for the deposited material.

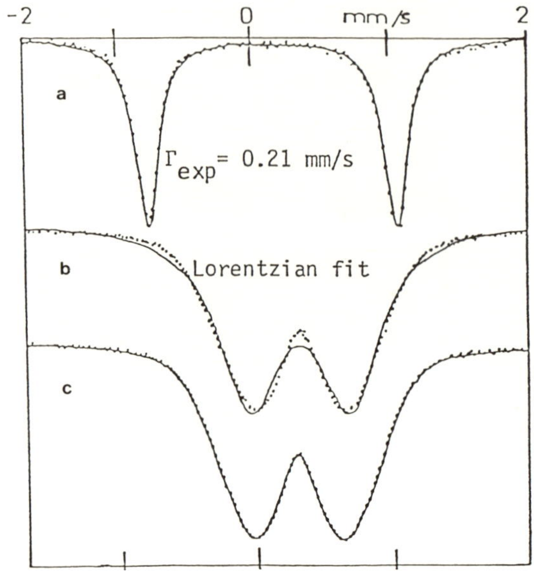

Figure 18. Room-temperature Mössbauer spectrum recorded on amorphous $KFeF_4$ with a Lorentzian fit (b) or a $P(\Delta)$ fit (c), compared with the spectrum of the crystalline compound (a), which has narrow Lorentzian quadrupole doublet lines. Reproduced from Ref. 73.

4.2. Paramagnetic Behavior

At room temperature, the Mössbauer effect hyperfine structures observed for all of these compounds show the presence of pure electric quadrupolar interactions.[65,67–69] The paramagnetic spectra consist of broad non-Lorentzian lines which reflect the diversity of local structural situations (see Fig. 18). These doublets, which are symmetric, are well reproduced by a discrete

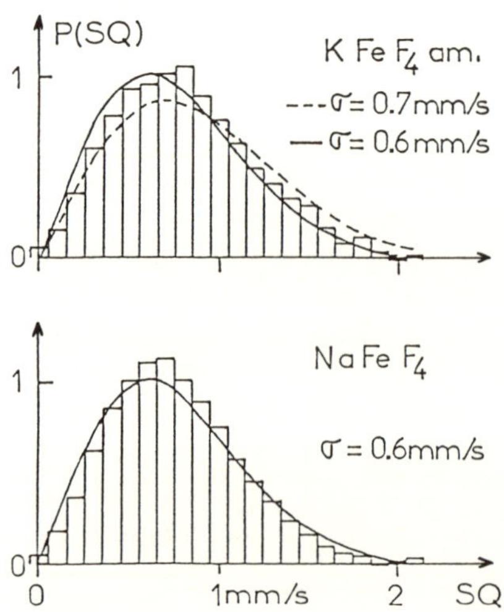

Figure 19. Examples of experimental quadrupole splitting distributions compared with the Gaussian model; see text. Reproduced from Ref. 78.

Table 3. Magnetic and Mössbauer Spectral Parameters for Some Deposited Amorphous Fluorides and for the Fluorinated Variety of FeF_3, in Parentheses[a]

	FeF_3	$NaFeF_4$	$KFeF_4$
IS_{cryst} (mm/s)	0.49[c]	0.46	0.45
$\langle IS \rangle_{am}$[b] (mm/s)	0.48 (0.45–0.47)	0.45	0.44
Δ_{cryst} (mm/s)	0.05[c]	0.69	1.46
$\langle \Delta_{am} \rangle$ (mm/s)	0.58 (0.51–0.56)	0.73	0.77
σ_{am} (mm/s)	~0.45 (~0.40)	~0.60	~0.65
H_{hyp}^{cryst} (T)	61.8[c]	56.2	54.0
$\langle H_{hyp}^{am} \rangle$ (T)	56.0 (55.0–56.0)	53.0	52.0
T_N^{cryst} (K)	362[c]	111.5	141.5
$T_{freezing}^{susc}$ (K)	29 (26)	12	<4.2
$T_{freezing}^{Möss}$ (K)	39 (36)	20	12
Θ_P (K)	−486 (−490)	−290	~248

[a] Adapted from Refs. 67, 79, 89, 90, 103.
[b] Isomer shift values relative to metallic iron at 300 K.
[c] Data concerning the r-FeF_3 phase.

distribution of quadrupole splittings, $P(\Delta)$, the isomer shift value being constant for all components. Histograms of $P(\Delta)$ are presented in Fig. 19 and the data are reported in Table 3.

It is worth noting that the isomer shift values of the amorphous and crystalline analogues are essentially the same and consistent with a trivalent iron state (high spin) in octahedral surroundings. Further, the $\langle \Delta \rangle$ values of FeF_3 and $AFeF_4$ are close to each other but differ from those obtained for the crystalline phases.

The quadrupole interactions were quantitatively evaluated with a fitting procedure which involves some arbitrary parameters but which leads to a discrete distribution of hyperfine parameters.[70–73] We have analyzed the influence of these arbitrary parameters for the case of a quadrupole doublet spectrum.[73] Systematic fits were performed by assuming various values of the total number of components, the increment value of the quadrupolar splitting, and the linewidth of the Lorentzian lines, Γ. The authors have concluded that the shape of the quadrupolar splitting distribution and the mean $\langle \Delta \rangle$ value are strongly influenced by the choice of Γ. In the study reported here, the value given to Γ was always the "instrumental" linewidth, 0.215 mm/s.

For any disordered system the relative width of the quadrupole splitting distribution $P(\Delta)$ can be characterized by the ratio $q = \langle \Delta^2 \rangle / \langle \Delta \rangle^2$,[73] a ratio which can then easily be compared with the predictions resulting from various models. A model proposed by Czjzek and co-workers,[74] based on an analytic approach developed from environment models of random ionic coordinations, permits the determination of the joint distribution function $P(V_{zz}, \eta)$ with V_{zz} and η, both of which were defined in Section 2. From a consideration of random environments in isotropic solids, they obtain

$$P(V_{zz}, \eta) = \frac{1}{\sqrt{2\pi}\,\sigma^5} V_{zz}^4 \eta (1 - \eta^2/9) \exp\left(\frac{-V_{zz}^2(1 + \eta^2/3)}{2\sigma^2}\right) \quad (1)$$

where σ represents the standard deviation of the Gaussian component. In the case of the iron nucleus, with a nuclear spin of 3/2; Eq. (1) becomes

$$P(\Delta) \propto \frac{\Delta^{d-1}}{\sigma^d} \exp - \frac{\Delta^2}{2\sigma^2} \quad (2)$$

where d is a dimensional parameter giving the number of independent random variables in the surroundings of the Mössbauer atom. These variables correspond to displacements of ions or angular distortions of a polyhedron formed by an ionic coordination shell.[75] The value of d is restricted to $d \leq 5$. The normalized and theoretical quadrupole distributions, deduced from Eq. (2) are represented in Fig. 20. A simple relationship between the "dimensional" d parameter and the relative width parameter q has been given in Ref. 73.

For the amorphous ferric fluorides prepared by the vapor-quenching technique, the nuclear quadrupole splitting distribution, $P(\Delta)$, is well reproduced by Eq. (2) with $d \simeq 2$.[73,76] In conjunction with the isomer shift values, Czjzek[75] showed that this result implies that the Fe^{3+} ions are surrounded by weakly and randomly distorted fluorine octahedra, the distortions of which can be described by two independent random variables. It should be noted that the same expression with $d = 2$ was previously established,[77,78] on the basis of a Gaussian model of three independent distribution functions, $P(V_{ii})$, $i = x, y, z$, coupled by the Laplace equation.[77]

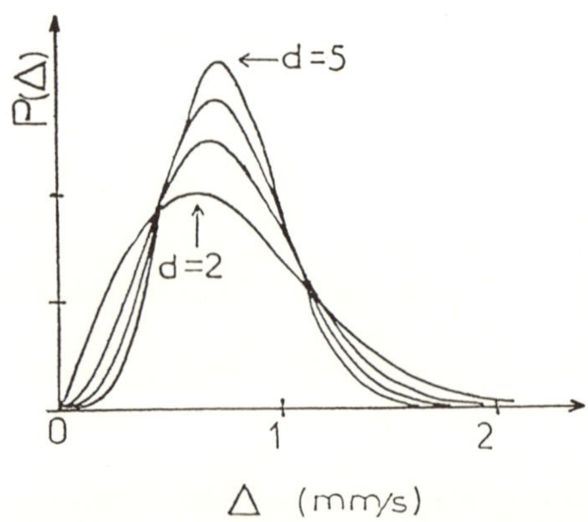

Figure 20. Theoretical quadrupole distributions obtained for different values of d. Reproduced from Ref. 73.

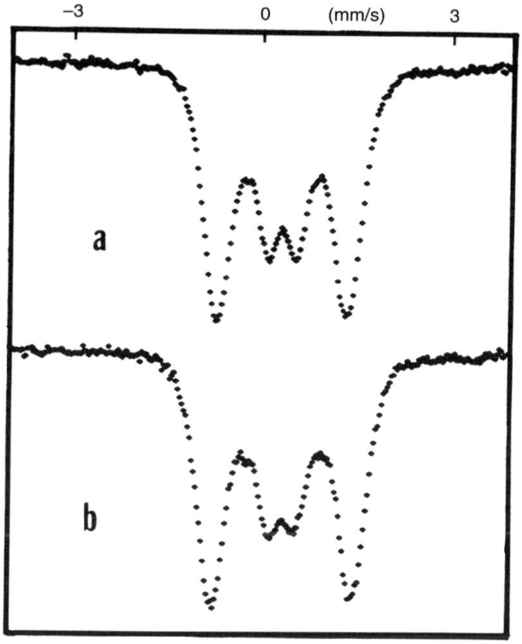

Figure 21. Mössbauer spectra of amorphous FeF_3, xHF (a) and deposited amorphous FeF_3 (b) recorded at 200 K in a 70-kG magnetic field applied parallel to the γ ray. Reproduced from Ref. 79.

In the case of the fluorinated form of FeF_3, $P(\Delta)$ is fitted with $d = 2.7$.[79] This higher d value is attributed both to the presence of a small $FeCl_3$ impurity and to the HF, which can induce a real change in the topology surrounding the Fe^{3+} ion.

High-field Mössbauer spectroscopy was carried out on the two forms of amorphous FeF_3 in the paramagnetic temperature range.[79] The rather symmetric shape of the spectra, as shown in Fig. 21, indicates an almost equal probability of finding positive and negative values of the EFGs at the iron sites. The fitted proportions of ~48(2) and 52(2)%, respectively, were obtained from a linear combination of simulated Mössbauer spectra composed of either pure positive or pure negative axial EFGs as in agreement with Eq. (2).[76,79]

These results were discussed by comparison with EFG calculations which used point charge summations on idealized atomic structures.[15] We considered only the monopolar contribution because of computing time limitations (see Section 2). The summations were performed on sets of successive spherical shells centered on a large number of iron sites located near the center of the structural model, the so-called "electric field gradient sphere".

The structural modeling was only undertaken in the case of FeF_3, because of its rather simple chemical composition. After analysis of the magnetic properties, as will be discussed in the next section, we established that the structure of the ideal amorphous FeF_3 could be described by a continuous

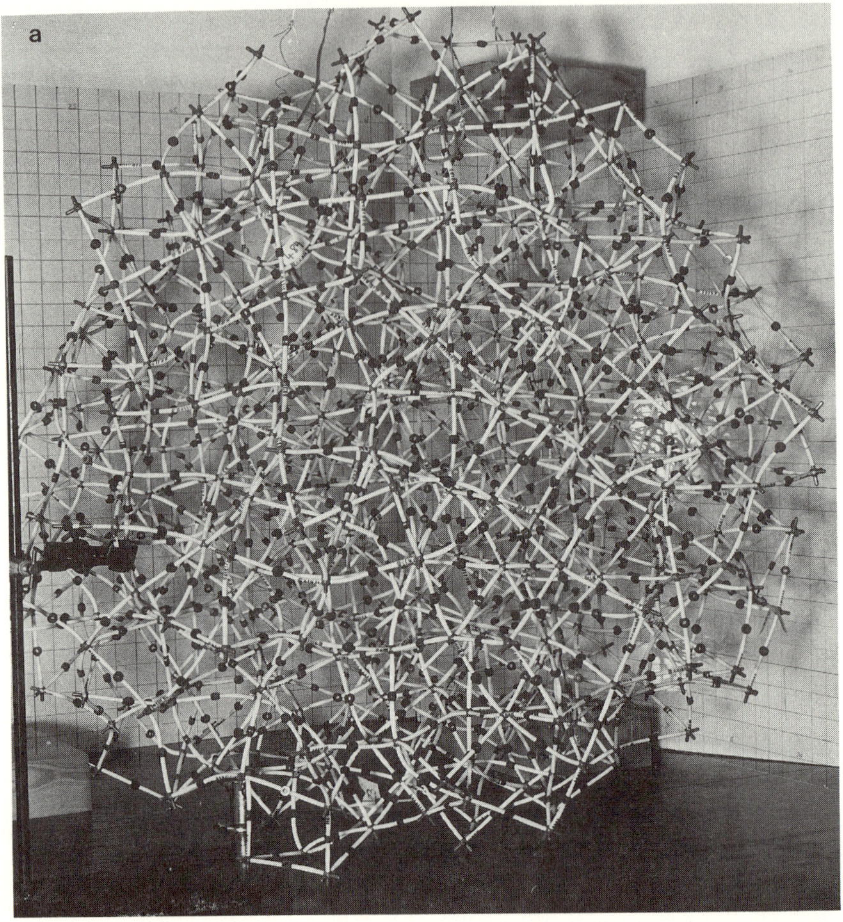

Figure 22. General view of hand-built random networks of (a) 2064 atoms or (b) 300 octahedra. Reproduced from Refs. 80 and 81.

dense random packing of corner-sharing octahedra with no dangling bonds.[44] Two hand-built models were proposed earlier. One model was composed of 2064 ball and spoke units, equivalent to 516 octahedra,[80] whereas the second model was composed of 300 rigid plastic octahedral units,[81] as illustrated in Fig. 22. More recently, we successfully generated a computer-built model containing ~4400 ions.[82] This network, as a first step, accounts only for the cationic positions and starts from a "nucleus" of some 200 cations adapted from the central part of the hand-built model.[80] Then, it is grown by computer, using an algorithm which adds new cationic positions, A, and creates bonds with the previous positions, by respecting the following criteria: (1) sixfold A–A coordination, (2) geometric considerations based on the steric en-

Figure 22. (*Continued*)

cumbrance of the two ions, and (3) the presence of odd-membered and even-membered cationic rings, in agreement with observed magnetic properties. The respective proportions of rings were deduced from an analysis of the correlation between the proportion of odd-membered rings and the magnetic ordering temperatures of the three crystalline phases. Then, anionic positions, B, are located at the middle of the cationic bonds, and finally a relaxation process, assuming pair harmonic potentials between B–B and A–B, is performed iteratively, leading to a so-called "relaxed model". The optimization of the growth process in the model is discussed in Ref. 82.

We evaluated this relaxed model in terms of the following structural and topological aspects.[82] The calculated radial distribution function is consistent with the available experimental data. Specifically, the Fe–F peak width corresponding to the first neighbors is close to that observed by EXAFS experiments. The density is found to be between those of the rhombohedral and the hexagonal tungsten bronze forms of FeF_3. The octahedral units are weakly distorted. The distribution of the Fe–F–Fe superexchange bonds agrees well with the steric encumbrance of Fe and F ions. The average value and full width at half-maximum of this distribution are estimated as 154° and 40°, respectively. We have checked that the ring statistics are consistent with the reduction of the spin freezing temperature of the different varieties of FeF_3 (see details in the next section).

The EFG distribution derived from the hand-built models showed poor accuracy, and more accurate results were obtained from the computer-built model. The monopolar EFGs were calculated over 495 iron sites, and a good quantitative and qualitative agreement was found with the observed quadrupole splitting distribution shape, $P(\Delta)$ (see Fig. 23), the average value of $\langle\Delta\rangle$, the proportions of positive and negative values of V_{zz}, and the value of d.[15] All of this supports the idea that the EFG distribution at the iron sites calculated in structural models is basically governed by the weakly distorted octahedra, through two independent random parameters.[75]

An alternative approach for computer modeling of the structure of amorphous FeF_3 was proposed by Lines.[83] A cluster containing 858 ions was obtained from a dense random packing involving the sequential introduction of two kinds of hard spheres of different radii. The relaxation procedure was carried out by including coulombic repulsion terms in order to separate similarly charged ions. Unfortunately, such a network does not satisfy the sixfold coordination of the cation which is clearly evidenced by Mössbauer spectroscopy and diffraction techniques. However, the EFG distribution calculated at the iron sites exhibits an almost equal probability of finding positive and negative V_{zz} values, a result in nice agreement with the Mössbauer data.[84]

We have recorded Mössbauer spectra of amorphous FeF_3 in an applied

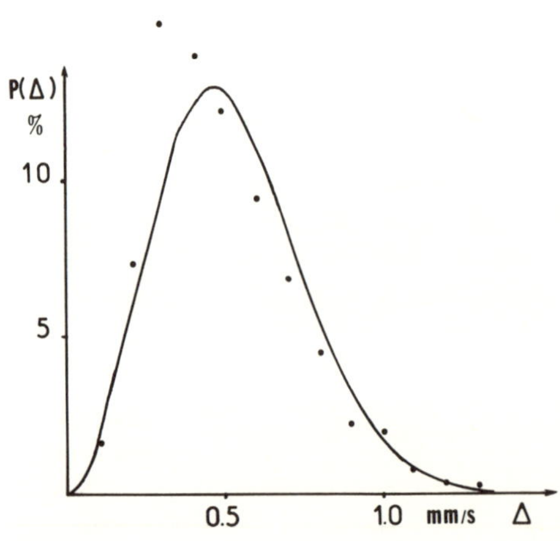

Figure 23. Quadrupole splitting distribution calculated from the point charge model (○) and the solid curve representing that deduced from the random packing model proposed by Czjzek *et al.*[74] with d = 2.9. Reproduced from Ref. 82.

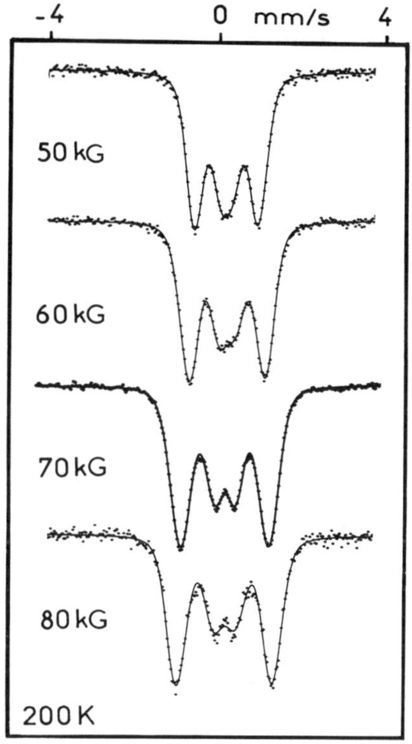

Figure 24. Mössbauer spectra of FeF_3, xHF recorded at 200 K as a function of an external magnetic field applied parallel to the γ ray. Reproduced from Ref. 85.

magnetic field as a function of temperature, in the paramagnetic temperature range.[85] As shown in Fig. 24, the spectra, which are composed of four lines with a Gaussian profile, are indicative of magnetic moments which are well aligned with the external field. At lower temperatures, the spectra recorded in an external magnetic field reveal that the magnetic moments are progressively canted (see Figs. 25 and 26).[85] Consequently, both the angles and the magnitudes of the hyperfine fields are distributed such that their quantitative analysis is far from simple. The determination of the combined distribution, $P(H_{hyp}, \Theta)$, is still an unresolved problem.

4.3. Magnetically Frozen Phase

The magnetic hyperfine splitting in amorphous FeF_3 progressively appears with decreasing temperatures.[67,86] The transition to the magnetically ordered state is described as a progressive spin freezing, because the corresponding temperature range (see the values reported in Table 3) depends on the various experimental techniques, such as Mössbauer spectroscopy or magnetic susceptibility; this behavior is similar to that of spin-glasses. The spin fluctuations

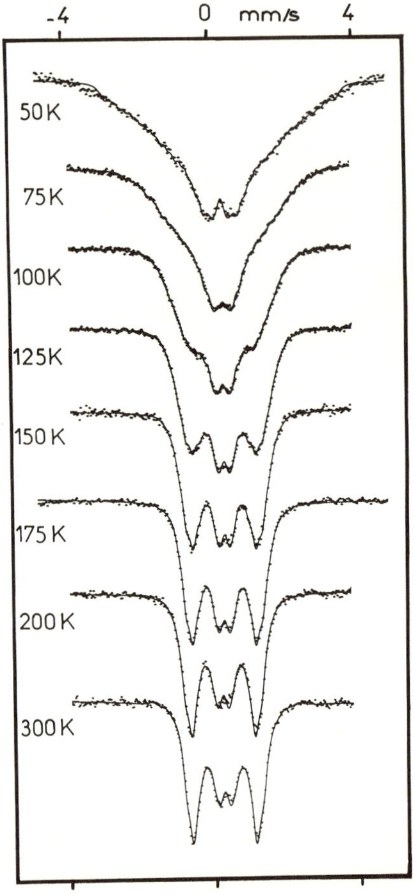

Figure 25. Mössbauer spectra of FeF_3, xHF recorded as a function of temperature, in a 60-kG external magnetic field applied parallel to the γ ray. Reproduced from Ref. 85.

have been followed by measuring the depth of the paramagnetic line by using the thermal scanning method. On a reduced scale, the same behavior is observed for the different amorphous fluorides, FeF_3, and $AFeF_4$, in which the freezing temperature depends on the nature of the cation (see Fig. 27). The magnetic susceptibility obeys the Curie–Weiss law with a large negative Θ_p, similar to that observed in crystalline fluorides, indicating the presence of strong antiferromagnetic interactions.[87] However, the value of the molar Curie constant is unusually low. This has been explained by Coey[88] as an artifact resulting from a distribution of exchange integrals, which should be related to the angular distribution of superexchange bonds postulated by the structural models. Further observations indicate a small magnetic irreversibility below the spin-freezing temperature.

The spin-freezing temperatures of approximately 35 to 45 K for the amorphous forms of FeF_3 are extremely low relative to those of the crystalline

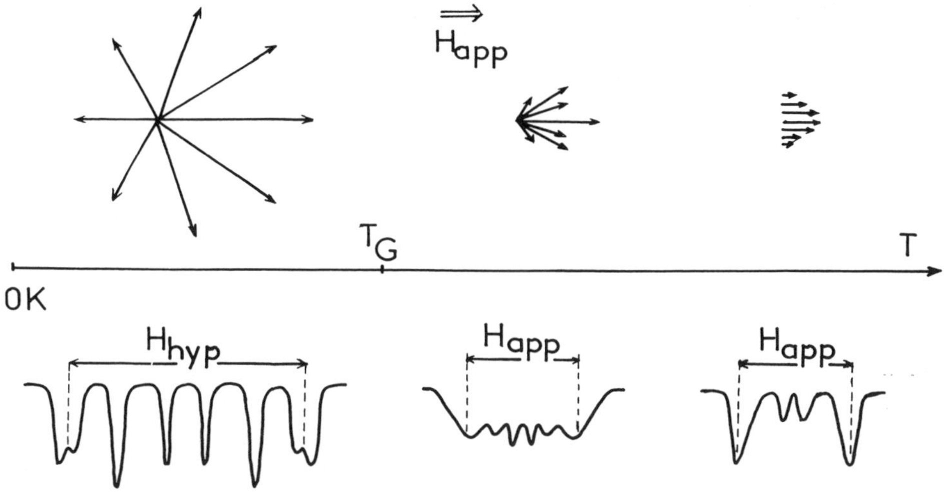

Figure 26. Schematic representation of the thermal evolution of a randomly distributed magnetic system in an external magnetic field and the corresponding Mössbauer spectra. Reproduced from Ref. 85.

rhombohedral form of FeF_3 (see Table 3). Ferey *et al.*[87] have attributed this reduction and the very large value of the $|\theta_p|/T_G$ ratio (see Table 3) to the frustration of superexchange bonds implied by the cationic topology. Their conclusion is consistent with the concept of frustration previously developed to explain the magnetic properties of spin-glasses.[91–93]

At 4.2 K, the Mössbauer spectra of the series of amorphous ferric fluorides are completely resolved, as shown in Fig. 28 for $FeF_3, x HF$. These

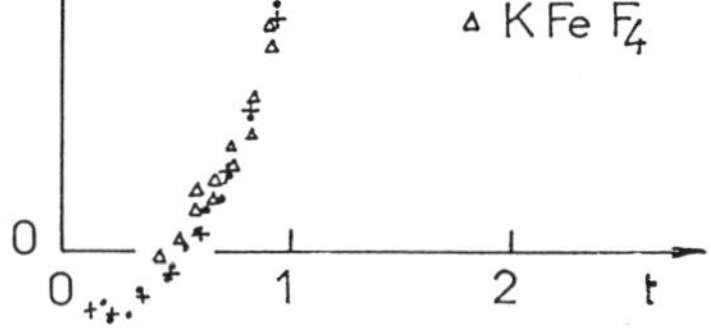

Figure 27. Thermal scan on reduced scale for several amorphous fluorides. F_P represents the Mössbauer absorption at a constant velocity chosen to coincide with a paramagnetic absorption line. Reproduced from Ref. 67.

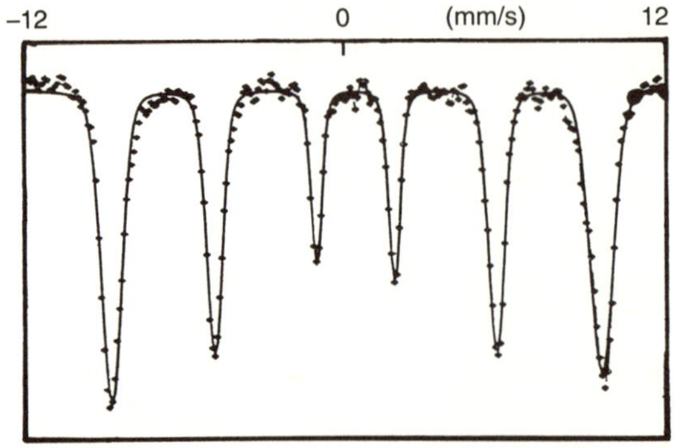

Figure 28. Mössbauer spectrum of FeF_3, xHF measured at 4.2 K. The theoretical curve results from the fitting procedure of Eibschutz and Lines.[95,96] Reproduced from Ref. 79.

spectra exhibit magnetic Zeeman sextets with weakly broadened lines[94] which have been analyzed with a method developed by Eibschutz and Lines.[95,96] Structural information on the statistical distribution of iron neighbors can be derived from a quantitative analysis of the mean positions, the root-mean-square widths, and the profiles of the six lines of the magnetic Zeeman sextets. The average values of the relevant hyperfine parameters and the correlations between them have been determined. Eibschutz and Lines have showed that, in the case of metallic glasses, the well-known linewidth asymmetry originates mainly from a correlation between the hyperfine field, the EFG, and their respective orientation. A secondary correlation occurs between the hyperfine field and the isomer shift.[97,98] For amorphous ferric insulators, the same authors have concluded that no preferred EFG sign was evidenced in the distribution, and that EFG fluctuations predominate over isomer shift distributions, confirming the previous observations in the paramagnetic temperature range. In addition, the distribution of quadrupole shifts in the magnetic sextets is well described by a symmetric Gaussian function whereas the isomer shift is quite constant, in agreement with an octahedral coordination for all of the iron sites.[94]

In all amorphous FeF_3 and $AFeF_4$ samples, the mean value of the hyperfine field is reduced relative to that of the crystalline systems, as shown in Table 3. In addition, small differences in the mean values of the hyperfine field were found in samples of amorphous FeF_3 prepared under different conditions. It is clear than the properties of the amorphous state depend on the deposition rate, the pressure, the temperature of the cooled substrate, and the purity of the initial crystalline FeF_3 in the case of the deposition technique, or the purity of the initial $FeCl_3$ for the fluorinated systems[79,99] (see Table 1 in Ref. 79). Additional structural information comes from the hyperfine field distribution which exhibits a small-amplitude, low field, tail.[79,94,100–102] In the case of amorphous FeF_3, this tail contains 5% of the iron sites at 4.2 K. Further, this tail should disappear in the limit as $T \rightarrow 0$. This low-field tail, which, according to Lines and Eibschutz,[101] only exists on hyperfine field distribution of amorphous speromagnets, is attributed by these authors to the presence of a low-energy exchange field, caused by the cancellation of the nearest-neighbor contributions within the frustrated spin network. The main narrow peak in the hyperfine field distribution does not reflect any significant variation in the magnitude of the ferric moments from one site to another, in agreement with a purely octahedral coordination environment.

High-field Mössbauer spectroscopy experiments were performed at 4.2 K on the different ferric amorphous fluorides.[44,67,79,87,103] All of the spectra exhibit an asymmetric broadening of the outer lines, as shown in Fig. 29. This broadening is consistent with the expected speromagnetic structure discussed in Section 3. This isotropic strong magnetic order, in agreement with the rather high spin-freezing temperature, is little affected by the presence of the applied

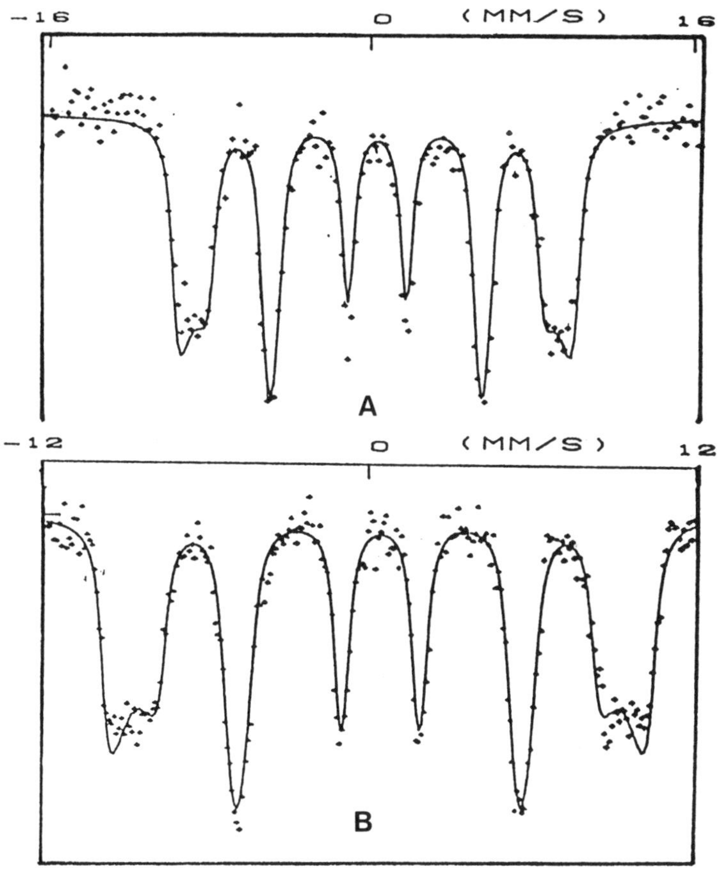

Figure 29. Mössbauer spectra recorded at 4.2 K in an external magnetic field applied parallel to the γ ray on (A) deposited amorphous FeF_3 (60 kG) and (B) fluorinated amorphous FeF_3, xHF (70 kG). Reproduced from Ref. 44.

field. Consequently, the conservation of the isotropic orientation of the magnetic moments in the presence of the external field induces a particular broadening of the outer sextet lines, as discussed previously (see, e.g., Ref. 71). To account for the magnetic properties, the cationic topology of these materials must involve both odd- and even-membered cationic rings. The presence of these rings frustrates the superexchange interactions and justifies the assumptions of the structural model discussed in Section 4.2.

The modeling of the magnetic properties of the amorphous FeF_3 materials is in progress. The basic features of the model are magnetic frustration, due to the topology of the lattice, and a distribution of superexchange integrals resulting from a range of Fe–F–Fe angles. The first quantitative description was given by Ferey *et al.*[87] who assumed a distribution of the effective exchange interaction, J_{eff}, at each site with the bond length and angle fluctuations expressed by

$$P(J_{\mathrm{eff}}) \propto J_{\mathrm{eff}}^{n} \exp(-J_{\mathrm{eff}}^{2}/2\sigma^{2})$$

We are currently applying a Monte–Carlo algorithm to the various simulated networks in an attempt to better describe the amorphous form of FeF_3. A first approach, proposed on a two-dimensional topologically disordered lattice, revealed a noncollinear magnetic configuration at 0 K.[104]

4.4. Crystallization of Amorphous Fluorides

The amorphous-to-crystal transformation, obtained through thermal annealing, represents the most specific characteristic of the amorphous state. This transformation is associated with an exothermic heat of crystallization which has been easily observed in the case of amorphous FeF_3 by differential thermal analysis. In addition, Mössbauer spectroscopy is an excellent tool for the qualitative and quantitative analysis of the crystallization products and the kinetics of the transformation.

Preliminary experiments performed on the amorphous form of $KFeF_4$ revealed that the crystallization mechanism is strongly influenced by the hydrolysis.[105] Consequently, isothermal and nonisothermal treatments were carried out on FeF_3 and vapor-deposited or fluorinated forms of FeF_3, by using a "cryofurnace" to heat the samples to 400°C. The Mössbauer spectra are recorded "*in situ*" at 77 K. These spectra permit a more complete characterization of the different crystalline phases, which are magnetically ordered at this temperature.

Annealings were performed at different temperatures on the fluorinated form of FeF_3. The main results indicate that this form always crystallizes directly into the rhombohedral phase without any formation of divalent iron, even at annealing temperatures below the transformation temperature of the hexagonal tungsten bronze FeF_3 phase into the rhombohedral FeF_3 phase.[106] In the case of the vapor-deposited material, isothermal high-temperature treatments led only to the rhombohedral form of FeF_3.[101] An early study by Lopez-Herrera *et al.* showed that the hexagonal tungsten bronze form of FeF_3 was present after annealing of vapor-deposited amorphous FeF_3 at 180°C for 2 days.[38] Such a metastable phase was discovered by "chimie douce,"[107] which is a low-temperature solid-state reaction. Further experiments have confirmed that both rhombohedral and hexagonal tungsten bronze forms of FeF_3 can be obtained by thermal treatment; the hydrolysis phenomena speeding up the transformation kinetics in the latter crystalline phase.[68]

The evolution of the Mössbauer spectra, shown in Fig. 30, permitted an estimate of the rate of crystallization of rhombohedral FeF_3 as a function of time for several annealing temperatures. The data followed a phenomenological law of Johnson–Mehl–Avrani such that

$$X_{\text{cryst}} = 1 - \exp[-(Kt)^n]$$

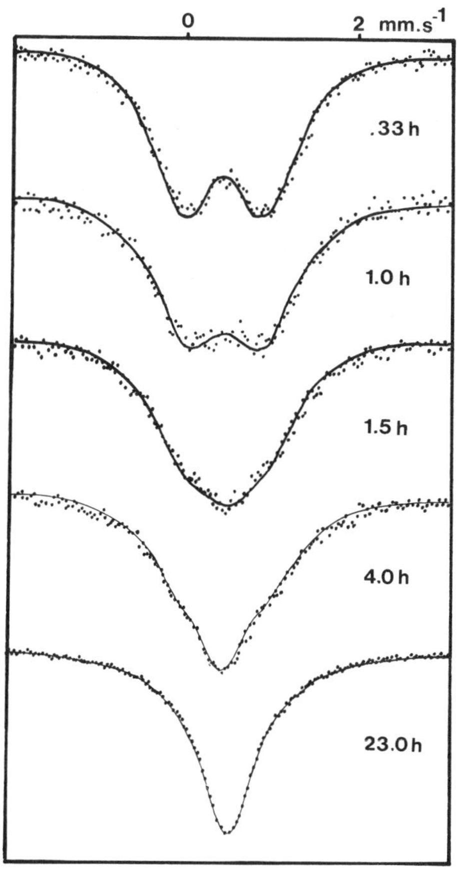

Figure 30. Room-temperature Mössbauer spectra measured after thermal treatment at 280°C and given as a function of annealing time. Reproduced from Ref. 106.

where X_{cryst} is the rate of crystallization, K is the kinetic constant, and n is the Avrami exponent. A first-order reaction, $n \simeq 1$, with an activation energy of ~2 eV was observed.[68] The activation energy compares well with that generally measured for metallic glasses or glasses, whereas the reaction order is lower than in these disordered systems, for which $\sim 1.5 < n < 5$ is usually found.

4.5. Glassy Fluorides

After the discovery of ZrF_4-based glasses in 1975,[108] related fluoride glasses have received a good deal of attention because of their potential applications as infrared optic fiber material (see, e.g., a review paper on fluoride glasses[109]). Numerous glass-forming systems were synthesized and investigated, with some transition metal elements introduced in order to change the optical, electric, and magnetic properties of the glass. Low concentrations of iron ions act as a coloring agent for fluoride glasses, and, consequently, are not favorable for ameliorating optical properties. In contrast, high concentrations of iron confer semiconductor or ferromagnetic properties to these materials. In spite of this technological interest, few iron-based fluoride glasses were prepared for Mössbauer spectroscopy studies. The sole objective of these few studies was mainly to determine the oxidation state and the coordination of the iron atoms, or the redox equilibrium and the structural role of the iron in the glass.

Kawamoto and co-workers[110–114] have focused their interest on materials deriving from the following ternary diagrams: ZrF_4–BaF_2–FeF_3, ZrF_4–BaF_2–FeF_2, ZrF_2–BaF_2–FeF_3, PbF_2–BaF_2–FeF_3, ZrF_4–BaF_2–CsF, and FeF_3–MnF_2–PbF_2, and the quaternary AlF_3–BaF_2–CaF_2–FeF_3 system. The resulting paramagnetic Mössbauer spectra recorded at room temperature result from the superposition of two doublets. Both of these quadrupolar components exhibit rather broad lines in agreement with the glassy state of these samples. To fit these spectra, the authors have commonly used two independent doublets with broad Lorentzian lines, whose linewidth range is from 0.50 to 0.70 mm/s. A disagreement between experimental and theoretical curves is observed in the centre of the spectrum and in the wings of the absorption lines. A better agreement could be obtained by using two independent distributions of the quadrupolar splitting. But assumptions have to be made due to the superposition of the lines and, consequently, the quadrupolar splitting distribution shapes should be rather difficult to determine unambiguously. As a matter of fact, the quantitative interpretation proposed by Kawamoto and co-workers does not drastically depend on the fitting assumptions.

The isomer shift values are found in the range of 0.40 to 0.50 and 1.20 and 1.30 mm/s and the quadrupole splittings in the range of 0.40 to 0.80 and 2.20 to 2.50 mm/s for both components. The authors have attributed these values to the presence of Fe^{3+} and Fe^{2+}, in the high-spin state coordinated to six fluoride ions, respectively. Similar results have been reported for sodium–iron fluorophosphate glasses.[115] Through the values of the ratio $[Fe^{2+}]/[Fe^{2+} + Fe^{3+}]$ estimated from the corresponding Fe^{2+} and Fe^{3+} Mössbauer spectral areas, the same authors have concluded that the ferric octahedra have the character of a network former and have characterized the $Fe^{3+} \rightarrow Fe^{2+}$ reduction reaction.

In the vitreous insulating ferric fluorides, $PbMnFeF_7$ and $PbMnFeF_9$, a spin-glass behavior was observed by magnetic measurements and Mössbauer spectroscopy.[116] Sharp cusps, indicative of spin-glass transitions, were observed at $T_G = 11.77$ and 5.26 K, respectively, by low-field ac susceptibility measurements. The progressive spin freezing, centered around 7.5 and 16.5 K, was confirmed by using a thermal scan of the Mössbauer absorption lines. The temperature discrepancy is explained by the differences in the measurement times of the two techniques (see Section 4.3).

5. CONCLUSIONS AND PERSPECTIVES

In this review, we have selected from among the large number of fluorine-based compounds, some crystalline and amorphous ferric fluorides which exhibit very attractive structural, electronic, and magnetic properties. We have shown that iron-57 Mössbauer spectroscopy, including high-field Mössbauer spectroscopy, can play a valuable role beyond that of materials characterization, in understanding such properties.

The observation and computation of the electric quadrupole interactions provide an efficient tool for structural investigation, a tool which has been tested successfully in crystalline fluorides. Because of the local character of the hyperfine interactions, this approach is of major interest for the study of less-ordered systems, such as amorphous structures or solid solutions. From the magnetic point of view, the superexchange interactions between paramagnetic centers via fluoride orbitals favor antiferromagnetic couplings in the transition metal fluorides. In addition, the cationic topology can produce frustrated antiferromagnetic interactions which lead to pathological magnetic structures which are easily detected by Mössbauer spectroscopy. Similarly, the Monte–Carlo simulation of magnetic structures, based on topologic considerations, has contributed to an understanding of their thermal evolution, particularly in the case of noncollinear magnetic structures. This procedure remains to be applied to disordered systems. Finally, the analysis of both electric and magnetic hyperfine interactions supports basic assumptions found in the structural description of amorphous fluorides.

The fundamental Mössbauer spectral hyperfine properties are well described for the bulk fluorides. This knowledge promotes their investigation as surface layers or nanoparticles. In other respects, some recent Mössbauer effect experiments on diluted[117] or mixed[63,118,119] crystalline fluorides should provoke renewed interest in the understanding of magnetic clusters.

Further, activity in chemical synthesis and crystal growth of new fluorides remains high and will provide new systems with outstanding properties for future studies.

ACKNOWLEDGMENT. We are indebted to the Fluoride Chemistry Group, URA CNRS No. 449, University of Le Mans, which has been deeply involved in most of the studies reported herein.

REFERENCES

1. D. Babel and A. Tressaud, in *Inorganic Solid Fluorides,* P. Hagenmuller, ed., p. 78, Academic Press, New York, 1985.
2. G. Ferey, in *Mixed Valency Systems: Applications in Chemistry, Physics and Biology,* p. 155, Kluwer Academic, Dordrecht, 1991.
3. J. M. Dance and A. Tressaud, in *Organic Solid Fluorides,* P. Hagenmuller, ed., p. 371, Academic Press, New York, 1985.
4. T. E. Cranshaw and G. Longworth, in *Mössbauer Spectroscopy Applied to Inorganic Chemistry,* Vol. 1, G. J. Long, ed., p. 171, Plenum Press, New York, 1984.
5. G. Ferey, M. Leblanc, R. De Pape, and J. Pannetier, in *Inorganic Solid Fluorides,* P. Hagenmuller, ed., p. 395, Academic Press, New York, 1985.
6. J. M. D. Coey, *Can. J. Phys.* **65**, 1210 (1987).
7. C. E. Johnson, in *Mössbauer Spectroscopy Applied to Inorganic Chemistry,* Vol. 1, G. J. Long, ed., p. 619, Plenum Press, New York, 1984.
8. H. Spiering, in *Mössbauer Spectroscopy Applied to Inorganic Chemistry,* Vol. 1, G. J. Long, ed., p. 77, Plenum Press, New York, 1984.
9. J. Teillet, Y. Calage, and F. Varret, *J. Phys. Chem. Solids* **43**, 863 (1982).
10. Z. M. Stadnik, *J. Phys. Chem. Solids* **45**, 311 (1984).
11. F. Bertaut, *J. Phys. Chem. Solids* **39**, 97 (1979).
12. Y. Calage and J. Pannetier, *J. Phys. Chem. Solids* **38**, 711 (1977).
13. C. K. Coogan, *Aust. J. Chem.* **20**, 2551 (1967).
14. H. C. Bolton, W. Fawcett, and I. D. C. Gurney, *Proc. Phys. Soc.* **80**, 199 (1962).
15. J. M. Grenèche, F. Varret, and J. Teillet, *J. Phys. (Paris)*, **49**, 243 (1988).
16. J. Teillet, G. Ferey, M. Leblanc, and F. Varret, *Solid State Commun.* **27**, 1083 (1978).
17. F. Menil, A. Tressaud, R. Sabatier, and G. Le Flem, *Mater. Res. Bull.* **12**, 983 (1977).
18. M. Eibschutz, H. J. Guggenheim, L. Holmes, and J. L. Bernstein, *Solid State Commun.* **11**, 457 (1972).
19. M. Eibschutz, G. R. Davidson, and H. J. Guggenheim, *Phys. Rev. B* **9**, 3885 (1974).
20. G. Heger and R. Geller, *Phys. Status Solidi B* **53**, 227 (1972).
21. J. D. Rush, A. Simopoulos, M. F. Thomas, and B. M. Wanklyn, *Solid State Commun.* **18**, 1039 (1976).
22. W. Kundig, A. B. Denison, and P. Ruegsegger, *Phys. Lett.* **42A**, 199 (1972).
23. J. Teillet and G. Courbion, *J. Phys. Chem. Solids* **46**, 559 (1985).
24. G. Courbion, C. Jacoboni, and R. De Pape, *J. Solid State Chem.* **45**, 127 (1982).
25. W. Viebahn, *Z. Anorg. Allg. Chem.* **413**, 77 (1975).
26. M. A. Hepworth, K. H. Jack, R. D. Peacock, and G. J. Westland, *Acta Crystallogr.* **10**, 63 (1957).
27. M. Leblanc, J. Pannetier, G. Ferey, and R. De Pape, *Rev. Chim. Min.* **22**, 107 (1985).

28. M. Leblanc, G. Ferey, P. Chevallier, Y. Calage, and R. De Pape, *J. Solid State Chem.* **47**, 53 (1983).
29. R. De Pape and G. Ferey, *Material Res. Bull.* **21**, 971 (1986).
30. L. A. Levinson, *J. Phys. Chem. Solids* **29**, 1331 (1968).
31. E. O. Wollan, H. R. Child, W. C. Koehler, and M. K. Wilkinson, *Phys. Rev.* **112**, 1132 (1958).
32. G. K. Wertheim, H. J. Guggenheim, and D. N. E. Buchanan, *Solid State Commun.* **5**, 537 (1967).
33. I. Dzialhozsinsky, *J. Phys. Chem. Solids* **4**, 241 (1958).
34. T. Moriya, in *Magnetism,* Vol. 1, G. T. Rado and H. Suhl, eds., Academic Press, New York, 1963.
35. M. Leblanc, R. De Pape, G. Ferey, and J. Pannetier, *Solid State Commun.* **58**, 171 (1986).
36. G. Ferey, R. De Pape, M. Leblanc, and J. Pannetier, *Rev. Chim. Min.* **23**, 474 (1986).
37. G. K. Wertheim, H. J. Guggenheim, and D. N. E. Buchanan, *Phys. Rev.* **169**, 465 (1968).
38. M. E. Lopez-Herrera, F. Varret, Y. Calage, and G. Ferey, *J. Magn. Magn. Mater.* **44**, 304 (1984).
39. Y. Calage, M. Leblanc, G. Ferey, and F. Varret, *J. Magn. Magn. Mater.* **43**, 195 (1984).
40. Y. Calage, M. Zemirli, J. M. Greneche, F. Varret, R. De Pape, and G. Ferey, *J. Solid State Chem.* **69**, 197 (1987).
41. G. A. Sawatzky and F. Van Der Woude, *J. Phys. (Paris) C6* **35,** 47 (1974).
42. J. Chappert, *J. Phys. (Paris) C6* **35**, 71 (1974).
43. J. Chappert, *Hyp. Interact.* **13**, 25 (1983).
44. J. M. Grenèche, A. Le Bail, M. Leblanc, A. Mosset, F. Varret, J. Galy, and G. Ferey, *J. Phys. C* **21**, 1351 (1988).
45. J. M. D. Coey and P. W. Readman, *Nature (London)* **246**, 476 (1973).
46. J. M. D. Coey and P. W. Readman, *Earth Planet. Sci. Lett.* **21**, 45 (1973).
47. Y. Laligant, Y. Calage, E. Torres-Tapia, J. M. Grenèche, F. Varret, and G. Ferey, *J. Magn. Magn. Mater.* **61**, 283 (1986).
48. P. Imbert, G. Jehanno, Y. Macheteau, and F. Varret, *J. Phys. (Paris)* **37**, 969 (1976).
49. J. M. Grenèche, J. Linares, F. Varret, Y. Laligant, and G. Ferey, *J. Magn. Magn. Mater.* **73**, 115 (1988).
50. Y. Laligant, J. Pannetier, and G. Ferey, *J. Solid State Chem.* **66**, 242 (1987).
51. P. Lacorre and J. Pannetier, *J. Magn. Magn. Mater.* **71**, 63 (1987).
52. C. M. Hurd, *Contemp. Phys.* **23**, 469 (1982).
53. X. M. Kan and J. M. D. Coey, *Am. Mineral.* **70**, 576 (1985).
54. E. Herdtweck, *Z. Anorg. Allg. Chem.* **501**, 131 (1983).
55. M. Leblanc, G. Ferey, Y. Calage, and R. De Pape, *J. Solid State Chem.* **53**, 360 (1984).
56. M. Guillot, A. Marchand, M. Leblanc, and G. Ferey, *J. Phys. C* **20**, 2405 (1987).
57. M. Leblanc, G. Ferey, P. Lacorre, and J. Pannetier, *J. Magn. Magn. Mater.* **92**, 359 (1991).
58. P. Lacorre, J. Pannetier, M. Leblanc, and G. Ferey, *J. Magn. Magn. Mater.* **92**, 366 (1991).
59. N. Metropolis, A. E. Rosenbluth, M. N. Rosenbluth, A. H. Teller, and E. Teller, *J. Chem. Phys.* **21**, 1087 (1953).
60. G. Courbion, R. De Pape, J. Teillet, F. Varret, and J. Pannetier, *J. Magn. Magn. Mater.* **42**, 217 (1984).
61. J. Kanamori, *J. Phys. Chem. Solids* **10**, 87 (1959).
62. J. B. Goodenough, *Magnetism and the Chemical Bond,* Wiley–Interscience, New York, 1963.
63. J. J. Maguer, G. Courbion, and J. M. Grenèche (submitted).
64. M. Tamine, Y. Calage, M. Leblanc, G. Ferey, and F. Varret, *Hyp. Interact.* **28**, 529 (1986).
65. G. Ferey, A. M. Leclerc, R. De Pape, J. P. Mariot, and F. Varret, *Solid State Commun.* **29**, 477 (1979).
66. M. Leblanc, G. Ferey, J. M. Grenèche, A. Le Bail, F. Varret, R. De Pape, and J. Pannetier, *J. Phys. (Paris) C8* **46**, 175 (1985).

67. M. Henry, F. Varret, J. Teillet, G. Ferey, O. Massenet, and J. M. D. Coey, *J. Phys. (Paris) C1* **41**, 279 (1980).
68. J. M. Grenèche, M. Leblanc, F. Varret, and G. Ferey, *Hyp. Interact.* **27**, 317 (1986).
69. M. Eibschutz, M. E. Lines, L. G. Van Uitert, H. J. Guggenheim, and G. J. Zydzik, *Phys. Rev. B* **24**, 2343 (1981).
70. J. M. Grenèche, M. Henry, and F. Varret, *Proc. Int. Conf. Appl. Mössbauer Effect (Jaipur)*, p. 908 (1982).
71. F. Varret, *Proc. Int. Conf. Appl. Mössbauer Effect (Jaipur)*, p. 129 (1982).
72. F. Varret, J. M. Grenèche, and J. Teillet, *Proc. Int. Workshop on Mössbauer Spectroscopy (Seeheim)*, p. 101 (1983).
73. M. E. Lopez-Herrera, J. M. Grenèche, and F. Varret, *Phys. Rev. B* **28**, 4944 (1983).
74. G. Czjzek, J. Fink, F. Gotz, H. Schmidt, J. M. D. Coey, J. P. Rebouillat, and A. Lienard, *Phys. Rev. B* **23**, 2513 (1981).
75. G. Czjzek, *Phys. Rev. B* **25**, 4908 (1982).
76. J. M. Grenèche, in *XXVI Zakopane School on Physics*, J. Stanek, ed., p. 129, World Scientific, Singapore, 1991.
77. J. M. D. Coey and G. Czjzek, unpublished results.
78. F. Varret and M. Henry, *Rev. Phys. Appl.* **15**, 1057 (1980).
79. J. M. Grenèche, F. Varret, M. Leblanc, and G. Ferey, *Solid State Commun.* **63**, 435 (1987).
80. J. M. Coey and P. J. K. Murphy, *J. Non-Cryst. Solids* **50**, 125 (1982).
81. J. M. Grenèche, J. Teillet, and J. M. D. Coey, *J. Non-Cryst. Solids* **83**, 27 (1986).
82. J. M. Grenèche, J. Teillet, and J. M. D. Coey, *J. Phys. (Paris)* **48**, 1709 (1987).
83. M. E. Lines, *Phys. Rev. B* **21**, 5793 (1980).
84. M. E. Lines, *J. Non-Cryst. Solids* **46**, 1 (1981).
85. J. M. Grenèche and F. Varret, *Hyp. Interact.* **42**, 935 (1988).
86. J. Teillet, A. Hauet, and J. M. Grenèche, *J. Phys. (Paris) C8*, **49**, 1095 (1988).
87. G. Ferey, F. Varret, and J. M. D. Coey, *J. Phys. C* **12**, L531 (1979).
88. J. M. D. Coey, *J. Appl. Phys.* **49**, 1646 (1978).
89. J. M. Dance, R. Sabatier, F. Menil, M. Wintenberger, J. C. Cousseins, G. Le Flem, and A. Tressaud, *Solid State Commun.* **19**, 1059 (1976).
90. R. Heger and R. Geller, *Phys. Status Solidi B* **53**, 227 (1972).
91. G. Toulouse, *Commun. Phys.* **2**, 115 (1977).
92. J. Vannimenus and G. Toulouse, *J. Phys. C* **10**, L537 (1977).
93. S. Kirkpatrick, *Phys. Rev. B* **16**, 4630 (1977).
94. M. Eibschutz and M. E. Lines, L. G. Van Uitert, H. J. Guggenheim, and G. J. Zydzik, *Phys. Rev. B* **29**, 3843 (1984).
95. M. Eibschutz and M. E. Lines, *Phys. Rev. B* **25**, 4256 (1982).
96. M. Eibschutz and M. E. Lines, *Phys. Rev. B* **26**, 2288 (1982).
97. M. Eibschutz, M. E. Lines, H. S. Chen, and T. Masumoto, *J. Phys. F* **14**, 505 (1984).
98. M. E. Lines and M. Eibschutz, *Solid State Commun.* **45**, 435 (1983).
99. J. M. Grenèche, J. Teillet, F. Varret, J. Litterst, and G. Ferey, unpublished results.
100. M. Eibschutz and M. E. Lines, *J. Magn. Magn. Mater.* **54–57**, 297 (1986).
101. M. E. Lines and M. Eibschutz, *Phys. Rev. B* **32**, 3067 (1985).
102. M. Eibschutz and M. E. Lines, *Hyp. Interact.* **27**, 47 (1986).
103. G. Ferey, J. M. D. Coey, M. Henry, J. Teillet, F. Varret, and R. Buder, *J. Magn. Magn. Mater.* **15–18**, 1371 (1980).
104. J. M. Grenèche and J. M. D. Coey, *J. Phys. (Paris)* **51**, 231 (1990).
105. M. E. Lopez-Herrera, Thesis, University of Le Mans, unpublished (1983).
106. J. M. Grenèche, F. Varret, M. Leblanc, and G. Ferey, *Solid State Commun.* **61**, 813 (1987).
107. A. R. West, *Solid State Chemistry and Its Applications*, p. 30, Wiley, New York, 1985.
108. M. Poulain and J. Lucas, *Mater. Res. Bull.* **10**, 243 (1975).
109. J. J. Videau and J. Portier, in *Inorganic Solid Fluorides*, P. Hagenmuller, ed., p. 309, Academic Press, New York, 1985.

110. Y. Kawamoto, T. Horisaka, K. Hirao, and N. Soga, *Chem. Lett.* **1984**, 1441 (1984).
111. Y. Kawamoto, T. Nohara, K. Hirao, and N. Soga, *Solid State Commun.* **51**, 769 (1984).
112. Y. Kawamoto and K. Nakashio, *Mater. Res. Bull.* **21**, 779 (1986).
113. Y. Kawamoto, J. Fujiwara, K. Hirao, and N. Soga, *J. Non-Cryst. Solids* **95–96**, 926 (1987).
114. Y. Kawamoto and D. Koumiyoji, *J. Electrochem. Soc.* **136**, 1816 (1989).
115. F. Menil, L. Fournès, J. M. Dance, and J. J. Videau, *J. Non-Cryst. Solids* **34**, 209 (1979).
116. J. P. Renard, J. P. Miranday, and F. Varret, *Solid State Commun.* **35**, 41 (1980).
117. J. Chadwick, D. H. Jones, J. A. Johnson, C. E. Johnson, and M. F. Thomas, *J. Phys. Condens. Matter* **1**, 6731 (1989).
118. M. Lahlou-Mimi, M. Leblanc, and J. M. Grenèche, *J. Magn. Magn. Mater.* **92**, 375 (1991).
119. M. Lahlou-Mimi, Y. Pennec, J. M. Bassat, M. Leblanc, and J. M. Grenèche, *J. Magn. Magn. Mater.* (submitted).

6

Mössbauer Spectroscopy Studies of Slow Paramagnetic Relaxation and Magnetic Interactions in Ferrocenium-Based Molecular Magnetic Materials

William M. Reiff

1. INTRODUCTION

Analogous to the quest for ever higher T_c organic and inorganic superconductors, the search for molecular magnetic materials with high Curie and Néel temperatures continues unabated. By molecular, we will mean truly organic, primarily carbon-based materials, or perhaps polar organic solvent-soluble organometallic species. They should exhibit spontaneous magnetization to something approximating a bulk magnetically ordered state where such three-dimensional ordering occurs in zero applied field. We like to believe that some of the impetus for the current interest in this area came from our discovery[1] that the charge transfer salt, decamethylferrocenium tetracyanoethyleneide, $\{DMFc^+\}\{TCNE\}^-$ (alternatively $[Fe(Cp^*)_2]^+[TCNE]^-$), shows bulk ferromagnetism, with a Curie temperature $T_c = 4.8$ K. This conclusion was confirmed by classical susceptibility, heat capacity,[2] neutron diffraction, and zero- and applied-field Mössbauer spectroscopy measurements.

William M. Reiff • Department of Chemistry, Northeastern University, Boston, Massachusetts 02115.

Mössbauer Spectroscopy Applied to Magnetism and Materials Science, Volume 1, edited by Gary J. Long and Fernande Grandjean. Plenum Press, New York, 1993.

A variety of paradigms have been proposed, tried, and recently discussed[3] vis-à-vis rational approaches to isolation of such materials. There are clearly challenging synthetic chemistry and structural characterization problems remaining that must be mastered for continued advances in this area.

Since the initial advances in this rapidly growing area of materials research have primarily involved compounds of iron and in particular charge transfer salts of low-spin iron(III) as the ferrocenium ion, we introduce this chapter with selected aspects of the electronic properties of this form of spin-doublet iron. As will be seen, the Mössbauer spectroscopy of this less common form of iron has proven extremely informative in both the synthetic effort to prepare new compounds and subsequent detailed characterization of magnetic properties. The power of Mössbauer effect spectroscopy in this context of materials research clearly makes it an appropriate subject for the present volume of this series.

2. ASPECTS OF ELECTRONIC STRUCTURE OF LOW-SPIN IRON(III)

2.1. Orbital Contributions

Since we are only interested in the ground state (magnetic properties), we take as our starting point the $^2T_{2g}$ state that arises for strong field ligation of iron(III) in O_h symmetry. Rather complete details of the theory for this state with emphasis on the interpretation of ESR spectra are found in the work of Griffith.[4,5] Subsequently, this theory was refined, enhanced, and enlarged to interpret the experimental temperature dependence of molar susceptibility,[6,7] the quadrupole interaction,[8] and finally paramagnetic relaxation phenomena of low-spin iron(III) compounds in seminal studies by Lang and Oosterhuis.[9,10] For now we give only qualitative generalizations appropriate to the present context.

A nominal $^2T_{2g}$ ground state arising from either the t_{2g}^1 configuration of d^1 in O_h symmetry (e.g., Ti^{3+} ion) or the t_{2g}^5 configuration of low-spin d^5 (Fe^{3+}) can suffer the lifting of its sixfold degeneracy via a combination of spin–orbit splitting effects ($\lambda Fe^{3+} \approx -460\,cm^{-1}$)[11] and low-symmetry (axial and/or rhombic) ligand field components.

The degeneracy removal effects are typically related to the experimental temperature dependence of various properties via parameterizations involving functions of Δ/λ where Δ is a distortion parameter and λ is the "free ion" spin–orbit coupling constant. One generally expects significant nonzero orbital angular momentum effects and in particular large g-factor anisotropy. As Carlin[12] has amply pointed out, the parent single-ion magnetic properties can strongly influence if not dominate the three- and lower-dimensionality magnetic exchange interactions of the related condensed magnetic systems. Iron(III) does not behave differently in this respect and several examples of these effects will be evident in the review that follows, particularly highly anisotropic 1D and 3D magnetic exchange effects.

More specific to Mössbauer spectroscopy, one should consider the various contributions to the effective internal hyperfine field H_{eff} given[13] in the equation: $H_{\text{eff}} = H_{\text{F}} + H_{\text{L}} + H_{\text{D}} + H_{\text{O}}$ and the consequences for paramagnetic relaxation phenomena and the limiting observed values of H_{n}. The large internal hyperfine fields (H_{n}) exerted on the nuclei of atoms in exchange-coupled cooperatively ordered or slowly relaxing systems ultimately arise from the paramagnetism of the atom's electrons and can be divided into three components:[13]

1. $H_{\text{F}}\alpha\{\psi\uparrow(0)^2 - \psi\downarrow(0)^2\}$ (the Fermi contact term)
2. $H_{\text{L}}\alpha\langle r^{-3}\rangle\langle L_z\rangle$ or $H_{\text{L}}\alpha\langle r^{-3}\rangle(g-2)\langle S_z\rangle$ (the orbital moment contribution)
3. $H_{\text{D}}\alpha\langle r^{-3}\rangle\langle 3\cos^2\Theta - 1\rangle\langle S_z\rangle$ or alternatively $H_{\text{D}}\alpha V_{zz}/e\langle S_z\rangle$ (the dipolar interaction)

It is seen that the first component is the result of an imbalance of "*s*" electron α and β spin density at the nucleus. This imbalance results from the polarization caused by the differential interaction of "*s*"(α) and "*s*"(β) with *unpaired* valence shell "*d*" and "*f*" electrons. On the other hand, the second component is directly related to the magnitude of orbital angular momentum (L) for a particular unpaired valence shell electron of radius r. It is clear that the third component is operative only for sites of less than cubic symmetry and is the result of a *through space* interaction of the valence shell spin angular momentum with that of the nucleus. The terms H_{L} and H_{D} can either oppose or add to H_{F}. For the case of iron, H_{F} is ~11 T per unpaired electron (1 T = 10 kG). Thus, for high-spin iron(III), ^6A ground term, in cubic symmetry where L, H_{L}, and H_{D} are all ~0, limiting low-temperature values of H_{n} ranging between 40 and 60 T can be observed. For high-spin iron(II), somewhat smaller values of H_{n} (to ~35 T) are observed because of a combination of the smaller value of $\langle S\rangle$, namely two, and the possibility of an orbital contribution H_{L} that generally opposes H_{F}. Reduction of H_{n} for any oxidation state is also related to variable covalency and delocalization effects depending on the ligands. This can lead directly to delocalization of metal ion "*s*" spin density and an increase of $\langle r^3\rangle$ for the valence shell electrons. These are effects that result in a decrease of any of H_{F}, H_{L}, or H_{D} in view of their respective equations. An additional effect leading to substantially reduced values of H_{n} is the so-called "zero point spin reduction" operative in certain low-dimensional magnetic systems.

2.2. Single-Ion Zero-Field Splitting and Slow Paramagnetic Relaxation versus Cooperative Three-Dimensional Order

Slow paramagnetic relaxation–hyperfine splitting is a dynamic single-ion effect[14] resulting in part from the *z*ero-*f*ield *s*plitting of the ground (electronic) spin manifold, e.g., the spin sextet of high-spin iron(III). Since this is a Kramers ion, the ground state of the *zfs* ion must be the doubly degenerate $M_s = \pm\frac{1}{2}$ $(D > 0)$ or $M_s = \pm\frac{5}{2}$ $(D < 0)$ in zero applied magnetic field. Relaxation in the former is rapid while that in the latter as well as $M_s = \pm\frac{3}{2}$ is slow. For the moment, we will consider only the case of high-spin iron(III) with the axial zero-field splitting, D, *large* and *negative* and at low temperatures. In this situation the dominant relaxation mechanism is spin–spin relaxation (interatom exchange of S_z values) as opposed to spin–lattice relaxation that is more important to $L \neq 0$ ions such as high-spin Fe^{2+} or, of special importance to this chapter, low-spin Fe^{3+}. If the metal ions are closely situated so as to allow for rapid interatom spin flips via direct interactions (but not close enough for direct magnetic exchange or superexchange), there will be no spectral broadening even though a "slowly" relaxing ($M_s = \pm\frac{5}{2}$) doublet is being populated. However, dilution[15] of the metal ions to distances typically ≥7.5 Å [for high-spin Fe(III)] leads to longer spin–spin relaxation times whose reciprocal corresponds to a dynamic, temperature-dependent frequency that eventually becomes comparable to the Larmor precession frequency of the nuclear moment. This leads to the *gradual* Zeeman splitting of the Mössbauer spectrum as the "slowly relaxing" $M_s = \pm\frac{5}{2}$ doublet is progressively populated. One observes a *single* six-line pattern in the limit of low T since the $M_s = \pm\frac{3}{2}$ at 4D is little populated when D is large and T is very low.

The limiting (zero field) low-temperature spectra corresponding to 3D magnetic ordering processes (antiferro- and ferromagnetism) are generally identical in appearance to those for slow relaxation. The difference is that H_n at the individual metal ion sites now corresponds to and is generally thought of as collinear with a spontaneous magnetization developing in the bulk sample. The latter originates from exchange interactions (either direct, e.g., α-Fe, or superexchange, e.g., α-Fe_2O_3) between the metal ions that become comparable to or greater than the *thermal* spin randomization effects as the temperature is decreased and a molecular exchange field develops. In general and in contrast to slow relaxation, the hyperfine splitting process occurs "suddenly" over a *small temperature interval*[13] reflecting the cooperative (usually second order) phase transformation nature of magnetic ordering. In addition, the internal hyperfine field is very temperature dependent in the vicinity of $T_{critical}$ and only levels off as magnetic saturation is reached at low temperatures, often as $T \to 0$ K.

It is clear that in strong contrast to typical high-spin iron(III) (^6A, electronic ground state), H_L may well now be significant for low-spin ferric especially for cases in which the local iron atom coordination and/or molecular orbital population are such that the electric field gradient is close to or actually zero in value. In these circumstances, effective g values can be considerably different[16,17] from 2 corresponding to $H_L \neq 0$. The latter is precisely the case for the $Fe(Cp)_2^+$ cation as well as $Fe(Cp^*)_2^+$ (Cp^* = pentamethylcyclopentadiene) which typically exhibit relatively narrow linewidth, essentially singlet Mössbauer spectra.[14] In this situation one has the possibility of H_n values that differ widely from the usual expectation of ~11 T from a Fermi contact contribution corresponding to the single unpaired electron of spin-doublet Fe(III). An example[18] of this effect in a coordination compound that is otherwise apparently a rapidly relaxing paramagnet at 4.2 K (in zero external field) is seen in the recent study of $[Fe(TICOH)_2](BF_4)_3$ (TICOH = tris(n-methylimidazol-2-yl)hydroxymethane, a highly anisotropic low-spin Fe(III) system. At 4.2 K, with $H_O = 0.22$ T, $H_n = 42.5$ T (~13.7 mm/s) for the polycrystalline powder while $H_n = 44$ T for a frozen CH_3CN solution in $H_O = 0.22$ T. In the molecular magnets that follow, we will see several examples with such large orbital angular momentum anisotropy effects, in fact one to such an extreme that the effects of g-factor anisotropy are clearly resolved in zero-field Mössbauer spectra of polycrystalline absorbers.

We close this section and bridge to the next by pointing out an additional important "orbital effect," namely the coupling of the spin system to the lattice phonons. This occurs through the agency of spin–lattice relaxation that generally necessitates low temperatures[9] (more so than for high-spin ferric) in order to observe slow paramagnetic relaxation broadening and eventual hyperfine splitting.

3. ZERO-FIELD AND FIELD-INDUCED SLOW PARAMAGNETIC RELAXATION FOR LOW-SPIN IRON(III)

In the context of this section we distinguish and give representative examples of three extreme types of low-spin iron(III) compounds vis-à-vis paramagnetic relaxation effects.

3.1. Spatially Nondilute Systems with Diamagnetic Counterions—$K_3Fe(CN)_6$

The archetypical example[9] of this kind of system is $K_3Fe(CN)_6$. This system is in fact *not self-dilute* (smallest Fe–Fe distances ≪7 Å) and has a relatively rapidly relaxing ground state in a diagmagnetic matrix of K^+ and CN^- ions. Thus, slow relaxation is *not observed* for the "neat" undiluted compound in zero field at 4.2 K and even lower temperatures. In Lang's studies of this system, it was found necessary to (1) dilute the Fe^{3+} moments by substitution in the isomorphous diamagnetic host salt $K_3Co(CN)_6$ and (2) cool to <20 K in order to observe resolved hyperfine splittings corresponding to the infinitely long slow paramagnetic relaxation time limit. Complete resolution of hyperfine splitting and the decoupling of nuclear spin–electron spin moments was achieved by the application of external magnetic fields, H_O.

3.2. Simple Ferrocenium Salts—Relatively Self-Dilute Cations with Diamagnetic Counteranions

Probably the most complete study[19] of systems of this type is the high-field (up to ~6 T) and variable-temperature Mössbauer spectroscopy study of Cohn *et al.* These investigators studied a variety of mono- and binuclear ferrocenium ions with examples of both trapped and detrapped monocations for mixed valence binuclear systems. While a wide range of paramagnetic relaxation rates were observed, *in no case* was relaxation broadening or hyperfine splitting reported for *zero-field spectra* at $T \sim 4.2$ K. Alternatively stated, the spectra in zero field are typical of rapidly relaxing paramagnets with the expected narrow linewidths.

3.3. $[Fe(Cp^*)_2]^+[DDQ]^-$ Self-Dilute $S = \frac{1}{2}$ Cations with Paramagnetic $S = \frac{1}{2}$ Counteranions

This class of compounds is essentially the primary subject of the remainder of this chapter. They are obtained as donor–acceptor charge transfer complexes via one-electron reduction of polycyanide acceptors, e.g., those shown in Fig. 1. Three basic structure types (Fig. 2) arise: type I—so-called segregated stacks; type II—mixed stacks; and type III—containing cationic dimers, $\{D^+\}_2$, alternating with dianionic forms of some of the acceptors from Fig. 1. Decamethylferrocene [DMFc or $Fe(Cp^*)_2$] has a lower ionization potential than ferrocene, $Fe(Cp)_2$. Since it exhibits complete transfer most readily with typical electron acceptors, it has undergone by far the most intense study in our laboratory. The reaction sequence of Fig. 3 of TCNE with various methylated ferrocenes suggests the possibility that ferrocene itself may not be that far from the neutral-to-ionic interface.[20,21] In any event, X-ray study[22] shows $Fe(Cp)_2$ to yield the diamagnetic *neutral mixed stack* (i.e., no formal charge transfer) structure shown at the top of Fig. 4 while $Fe(Cp^*)_2$ leads to the paramagnetic, *ionic mixed stack motif*[1] shown in the center of Fig. 4.

Our initial result suggesting that the DDQ species is more appropriately formulated as the $S = \frac{1}{2}[DDQ]^{\cdot-}$ anion in $[Fe(Cp^*)_2]^+[DDQ]^-$ than as the diamagnetic $[DDQH]^-$ arose from the observation of a novel fully resolved slow paramagnetic relaxation broadening of the zero-field Mössbauer spectra[23] (Fig. 5). Slow (zero-field) paramagnetic relaxation is not common for low-spin iron(III). As mentioned previously, one of the few and better-documented

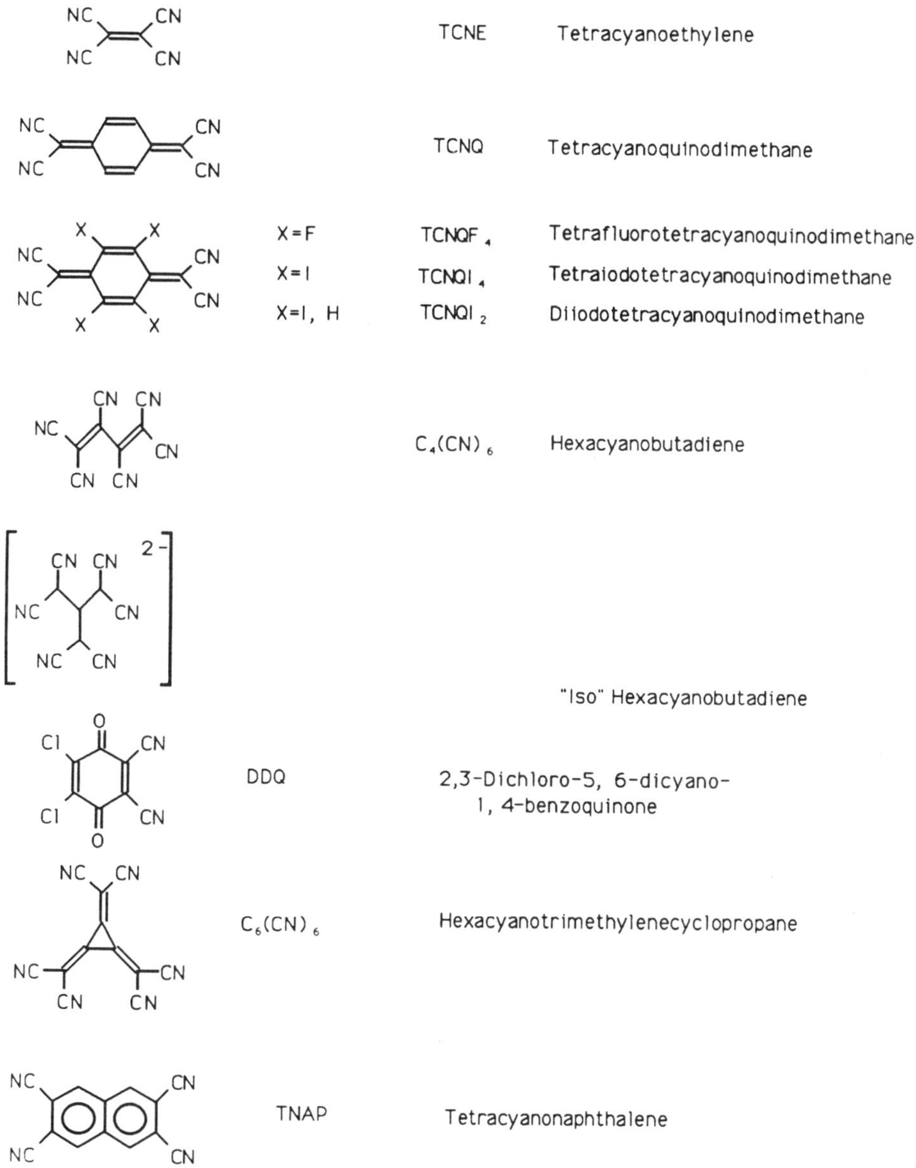

Figure 1. Acronyms and schematic structures of some polycyanide electron acceptors.

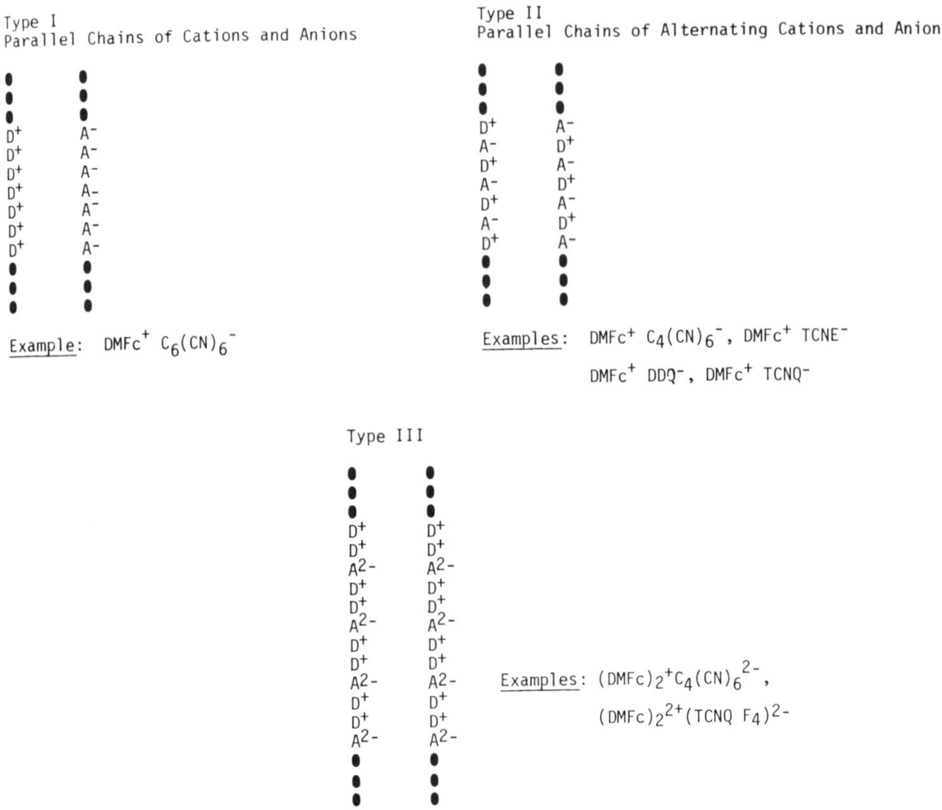

Figure 2. Structure archetypes for linear-chain charge transfer salt magnets.

examples[9] is the $K_3[Fe(CN)_6]$ complex isomorphously diluted with the diamagnetic Co^{III} analogue. The combination of dilution (leading to longer spin–spin relaxation times) and very low temperatures (decreasing the rate of spin–lattice relaxation) can ultimately lead to fully resolved magnetic hyperfine splitting of zero-field Mössbauer spectra. The normal absence of slow paramagnetic relaxation broadening for undiluted low-spin iron(III) compounds (save for the effects of solitons; *vide infra*) is a consequence of the fact that it is a Kramers ion and this ground state must be at least a spin doublet. There is then the fully allowed nature of transitions (rapid relaxation) between the $m = \pm\frac{1}{2}$ components of such a doublet. The gradual hyperfine splitting observed in Fig. 5 such that dH_n/dT is not large is more typical of slow relaxation as opposed to cooperative, extended 3D magnetic order. We find no evidence of slow paramagnetic relaxation broadening effects in the zero-field Mössbauer spectra of a variety of ferrocenium or decamethylferrocenium salts of simple diamagnetic anions (e.g., $[I_3]^-$, $[PF]_6^-$, $[BF_4]^-$) above 1.7 K. This is also true for the diamagnetic ($S = 0$) $[C_3(CN)_5]^-$ anion salt (Fig. 6) which

FERROCENE
$Fe(Cp)_2$

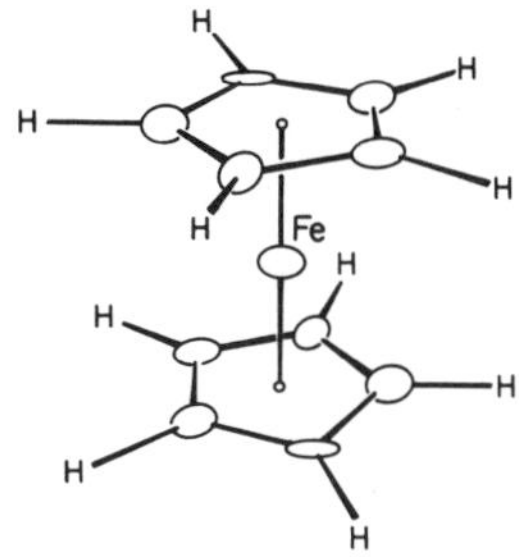

$Fe(Cp)_2$ + TCNE → $Fe(Cp)_2{}^0$ $TCNE^0$
(Ferrocene)

$Fe(Cp^*)_2$ + TCNE → $Fe(Cp^*)_2{}^+$ $TCNE^-$
(Decamethylferrocene)

$Fe(CpCp^*)$ + TCNE → $Fe(CpCp^*)_2{}^+$ $TCNE^-$
(Pentamethylferrocene)

$Fe(1\text{-}CH_3\text{-}CpCp)$ + TCNE → $Fe(1\text{-}CH_3\text{-}CpCp)_2{}^0$ $TCNE^0$
($1\text{-}CH_3$-ferrocene)

Cp* = pentamethylcyclopentadiene

Figure 3. The reaction of ferrocene and some of its methylated derivatives with tetracyanoethylene (TCNE).

possesses the same linear chain mixed stack structure that is observed for the $[Fe(Cp^*)_2]^{+\cdot}[DDQ]^{\cdot -}$ salt.[23] We and others[19] have found that external fields, H_O, can induce varying degrees of slow paramagnetic relaxation for the case of the $[Fe(Cp^*)_2]^{\cdot -}$ cation with diamagnetic anions, where the rate of relaxation is related to a complex combination of spin–spin and spin–lattice effects and the magnitude of H_O and is also dependent on the specific type of ferrocenium cations involved. *In any event, it seems clear that the slow relaxation observed for the present system in zero field is consistent with a small Zeeman splitting of the ground-state Kramers doublet of the* $[Fe(Cp^*)_2]^{\cdot -}$ *cation by dipolar fields arising from electron spin on the* $[DDQ]^{\cdot -}$ *species in the lattice. This combined with the already large* Fe^{III}–Fe^{III} *distances (shortest Fe–Fe distances >8.6 Å) in the undiluted solid leads to long relaxation times.* The shape of the limiting relaxation spectrum (infinitely long relaxation time), Fig. 5, is exactly that

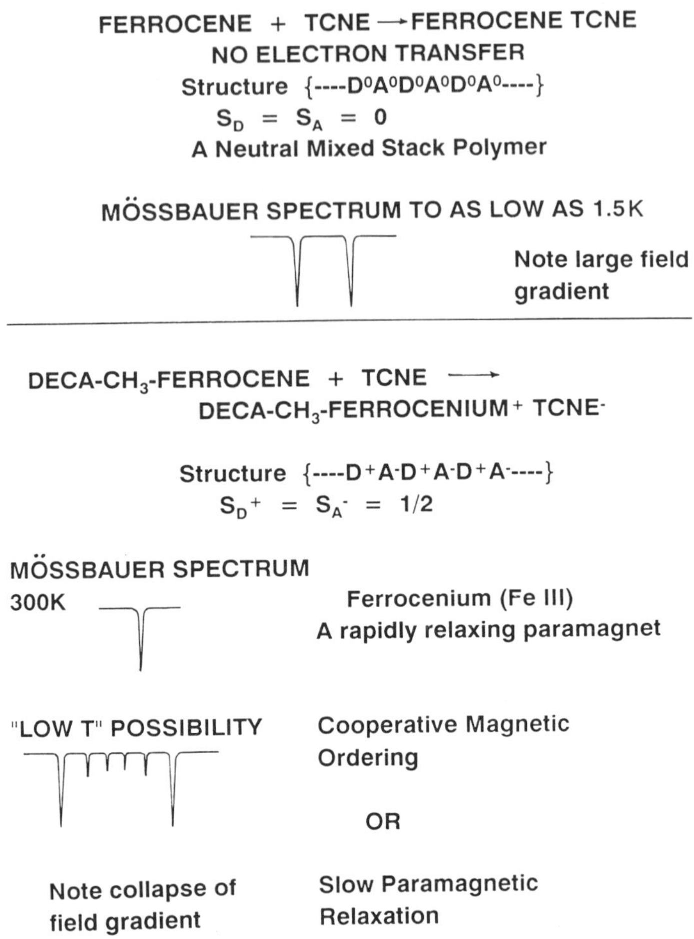

Figure 4. Basic Mössbauer spectra for neutral ferrocene and ionic decamethylferrocene TCNE chains.

calculated theoretically by Lang and Oosterhuis[10] for the case of a large positive axial field and an orbital doublet (^2E) ground term in the crystal field approximation with the scaling factor for quadrupole interaction very nearly zero owing to electron delocalization and/or a canceling lattice contribution to the electric field gradient. *To our knowledge this is the first observation of this limiting form of slow (zero-field) relaxation behavior for low-spin Fe^{III}.* The preceding ground state is suggested by previous susceptibility[24] and spectroscopy[16,17,25] studies and corresponds to the $a_{1g}^2 e_{2q}^3[(d_{z2})^2(d_{x2y2})^2 d_{xy}^1]$ configuration. This configuration is expected to lead to both a substantial nonzero field gradient and quadrupole interaction while nearly zero values are typically observed for various ferrocenium ions. This inconsistency is avoided

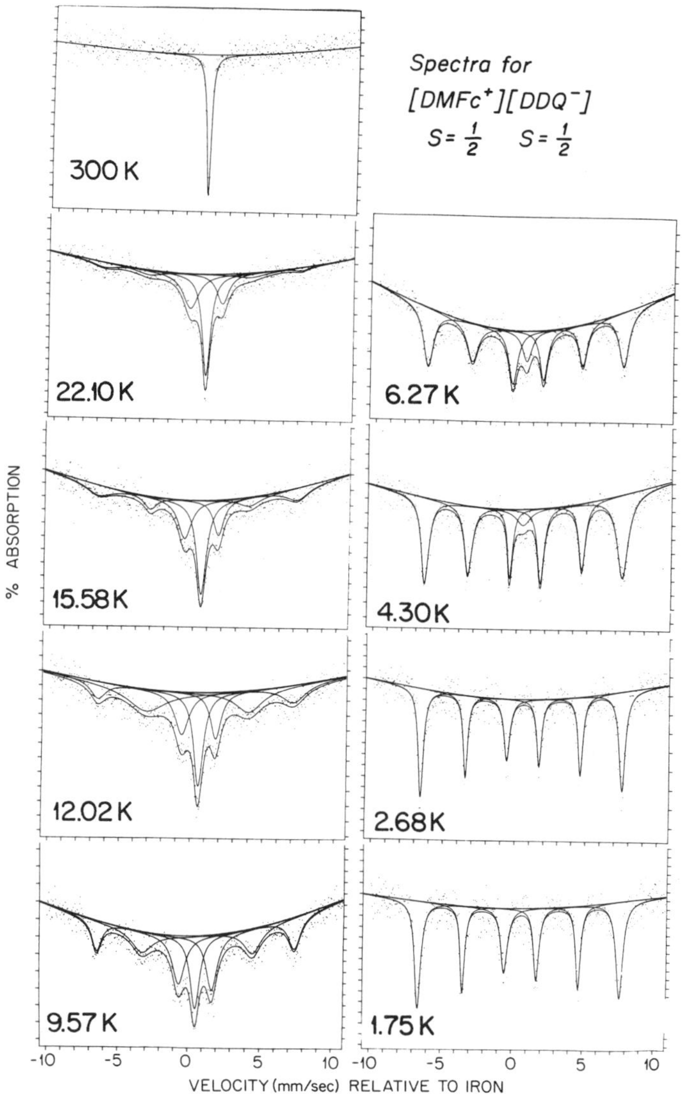

Figure 5. Evolution of hyperfine split Mössbauer spectra for $[Fe(Cp^*)_2]^+[DDQ]^-$ via slow paramagnetic relaxation.

in molecular orbital treatments[26] of the ferrocenium ion which predict essentially zero quadrupole interaction and at the same time allow for electron delocalization.

Actually, more recent susceptibility data[27] for $[Fe(Cp^*)_2]^+[DDQ]^-$ suggest that it either undergoes 3D ordering (probably as a low-field metamagnet) near ~3 K or exhibits a broad 1D maximum in χ_M at this temperature. In any

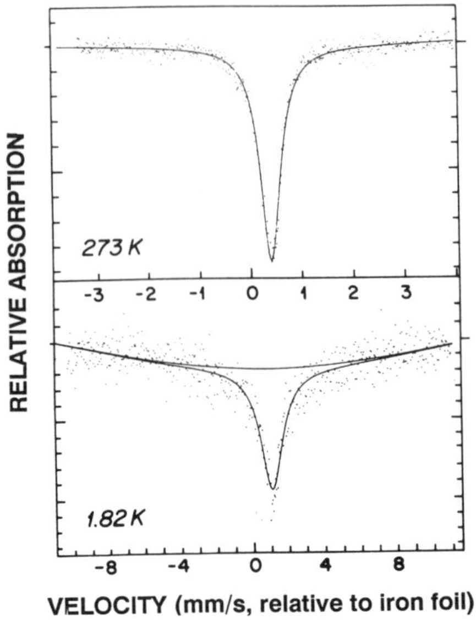

Figure 6. Zero-field Mössbauer spectra of $[Fe(Cp^*)_2]^+[C_3(CN)_5]^-$, a diamagnetic $S = 0$ anion system.

event the limiting low-temperature (zero-field) Mössbauer spectra of Fig. 5 cannot distinguish these possibilities.

4. LOW-TEMPERATURE COOPERATIVE MAGNETIC ORDERING

4.1. $K_3Fe(CN)_6$ (Ultralow Temperature)

Spin-doublet iron(III) is rare in nature simply because most naturally occurring ligands do not present large enough ligand fields to induce ferric ion electron spin pairing. In addition to this and the rarely observed slow relaxation of low-spin iron(III) mentioned earlier, cooperative ordering of such compounds is perhaps even more uncommon. In addition to those systems to be considered in detail herein, one of the few examples that comes to mind is again $K_3Fe(CN)_6$. Adiabatic demagnetization cryostat[28] and He-3,4-dilution refrigerator Mössbauer spectroscopy studies[29] show it to order antiferromagnetically as a neat (i.e., undiluted) material at 0.129 K with an internal field (H_n) of 19.4 T. This value is still far from that expected for an $S = \frac{1}{2}$ Fermi contact value, i.e., 11 T/spin. In this light, the "high" ordering temperatures and observed values of H_n (>40 T) for the present molecular magnets (*vide infra*) are extraordinary with the latter implying extremely large orbital contributions to H_{Int}. $K_3Fe(CN)_6$ is significantly different from the $Fe(Cp^*)_2^+$ systems considered herein in that it exhibits a substantial quadrupole splitting effect [$\Delta E(77\,\mathrm{K}) = 0.47\,\mathrm{mm/s}$, for the neat polycrystalline powder].[9] This distortion can measurably alter (likely partially quench) the ground-state orbital angular momentum of the Fe^{3+} ion and in turn the corresponding contributions of H_L and H_D to H_n. Hence, the very different values of H_n for $K_3Fe(CN)_6$ and $Fe(Cp^*)_2^+$ salts, i.e., ~20 T versus 40 T, are not unexpected.

4.2. Donor–Acceptor Charge Transfer-Type Molecular Magnets Based on the Decamethylferrocenium Ion $[Fe(Cp^*)_2]^+$

4.2.1. Ferromagnets

4.2.1.1. Soft Materials: $[Fe(Cp^*)_2^+[C_4(CN)_6]^-$. The change of the electron acceptor from DDQ to hexacyanobutadiene[30] appears to lead to a 3D-ferromagnetically ordered type II structure material (Fig. 2) with appropriately characteristic susceptibility, moment (field-dependent demagnetization effects, Fig. 8), and magnetization (rapid non-Brillouin rise in σ versus H, not shown). Some structure details are given in Fig. 7. The relatively sudden onset of Mössbauer spectra magnetic hyperfine splitting behavior (Fig. 9) suggests

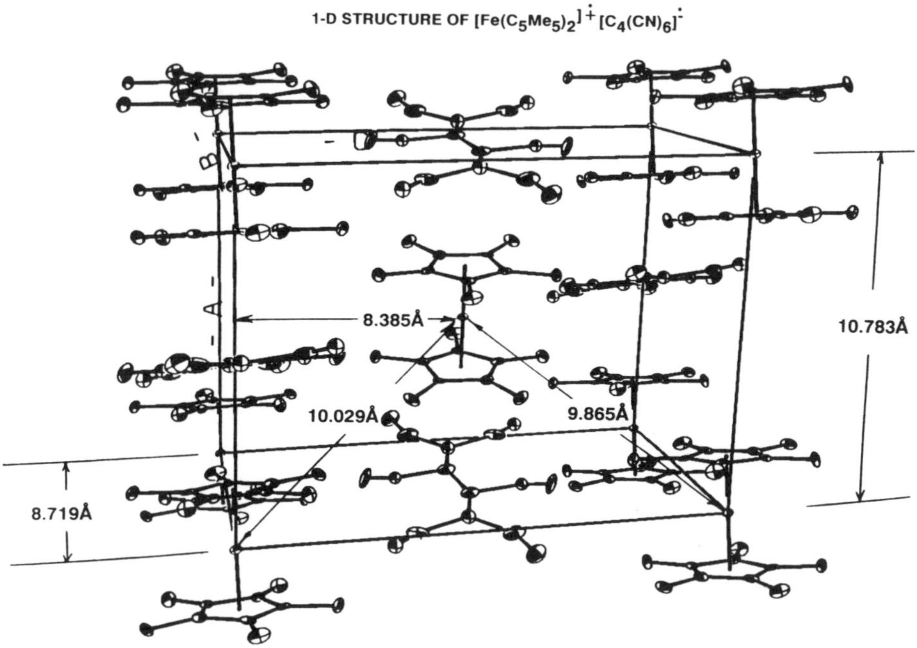

Figure 7. Packing structure and some important interatomic distances for the molecular ferromagnet $[Fe(Cp^*)_2]^+[C_4(CN)_6]^-$.

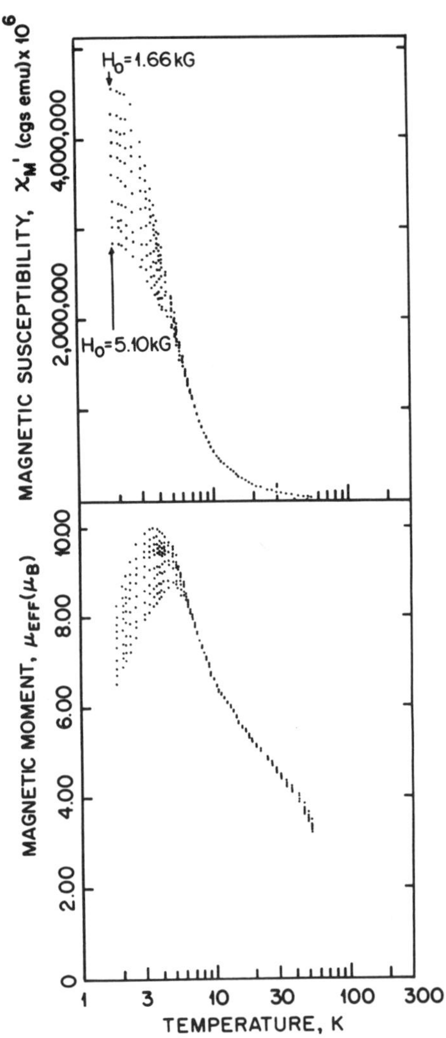

Figure 8. Field dependence of the molar susceptibility and effective magnetic moment of $[Fe(Cp^*)_2]^+[C_4(CN)_6]^-$ for ten values of applied field above and below T_c (7.5 ± 0.5 K).

$T_{Curie} \sim 8$ K, fully consistent with the susceptibility data for this system. The value of H_n at 1.68 K is 45 T. The field dependence of magnetization[31] for the material gives no evidence[27] of hysteresis effects, indicating that it is a soft ferromagnet.

4.2.1.2. Hard Materials: $[Fe(Cp^*)_2]^+[TCNE]^-$. The potential for the one-electron reduction of TCNE by DMFC is somewhat more negative than for $C_4(CN)_6$. Nevertheless, $[Fe(Cp^*)_2]^+[TCNE]^-$ can be isolated[1] and has a 1:1 type II structure and bulk ferromagnetic properties[32,33] generally similar to those of the hexacyanobutadiene compound whose packing structure is shown in Fig. 7. In contrast, however, hysteresis and spontaneous zero-field mag-

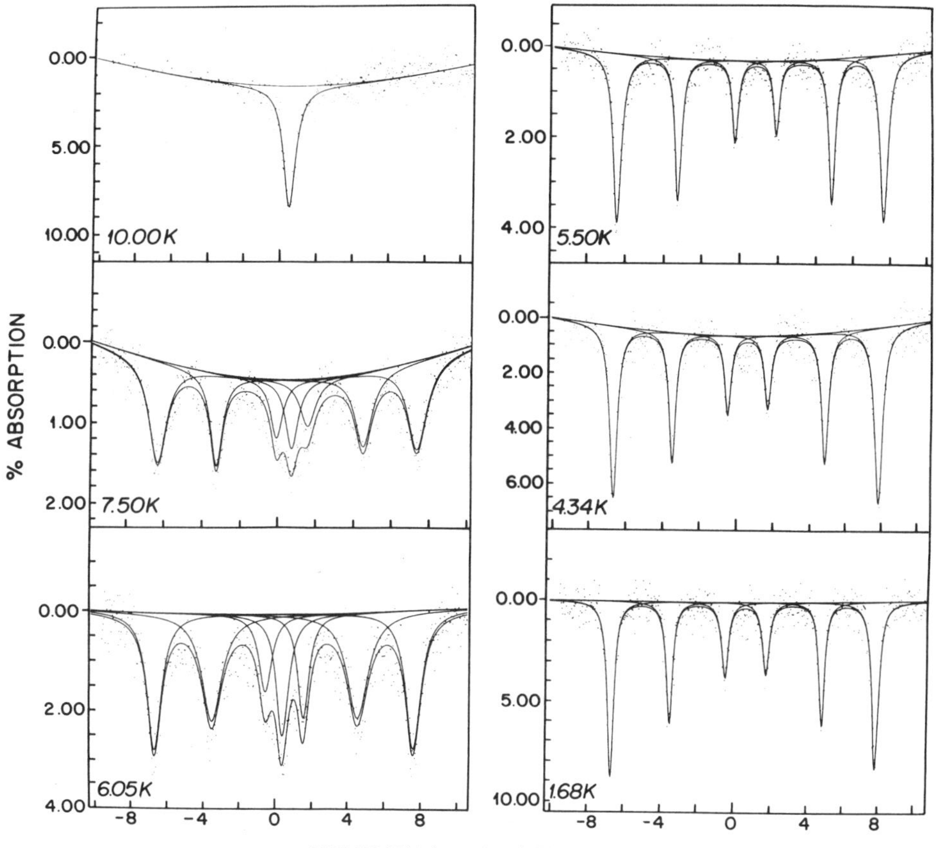

Figure 9. Zero-field Mössbauer spectra of $[Fe(Cp^*)_2]^+[C_4(CN)_6]^-$ in the vicinity of T_{Curie} (7.5 ± 0.5 K).

netization have been demonstrated[34] for the TCNE analogue (Fig. 10), confirming that it is a "hard" magnetic material. The origins of the difference in magnetic behavior for the $[C_4(CN)_6]^-$ and $[TCNE]^-$ salts, i.e., soft versus hard, are not clear at this time. The 3D-ferromagnetic ground state of this material is further elegantly demonstrated by its zero-field AC susceptibility[35] (Fig. 11) wherein the characteristic out-of-phase signal (χ'') of a ferromagnet is evident at 4.8 K. This value of T_{Curie} is in general agreement with the results of zero-field Mössbauer[1] and magnetic heat[2] capacity studies of $[Fe(Cp^*)_2]^-[TCNE]^-$.

As mentioned earlier, large orbital contributions to magnetic moments generally accompany highly anisotropic single-ion (paramagnetic state) and ordered-state behavior. One can elegantly probe the latter by the determination of Mössbauer spectra in external fields. Thus, for magnetically isotropic

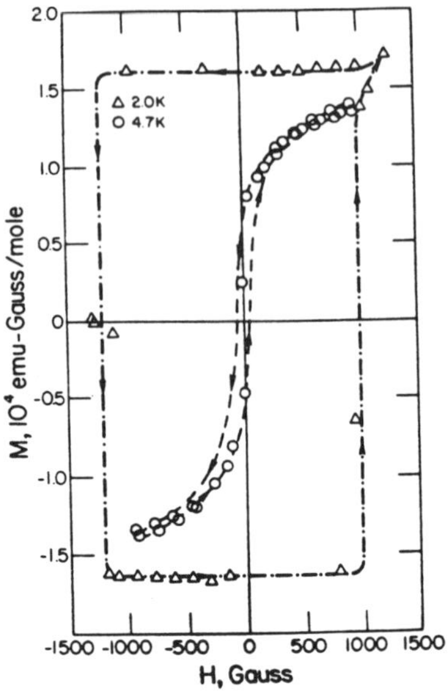

Figure 10. Magnetization hysteresis loops for $[Fe(Cp^*)_2]^+[TCNE]^-$ at 4.7 and 2.0 K.

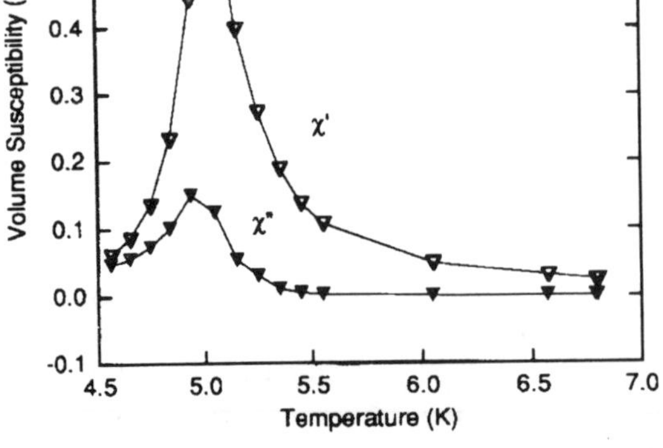

Figure 11. The temperature dependence of the real in-phase component (χ') and imaginary out-of-phase component (χ'') of the AC susceptibility of $[Fe(Cp^*)_2]^+[TCNE]^-$.

single-ion ground term-based systems [e.g., high-spin Fe(III), ^6A], the moments of a ferromagnetically ordered material are relatively easily polarized to the direction of an applied field. On the other hand, anisotropic systems (E or T irreducible representation single-ion ground states having orbital angular momentum) will resist such polarization. The Mössbauer spectra will reflect these extremes (even for polycrystalline samples) in terms of a *response* or *lack thereof* of the intensity pattern of the Zeeman split spectrum to an applied field. Specifically, for the case of a *ferromagnet* in longitudinally applied fields ($H_O \parallel E\gamma$), transitions two and five of a typical six-line Zeeman pattern (i.e., $\Delta M_I = 0$ transitions) will undergo diminution as the domains are polarized parallel to $E\gamma$. This is the result of the $\sin^2\theta$ dependence of their intensity where θ is the angle between $E\gamma$ and H_n and thus the axis of magnetization for the ordered state. For the ferromagnet α-iron, the moments of a thin foil are fully polarized by $H_O \approx 3\,T$[36] at which field the intensity of the $\Delta M_I = 0$ transitions has vanished. In strong contrast, we find[37] that the molecular ferromagnet $[DMFc]^+[TCNE]^-$ is highly anisotropic with little or no observable polarization to as high as $H_O = 9\,T$. A spectrum at $H_O = 7.5\,T$ is shown in Fig. 12. This is consistent with the highly anisotropic nature of this magnet. A remarkable feature of this spectrum that first becomes evident for $3\,T < H_O < 4.5\,T$ is the appearance of "doublets" at the extreme negative and positive velocity $\Delta M_I = \pm 1$ transitions. This is precisely the behavior predicted from theoretical simulations of applied-field Mössbauer spectra.[38–40] The

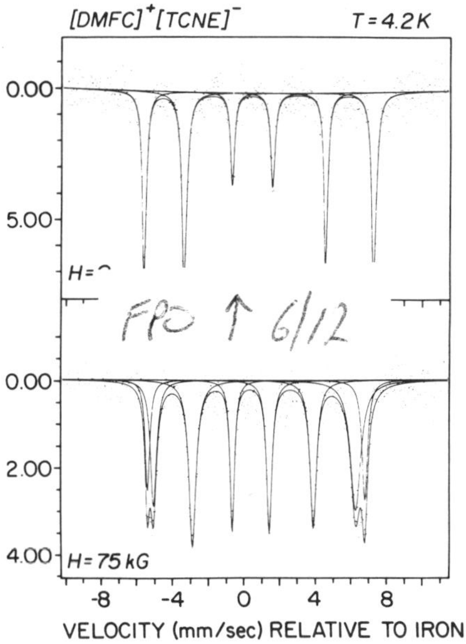

Figure 12. Zero-field and 7.5-T spectra of $[Fe(Cp^*)_2]^+[TCNE]^-$ at 4.2 K.

specific case is that of a random polycrystalline absorber whose anisotropy energy is so large that its magnetic structure is essentially unaffected by large external fields.

4.2.2. Antiferromagnets

4.2.2.1. Mixed Stack Polymers: $[Fe(Cp^*)_2]^+[TCNQ]^-$. Both of the preceding $TCNE^-$ and $C_4(CN)_6^-$ systems exhibit dominant 1D-ferromagnetic intrachain effects ($\theta \approx +30\,K$) along with 3D-ferromagnetic ground states. We now consider $[Fe(Cp^*)_2]^+[TCNQ]^-$, likewise a 1D ferromagnet along the $D^+A^-D^+A^-$ chains. However, this type II material has a 3*D antiferromagnetic* ground state ($T_{Néel} \approx 3\,K$) that exhibits a low-field (0.15 T) metamagnetic transition to a "paramagnetic" state.[41] In addition, an unusual magnetostructural transformation leading to inequivalent iron sites occurs[37,41] at ~2.5 K, i.e., just below the 3D-ordering temperature of ~3.0 K. This *zero-field* transition is apparent in the sample Mössbauer spectra of Fig. 13. The transformation is highly reproducible, fully reversible, and results in an intensity ratio for the "inequivalent" Fe(III) sites of ~2:1 with internal fields of 39.0 and 43.5 T for the more and less intense hyperfine patterns, respectively. It seems likely that the observed transformation is in some way the result of a soft-phonon mode along the D^+A^- chain direction. We *speculate* one possibility, i.e., the trimerization $\ldots D^+A^-D^+A^-D^+A^- \ldots \rightarrow \ldots \{D^+A^-D^+\}\{A^-D^-A^-\}\ldots$, which would result in 2:1 Fe(III) site inequivalence and different internal fields at these sites. Incidentally, an additional anomaly[27] is observed in the temperature dependence of the magnetic susceptibility (χ_M) coincidental with the aforementioned Mössbauer behavior. The full details of the behavior for this and related systems will be published elsewhere. Suffice it to say for now that low-field susceptibility studies[42,43] of the related $[Mn(Cp^*)_2]^+[DDQ]^-$ reveal similar anomalies in their low-field χ_M versus T behavior that have been correlated with spin-Peierls-like transitions.

In addition to the foregoing (kinetically favored) 1:1 metamagnetic phase of $[Fe(Cp^*)_2]^+[TCNQ]^-$, there is a thermodynamically favored rapidly relaxing dimeric phase,[41] $[Fe(Cp^*)_2]_2^+[TCNQ]_2^{2-}$, whose structure is shown in Fig. 14. The material corresponds to strongly antiferromagnetically coupled, effectively diamagnetic $(TCNQ)_2^{2-}$ dimers which separate the paramagnetic $[Fe(Cp^*)_2]^+$ units. This structure points to the crucial importance of A^- spin and paramagnetism as a necessary but not sufficient condition for the rich variety of magnetic behavior observed for $[Fe(Cp^*)_2]$ charge transfer salts. That is, with intervening diamagnetic A_2^{2-} units, the material is a simple (relatively uninteresting) rapidly relaxing paramagnetic as indicated by the spectra of Fig. 15.

4.2.2.2. Segregated Stack Polymers $[Fe(Cp^*)_2]^+[C_6(CN)_6]^-$ and $[Fe(Cp)_2]^+[TCNQF_4]^-$. There are two segragated chain (type I structure)

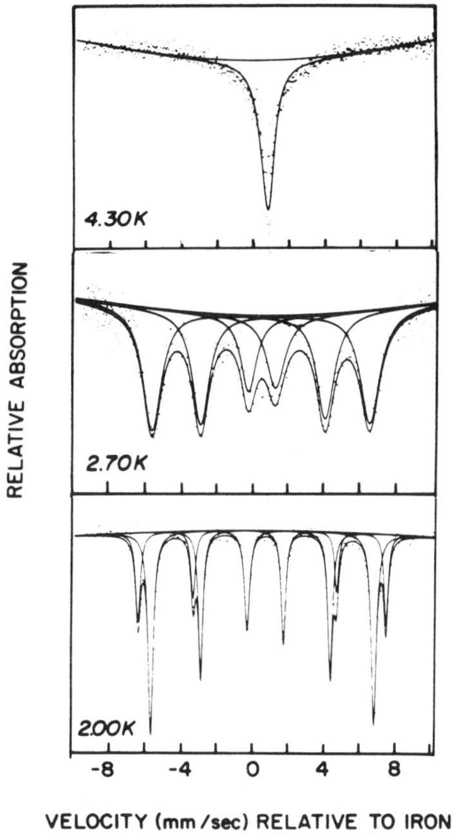

Figure 13. Sample Mössbauer spectra for the 1:1 linear chain form of $[Fe(Cp^*)_2]^+[TCNQ]^-$ above and just below T_N (~3.0 K) suggesting a spin-Peierls-type or energy gap transition.

charge transfer salts to be considered herein, one of $[Fe(Cp^*)_2]^+$ and one of $[Fe(Cp)_2]^+$. The structure of the former[44] and its low-field magnetic susceptibility are shown in Fig. 16. The anion chains are already diamagnetic (at ambient temperatures) singlet ground state by virtue of very strong antiferromagnetic exchange coupling along the chains. On the other hand, the cations of this system are paramagnetic and exhibit distinctly weaker antiferromagnetic exchange interactions along their respective chains leading to a (1D) maximum in χ_M versus T at ≈4 K for fields ≤50 G. However, for only slightly larger fields, ≈100 G, the maximum is largely suppressed and ultimately disappears with increasing field to give a high-moment state, as seen in Fig. 16.

Diamagnetic $[C_6(CN)_6]^-$ chains have been observed[45] in $[(CH_3)_4N]^+[C_6(CN)_6]^-$ which has short interplanar anion–anion separations (3.25 Å) similar to those in $[Fe(Cp^*)_2]^+[C_6(CN)_6]^+$. At 350 K, $[(CH_3)_4N]^+[C_6(CN)_6]^-$ undergoes a magnetostructural transition (likely spin-

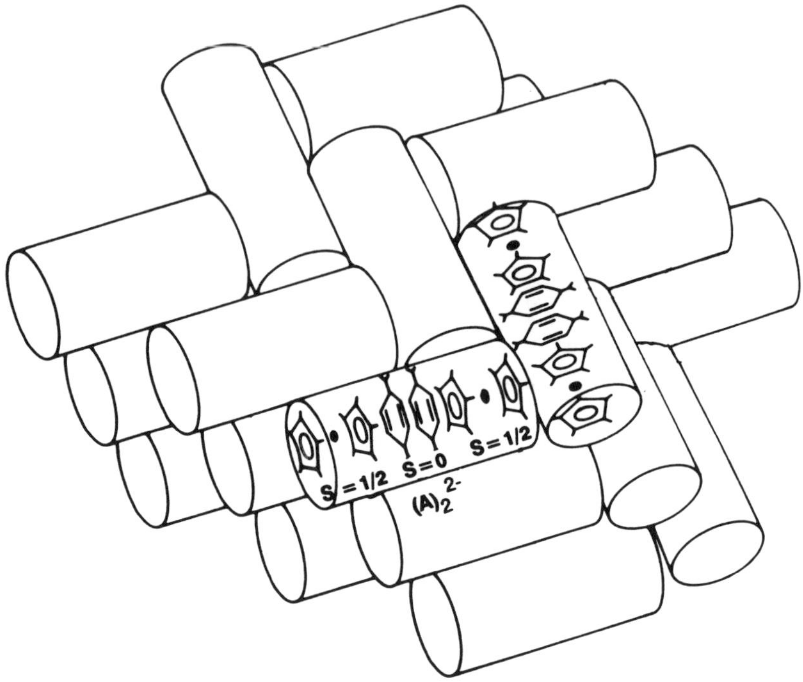

Figure 14. Schematic illustration of the crystal structure of the thermodynamically stable rapidly relaxing paramagnetic dimer form of $[Fe(Cp^*)_2]^+[TCNQ]^-$. Each cylinder represents a dimer. The ground state of the $(A)_2^{2-}$ dimer is $S_{total} = 0$.

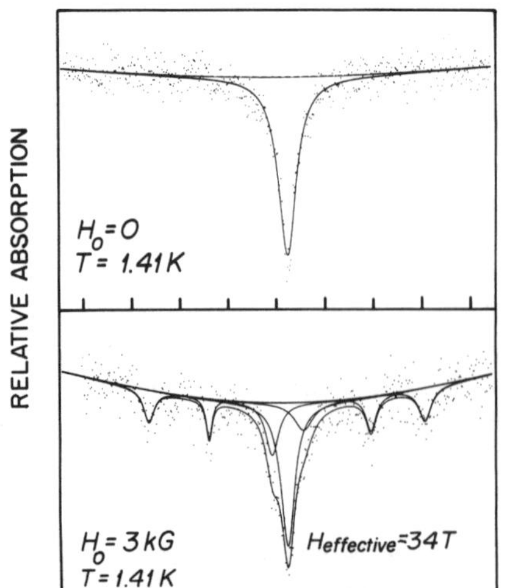

Figure 15. Zero- and small-applied-field (H_O = 0.3 T) Mössbauer spectra of the rapidly relaxing paramagnetic dimer form of $[Fe(Cp^*)_2]^+[TCNQ]^-$.

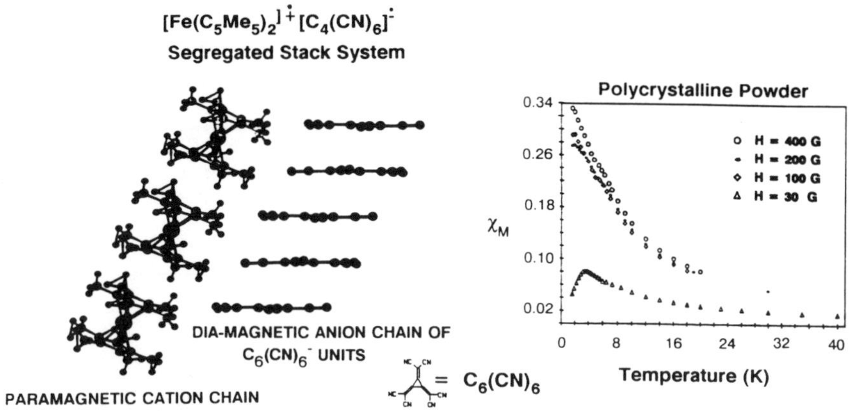

Figure 16. Schematic structure and temperature dependence of the molar susceptibility of monoclinic $[Fe(Cp^*)_2]^+[C_6(CN)_6]^-$.

Peierls) such that it exhibits strong antiferromagnetic exchange interaction above 315 K and is diagmagnetic at low temperatures. In this system, the spin on each anion is paired with the spin on an adjacent anion owing to short interplanar separation, which leads to a uniform diagmagnetic anion chain. Any paramagnetism for this system necessarily resides in the anion chains since the $[(CH_3)_4N]^+$ $S = 0$ cation units are diamagnetic.

$[Fe(C_5H_5)_2]^+[TCNQF_4]^-$ is the other example[46] considered herein that is comprised of segregated chains of $S = \frac{1}{2}$ cations and $S = \frac{1}{2}$ anions. Note that this system contains unmethylated ferrocene. Its short interplanar anion–anion separation (3.225 Å) results in the formation of chains of dimers, $[(TCNQF_4)_2]^{2-}$, and an $S_{tot} = 0$ ground state for anion chains. Unlike the $[C_6(CN)_6]^-$ system and for reasons not at all clear, the paramagnetic cation chains of this system exhibit neither slow paramagnetic relaxation nor magnetic exchange effects. In relation to the latter, it is conceivable that there is simply less delocalization of electron spin density to Cp ligands than to Cp* and thus J intrachain is smaller.

The zero-field ^{57}Fe Mössbauer spectra for monoclinic $[Fe(C_5Me_5)_2]^+\{C_3[C(CN)_2]_3\}^-$ as a function of temperature between 0.36 and 78 K are shown in Figs. 17 and 18. The zero-field Mössbauer spectrum at ambient temperature for a velocity sweep ±4 mm/s is fit to a singlet with isomer shift, δ, of 0.444 mm/s and linewidth, Γ, of 0.414 mm/s, values characteristic of the decamethylferrocenium cation. This narrow singlet varies little between 100 and 300 K. However, gradual and *unprecedentedly high-temperature* broadening and hyperfine splitting below 100 K are clearly evident. This gradual increase in hyperfine splitting with decreasing temperature is similar to that observed for the mixed stack linear chain complex

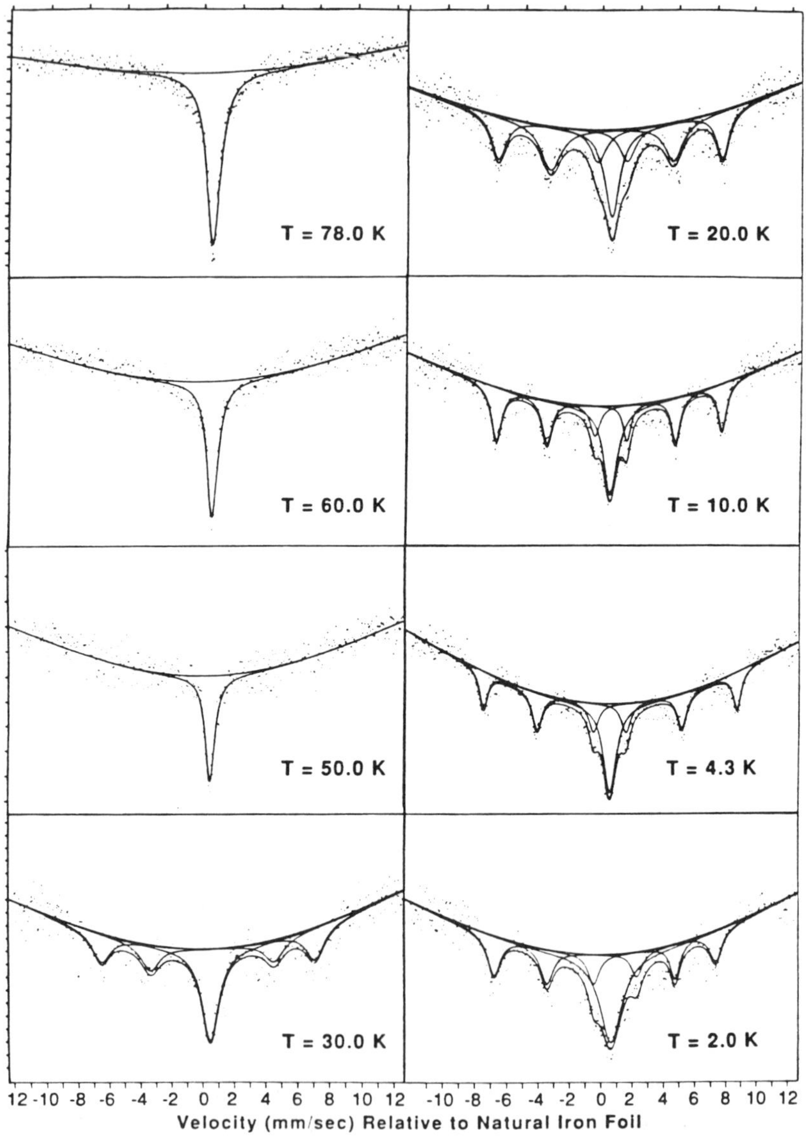

Figure 17. Evolution of magnetic hyperfine splitting via slow paramagnetic relaxation for $[Fe(Cp^*)_2]^+[C_6(CN)_6]^-$ in the range 78 to 2 K.

$[Fe(C_5Me_5)_2]^+[DDQ]^-$ below 25 K (Fig. 5), and is typical of slow paramagnetic relaxation as opposed to the onset of cooperative, extended 3D magnetic ordering. There is no evidence of 3D magnetic ordering or weak exchange from the susceptibility data for this temperature range. The slow paramagnetic relaxation observed for the present complex in the zero field is now attributed

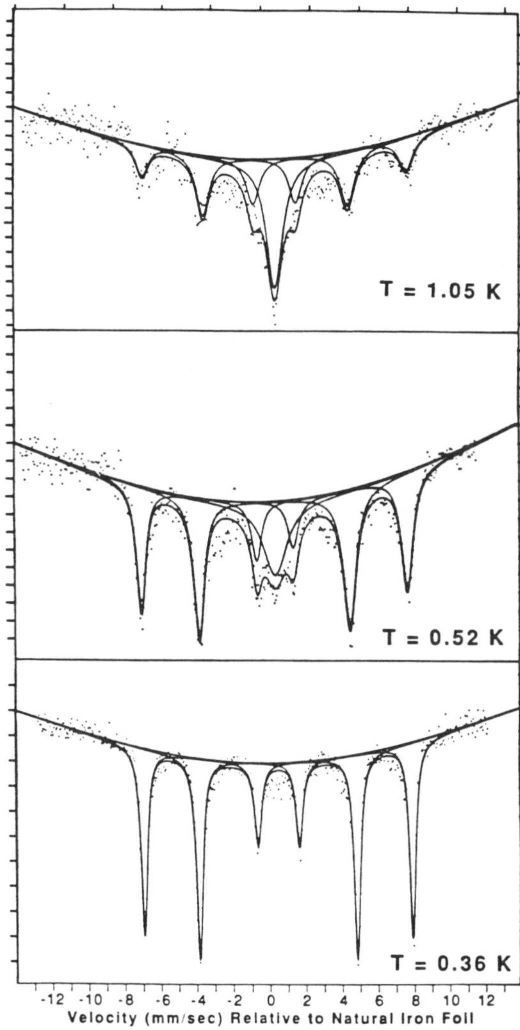

Figure 18. Vanishing of the rapidly relaxing paramagnetic phase and full resolution of hyperfine splitting for $[Fe(Cp^*)_2]^+[C_6(CN)_6]^-$.

in part to a small Zeeman splitting of ground-state Kramers doublet of the $[Fe(C_5Me_5)_2]^+$ cation by dipolar fields arising from electron spin on the neighboring cations. The Fe(III) cations are self-dilute with Fe–Fe separations > 8 Å and this leads to long spin–spin relaxation times. The decrease in temperature again leads to longer spin–lattice relaxation times as the lower-energy phonon states are populated. The effects along with the aforementioned Zeeman splitting induced via local dipolar fields result in unprecedented (relatively high temperature and in zero external applied field) slow relaxation in the $S = \frac{1}{2}$ ground doublet of low-spin Fe(III).

Remarkably the intensity from the central singlet corresponding to rapidly relaxing paramagnetic phase persists to ~0.5 K at which the hyperfine splitting is still not fully resolved. At 0.36 K the spectrum exhibits the fully resolved characteristic six-line Zeeman pattern with narrow linewidths (Fig. 18, bottom). *This limiting spectrum could arise from either slow paramagnetic relaxation combined with a 3D magnetic ordering* process as observed in the $[Fe(C_5Me_5)_2]^+[DDQ]^-$, or more likely only slow paramagnetic relaxation in the limit of infinitely long relaxation time between the components of its ground Kramers doublet. Zero-field Mössbauer spectroscopy studies cannot readily distinguish these two situations as the limiting zero-field, low-temperature spectra corresponding to 3D magnetic ordering processes are identical to those for slow relaxation. To clarify this situation, susceptibility measurements at these very low temperatures are necessary. The internal field at the limiting temperature of 0.32 K is 455 kG, which is comparable to that found for other $[Fe(Cp^*)_2]^+$ salts consisting of alternating cation–anion chains.

In a 5.2-kG applied field at 0.52 K, the spectrum is only slightly changed relative to $H_O = 0$ (Fig. 18, middle), and intensity from the central peak is still evident. This strongly suggests that a fraction of the material is still in a rapidly relaxing paramagnetic state. In a transverse applied field of 15.1 kG at 0.52 K (Fig. 19), the material is nearly fully polarized such that it exhibits a hyperfine splitting intensity pattern of 2.2:2.6:1:1:2.6:2.2, i.e., approaching intensity pattern of 3:4:1:1:4:3. This latter is precisely the intensity profile that one expects to induce via a transverse applied field ($H_O \perp E\gamma$) for either a paramagnet (that has been induced to slowly relax by an external field) or a ferromagnetically ordered species. The latter ground state is, however, clearly ruled out by the magnetic susceptibility that shows that the dominant magnetic interaction along the cation chain is antiferromagnetic, i.e., the paramagnetic Curie temperature, θ, is negative.

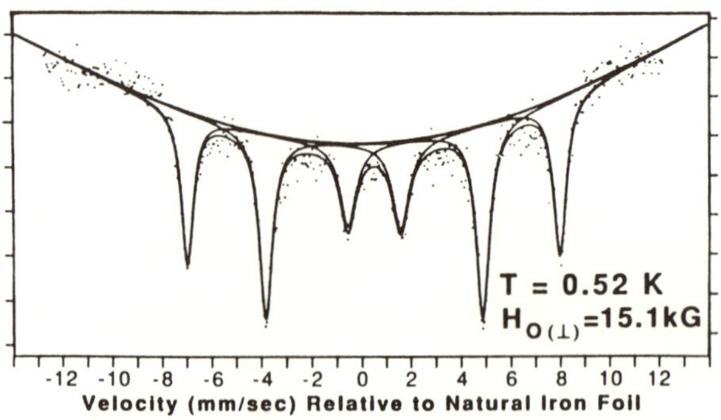

Figure 19. Mössbauer spectrum of $[Fe(Cp^*)_2]^+[C_6(CN)_6]^-$ in a transverse field of 1.51 T at 0.52 K.

4.2.3. Effects of Systematic Change of Charge Donor

To date, the effects of variation of donor [within the context of low-spin iron(III) chemistry] have been disappointing. The 1:1 charge transfer salts of both $[Fe(C_5Et_5)_2]$[47] and $[Fe(C_5Me_4H)_2]$[48] with TCNE and TCNQ have the desirable mixed stack chain (type II, Fig. 2) structure that is apparently requisite for interesting low-temperature magnetism. However, none of these materials exhibits[48] slow paramagnetic relaxation or extended magnetic order on a Mössbauer spectroscopy time scale for temperatures as low as 1.5 K. The ethyl system shows the classical characteristic sharp singlet typical of ferrocenium ions. Interestingly, the octamethyl compound $[Fe(C_5Me_4H)_2]^+[TCNE]^-$ exhibits a small but well-resolved quadrupole splitting effect (Fig. 20) reflecting the lower symmetry (C_{2h}) versus nominal D_{5h} for $Fe(Cp^*)_2$, i.e., $Fe(C_5Me_5)_2$. The field dependence of χ_M for $[Fe(C_5Me_4H)_2]^+[TCNE]^-$ exhibits behavior characteristic of either metamagnetism or ferromagnetism for $T \leq 6$ K, suggesting the possibility that magnetic hyperfine splitting of its Mössbauer spectra may occur below 1.5 K. However, it is also possible that such splitting is not evident and will not occur owing to fortuitous cancellation of the various components (H_F, H_L, H_D) to H_n. Precedence[40] for this unusual situation arises in the example of anhydrous ferrous chloride, $FeCl_2$, whose susceptibility, magnetization, and neutron diffraction clearly indicate that it is a 3D ordered metamagnet[12] ($T_{Néel} \approx 24$ K) and yet little or no hyperfine splitting of its Mössbauer spectra is evident at 4.2 K.

4.2.4. Effects of Systematic Change of Charge Acceptor

This kind of study was performed by detailed investigation[50] of the series $[Fe(Cp^*)_2]^+[TCNQR_2]^-$ where R is a Cl, Br, I, Me, OMe, or OPh 2,5-substituent on the TCNQ acceptor. These materials are again 1:1 type II mixed stack linear chains. Unfortunately, none of the systems gives clear evidence of

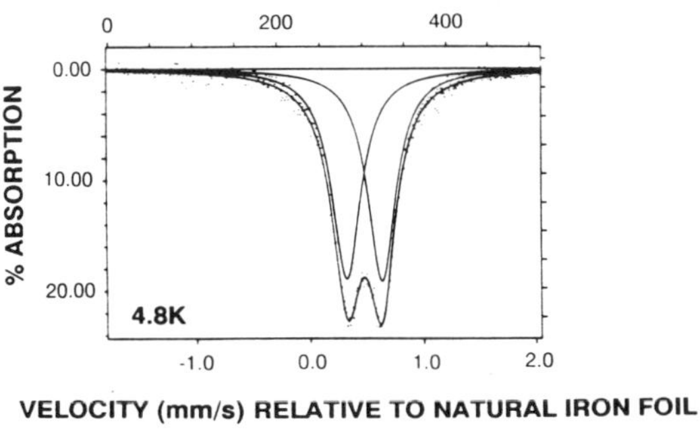

Figure 20. Zero-field Mössbauer spectrum of octamethylferrocenium tetracyanoethylenide $[Fe(C_5Me_4H)_2]^+[TCNE]^-$ at 4.8 K.

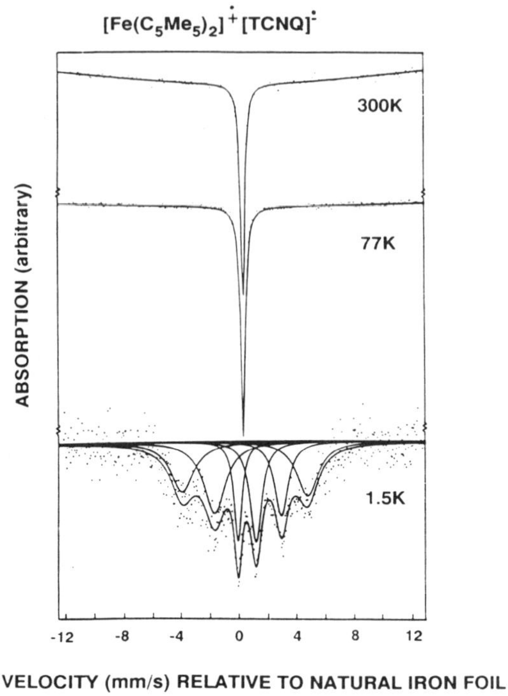

Figure 21. Some zero-field Mössbauer spectra for $[Fe(Cp^*)_2]^+[TCNQI_2]^-$ indicating incipient magnetic order below ~2 K.

interesting cooperative magnetic behaviour except for $[Fe(Cp^*)_2]^+[TCNQI_2]^-$, which exhibits incipient magnetic hyperfine splitting at 1.5 K (Fig. 21) and may fully order in the helium-3 cryogenics range.

4.2.5. Ferromagnetically Coupled Dimers

We conclude with observations[14,51,52] for some new 2:1 cation:polycyanide dianion phases whose charge transfer polymer structure is shown in Fig. 22 for the example of $[Fe(Cp^*)_2]_2^+[C_4(CN)_6]^{2-}$. The salt contains a dianionic polycyanide species. The temperature dependence of its Mössbauer spectra is clearly unprecedented and to our knowledge is the first example in which hyperfine splitting grows, reaches a maximum intensity, and then, with further decreasing temperature, totally vanishes. $[Fe(Cp^*)_2]_2^+[C_4(CN)_6]^{2-}$, as pictured in Fig. 22, corresponds to parallel chains of *pairs* of $[Fe(Cp^*)_2]^+$ cations alternating with single $C_4(CN)_6^{2-}$ dianions (type III, Fig. 2). Its characteristic *singlet* Fe-57 Mössbauer spectrum at 300 K (Fig. 24, top) *incontrovertibly* confirms the fact that all iron is present as low-spin Fe(III), i.e., decamethylferrocenium cations [ferrocene and decamethylferrocene formally contain iron(II) and always exhibit well-resolved quadrupole *doublet* spectra $\Delta E > 2$ mm/s]. Thus, from

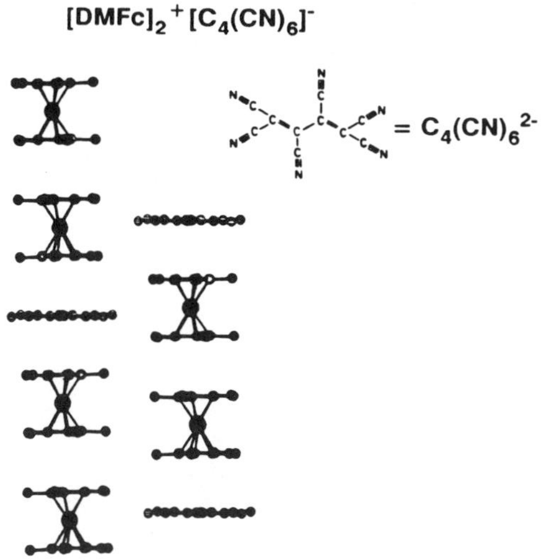

Figure 22. Schematic structure of the 2:1 charge transfer polymer $[Fe(Cp^*)_2]_2^+[C_4(CN)_6]^{2-}$.

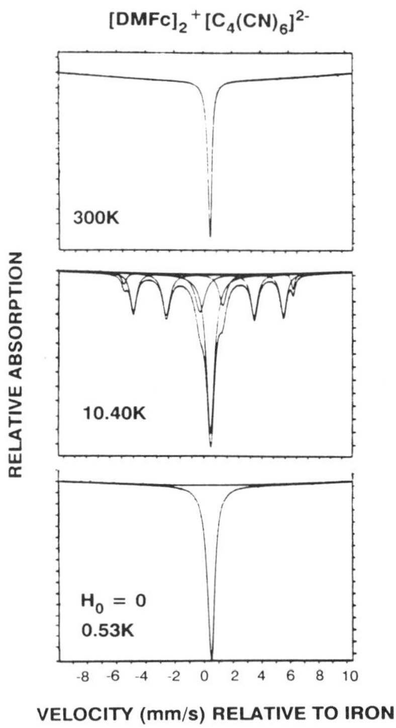

Figure 23. Sample Mössbauer spectra showing the growth and vanishing of hyperfine splitting for $[Fe(Cp^*)_2]_2^+[C_4(CN)_6]^{2-}$.

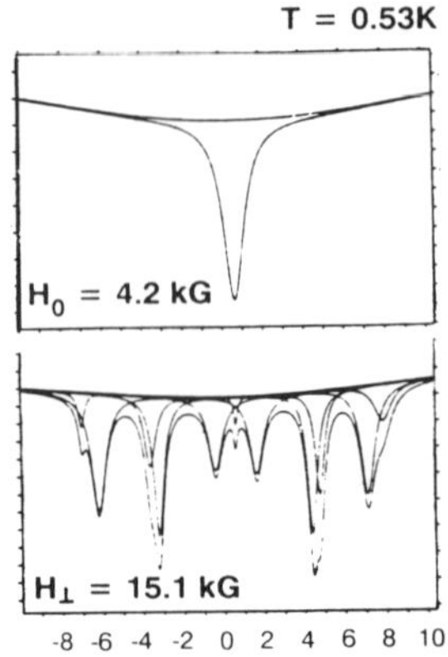

Figure 24. Some transverse applied-field Mössbauer spectra of $[Fe(Cp^*)_2]^+[C_4(CN)_6]^{2-}$.

electrical neutrality considerations and the 2:1 stoichiometry confirmed by X-ray structure determinations, the hexacyanobutadiene moiety must be present as a dianionic species, $C_4(CN)_6^{2-}$.

The Mössbauer spectra that we now present indicate unusual dynamic behavior and the presence of a novel triplet species in $[Fe(Cp^*)_2]_2^+$-$[C_4(CN)_6]^{2-}$ for *ferromagnetically coupled dimeric cation pairs*. Sample spectra are shown in Figs. 23 and 24. On gradual cooling, slow paramagnetic relaxation broadening and hyperfine splitting develop with maximum intensity and resolution for the hyperfine background occurring at ~10 K. However, and to our surprise, below this temperature the strong central singlet corresponding to the rapidly relaxing paramagnetic phase begins to *grow* again until a singlet reminiscent of 300 K is all that remains at 0.53 K. Closer examination of the spectra (not shown) in the vicinity of ~25 K shows the emergence of two hyperfine patterns. The multiple Zeeman splitting patterns are observed as the result of highly anisotropic magnetic hyperfine interactions. These interactions undoubtedly arise from the large g-factor anisotropy of the (single) low-spin Fe(III) ions and the strongly anisotropic (ferro) magnetic exchange between them. We emphasize that the observed behavior has been reproduced and is entirely reversible. In addition, it is also found for other dianion species such as $C_6(CN)_6^{2-}$ and *iso*-$C_4(CN)_6^{2-}$ (Fig. 1) as well as the $S = 0$ Co(III) partially diluted analogues. The "inequivalence" (not resolved in the *broadened* singlet at 0.53 K; compare the tops of Figs. 23 and 24) is clearly resolved again in a *transverse applied field* spectrum at 0.53 K as shown in Fig. 24 bottom. The limiting gross intensity pattern (H_O = 15.1 kG) is exactly as predicted for a *rapidly relaxing paramagnet* or a *diamagnet* whose sites have been induced to slowly relax by the applied field and polarized so that their magnetization is perpendicular to the direction of gamma-ray propagation. The susceptibility of the present 2:1 system has been measured at 30, 300, and 3000 G with the same result, "*simple*" *paramagnetism* but very non-Curie–Weiss behavior. We observe *no maximum* in the temperature dependence of the magnetic susceptibility for any value of applied field to as low as 30 G and temperatures as low as 1.4 K.

The explanation of our observations for the 2:1 cation:dianion phases is suggested by the structure of these systems, i.e., one containing *cation pairs* (Fig. 22). At sufficiently low temperatures, such pairs can act as magnetic exchange-coupled *dimers* (very anisotropic along the cationic dimer axis). The "magnetic" literature is replete with such $S = \frac{1}{2}$, $S = \frac{1}{2}$ dimers, particularly $S_{\text{total}} = 0$ (singlet) ground-state *a*nti*f*erromagnetically coupled dimers of Cu(II). For such AF dimers, a *maximum* in the temperature dependence of the magnetic susceptibility (χ_m) is typically observed (for samples free of paramagnetic impurities) and is theoretically predicted.[12] The exchange interaction J is negative with the singlet–triplet separation being $2J$. There are, however, well-characterized examples of $S_{\text{total}} = 1$ ground-state ($J > 0$) ferromagnetically coupled dimers for which a simplified (spin only) energy level diagram is given in Fig. 25. This is undoubtedly the case for the present $[Fe(Cp^*)_2]_2^+$ pairs. Such a dimer's $S_{\text{total}} = 1$ (nominal) ground state can be (positive) zero field split to an excited $m_{s(\text{total})} = \pm 1$ and a true $m_{s(\text{total})} = 0$ ground state, as shown on the right side of Fig. 25. Large zero-field splittings of this type can result from the highly anisotropic magnetic exchange operative in these systems. Decreasing the temperature for such a system would then *initially* lead to population of the *slowly relaxing* $m_{s(\text{total})} = \pm 1$ (since $\Delta m_{s(\text{total})} = \pm 2$ electronic transitions are highly forbidden) and hyperfine splitting via slow (intra- and inter-) dimer paramagnetic relaxation. Ultimately, exclusive population of the *nonmagnetic* $m_{s(\text{total})} = 0$ *ground state* would occur at very low T. Concomitant with the latter would be the vanishing of all hyperfine splitting effects as observed. *Finally, and consistent with our observations, J positive dimers* are *not expected*[12] *to exhibit a maximum in* χ_m. Good fits to the variation of the effective magnetic moment μ versus T for the combined effects of ferromagnetic intradimer exchange and zero-field splitting are obtained for

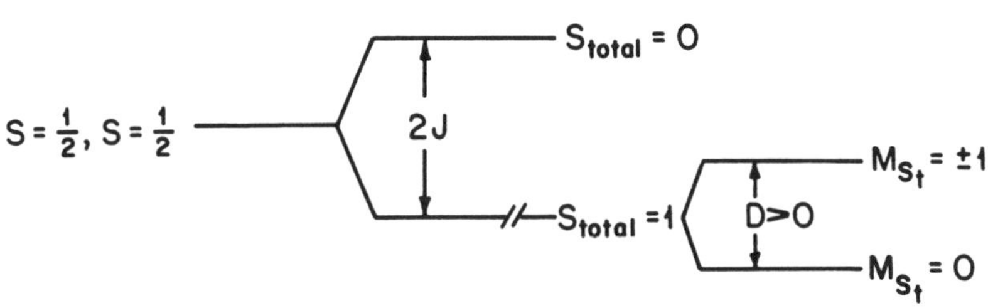

Figure 25. Energy level diagram for a ferromagnetically coupled ($J > 0$) dimer including positive axial zero-field splitting ($D > 0$) of the triplet state, to give a non-magnetic $M_{\text{st}} = 0$ ground state.

$2J$ and D values of $+180\ \mathrm{cm}^{-1}$ and $+6\ \mathrm{cm}^{-1}$, respectively, where D positive is clearly confirmed by the Mössbauer data, i.e., a nonmagnetic, nonhyperfine split ground state. A value of $D = 6\ \mathrm{cm}^{-1}$ is equivalent to $\sim 9\ \mathrm{K}$. This is consistent with the observed temperature, $\sim 10\ \mathrm{K}$, of *maximum intensity* for the hyperfine background that eventually vanishes.

5. NONLINEAR LOW-DIMENSIONALITY EFFECTS—SOLITONS

It is clear that structurally induced magnetic one-dimensionality is a hallmark of many of the systems discussed herein. This effect reinforces the already significant anisotropic single-ion magnetic behavior typical of low-spin Fe(III) (*vide supra*), particularly ferrocenium ions, and can potentially lead to highly anisotropic 1D and 3D exchange behaviors intermediate to the Heisenberg and Ising extremes. It is also known that the motions of domain walls (kinks) in antiferro- or ferromagnetic chains obey the sine-Gordon equation and thus can be viewed as soliton excitations.[53–55] The spin flips accompanying domain wall boundary movement in a highly 1D magnet can lead to reversals of the internal hyperfine field at Mössbauer active nuclei with observable linewidth broadening effects.[56] Their observation depends on the rate of soliton passage versus the reciprocal of the excited-state lifetime of the Mössbauer effect nuclide. Such soliton broadening effects have in fact been observed in detailed linewidth analysis[57] (just above and below T_{critical}) for several chain magnets based on high-spin Fe(II). In these systems the spin resides exclusively on the ferrous ions.

It seems possible that the broadening effects observed for some of the compounds discussed here are soliton-related rather than solely single-ion or superparamagnetic relaxation effects. In any event, this is certainly a rich area for further physics study of donor–acceptor chain magnetic materials. Such investigations are expected to be more complicated for the low-spin Fe(III) systems because of larger, highly sensitive orbital contributions to the internal field and the fact that both the *anion* and the cation of the donor–acceptor systems possess spin. In this situation, simple dilution experiments with diamagnetic $S = 0$ species such as decamethylcobalticenium may not clearly distinguish single-ion versus cooperative soliton effects.

6. CONCLUSIONS AND PROSPECTS FOR FURTHER RESEARCH

It is clear that the charge transfer salts of decamethylferrocene have provided a rich area for significant new chemical and physical research. I conclude this chapter by briefly outlining what I believe to be some novel, nontrivial, logical extensions of this effort and that in part have already been initiated in our laboratory.

6.1. Charge Transfer Salt Magnets Based on Completely Open and Half-Open Pentadienyl Iron Systems

The noncyclic or open pentadienyl system, C_5H_8, can be stabilized in η^5-bonded metallocenes as the 2,4-di-CH_3 substituted version in both open and half-open ferrocene-like complexes, $[Fe(\eta^5\text{-}2,4\text{-}C_7H_{11})_2]$ and $[Fe(\eta^5\text{-}C_5H_5)(\eta^5\text{-}2,4\text{-}C_7H_{11})]$, respectively (Fig. 26). The pioneering chemistry in this area was initiated by Ernst and co-workers.[58] While the completely open and half-open systems appear to oxidize more readily than the completely closed [ordinary ferrocene-$Fe(\eta^5\text{-}C_5H_5)_2$] analogue, matters are apparently complicated by pentadienyl–pentadienyl coupling reactions for the radical cation forms in solution. However, stable charge transfer linear chains (of any of the structure types shown in Fig. 2) may result under preparative conditions leading to rapid precipitation of the solid. We are proceeding with investigations in this area with this hope in mind.

6.2. Charge Transfer Salt Magnets Based on Fully Detrapped Valence Averaged Binuclear Aggregates of Iron

We have recently shown[59,60] that the monocationic forms of certain *fused* hydrocarbon, μ-bridging ligand, binuclear iron systems are fully valence detrapped to as low as 1.5 K. They contain bridging pentalene or symmetric indacene units and can be formally viewed as corresponding to $Fe^{2.5+}$ as for example in $[(Cp^*Fe)_2\text{pentalene}]^+$ (Fig. 27). We have recently discovered that the reaction of the neutral $[(Cp^*Fe)_2\text{pentalene}]$ with some of the acceptors of Fig. 1 goes smoothly, potentially leading to type I (segregated) or type II

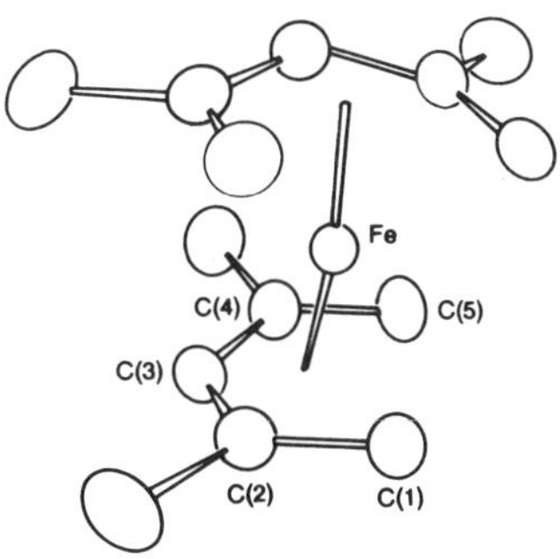

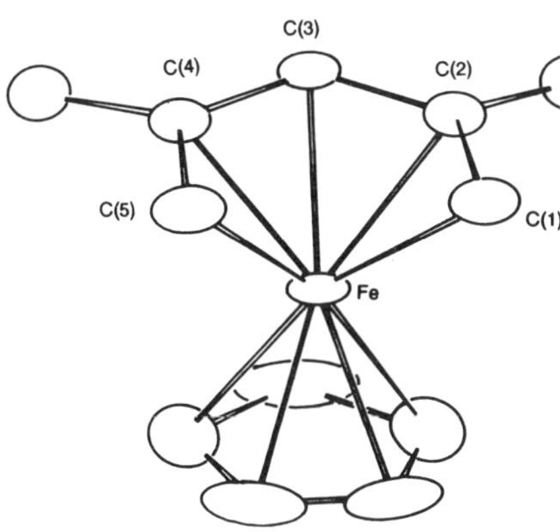

Figure 26. Fully and half-open η^5-bonded pentadienyl complexes of iron.

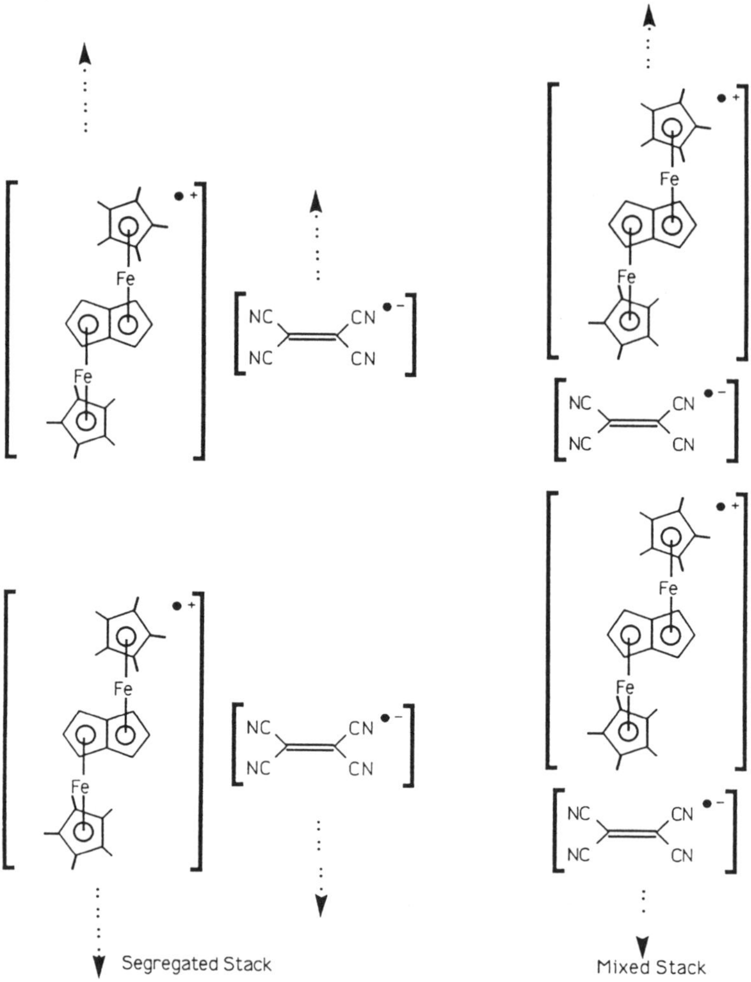

Figure 27. The $[(Cp^*Fe)_2pentalene]$ dimer and its mixed and segregated stack charge transfer salt chain polymers with the electron acceptor TCNE.

(mixed stack) chains as pictured in Fig. 27. The novel aspect here is that one of the basic building blocks of the system is now a fully delocalized valence averaged binuclear unit. The presence of such units may lead to novel magnetic properties for type II structures and both interesting magnetism and electron transport behavior for the segregated type I systems.

6.3. Charge Transfer Salt Magnets of Ferrocene with Halogenated Buckminsterfullerene (C_{60})

We have recently initiated a program of synthetic chemistry with the goal of preparation of new molecular magnetic materials based on unsubstituted ferrocene charge transfer salts of C_{60}. We are encouraged in this effort through the findings of Wudl *et al.*[61,62] who have prepared the 1:1 phase, $\{TDAE\}^+C_{60}^-$. TDAE is the very strong organic electron donor tetrakis-dimethyl aminoethylene. $\{TDAE\}^+C_{60}^-$ orders as a soft ferromagnet at 16.1 K. We find that ordinary ferrocene is simply not "strong enough" to reduce C_{60}. However, our work indicates that halogenated versions of C_{60} undergo facile reduction leading to ferrocenium charge transfer salts. In particular, we have found that the reaction of ferrocene with $C_{60}Br_{24}$[63] (an acceptor whose structure has recently been determined) leads to a new magnetic material whose Mössbauer spectra show 3D ordering at ~14 K. We are actively pursuing investigations in this fast-moving area.

ACKNOWLEDGMENTS. The research reviewed herein has extended over the period ranging from 1986 to the present. At various times it has received the support of the NSF Division of Materials Research Solid State Chemistry Program. More recently, support has come through PRF type AC-funding (Grant ACS-PRF 2436-AC3) and the NSF Division of Materials Research Instrumentation Program in the form of funding for an automated AC-susceptometer and AC–DC resistivity equipment. Much of this effort corresponds to a continually fruitful collaboration with Dr. Joel S. Miller and his colleagues at E. I. Dupont de Nemours & Co., Central Research and Development at the Experimental Station in Wilmington, Delaware. Finally, the author is pleased to acknowledge his debt of gratitude to Dr. Jian Hua Zhang for his devoted efforts in this research area.

REFERENCES

1. J. S. Miller, J. C. Calabrese, H. Rommelmann, S. R. Chittipeddi, J. H. Zhang, W. M. Reiff, and A. J. Epstein, *J. Am. Chem. Soc.* **109**, 768 (1987).
2. M. Nakano and M. Sorai, *Chem. Phys. Lett.* **169**, 27 (1990).
3. J. S. Miller and D. Dougherty (eds.), *Proceedings of the International Symposium on High Spin Molecular Based Materials, Mol. Cryst. Liq. Cryst.* **176** (1989).
4. J. S. Griffith, *Nature* **180**, 30 (1957).
5. J. S. Griffith, *Mol. Phys.* **21**, 135 (1971).
6. B. N. Figgis, *Trans. Faraday Soc.* **56**, 1553 (1960).
7. R. M. Golding, *Applied Wave Mechanics,* Van Nostrand, Princeton, N.J. (1969).
8. R. M. Golding, *Mol. Phys.* **12**, 13 (1967).
9. W. T. Oosterhuis and G. Lang, *Phys. Rev.* **178**, 439 (1969).
10. G. Lang and W. T. Oosterhuis, *J. Chem. Phys.* **51**, 3608 (1969).
11. B. N. Figgis, *Introduction to Ligand Fields,* Interscience, New York (1966).
12. R. L. Carlin, *Magnetochemistry,* Springer-Verlag, Berlin (1986).
13. N. N. Greenwood and T. C. Gibb, *Mössbauer Spectroscopy,* Chapman & Hall, London (1971).
14. W. M. Reiff, *Mol. Cryst. Liq. Cryst.* **176**, 391 (1989).
15. J. W. G. Wignall, *J. Chem. Phys.* **44**, 2462 (1966).

16. R. Prins, *Mol. Phys.* **19**, 603 (1970).
17. A. H. Maki and T. E. Berry, *J. Am. Chem. Soc.* **87**, 4437 (1965).
18. K. S. Back, P. G. Debrunner, Y. M. Xia, F. J. Wu, and D. M. Kurtz, *Hyp. Interact.* **54**, 447 (1990).
19. M. J. Cohn, M. D. Timken, and D. N. Hendrickson, *J. Am. Chem. Soc.* **106**, 6683 (1984).
20. J. B. Torrance, J. E. Vazquez, J. J. Mayerle, and V. X. Lee, *Phys. Rev. Lett.* **46**, 253 (1981).
21. J. B. Torrance, A. Girlando, J. J. Mayerle, J. I. Crowley, V. Y. Lee, P. Batail and S. J. LaPlaca, *Phys. Rev. lett.* **47**, 1747 (1981).
22. E. Adman, M. Rosenblum, S. Sullivan, and T. N. Margulis, *J. Am. Chem. Soc.* **89**, 4540 (1967).
23. J. S. Miller, P. J. Krusic, D. A. Dixon, W. M. Reiff, J. H. Zhang, E. C. Anderson, and A. J. Epstein, *J. Am. Chem. Soc.* **108**, 4459 (1986).
24. D. N. Hendrickson, Y. S. Sohn, and H. B. Gray, *Inorg. Chem.* **10**, 1559 (1971).
25. D. M. Duggan and D. N. Hendrickson, *Inorg. Chem.* **14**, 955 (1975).
26. R. L. Collins, *J. Chem. Phys.* **42**, 1072 (1965).
27. W. M. Reiff and J. S. Miller, unpublished results.
28. M. Shinohara, A. Ishigaki, and K. Ono, *Jpn. J. Appl. Phys.* **7**, 170 (1968).
29. J. L. Groves, A. J. Becker, L. M. Chirovsky, W. P. Lee, G. W. Wang, and C. S. Wu, *Hyp. Interact.* **4**, 930 (1978).
30. J. S. Miller, J. H. Zhang, and W. M. Reiff, *J. Am. Chem. Soc.* **109**, 4584 (1987).
31. J. S. Miller and W. M. Reiff, unpublished results.
32. J. S. Miller, J. C. Calabrese, A. J. Epstein, R. W. Bigelow, J. H. Zhang, and W. M. Reiff, *J. Chem. Soc. Chem. Commun.* **1986**, 1026 (1986).
33. J. S. Miller, A. J. Epstein, and W. M. Reiff, *Acc. Chem. Res.* **21**, 114 (1988).
34. J. S. Miller, A. J. Epstein, and W. M. Reiff, *Chem. Rev.* **88**, 201 (1988).
35. W. M. Reiff, EMU 1, No. 1, 1 (1992), Lakeshore Cryotronics Inc., Westerville, Ohio.
36. N. Blum, S. Foner, and R. B. Frankel, *Phys. Rev.* **181**, 863 (1969).
37. W. M. Reiff, *J. Appl. Phys.* **63**, 2957 (1988).
38. Y. Calage, M. Zemirli, J. M. Grenèche, F. Varret, R. DePape, and G. Ferey, *J. Solid State Chem.* **69**, 197 (1987).
39. F. Varret, *Proceedings of the International Conference on the Applications of the Mössbauer Effect,* Jaipur, India, p. 129 (1981).
40. J. Chappert, J. Teillet, and F. Varret, *J. Magn. Magn. Mater.* **11**, 100 (1979).
41. J. S. Miller, J. H. Zhang, W. M. Reiff, A. D. Dixon, L. D. Preston, A. H. Reis, H. Gebert, M. Extine, J. Troup, A. J. Epstein, and M. D. Ward, *J. Phys. Chem.* **91**, 4344 (1987).
42. J. S. Miller, R. S. McLean, C. Vazquez, G. T. Yee, K. S. Narayan, and A. J. Epstein, *J. Mater. Chem.* **1**, 479 (1991).
43. K. S. Narayan, O. Heres, and A. J. Epstein, *J. Magn. Magn. Mater.* **110**, L6 (1992).
44. J. S. Miller, M. D. Ward, J. C. Calabrese, J. H. Zhang, and W. M. Reiff, *Inorg. Chem.* **29**, 4063 (1990).
45. S. C. Abrahams, H. E. Bair, F. J. DiSalvo, P. Marsh, and L. A. Deuring, *Phys. Rev.* **3**, 1258 (1984).
46. J. S. Miller, J. H. Zhang, and W. M. Reiff, *Inorg. Chem.* **26**, 600 (1987).
47. K. M. Chi, J. C. Calabrese, W. M. Reiff, and J. S. Miller, *Organometallics* **10** (1991).
48. J. S. Miller, D. T. Glatzhofer, D. M. O'Hare, W. M. Reiff, A. Chakraborty, and A. J. Epstein, *J. Inorg. Chem.* **28**, 2930 (1989).
49. K. Ono, A. Ito, and T. Fujita, *J. Phys. Soc. Jpn.* **19**, 2119 (1964).
50. J. S. Miller, J. C. Calabrese, R. L. Harlow, D. A. Dixon, J. H. Zhang, W. M. Reiff, S. Chittipeddi, M. A. Selover, and A. J. Epstein, *J. Am. Chem. Soc.* **112**, 5496 (1990).
51. J. H. Zhang, W. M. Reiff, and J. S. Miller, *Hyp. Interact.* **42**, 1099 (1988).
52. W. M. Reiff, *Hyp. Interact.* **40**, 195 (1988).
53. E. Enz, *Helv. Phys. Acta* **37**, 245 (1964).

54. H. J. Mikeska, *J. Phys.* (*Paris*) **C13**, 2913 (1980); H. J. Mikeska, *J. Appl. Phys.* **52**, 1950 (1980).
55. J. Villain, *Physica* **79B**, 1 (1975).
56. R. C. Thiel, H. deGraaf, and L. J. deJongh, *Phys. Rev. Lett.* **47**, 1415 (1981).
57. H. J. M. deGroot, L. J. deJongh, M. Elmassalami, H. H. A. Smit, and R. C. Thiel, *Hyp. Interact.* **27**, 93 (1986).
58. R. D. Ernst, *Chem. Rev.* **88**, 1255 (1988).
59. W. M. Reiff, J. M. Manriquez, and J. S. Miller, *Hyp. Interact.* **53**, 397 (1990).
60. W. R. Reiff, J. M. Manriquez, M. D. Ward, and J. S. Miller, *Mol. Cryst. Liq. Cryst.* **176**, 423 (1989).
61. P. M. Allemand, K. C. Khemani, A. Koch, F. Wudl, K. Holczer, S. Donovan, G. Gruner, and J. D. Thompson, *Science* **253**, 301 (1991).
62. P. W. Stephens, D. Cox, J. W. Lauher, L. Mihaly, P. M. Allemand, A. Hirsch, K. Holczer, Q. Li, J. D. Thompson, and F. Wudl, *Nature* **355**, 331 (1992).
63. F. N. Tebbe, *Science* **256**, 822 (1992).

7

Mössbauer Effect Studies of Nanostructured Materials

S. J. Campbell and H. Gleiter

1. INTRODUCTION

1.1. Mössbauer Spectroscopy

Mössbauer effect spectroscopy has a high-caliber pedigree, of over 30 years standing, as an incisive technique in the delineation of structural details of condensed matter. Nowhere have the contributions of Mössbauer spectroscopy been more effective than in the fields of solid-state physics and materials science. Examples of applications in which measurements of nuclear hyperfine interactions by Mössbauer spectroscopy provide insight into the local atomic environment of the Mössbauer isotope atom, abound in the literature. The hyperfine interactions of prime interest are: the electric monopole interaction (the isomer shift), the electric quadrupole interaction (quadrupole splitting) and the magnetic dipole interaction (nuclear Zeeman effect; magnetic hyperfine splitting).[1,2] Just one example of an early application of Mössbauer spectroscopy in this way is the investigation of the cubic Laves phase intermetallic compound $ZrFe_2$. Wertheim *et al.*[3] were able to account fully for their Mössbauer spectrum in terms of the directions of easy/hard magnetization and the principal axis of the electric field gradient, consistent with the number of iron atoms in the different locations on the corners of tetrahedra.

S. J. Campbell • Department of Physics, University College, The University of New South Wales, Australian Defense Force Academy, Canberra, ACT 2600, Australia.
H. Gleiter • Werkstoffwissenschaften, University of Saarland, D-6600 Saarbrücken, Germany.

Mössbauer Spectroscopy Applied to Magnetism and Materials Science, Volume 1, edited by Gary J. Long and Fernande Grandjean. Plenum Press, New York, 1993.

Besides its ready application in the study of crystalline materials, Mössbauer spectroscopy has also been applied successfully for many years in the investigation of the amorphous state[4] or indeed other artificially structured materials such as superlattices, multilayers, quasicrystals, metastable crystalline alloys, and granular metals.[5] Indeed, any advances in the development of new materials or new solid-state structures quickly attract and benefit greatly from the microstructural insight afforded by Mössbauer spectroscopy. This is reflected in the recent analysis by Stevens[6] of trends in the applications of Mössbauer spectroscopy as evidenced by the number of publications in the different fields in which Mössbauer spectroscopy plays an important part. Since 1984, about a half of the growth of about 6–7% in the number of Mössbauer papers published each year can be accounted for by Mössbauer effect studies of amorphous materials and, since 1987, high-temperature superconductors.

Interest in the present chapter centers around the application of Mössbauer spectroscopy in the study of the new solid-state structure afforded by nanostructured materials.[7] However, a distinction should be drawn between materials that are of a nanoscale and those that have been designed with the nanostructured state in mind.[7] The study of particles in the nanoscale is well known and well established in the scientific literature with Hess and Parker,[8] for example, reporting a process of preparing metallic cobalt particles of uniform size in the range 1–100 nm. Evidence for materials of a nanoscale even extends to other planets in the solar system: evidence for nanocrystalline hematite being present in the soils of Mars has been reported by Morris *et al.*[9] Their conclusions were based on features resulting from ferric iron in the Martian spectral data and the magnetic nature of soil at the landing sites of the Viking Mission crafts. Given that a Mössbauer spectrometer is included in the preliminary payload of a rover to be placed on the surface of Mars in the mission to the planet scheduled for 1996,[10] with one of the scientific objectives being to study the properties of magnetically ordered phases in the Martian soil, including particle size, it may be that Mössbauer spectroscopy could provide direct evidence for the presence of such nanoscale minerals on Mars. By comparison, the systematic investigation of designed and synthesized nanostructured materials has only taken place in the last dozen or so years.[7]

The basic ideas surrounding the design and preparation of nanostructured materials are outlined below. The key point is that a new type of solid can be generated by incorporating many defects so that a disordered heterogeneous structure as opposed to, say, the homogeneously disordered structure of glass, can be obtained. Nanostructured materials offer a range of remarkable properties and corresponding potential applications beyond those available for crystalline or amorphous materials[11,12] and, as such, their physical properties have been investigated thoroughly (for recent reviews see, e.g., Refs. 13–15). This interest and activity reflect the systematic studies of nanostructured materials carried out by Gleiter and co-workers over the past dozen years or so since introducing the nanostructured state.[7] The present overview of

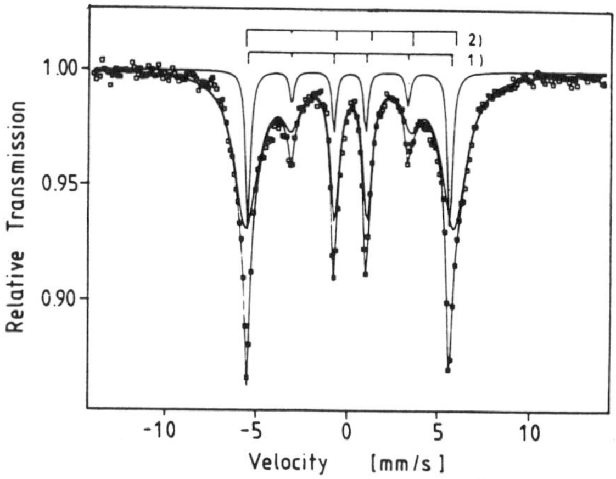

Figure 1. The 77 K Mössbauer spectrum of nanostructured Fe (see Fig. 15, Section 4.1). (After Herr *et al.*[16])

Mössbauer effect studies of nanostructured materials is therefore concerned primarily with materials purposely designed with the unique structure of the new nanostructured state in mind, rather than those materials that merely happen to be of the nanoscale. The merits of the technique of Mössbauer spectroscopy in the study of new states of matter such as nanostructured material lie not so much in the delineation of the intrinsic structure itself (for which more conventional methods of structure determination such as X-ray and neutron diffraction and electron microscopy are required) but rather in elucidation of structural details at the microscopic level thereby confirming, extending, or showing agreement with the structural information available from these other techniques. This is demonstrated well in the present case by the analysis reported by Herr *et al.*[16] on the 77 K spectrum of nanostructured Fe shown in Fig. 1. The work of Herr *et al.* represents the first Mössbauer effect study of purpose designed nanostructured material. Their analysis and conclusions are discussed in Section 4.1.

1.2. Nanostructured Materials: Basic Ideas

In the past, new fields of research in materials science and condensed matter physics have been stimulated in many instances by the discovery of new methods of investigation or by the synthesis of new types of materials. The discovery of tunneling microscopy or the synthesis of high-temperature superconductors are examples of these two kinds. In the case of nanostructured materials, an attempt has been made to synthesize new materials by utilizing the atomic structure in the cores of lattice defects. In order to explain this approach and relate nanostructured materials to other solid-state structures, it is simplest to start from the atomic arrangement in a perfect crystal. In the early part of this century, it was recognized that deviations from the perfect crystal structure result in disordered solids with new and interesting properties. Basically, there are two ways to introduce disorder into an initially perfect crystal; each way results in a different class of disordered materials with different structural features and properties.

The first class is obtained by freezing-in the disorder that is generated by enhancing the energy of a crystal homogeneously everywhere in the volume (or at least within a volume of macroscopic dimensions). The best-known examples of this kind are glasses produced by quenching the melt, i.e., the thermally disordered state. The characteristic features of disordered materials belonging to this category are that their density and nearest-neighbor coordinations deviate little (typically a few percent) from the corresponding quantities of the perfect crystal; the major difference between a perfect crystal and a glass is the loss of long-range structural order.

In the last decade, attention has been focused on the second class of disordered materials. In these materials—termed nanostructured materials[12]—disorder is introduced heterogeneously into a formerly perfect crystal by incorporating a high density of lattice defects such as dislocations or grain boundaries.[7] The atomic arrangements in the cores of lattice defects (Fig. 2a,b) are characterized by locally reduced densities and changes of the nearest-neighbor coordinations, relative to the perfect crystal. For example, the core of the boundary[17] displayed in Fig. 2b consists of a periodic sequence of two pentagonal units, whereas the crystal lattices on both sides of the boundary have cubic symmetry. In fact, the atomic density in the boundary shown in Fig.

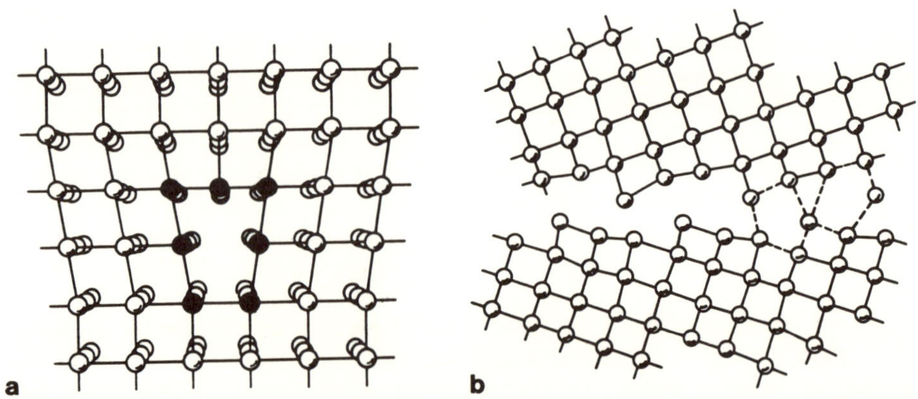

Figure 2. (a) Edge dislocation in a simple cubic crystal. In the dislocation core, the atomic density and coordination are changed relative to the perfect cubic lattice. In fact, the core of the dislocation may be interpreted as a channel formed by seven atoms, as shown. (b) Atomic structure in the core of grain boundary between two crystals tilted relative to one another by 36.9° about a common [100] direction. The structure was deduced from the high-resolution electron micrograph shown in Fig. 6 (Section 3.1). The boundary core structure may be described as a two-dimensional periodic array of two different pentagonal polyhedra (indicated on the right side of the boundary).

2b was found experimentally to be about 75% of the crystal density[17] whereas glasses have typical densities of 95% or more. The densities and the nearest-neighbor coordinations in boundary cores depend on the misfit between the two lattices of different crystallographic orientations joined together at the boundaries. In other words, the interatomic forces between the atoms in the crystals on either side of a boundary and the atoms in the boundary cores, lead to the formation of atomic arrangements in the boundary (defect) cores that are characterized by densities and nearest-neighbor coordination numbers that differ from the densities and the coordination numbers in crystals and glasses with the same chemical composition. In the cores of defects a (constrained) state of solid matter exists that does not exist anywhere else. In terms of this picture, the existence of two classes of disordered solids has the following physical origin. The disorder in glasses originates from thermal fluctuations in the melt and requires no constraints, whereas the disorder in boundary cores (defect cores) is caused by constraints and requires no thermal fluctuations. Hybrid-disordered structures, resulting from constrained thermal fluctuations, lead to materials that will be discussed in Section 3.3.

It is the basic idea of nanostructured materials that a new type of solid can be generated by incorporating so many defects, e.g., grain boundaries, into an initially perfect crystal that the fraction of atoms situated in the cores of these defects become comparable to the fraction of atoms in the residual crystals. Depending on which type of defect (e.g., dislocations, grain boundaries, interphase interfaces) is incorporated, different kinds of nanostructured materials result. However, all of these materials have the following micro-

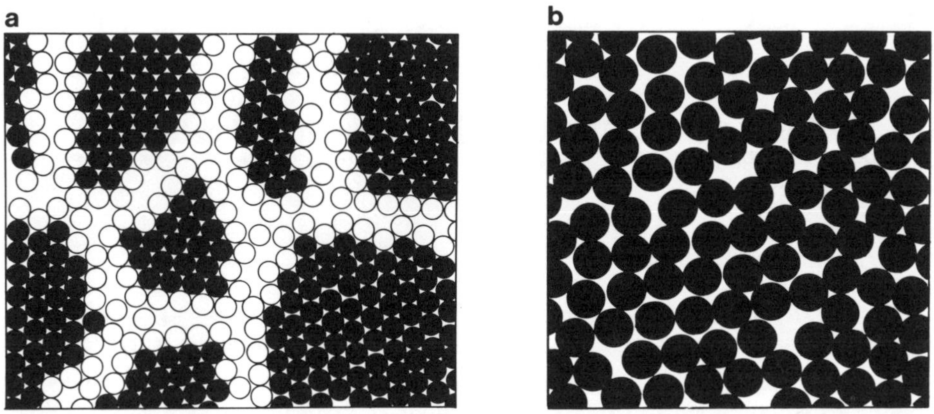

Figure 3. (a) Atomic structure of a two-dimensional nanostructured material. It consists of small crystallites (of different crystallographic orientations) connected by grain boundaries. For the sake of clarity the atoms in the centers of the "crystals" are shown as solid circles. The ones in the boundary core regions are represented by open circles. Both types of atoms are assumed to be chemically identical. (b) Atomic arrangement in a two-dimensional glass.

structural feature in common. They consist of numerous defect cores that are embedded into (an elastically distorted) crystal lattice or lattices. It is this heterogeneous structure that distinguishes nanostructured materials from homogeneously disordered solids such as glasses. In order to illustrate this difference, Figure 3 displays the atomic arrangements in a two-dimensional nanostructured material relative to a two-dimensional glass. In the grain boundaries between the differently oriented hexagonal arrays of atoms (representing the crystallites) a network of boundaries is formed. Because of the misfit between adjacent crystallites, the boundaries exhibit reduced densities and a wide distribution of interatomic spacings.

The synthesis of nanostructured materials is central to any discussion about the properties and behavior of nanostructured materials and an outline of the preparation techniques is presented in Section 2. In Section 3 we highlight the basic features of the atomic structure of nanostructured materials. This provides the background for the discussion presented in Section 4, which outlines the structural and magnetic information that has been obtained so far about nanostructured materials by means of Mössbauer spectroscopy.

2. SYNTHESIS

So far, the synthesis of nanostructured materials has been carried out most frequently by assembling pregenerated small clusters by means of *in situ* consolidation and sintering. We shall first review the various techniques belonging to this category. The subsequent sections will be devoted to the procedures that do not require pregenerated clusters. For reviews of the synthesis of small crystals, see Refs. 18–25.

2.1. Generation of Nanometer-Sized Clusters

The techniques for the generation of nanometer-sized clusters may be divided into three broad categories[18]: vacuum, gas-phase, and condensed-phase synthesis.

2.1.1. Vacuum Synthesis

2.1.1.1. Sputtering. When ions of a suitable substance (e.g., those of Ar or Kr) are accelerated to high energies and directed toward a surface, atoms and clusters, both neutral and ionic, are ejected. This method of vaporizing materials is called sputtering. Today, primarily two types of sputtering sources are in use. Energetic ions are produced in an ion gun and accelerated toward the surface of the target material.[26] Clusters are ejected directly from the target surface as a small fraction (typically 10^{-4} or less) of the overall material sputtered.

Alternatively, hollow cathode sputtering may be utilized.[27] The material to be vaporized is fabricated into a cylindrical cup perforated at one end. A coaxial electrode is placed within the cup and a high voltage is imposed between the cylinder and the internal electrode so as to make the former the cathode. When gas is passed through the system, a discharge is struck and the ionized gas (largely in the form of cations) is accelerated toward the inner surface of the cylinder, vaporizing a portion of it. The vapor is confined for a period of time within the hollow cathode and, if it becomes supersaturated, cluster growth occurs. A drawback to sputtering is that it usually produces small amounts of clusters and cluster intensity distributions that decrease exponentially with increasing cluster size.

2.1.1.2. Laser Ablation. This technique uses high-power pulsed lasers to vaporize cluster ions from solid surfaces. The wavelength of the laser light has to be adjusted to the material. For example, the ablation of metals requires UV lasers (e.g., excimer lasers).

2.1.1.3. Liquid-Metal Ion Sources. Liquid-metal ion sources are typically formed by a small tungsten wire wetted with another metal above its melting point. When this tip is brought to a potential of a few kilovolts with respect to a grounded aperture, the liquid is pulled up into a cone by the electrostatic forces. With the very high electric field at the end of the cone, ions, ionized clusters, and even ionized droplets are emitted by field evaporation. These sources work for those metals whose vapor pressures are sufficiently low at their melting points so as not to interfere with maintaining the extraction voltage, e.g., Au, Ga, and In.

2.1.2. Gas-Phase Synthesis

2.1.2.1. Inert Gas Condensation. In the case of inert gas condensation the aggregation of volatilized monomers into clusters is achieved by: (1) establishing a monomer population, (2) "cooling" the monomers by collisions with "cold" inert gas atoms, and (3) growth of clusters both by addition of monomer to individual clusters and by aggregation resulting from collisions between clusters. So far, oven and sputtering sources, electron gun or laser evaporation, pyrolysis, hydrolysis, or supersonic expansion have been utilized.

2.1.2.2. Oven Sources. The simplest technique for establishing a monomer population is by means of a heated crucible or oven. Near the source, small clusters of fairly uniform size are observed. Farther from the source, the clusters become larger with a broader size distribution. Finally, at some limiting distance from the oven, which depends on inert gas pressure and evaporation rate, the clusters reach a limiting size that increases with enhanced evaporation rate and with the atomic weight of the inert gas. The mean cluster size in such a source can be controlled by varying the evaporation rate from the oven and the pressure of the inert gas in the evaporation chamber. The mean cluster size can be reduced and the size distribution sharpened by imposing a forced convective flow on the inert gas. Several sources based on this concept have been developed.[28–31] Short residence times of the clusters close to the source reduce cluster–cluster aggregation resulting in smaller clusters and narrower size distributions.

Naturally, sublimation may also be utilized instead of evaporation for solids with high vapor pressures. This applies, for example, to MgO, which has been heated in He at a pressure of 200 Pa to temperatures around 1600°C (MgO melts at 2852°C).[32] The material that sublimes was found to be oxygen deficient, but is fully converted to stoichiometric MgO by subsequent exposure to oxygen introduced into the vacuum chamber.

Oven-based sources suffer from the following drawbacks: the temperature limitations imposed by the oven or crucible material; chemical reactions occur between many metals and the frequently used refractory metal crucibles; inhomogeneous temperature distributions exist in the molten metal leading to unsatisfactory control and reproducibility; the production of alloy clusters is restricted by the different activities of the constituents, and evaporation of technologically interesting ceramics such as Al_2O_3 is limited by the oxidation of refractory boats.

The preparation of oxide ceramics by oven-based sources has been achieved by evaporating the metallic component (Ti, Fe) in He so that a nanometer-sized metal powder was obtained. The loose metallic powder was subsequently oxidized by introducing oxygen (typical pressure 2 kPa) into the chamber.[33–36]

Naturally, postoxidation is not limited to fully oxidized particles. For example, oxidation of small Al, Zn, or Mg crystals has been utilized to generate nanometer-sized particles with metallic cores and oxidized surfaces. The method of reacting small metallic particles with a gaseous phase has been applied to generate nanometer-sized hydride particles (e.g., TiH_2) by evaporating the metallic component in a H_2 atmosphere of 50 kPa pressure.[37] Subsequent heating of the TiH_2 particles (400°C, 1 min, in vacuum) converted them into Ti of hcp structure. Rehydrogenation into TiH_2 was achieved by annealing the particles at 250°C for 5 min at 10 kPa hydrogen pressure. Iwama and co-workers[38] have succeeded in producing a variety of transition-metal nitrides in ultrafine powder form by evaporating metals in either N_2 or NH_3 gas.

Recent experiments have revealed that, at least in one case, control of the inert gas pressure affects not only particle size, but also the phase of the resulting material.[39] Ultrafine powders of Ti were found to react with oxygen to form rutile (TiO_2) particles if He pressures of greater than 500 Pa were used during evaporation. However, if the He pressure used was less than 500 Pa and greater than about 10 Pa, small Ti particles were still formed, but these particles did not form rutile when exposed to oxygen. Instead, in this case, an unexpected amorphous phase was formed. Unexpected new phases of erbium oxides have also been produced via the same processing route.[40] In fact, compacted nanometer-sized erbium oxide (10-nm crystal size) particles were found to be composed of two fcc forms of erbium oxides, one of which is a new fcc erbium sesquioxide with a lattice constant of 0.374 nm, and a new Er_2O_3 phase with a monoclinic structure.

2.1.2.3. Sputtering. dc and rf magnetron sputtering seem in principle suitable for the production of nanostructured materials. The normal operating pressure of sputter sources (10^{-1} to 10^{-2} Pa), however, is several orders of magnitude lower than the pressure range required for particle formation during inert gas condensation in laboratory-size evaporators (10^2–10^3 Pa). Some pioneering work[41] has shown that simple diode sputtering could be used to produce ultrafine particles in the desired size range. In addition to its wide applicability, sputtering has other advantages over most thermal evaporation techniques: (1) the composition of the sputtered material is the same as that of the target, making alloy synthesis possible, (2) the sputtering conditions are stable and readily controllable by means of the plasma current, and (3) the heat load on the chamber walls is far smaller than during thermal evaporation, reducing outgassing and subsequent impurity incorporation by the small particles.

The application of a commercially available magnetron sputtering device for preparation of metallic and ceramic clusters (Al, Mo, W, $Cu_{91}Mn_9$, $Al_{52}Ti_{48}$, TiO_2, NiO, and ZrO_2) with diameters in the range 7–50 nm has been studied recently.[42,43]

2.1.2.4. Laser Ablation. In the case of laser ablation, a high-power laser beam ablates a sample target located in an inert gas chamber. The resulting plume of metal is entrained and cooled by the carrier gas, yielding the required supersaturation and cluster growth. Virtually any material that can be fabricated into an appropriate target may be used. Clusters of semiconductors,[44] transition metals,[45] and main group metals[46] have been generated in this way. Alloy targets have been used and, in general, the clusters show the same chemical composition as the target.[47] Another procedure combines laser vaporization of one metal with simultaneous laser pyrolysis of a gaseous precursor of another metal.[48] This precursor is mixed with the carrier gas so that both types of monomers are present in the clustering region. In this way, mixed clusters of elements that do not normally alloy can be prepared.

2.1.2.5. Pyrolysis. Laser pyrolysis is a technique designed to synthesize ultrafine refractories by rapidly heating with a laser a flowing reactant gas mixed with an inert carrier. This leads to rapid gas-phase decomposition of the reactants and produces a saturated vapor of the desired constituent atoms. Nucleation and growth of clusters then occur as the decomposition products are quenched by collisions with the atoms (molecules) of the carrier gas. The technique has been used to synthesize ultrafine powders of Si, Si_3N_4, and SiC. ZrB_2, $FeSi_2$, B_4C, TiB_2, TiO_2, Al_2O_3, and TiO_2 have been obtained by a similar approach.[49–51] The synthesis of nonmetals uses volatile hydrides, with

H_2 as the by-product. Compounds of nonmetals are produced by adding ethylene to the gas mixture to yield the carbide, ammonia to yield the nitride etc.

Metals and compounds of metals, as well as complex compounds of nonmetals, require a different approach. The metals are incorporated into compounds that are volatile under the experimental conditions. This is usually done with organometallic compounds, or with metal carbonyls. Rice and Woodin[52] have synthesized Si/N/C powder of 100-nm-diameter spheres from the pyrolysis of $Me_3SiNHSiMe_3$. WC and W/C/O as well as MoS_2 and MoS_2C_x have been synthesized by this approach; doping these materials with Co or Zn was also achieved. Unique FeC catalysts have also been synthesized by this procedure.[53] Laser pyrolysis has the advantage of being a continuous process, which is scalable with laser power and reactant flow rate. The cost of materials produced by this approach is primarily determined by the cost of the chemicals.

2.1.2.6. Flame Hydrolysis. Flame hydrolysis is the reaction of volatile compounds (e.g., $TiCl_4$ or $SiCl_4$) in an oxygen–hydrogen flame. It leads to highly dispersed oxide clusters and has been applied to a wide variety of compounds[54–57] such as SiO_2, Al_2O_3, TiO_2, ZrO_2, Bi_2O_3, CrO_3, Fe_2O_3, NiO, GeO_2, and V_2O_5. The major advantages of this technique are high purity, chemical flexibility, and the possibility to synthesize mixed oxides. Depending on the process conditions, cluster sizes between 5 and 50 nm are obtained.

2.1.3. Condensed-Phase Synthesis

2.1.3.1. Metals. If a reducing agent is added to an acidic aqueous solution of metal ions, then small neutral metal clusters will form. Dialysis can be used to remove remaining ions and a thickener, such as gelatin, can be added to prevent aggregation of the particles. Relatively narrow size distributions have been achieved with this technique (e.g., Ag,[58] Au,[59] Pt[60]) similar to those obtained with inert-gas condensation.

The growing of small metal clusters by means of the microelectrode method has been utilized in recent years to generate small Ag clusters.[61,62] A solution containing about 10^{-4} mol/litre Ag^+ ions was exposed to a short pulse of radiation producing a known amount of hydrated electrons in the 10^{-6} molar range. The latter rapidly react with silver ions to form atoms, $Ag^+ + e^-_{aq} \rightarrow Ag^0$, which subsequently agglomerate to larger particles. After a certain time of agglomeration, a metallic silver colloid was present, recognizable by the appearance of the 380-nm plasmon absorption. In a second study, utilizing the same technique,[63] sulfonatopropylviologen (SPV) was used. The Ag atoms agglomerate into silver particles and grow further by the reaction: $SPV^- + Ag^+ + Ag_n \rightarrow SPV + Ag_{n+1}$. Small gold clusters have been grown by a closely related method.[64] In all systems studied, the small metal particles were noted to agglomerate if the suspension was stored under ambient conditions. Hence, stabilization of intermediate cluster sizes would be of interest for the controlled growth of metal clusters of a predetermined size. Such a stabilization has recently been achieved in the case of silver clusters by sodium polyphosphate.[65]

Colloidal suspensions of metal clusters in organic solvents have also been generated by depositing metal atoms into low-temperature organic solvents.[66] Cluster growth is eventually inhibited by strongly bound solvent molecules.

Furthermore, spark erosion has been utilized as a method for producing fine powders of metals, alloys, semiconductors, and compounds.[67] The technique involves maintaining repetitive spark discharges among chunks of material immersed in a dielectric liquid. As a result of the spark discharge, there is highly localized melting and/or vaporization of the material. The powders are produced by the freezing of the molten droplets or the condensation and freezing of the vapor in the dielectric liquid. The powders are quenched *in situ* to the temperature of the liquid. Particles can be produced in sizes ranging from 5 nm to 75 μm. The average powder size and production rate depend on the power parameters, the material used, and the dielectric liquid.

The generation of metallic clusters by precipitation from a solid crystalline or glassy matrix has been used for numerous studies on isolated clusters.[68–70] In a few cases it has been possible to dissolve the matrix selectively without dissolving the cluster, resulting in a colloidal suspension of clusters in the solvent.

2.1.3.2. Semiconductors. By control of temperature, concentration, and solvent during liquid-phase precipitation of compound semiconductors, it has become possible to make and stabilize crystallites with diameters <5 nm.[71] Better size control and colloid stability can be achieved through the use of microscopically structured liquid mixtures, such as water–soap–heptane, in which the crystallites are stabilized in small (2–10 nm) water pools[72] termed "inverse micelles." Small semiconductor clusters have also been prepared inside ionomers. Ionomers are a class of copolymers containing ionic side-chain groups such as COO^- and SO_3^-. These ionic groups tend to aggregate together forming domains, analogous to the formation of ionic pools in micelles. Metal ions, such as Cd^{2+} and Pb^{2+}, can be easily exchanged into these ionic domains where synthesis of desired semiconductor clusters can be conducted. For example, stable PbS clusters, ranging in sizes from monomolecular to bulk, have been synthesized in ethylene–15% methacrylic acid copolymer.[73]

Silicate glasses have been used as hosts for the high-temperature synthesis of semiconductor clusters such as CdS_xSe_{1-x} and CuCl. In the case of CdS_xSe_{1-x} doped glasses, which are available as commercial color filters, elementary Cd, S, and Se are first dissolved in the molten glass in the presence of excess amounts of Se. After the glass is formed upon cooling, CdS_xSe_{1-x} crystallites of various sizes can then be produced by controlled annealing.[74]

2.1.3.3. Ceramics. Decomposition and precipitation reactions in ionic materials have been utilized to generate nanometer-sized clusters. For example, decomposition of $Mg(OH)_2$ and $MgCO_3$ yielded MgO clusters with diameters of about 2 nm.[75–77] The sol–gel method has been used in a variety of systems to generate small (<10 nm) clusters of SiO_2, Al_2O_3, TiO_2.

Hydrothermal reactions (reactions of the water above its boiling point) have been proposed to be applied to synthesize small clusters. The two reactions tested so far are hydrothermal precipitation and hydrothermal oxidation.[78] Both reactions yield suspensions of crystalline metal oxides in water. Simple oxides (ZrO_2, SiO_2, Al_2O_3, TiO_2, MgO, CaO) as well as mixed oxides (ZrO_2–Y_2O_3, ZrO_2–Y_2O_3–MgO, ZrO_2–MgO, ZrO_2–Al_2O_3, $BaFe_{12}O_{19}$, $BaTiO_3$) have been prepared in the size range 10–100 nm.

2.1.4. Capped Clusters

In order to isolate macroscopic amounts of clusters from the liquid-phase reaction media, it is necessary to modify chemically the bare surface to prevent cluster fusion. Polymeric, highly ionic, inorganic phosphates have been used[79] to protect powders of Cd_3P_2, CdS, etc., which could then be redispersed in water. A direct organometallic synthetic method for surface reactions on clusters has recently been achieved by means of inverse micelle colloidal preparations.[80] This method chemically bonds organic groups to chalcogenide atoms on the cluster surface. The surface changes from hydrophilic to hydrophobic and, as a result, the clusters precipitate as a pure, capped, cluster powder.

Extremely small monodisperse metal and semiconductor cluster molecules, consisting of 30 atoms or less, can be prepared and crystallized via direct chemical synthesis. Examples such as iron sulfur clusters[81] and gold–silver clusters[82] have been reported in the literature.

2.1.5. Cluster Arrays

Here, the goal is to assemble clusters in well-defined three-dimensional arrays. In principle, the existence of solids of this type has been demonstrated recently with the synthesis of CdS superclusters inside zeolites. In zeolite Y,[83] discrete $(CdS)_4$ cubes with a CdS bond length of 0.247 nm have been synthesized within the 0.5-nm sodalite cages. Cubes in the adjacent sodalite units interact with each other, via superexchange through the host lattice, to form an interlocking three-dimensional "supercluster" structure. The optical absorption spectrum of these superclusters lies between those of clusters and the bulk. A novel orientational effect has also been observed. By changing the host from zeolite Y to A, the relative orientation of the cubes is changed from vertex-to-vertex to face-to-face. This results in a 0.3-eV blueshift in the supercluster absorption spectrum, reflecting the change in the geometrical and electronic structures of the superclusters as imposed by the zeolite host. Using different zeolites as templates, superclusters with different structures and electronic properties can be built.

Another approach toward the same goal was proposed by Umemura.[84] In order to generate a regular array of small particles, a two-step process has been developed. In the first step, a square lattice of carbon was deposited on the Si by means of a narrow electron beam. By depositing subsequently an Au or a KCl particle into every square, a regular array was obtained. The accuracy of the lattice is still inferior relative to the quality achieved by modern microfabrication methods. A somewhat related approach is the generation of small particle devices called "superatoms."[85–87] These particles are composed of two semiconductor substances arranged regularly as a core and a shell which resembles functionally the core and electron shell of an atom, respectively. The application of such structures requires the availability of monodisperse small particles.

2.2. Cluster Consolidation

One approach to fabricate a bulk piece of a nanostructured material is to consolidate the small clusters generated by one of the methods discussed in the previous section. The following three methods of cluster consolidation have been studied so far.

2.2.1. Consolidation by High-Speed Deposition

Kashu *et al.*[88] have deposited nanometer-sized clusters of Ag, Fe, Ni, Cu, Au, Ti, Pb, Zn, TiN, and RuO_2 by using clusters entrained in a gas jet. The clusters collide with a glass, metal, or plastic substrate surface located close and normal to the jet. With a sufficiently high velocity (>100 m/s) of jet flow against the substrate, the clusters have enough energy that is released during the collision to make them stick to the substrate and to one another. The deposit exhibited a green density (i.e., the density after consolidation of the powder at ambient temperature) of about 55% for Ag, Cu, Ni, etc., which corresponds to a conventional powder compact obtained by high-pressure consolidation. If a conventional millimeter-sized fine powder was used instead of the nanometer-sized clusters, no solid deposit was formed. Codeposition of different materials, e.g., Ag–Fe, Pb–Zn, from a single nozzle was possible, regardless of the specific density of each component. The microanalytical studies indicated uniform distribution of the components. In classical metallurgical terms, this process seems to be based on low-temperature ballistic consolidation.

2.2.2. Deposition by Ionized Cluster Beams

Takagi[89] and Yamada and co-workers[90] have pioneered thin-film deposition by "ionized cluster beams." These beams are generated by expanding a vapor beam adiabatically through a nozzle. During expansion, a fraction of the atoms condense in the form of small clusters containing typically 1000 atoms. An electron beam post ionizes a small fraction of the clusters which can then be accelerated to a deposition substrate with any desired impact energy, typically about 1 eV per atom of the cluster. There is no mass selection; atoms, small clusters, and neutral large clusters as well as the ionized clusters strike the substrate. The ionized, accelerated clusters are pictured as dissociating at the moment of impact on the substrate, desorbing loosely bonded atoms and impurities and giving mobility to atoms that have already landed at the surface as well as those of the dissociated clusters. In one sense, the ionized cluster beams are an "ion-assisted" process, since only a few percent of the total amount of material being deposited is believed to consist of ionized clusters. Ionized cluster beam sources have been utilized to generate thin films, e.g., epitaxial aluminum on silicon[90] or gold mirror coatings.[91]

2.2.3. Consolidation

The first attempt to generate and study bulk nanostructured materials was based on the collection and consolidation of small clusters produced by inert gas evaporation or codeposition of a metal vapor and a rare gas on a cold substrate followed by the sublimation of the rare gas.[7]

A schematic illustration of an ultrahigh vacuum apparatus developed for this purpose is shown in Fig. 4.[92] The nanometer-sized crystals are produced by a two-step process. The substance (e.g., Cu) is evaporated by resistive heating from a refractory metal boat. The evaporated atoms transfer their thermal energy to the inert gas and condense in the form of small crystals. The crystals are carried by a convective current of inert gas to a liquid-nitrogen-cooled cold finger. Subsequently, the inert gas is removed and the particles are scraped from the cold finger and funneled into a piston and anvil-like device where they are compacted using pressures of up to 5 GPa. Cooling or heating of the specimen during compaction is readily achievable. As a consequence of the clean handling conditions, the powders become partially sintered during compaction and have densities of about 70–90% of the bulk. Clearly, the processing technique allows for great versatility: composite materials can be produced by using two or more evaporation sources and oxide or other ceramic materials can be obtained by mixing or replacing the inert gas with a reactive gas. The particle size can be controlled by varying the evaporation rate and the inert gas pressure.[92]

Although the general features of the synthesis seem to be clear, the detailed processes that take place during compaction and the resulting structure of the material are not yet well characterized. During the collection of the powder on the cold finger, some of the particles agglomerate and form grain boundaries; in compound powders, both grain boundaries and heterogeneous interfaces result. Most interfaces, however, are formed during the

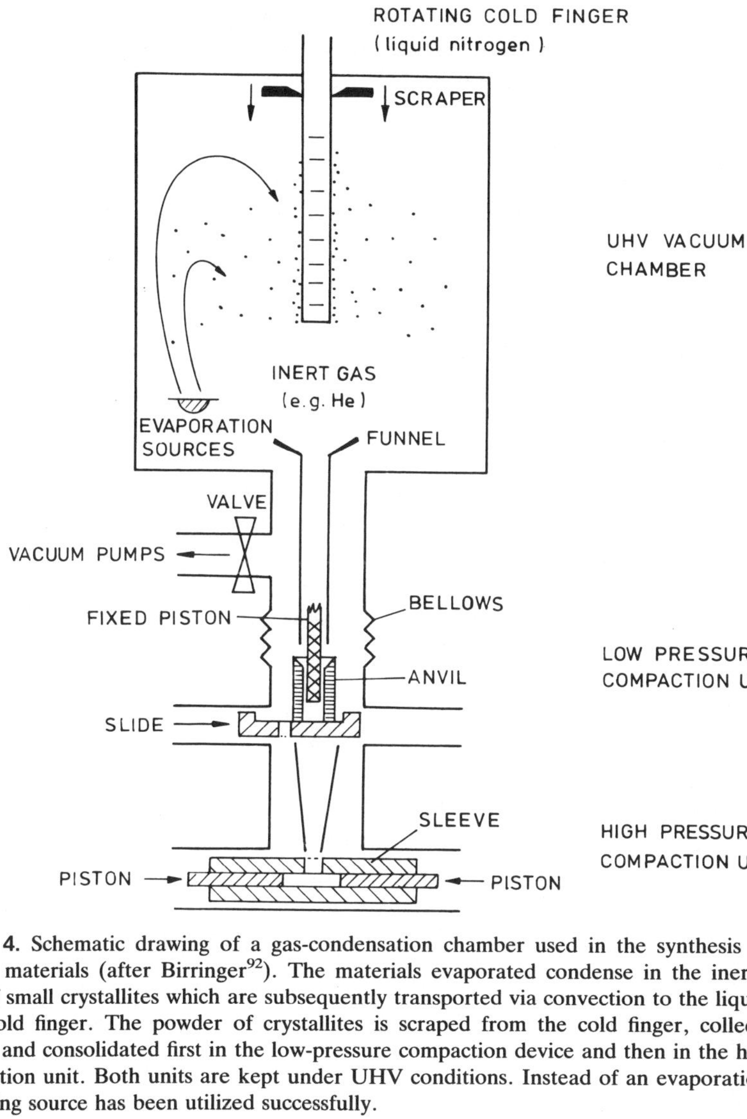

Figure 4. Schematic drawing of a gas-condensation chamber used in the synthesis of nanocrystalline materials (after Birringer[92]). The materials evaporated condense in the inert gas in the form of small crystallites which are subsequently transported via convection to the liquid-nitrogen-filled cold finger. The powder of crystallites is scraped from the cold finger, collected via the funnel, and consolidated first in the low-pressure compaction device and then in the high-pressure compaction unit. Both units are kept under UHV conditions. Instead of an evaporation device, a sputtering source has been utilized successfully.

compaction process. Before two particles (nanometer-sized crystals) are brought into contact, their atoms lie on regular lattice sites, with possible relaxations at the free surfaces. As contact is made between the two crystals, an interface is formed with an atomic structure controlled by the parameters discussed in Section 1.2.

Further improvements in powder consolidation seem to be possible by utilizing the "core extrusion process" which has been successfully applied to the consolidation of metallic glasses.[93] When a powder of glassy particles is packed into a cylindrical container and is extruded, an inhomogeneous, porous bulk glassy material commonly results. The homogeneity of the extruded material may be improved by placing a core of a suitable material in the central axis of the container. The published observations[93] demonstrate that the coextrusion of the core and the container results in a homogeneous glassy body with a residual porosity of less than 0.1%. The same technique seems applicable to the processing of nanostructured materials.

2.3. Other Methods

2.3.1. High-Energy Milling

The reduction of grain size in powder samples to a few nanometers during heavy mechanical deformation has been investigated by several workers.[94–98] Using high-energy ball-milling, the grain size of pure bcc metals (Cr, Nb, W), metals with hcp structure (Zr, Hf, Co, Ru), and intermetallic compounds with CsCl structure (CuEr, NiTi, AlRu, SiRu) and immiscible systems (FeAl) can be reduced to nanometer scales. For longer durations, ball-milling results in an extremely fine-grained microstructure with randomly oriented grains (5–13 nm in diameter) separated by high-angle grain boundaries. Thermal analysis of these samples enabled a quantitative determination of the enthalpy stored in these materials and an estimation of the related grain boundary energies to be made. Excess enthalpies of up to 40% of the heat of fusion and excess heat capacities of up to 20% in comparison with the underformed state have been achieved, exceeding by far any values determined for conventional deformation processes. These values also exceed the typical specific energies of grain boundaries in fully equilibrated polycrystals. The values are comparable, however, to the excess enthalpies and the excess heat capacities reported for nanostructured materials prepared by consolidation of nanometer-sized clusters using the method discussed in Section 2.2.3.

Calka and Radlinski[99] have recently developed and studied a new design of ball-mill. This mill has the novel feature that the ball movement is controlled partially by an external magnetic field. In particular, various modes of operation are used to control the paths for the solid-state reactions between the components during the milling process. As a consequence, the microstructure of the final product can be manipulated by suitably selecting the milling parameters as well as by adding small amounts of surface-active substances.[100] The new mill has been applied to metal/metal (e.g., Al–Mg, Ni–Zr) as well as metal/metalloid (e.g., Ni–Si, Ti–Si, Ti–B, Fe–B) mixtures. In both cases, the formation of amorphous and nanostructured materials was achieved, depending on the milling conditions and the subsequent annealing treatment.

High-energy ball-milling also seems suitable for generating nanostructured alloys. For example, using a planetary-ball-mill or an attritor, an equiaxed crystalline structure was produced with a grain size of <10 nm in an Fe–Ta–W alloy.[66] Nanostructured cermets such as Ti–Ni–C could be generated by the same method. Similar results were obtained[95] for Al–Fe and Ag–Fe alloys. In both systems, the formation of a nonostructured material competed with the formation of an amorphous atomic arrangement. In fact, it turned out that the chemical composition strongly influenced the amorphization. For example, a small amount of Ti in the Al–Fe system favored the formation of the amorphous phase. Cryomilling (i.e., high-energy ball-milling at liquid-nitrogen temperature) leads to a matrix of aluminum (for example) with a fine grain size (<50 nm) of dispersed AlN particles.[97] A major advantage of the mechanical alloying approach to production of nanostructured materials is that it is a relatively large-scale process permitting several kilograms of material to be produced in times of up to 100 h.[96]

2.3.2. High-Strain-Rate Deformation

The reduction of the grain size into the submicron regime (0.1 μm) by means of a high-strain deformation process under high pressure has been demonstrated by Valiev *et al.*[101] The obvious advantages of this method are that it permits the generation of large quantities of fine-grained materials and does not require special precautions to prevent contamination effects as do the methods based on the consolidation of small, initially isolated particles (Section 2.1.).

2.3.3. Mixalloy Processing

The mixalloy processing utilizes the turbulent mixing of impinging alloy streams and *in situ* chemical reactions.[102–104] Two or more turbulent streams of molten metal alloys are directed into a mixing chamber. There the streams impinge at high velocities. Because of the turbulent flow, small-scale eddies provide rapid and effective mixing and permit chemical reactions on a nanometer scale. For example, if one of the molten streams is a Cu–B alloy and the other stream is a copper alloy containing a strong boride-forming element such as Ti, nanometer-sized TiB_2 crystals are formed in the zone of mixing. Subsequent rapid solidification results in a fine-grained material containing nanometer-sized thermally stable dispersions.

2.3.4. Deposition Methods

During thin-film generation, the formation of nanostructured materials has been reported by numerous authors. For example, thin films of $Fe_{60}Co_{40}$

produced by electron beam evaporation and ultrahigh vacuum conditions led to crystal sizes of about 8 nm.[105] The films were shown to consist of the small crystallites and (about 30 vol.%) grain boundaries. Thin films produced by laser breakdown chemical vapor deposition from nickel and iron carbonyls had grain sizes smaller than 10 nm.[94] Rapid solidification has been demonstrated—for several pure metals as well as alloys—to result in nanostructured materials.[106–108] Furthermore, electrodeposition of Ni–P alloys under suitable conditions leads to nanostructured materials.[109–112]

In the case of silicon, chemical transport in a low-pressure hydrogen plasma has been utilized.[113] The crystal size (and hydrogen content) varied from 5 nm (12%) to 15 nm (1%) as the deposition temperature was increased from 110 to 400°C. A chemical transport process for generating nanostructured, high-strength iron whiskers (crystal size between 5 and 30 nm) was developed more than two decades ago.[114,115]

Chemical vapor deposition (CVD) has been utilized to synthesize nanostructured materials based on the nitrides and carbides of titanium and silicon.[116] The CVD reactor used was a computer-controlled hot-wall reactor. Layered deposits were produced by pulsing the reactant gases judiciously under software control. The development of a columnar structure which is endemic to most CVD materials was suppressed by the above procedure. The same procedure was applied to grow nanocrystalline carbon films and filaments from methane–hydrogen mixtures by rf-plasma-assisted CVD. Crystal sizes as small as 2 to 3 nm were obtained.

2.3.5. Sol–Gel Method

Nanometer-sized ceramic structures have been generated by means of a sol–gel technique. Both structural and compositional nanometer structures are obtained by "seeding" of a ceramic precursor with crystalline sols of the final equilibrium phase to catalyze nucleation.[117–127] This process, which consists of introducing crystalline nuclei in a matrix in order to lower the nucleation energy required to form the expected phase, has been known for a long time in the case of liquids and glasses.

A significant advantage of the sol–gel technique relative to methods involving high temperatures (e.g., calcination, evaporation) is the low temperature of the method. The preparation of stoichiometric compounds containing one or more components with a high vapor pressure, e.g., Pb in $(BaPb)TiO_3$ ferroelectrics, poses serious problems because of the Pb loss. This difficulty can be avoided by the sol–gel approach. In fact, the specific surface area of $(BaPb)TiO_3$ sol–gel-derived powders has been found to be about $50\ m^2/g$ in comparison with $4\ m^2/g$ for powders prepared by calcined mixed oxides.[128]

3. STRUCTURE

In this section, we aim to highlight the basic features of the atomic structure of nanostructured materials in order to provide the background for the subsequent discussion of the structural information that has been obtained so far by means of Mössbauer spectroscopy.

3.1. Structure of Pure Nanostructured Materials

A two-dimensional hard-sphere model of a nanostructured solid is shown in Fig. 3a. For the sake of clarity, the atoms in the various crystals are indicated as solid circles whereas those in the boundary cores are represented by open circles, although both types of atoms are assumed to be chemically identical. Structurally, the material consists of the following two components: the crystals with different crystallographic orientations but identical atomic arrangements and a network of boundaries the structure of which varies from boundary to boundary and is characterized by the following two features.

1. The average atomic density in the boundary is reduced relative to the crystal density because of the misfit between adjacent crystallites. The boundary density cannot be enhanced by inserting additional atoms into the vacant sites existing between the atoms in the boundaries because the local free volumes of these sites are too small. In fact, interatomic penetration would result if atoms were inserted into these sites.
2. The boundaries exhibit a broad spectrum of interatomic spacings ranging from closely packed atoms to widely spaced ones.

As far as the volume ratio between boundaries and crystals in nanostructured materials is concerned, the situation is complementary to the one in coarse-grained polycrystals. In coarse-grained polycrystals, the crystal lattice represents the basic structural component and the lattice defects (e.g., grain boundaries) are perturbations limited to a small (typically 10^{-2} or less) volume fraction of the solid. In nanostructured materials, the priorities are inverted. The defect cores represent the essential structural component and the residual crystals are primarily required for generating the boundary cores. Here we focus attention on the discussion of the basic features of the structure of nanostructured materials. Many of the experimental and technical details have been documented in several recent reviews.[13,15,129–131] As the field has only been established for about a dozen years[7,132] and finds itself in a state of rapid expansion, many of our present views may have to be modified in the years ahead.

Studies by transmission electron microscopy and scanning tunneling microscopy have confirmed the structure proposed in Fig. 3a, consisting of

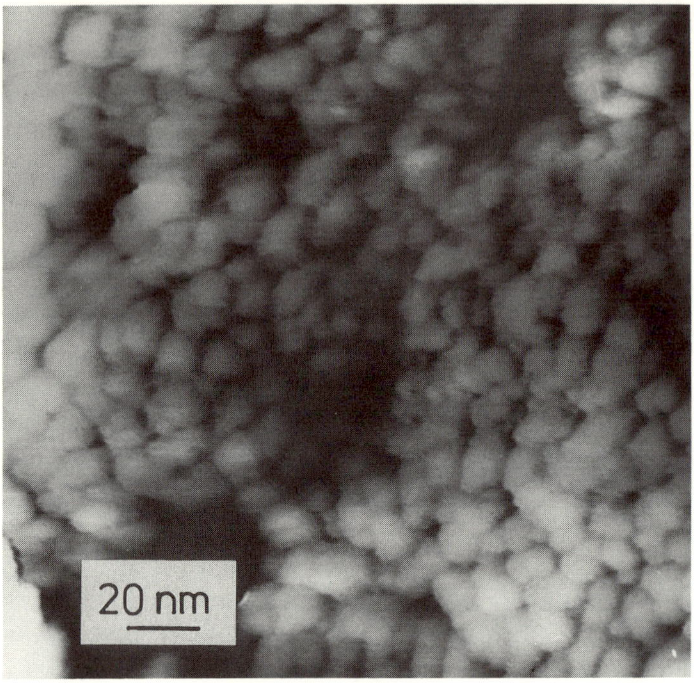

Figure 5. Tunneling micrograph of nanostructured Pd. The bright regions represent the crystals. The network of dark lines are the grain boundaries.

small equiaxed crystallites (i.e., crystallites of comparable dimensions in three orthogonal directions) interconnected by incoherent interfaces.[133,134] For example, Fig. 5 shows a tunneling micrograph of nanostructured Pd consisting of small crystallites (bright areas) joined together along a network of grain boundaries (pattern of dark lines).

The hypothesized low density in the boundary cores agrees with the following findings.

1. Precise measurements of the density of nanostructured Pd (crystal size about 11 nm) yield a density of about 96% of the pore-free material.[135] This indicates an average density in the boundaries of about 70% provided the small crystallites have the theoretical density and the boundary cores extend over a region that is about 1 nm wide. (All densities are given relative to the density of a perfect crystal with the same chemical composition.)
2. The measured scattering cross sections of nanostructured materials in small-angle neutron scattering experiments indicate an average density in the boundaries of about 70%.[136,137]

Figure 6. Electron micrograph of a symmetric 36.9° [100] tilt grain boundary in NiO. The black regions represent the atomic positions. In the boundary core the atomic density and coordination are changed relative to the perfect lattice. The boundary structure in the facets A and B are different because of the vertical displacement of the boundary plane relative to the two crystals. (After Merkle *et al.*[17])

3. High-resolution electron micrographs of special high-coincidence grain boundaries in NiO (Fig. 6) suggest an average boundary density of about 75%.[17]
4. The behavior of the hyperfine parameters of the Mössbauer spectrum of nanostructured Fe is consistent with a reduced density in the interfacial region[16] (see Section 4.1.).
5. A strongly (several orders of magnitude) enhanced solubility for solute atoms in the boundaries of a nanostructured material (H in Pd, Bi in Cu, Ga in W) relative to the solubility of crystals and/or glasses with the same chemical composition[138–140] suggests an enhanced free volume in the boundary regions.
6. In all nanostructured materials studied so far, the self and substitutional solute diffusion in the boundaries was found to be enhanced up to 20-fold relative to that in crystals and/or glasses with the same chemical composition. These high diffusivities suggest an enhanced free volume (reduced density) in the boundary cores.[141–144] This conclusion agrees with the fact that hydrostatic pressure reduced the enhanced diffusion rate[143,144] significantly.
7. Recent computations of the free volume of grain boundaries[145–147] suggest average boundary densities on the order of about 80% of the crystal density.
8. Recent studies of the pressure-induced changes in the Mössbauer spectrum,[148] the diffusivity,[143] and the positron lifetime[149] of nanostructured Fe or Pd suggest that the compressibility of the boundary cores is much higher than that of the crystals of the same chemical composition. This result is consistent with the proposed enhanced free volume in the boundary cores.

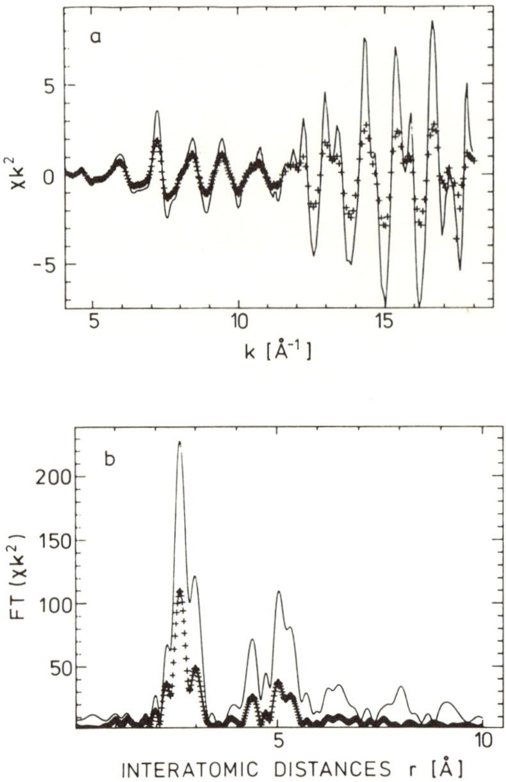

Figure 7. (a) The weighted EXAFS, χk^2, and (b) the Fourier transforms, FT(χk^2) (phase shift not included), of a nanocrystalline W sample (crosses) relative to polycrystalline W (solid line). In the nanocrystalline sample the amplitude of the EXAFS oscillations and the FT(χk^2) are weaker than in the polycrystal. The average diameter of the crystallites was 10 nm. (After Haubold *et al.*[151])

For comparison, the typical densities in metals and glasses are about 95% or more. As many properties of a solid depend primarily on their densities, the properties of nanostructured materials are expected to differ from those of crystals and/or glasses with comparable chemical compositions, as observed previously.[13,15,129–131]

The expected wide distribution of interatomic spacings in the boundary cores agrees with the following experimental results.

1. The amplitude of the EXAFS oscillations of a nanostructured material is reduced relative to that of a coarse-grained polycrystal with the same chemical composition.[150,151] An example is shown in Fig. 7. The reduction of the EXAFS amplitude originates primarily from the grain boundaries. This result indicates a broad (nearly random) distribution of interatomic spacings in the boundary cores.
2. The enhanced width of the distribution of the quadrupole splitting in the Mössbauer spectrum of nanostructured FeF_2 in comparison with a coarse-grained polycrystal of the same chemical composition suggests that a broad distribution of nearest-neighbor configuration exists in the boundary cores.[152]

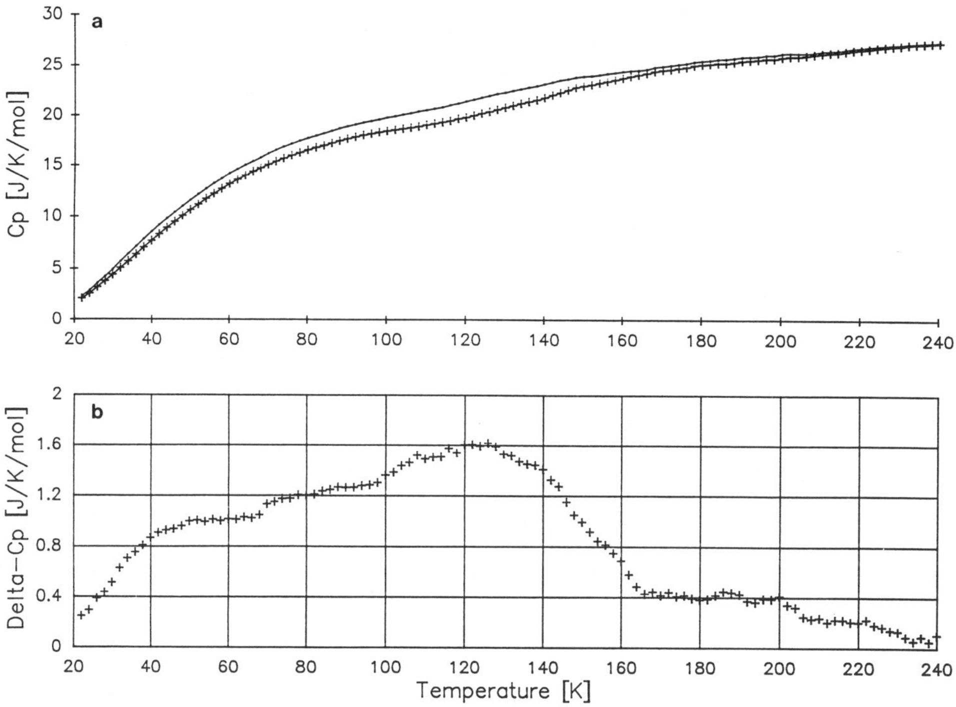

Figure 8. (a) Specific heat, C_p, of a nanostructured Y–7 at.% Fe alloy (solid line) relative to the same alloy in polycrystalline form (crosses). (b) The difference in C_p data between the nanostructural and polycrystalline forms. The enhancement of C_p in the temperature regime 40–160 K indicates a reduced Debye temperature for nanostructured Y–Fe alloy. (After Klein.[157])

3. The Debye temperatures of nanostructured materials are reduced relative to the values for the corresponding crystals. For example, the Debye temperatures of nanostructured Fe and $FeFe_2$ are reduced from 467 K to 345 K and 298 K to 86 K, respectively.[16,153] This reduction agrees with a wide distribution of interatomic spacing.[154]
4. The enhanced specific heat of nanostructured materials (Fig. 8[155–157]) shows that the vibrational modes of these solids differ from the modes of the corresponding crystals or glasses. A reduction of the density and a broad distribution of the interatomic spacings of a solid are known[154] to change the vibrational modes in such a manner that an enhanced specific heat results, as was indeed observed.
5. The thermal expansion (Fig. 9) of nanostructured Pd was found to be enhanced relative to that of Pd single crystals indicating an enhanced anharmonicity of the vibrational modes in the nanostructured materials. This is to be expected if broad distributions of interatomic spacings exist in the interfaces.[135,154] The interpretation of the enhanced thermal expansion of a nanostructured material in terms of an

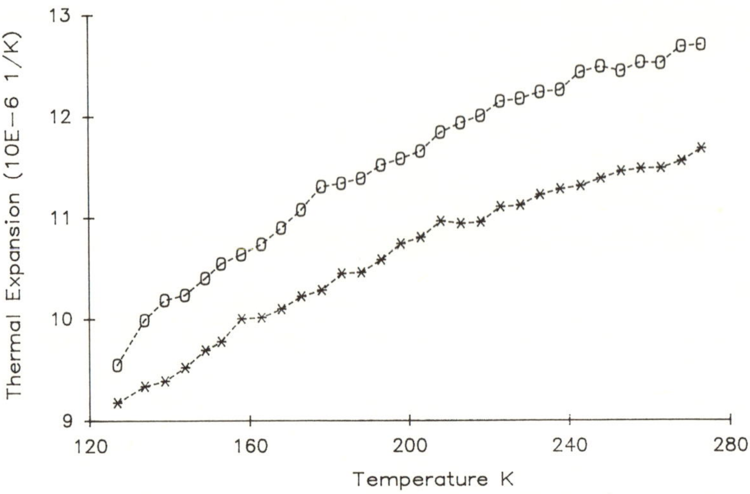

Figure 9. Thermal expansion coefficient of nanostructured Pd (average crystal size 11 nm, open symbols) and a coarse-grained polycrystal of Pd (crosses). The nanostructured Pd specimens were prepared by consolidating nanometer-sized Pd crystals at 20°C and 3.5 GPa. Subsequently the resulting material was hot-pressed (24 h at 100°C, 2 GPa). The final porosity was 2% or less.

enhanced expansion of the boundary regions agrees with recent measurements of the thermal expansion of grain boundaries in bicrystals and polycrystals.[158,159] In both cases the enhanced thermal expansion was suggested as originating from the modified vibrational modes in the boundary regions.

6. Nanostructured materials are found to exhibit positron lifetimes that indicate the presence of regions with enhanced free volumes,[160] i.e., regions in which the density is reduced and the interatomic spacings are increased relative to those in crystals and glasses with the same chemical compositions.
7. The results of recent molecular dynamic studies[146] indicate a broad distribution of interatomic spacings in grain boundary cores (Fig. 10).

A comprehensive discussion of these results and further references to the relevant original publications are given in Refs.[13,15,129–131] In summary, all available results seem to be basically consistent with a structural hard sphere model as indicated in Fig. 3a.

Nonetheless, it is clear that a hard sphere model such as the one shown in Fig. 3a cannot account for the details of the atomic arrangements in the cores of the boundaries in different materials. In recent years, evidence seems to have accumulated suggesting that such differences in the boundary core structures do in fact exist. For example, small-angle scattering studies[161] of nanostructured materials of different chemical compositions indicate a correlation between the chemical composition (the interatomic forces) and the

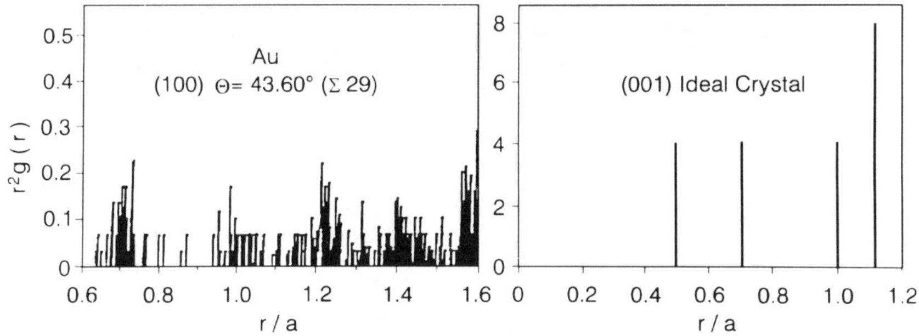

Figure 10. Radial distribution function $r^2g(r)$ for the atoms in the core of a 43.6° (Σ29) boundary. The interaction between the atoms was described by an embedded atom potential for gold (left panel). The panel on the right displays the radial distribution function of an ideal crystal in the (001) plane. (After Phillipot *et al.*[146])

boundary core structure. In nanostructured Pd an abrupt (atomically sharp) transmission was observed between the regions of low density (i.e., the boundary cores). This result agrees with a hard sphere model of the structure (Fig. 3a). However, such a model does not seem to apply to other nanostructured materials. For example, in a nanostructured $Si_{77}Au_{23}$ alloy, the density seems to vary gradually if one proceeds in the direction normal to the plane of the boundaries. In fact, the transition zone between the regions of low and high densities seems to be several interatomic spacings wide. A similar result emerged from a comparison of the wide-angle X-ray scattering data and the EXAFS oscillations of nanostructured Fe and Pd or W and Cu.[150,151,161,162] In both metals with an fcc lattice structure (Cu, Pd), a hard sphere model (Fig. 3a) accounts well for the wide-angle scattering behavior. In other words, the atoms in the boundary cores are displaced little from the corresponding lattice sites.[163] This does not seem to apply, however, to Fe and W. In these two materials, the atoms in the boundary cores were found to be displaced from lattice sites by distances comparable to an interatomic spacing.[162] This means that these materials consist structurally of the following two components: the crystallites in which the atoms occupy lattice sites and the boundaries in which the atoms are displaced from these sites (Fig. 11). A comparable correlation between the core structure of defects and the interatomic forces (chemical composition) has been demonstrated for the core structures of screw dislocations in fcc and bcc metals.[164]

The atomic arrangements formed in the cores of the boundaries seem to depend not only on the chemical composition (interatomic forces) but also on the time–temperature history of a sample and/or on the method of preparation. For example, the energy stored in nanostructured Pd (average crystal size 11 nm) was found to be reduced by about 50% when the specimen was annealed for several hours at 110°C although no measurable grain growth occurred.[165] Similar effects were noted for the enhanced thermal expansion

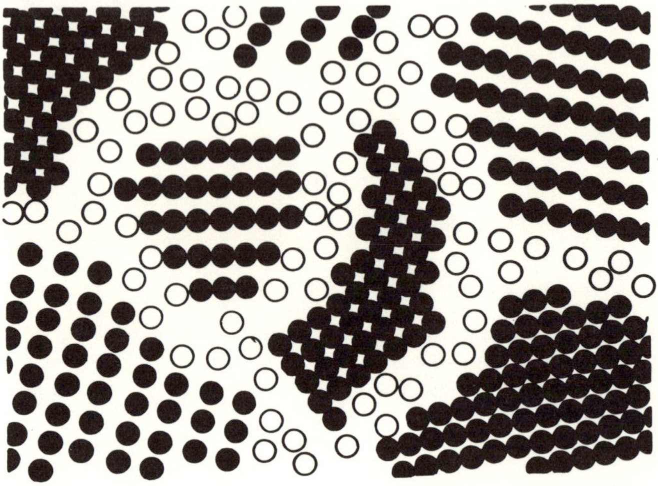

Figure 11. Schematic model of a nanostructured material if the atoms in the grain boundary core do not remain on lattice sites as was assumed in Fig. 3a. The atoms represented by open circles have been displaced in nonlattice site positions in order to minimize the total energy of the system. All atoms (solid and open circles) are assumed to be chemically identical.

and the enhanced specific heat. Both properties changed when the specimens were annealed at temperatures at which no measurable grain growth occurred[135,155,157] (see Fig. 8).

3.2. Nanostructured Alloys

So far, we have limited ourselves to nanostructured materials consisting of chemically identical atoms or molecules. The way to a new family of nanostructured materials called "nanostructured alloys" was open when this limitation was abandoned.[166] The preparation of nanostructured alloys follows a procedure similar to the one described in Section 2.2.3. Instead of using one evaporation unit (Fig. 4), several units are used for the substances A, B, C, . . . , which are evaporated (or sputtered) simultaneosuly. The small crystals of A, B, C, . . . , formed in the He atmosphere accumulate on a rotating cold finger as a loose powder. The powder was found to consist of a random mixture of small crystals of A, B, C, ... Subsequent consolidation of the stripped-off powder led to a nanometer-sized polycrystal, the individual crystallites of which consist of A, B, C, ... This product was called a "nanostructured alloy." Compared with single-component nanostructured solids, nanostructured alloys contain a high density of interphase as well as grain boundaries.

The atomic structure of nanostructured Ag–Fe alloys has been studied recently.[166] Ag–Fe alloys were selected as one of the model systems for study

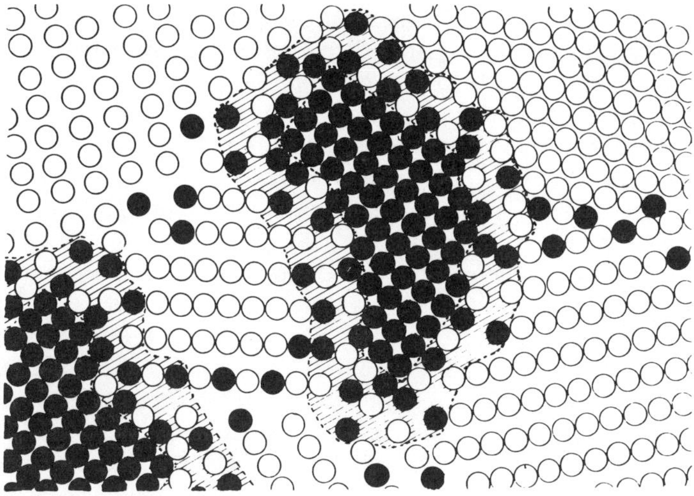

Figure 12. Schematic model of nanostructured Ag–Fe alloy based on analysis of X-ray and Mössbauer spectroscopy data (see Section 4.2). The alloy consists of a mixture of nanometer-sized Ag and Fe crystals. In the (strained) interfacial region between Ag and Fe crystals, solid solutions of Fe atoms in Ag and Ag atoms in Fe are formed although both components are immiscible in the liquid and/or solid state. Similar effects may occur in the grain boundaries between adjacent Fe and Ag crystals.

for the following reasons: Ag and Fe are insoluble in the solid and liquid states and the atomic structure can be investigated conveniently by X-ray diffraction as well as Mössbauer spectroscopy. X-ray diffraction and Mössbauer spectroscopy identified the following four components: bcc α-Fe crystals, bcc Fe–Ag solid solutions, fcc Ag–Fe solid solutions and pure Ag crystals (see Section 4.2). The atomic arrangement in a nanostructured Ag–30 at.% Fe alloy deduced from these observations is shown in Fig. 12. Apart from pure bcc α-Fe and fcc Ag crystals, solid solutions are formed in the vicinity of the Ag/Fe interphase boundaries and/or at the Fe/Fe, Ag/Ag grain boundaries, respectively. Clearly, if alloying occurs in a system such as Ag–Fe which is insoluble in the crystalline as well as in the molten state, one might expect alloy formation to occur in other (less insoluble) systems as well. This speculation seems to be confirmed by other experiments with alloying having been revealed recently in nanostructured Cu–Bi, W–Ga, Ir–Cu, Si–Fe, and Si–La.

All of these pairs of chemical elements exhibit virtually no mutual solute solubility if prepared in the form of coarse-grained polycrystals. The processes controlling the formation of alloys in nanostructured materials have not been elucidated in detail. Nonetheless, the following effects presumably play a role: (1) the enhanced entropy of mixing, if a solid solution is formed in the boundary regions; (2) a reduction of the energy stored in the strain fields of the boundaries. The energy of the strain field is known to be reduced if large (small) solute atoms are incorporated in regions of dilatation (compression). Such regions exist in the vicinity of boundaries due to the misfit in the boundary cores; (3) electronic screening effects due to different Fermi energies on both sides of interphase boundaries. All three effects have been revealed to exist at interfaces between chemically different crystals. In coarse-grained polycrystals, these effects change the solute solubility insignificantly relative to single crystals due to the small ($\leq 10^{-2}\%$) percentage of boundary regions. However, the opposite may apply to nanostructured alloys in which the boundaries occupy about 50% of the volume.

A second group of nanostructured alloys may be generated by modifying the surface regions of the small crystallites prior to consolidation. An example is the oxidation of the surface region of nanometer-sized Al crystals to Al_2O_3. Subsequent consolidation results in a nanostructured material with Al_2O_3-grain boundaries whereas the interior of the crystallites remains metallic Al. Similarly, the free surface of the crystallites, e.g., Al, may also be modified by depositing a thin layer of another material (e.g., NaCl) on it prior to consolidation. A further modification along the same lines is obtained if a second material is implanted into the small crystallites prior to consolidation. The advantages of this approach, in comparison with implantation in bulk pieces of materials, are apparent: the implantation energies for nanometer-sized crystals are small because the implantation depth is only a few atomic spacings. The material resulting from the consolidation of the implanted crystallites has the implant uniformly distributed in the entire volume and not just near the free surface. A first technological application of a nanostructured alloy with chemically modified interfacial regions seems to be Fe–Si–B alloys doped with Cu and Nb. Their magnetic properties (magnetic losses $\leq 200\,mW/cm^3$, effective permeabilities ~ 105) are comparable or better than the best values achieved up to now in metallic glasses.[167]

3.3. Imperfect (Distorted) Crystallites: Nanoglasses

So far, the nanometer-sized crystals have been assumed to be free of defects. A new structural "degree of freedom" opens up, however, if one incorporates defects into the crystallites prior to the compaction process. In other words, the resulting nanostructured material now contains two types of defects: the grain and/or interphase boundaries (as in all cases discussed so far) and, in addition, the defects inside the crystallites. An example of this kind of nanostructured material is illustrated schematically in Fig. 13. The crystallites contain large solute atoms that distort their lattices elastically.

This distortion reduces the atomic order relative to a nanostructured material made up of undistorted or less-distorted crystallites. In fact, this reduction is borne out by experimental results. Figure 14 displays X-ray scattering curves and the computed pair-correlation functions for nanostruc-

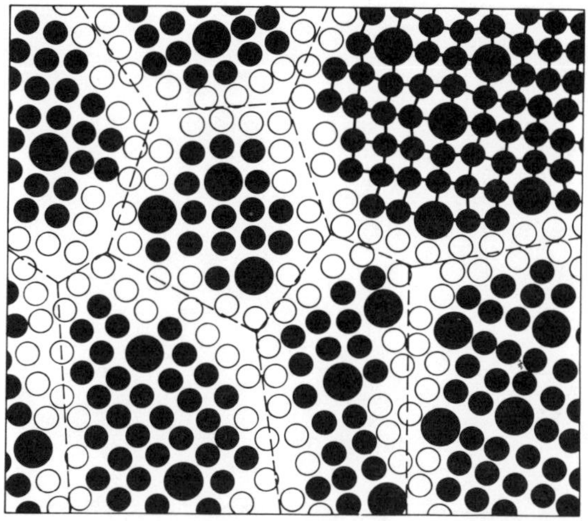

Figure 13. Schematic model of a nanostructured material consisting of small crystallites into which large solute atoms are incorporated substitutionally. These large solute atoms distort the lattice in their vicinity.

tured Si, a nanostructured Si–Fe-solid solution (3 at.%Fe), and a nanostructured Si–Au-solid solution (25 at.% Au). Fe and Au atoms differ in size by about 10 and 20 at.%, respectively, from Si atoms. In the case of the nanostructured Si–Fe, the atomic correlation was found to extend out to third-nearest neighbors whereas in nanostructured Si–Au (high content of large solute atoms), the atomic order is limited to the nearest neighbors only.[169] In other words, by varying the size and concentration of the solute atoms, the order of the material may be tuned in addition to the structural changes resulting from the variation of the crystal size, considered in the previous sections. It is the common feature of all nanostructured materials considered in this section that one first generates isolated small crystallites, the lattice of which is distorted by incoporating defects. These distorted crystals are consolidated in the second step of the production process into a nanostructured material. Obviously, if the distortion of the lattice of the crystallites is increased, it becomes more and more difficult to distinguish them structurally from nanometer-sized spheres with glassy structures. In fact, historically, the first nanostructured materials with distorted atomic arrangements in the nanometer-sized particles were prepared by evaporating glass-forming alloys (Pd–Si, Au–Si).[169–170] The resulting materials were described in terms of consolidated nanometer-sized glassy particles and were thus called "nanoglasses." It now appears that there is a continuous transition between nanocrystalline and nanoglassy materials. In fact, a continuous sequence of structures seems to exit extending (on the one side) from nanostructured materials that are generated by consolidating (small) defect-free crystals, to

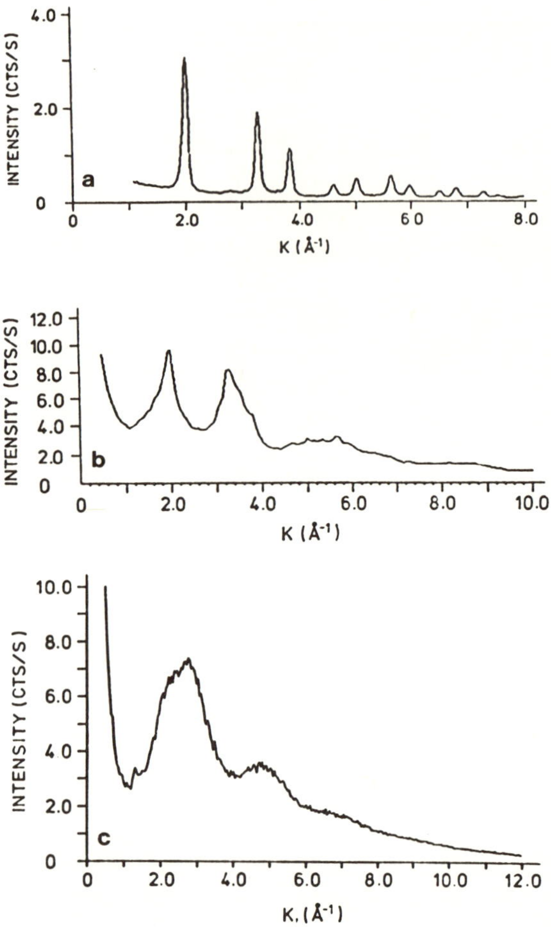

Figure 14. (a) X-ray diffraction curve of nanostructured Si (5-nm crystal size) (K is the scattering vector) (after Karch[168]). (b) X-ray diffraction curve of nanostructured Si–5 at.% Fe. (c) X-ray diffraction curve of nanostructured Si–25 at% Au (after Weissmueller *et al.*[169]). The decreasing number of maxima in the curves of panels (a)–(c) indicate a reduction in the range of atomic order with increasing alloy content. The materials of (b) and (c) have been produced in the same way as the nanostructured Si of (a).

nanostructured materials (on the other side) that are synthesized from nanometer-sized particles with highly distorted atomic bonds. In other words, it seems possible to tune the atomic arrangements in nanostructured materials from a nanocrystalline to a nanoglassy structure. In a nanocrystalline material, the low-energy structure of the crystallites is maintained by localizing the misfit into the boundaries. However, if the crystallites become more and more distorted, their energy increases and, thus, the driving force for localizing the misfit into narrow boundary regions (Fig. 3a) is reduced. Hence, in nanoglasses the regions where neighboring particles are joined together may become wider (more and more delocalized) as the particle distortion increases. Experimental observations by small-angle X-ray scattering supporting this view have been reported recently.[161]

4. MÖSSBAUER EFFECT STUDIES

4.1. Nanostructured Metals: Fe, Ni

A clear insight and understanding of the Mössbauer spectra of a range of nanostructured materials has been gained from the combined investigations of the groups of Gleiter and Gonser at the University of Saarland.[e.g.,16,170,171] A range of different classes of high-quality samples have been prepared by Gleiter and co-workers by consolidation of material produced by the inert gas condensation technique (Section 2.2.3).

^{57}Fe is of course the archetypal Mössbauer isotope with currently about 80% of Mössbauer effect studies being carried out with this isotope.[6] Fe was therefore chosen as the ideal sample for the first systematic investigation of the nanostructured state by Mössbauer effect spectroscopy.[16] Nonetheless, the relatively simple spectrum of nanostructured Fe shown in Fig. 1 (and repeated here for convenience as Fig. 15) belies the detailed information that can be obtained about the boundary regions between the crystals on analysis of this spectrum and related variable temperature spectra. The 77 K spectrum of Fig. 1 is well described by a fit to two sextets. Identification of component 1 with the fraction of crystalline Fe in the sample follows readily from the relative sharpness of the six lines and its hyperfine parameters. Identification of component 2 in terms of a contribution from the interfacial component, formed by the atoms located either in the crystals or in the interfacial region between them, is well based on the following grounds:

1. The enhanced value of the isomer shift of component 2 compared with that of component 1 reflects a reduction in the electron density of the interfacial component which in turn is consistent with the measured lower mass density, indicating a volume expansion in the interfacial

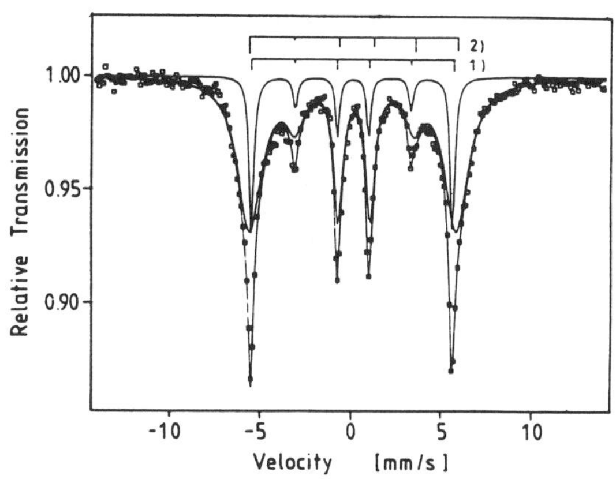

Figure 15. The 77 K Mössbauer spectrum of nanostructured Fe. The two-component fit to the spectrum is discussed in the text. (After Herr *et al.*[16])

component. Herr *et al.*,[16] in noting the results of optical and electron microscopy studies as well as X-ray diffraction measurements on nanostructured samples, concluded that the reduction in the mass density for nanostructured Fe compared with polycrystalline Fe results primarily from a lower density of the interfacial component (see also Section 3.1), in agreement with the findings of the Mössbauer experiments.

2. Likewise, the increased value of the average magnetic hyperfine field of component 2 ($\bar{H}_2 = 351$ kOe) compared with that of component 1 ($\bar{H}_2 = 343$ kOe) was taken as further evidence for volume expansion of the interfacial component. According to the Bethe–Slater curve, an increase in the near-neighbor distance would, in the case of Fe, lead to an increased value in the magnetic moment per atom and hence an increased magnetic hyperfine field.[16,171]
3. The large width of the lines comprising the sextet of component 2 reflects the broad distribution of magnetic hyperfine fields associated with this component. This is consistent with the broad spectrum of interatomic spacings found to exist in the interfacial component as discussed fully in Section 3.1.
4. The behavior of the Mössbauer spectrum of nanostructured Fe as a function of temperature, as also examined by Herr *et al.*,[16] gave further support to the above conclusions. As shown by Fig. 16, the temperature dependences of $\bar{H}_1(T)$ and $\bar{H}_2(T)$ differ significantly, reflecting the essential difference in the two states of matter identified for each of these components. Further, a calculation for the Debye temperature θ_D of the sample as a whole, based on a fit to the temperature-dependent total relative resonant absorption, led to the value $\theta_D^{nano} \sim 345$ K compared with $\theta_D^{Fe} \sim 467$ K for bulk Fe. Within the limitations of the Debye model of solids, and noting the good fit that the model gave to the temperature-dependent data, it is clear that the significant reduction in the θ_D^{nano} value is consistent with a significant fraction of a

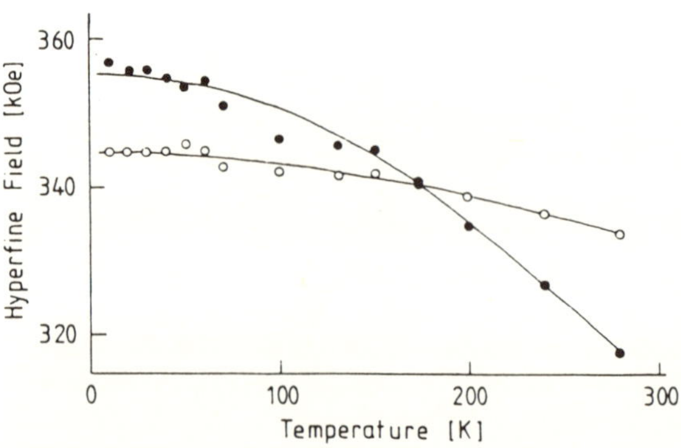

Figure 16. The temperature dependence of the mean magnetic hyperfine field values for the crystalline component (open circles) and the interfacial component (solid circles) of consolidated nanostructured Fe. (After Herr *et al.*[16])

noncrystalline component, i.e., the interfacial component (component 2), being present in the nanostructured Fe sample. Also, as noted previously (Section 3.1), this reduction in Debye temperature agrees with the existence of a wide distribution of interatomic spacings.[154]

The firm conclusion from these Mössbauer findings therefore is the existence of two components of comparable volume fraction in nanostructured Fe: the expected crystalline component associated with the crystals of sizes typically 1–10 nm and an interfacial component that exhibits distinctive behavior consistent with a broad spectrum of interplanar spacings for atoms in the boundary cores between the crystals.

Similar experiments carried out on nanostructured Ni doped with 2 at.% ^{57}Fe revealed essentially the same behavior[172] with the most important point again being the two-component nature of the spectrum. Unlike Fe, the exchange interaction in Ni does not, according to the Bethe–Slater curve, depend critically on the interatomic distances. This again agrees with the temperature dependence for the mean magnetic hyperfine fields for nanostructured Ni(^{57}Fe) in that no enhancement of the hyperfine field of the interfacial component over that of the crystalline component was observed, unlike the bahavior of nanostructured Fe.

An important qualification in the discussion of the structural and magnetic properties of nanostructured samples prepared by consolidation of gas condensed particles is to check the extent to which changes may occur as a result of the compacting process which can involve pressures of up to 5 GPa (Section 2.2.1). Hu *et al.*[173] have recently used polarized γ rays to examine the distribution of magnetic moments in nanostructured Fe. Their preliminary investigation revealed a different average value for the orientation of the magnetic moments with respect to the γ-ray direction in nanostructured Fe compared with a standard α-Fe sample. Such preferred orientation of the magnetic moments could be associated with the stress resulting from consolidation. This behavior, including the question of the possible influence of anisotropy resulting from the interaction between the interfacial components and the crystals, will require fuller investigation. Current *in situ* Mössbauer effect measurements of consolidated nanostructured Fe and *Fe*Pd subjected to external pressures[148] or high magnetic fields (0–16 T)[174] will help to clarify further the structural and magnetic behavior of the interfacial regions of the nanostructured state.

Interesting comparisons can be drawn between the behavior of nanostructured Fe as outlined above and that found for Fe with a submicrometer grain (SMG) structure of mean grain size on the order of 0.1 μm (100 nm). Using a method of high-strain-rate deformation (Section 2.3.2) based on intensive torsion straining under pressure, Valiev *et al.*[175] prepared SMG Fe. The sample was annealed at 440°C for 1 h and had a mean grain size of 0.22 μm.

Analysis of the room-temperature spectrum of this SMG Fe (Fig. 17)

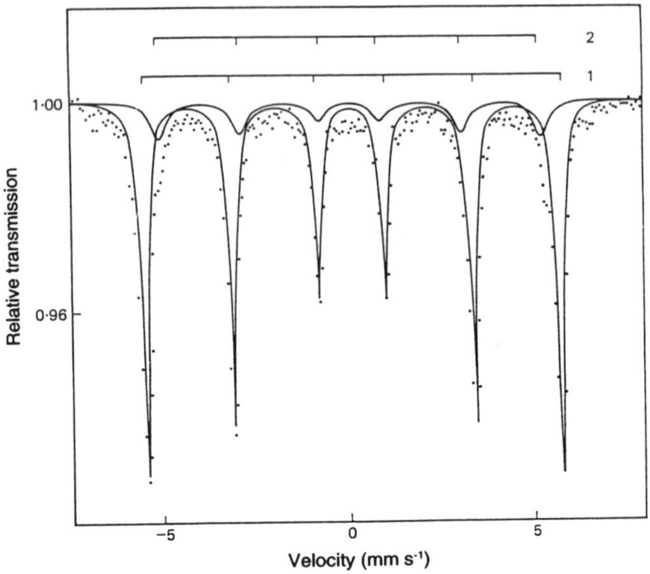

Figure 17. The room-temperature Mössbauer spectrum of Fe with a submicrometer-grained structure. The behavior of the two subspectra derived from a fit to the spectrum is discussed in the text. (After Valiev *et al.*[175])

again reveals two components consistent with the existence of two states for the Fe atoms. One component (subspectrum 1) shows the behavior expected for coarse-grained polycrystalline Fe while the other component, unlike the behavior observed in nanostructured Fe, exhibits a reduced value for the mean magnetic hyperfine field and a negative shift of the corresponding subspectrum (subspectrum 2). It was concluded that this second state (subspectrum 2) corresponds to a grain boundary phase (GP), the parameters of which are related to the physical width of the intercrystalline boundaries. Unlike the interfacial component of the nanostructured state, the change in isomer shift of the GBP subspectrum cannot be accounted for by an expansion of the lattice in the region of the intercrystalline boundaries, but rather is explained as a decrease in the number of conduction electrons in the GBP. This lower concentration of conduction electrons could also explain, in part, the decrease in the mean magnetic hyperfine field for this component although other electronic structure factors have also to be taken into account.[175] The overall differences in behavior of nanostructured Fe and SMG Fe are considered to be linked with the absence of local free volumes in SMG samples.

The similarities and differences between compacted nanostructured Fe and fine Fe particles (nanocrystals) suspended in silicon oil or paraffin can be gauged from a recent Mössbauer effect study[176] (see also Griffiths *et al.*,[177] for a discussion of the structural and magnetic properties of colloidal Fe dispersions for particles of median sizes in the range ~5–20 nm). The Fe particles of average crystal sizes ~6 nm and ~10 nm were prepared by the inert-gas condensation method and suspended in silicon oil or paraffin under vacuum to prevent oxidation. Inspection of such materials using electron microscopy reveals that not all of the crystals are isolated. Rather, most of the crystals form porous agglomerates (likely to occur when the crystals accumulate at the cold finger), comprising between a few and many thousands of crystals. Correspondingly, there is a fraction of the atoms inside the crystallites, a fraction of the atoms at free surfaces, and a further fraction at grain boundaries. The Mössbauer spectra of both samples, however, showed essentially the same two-component nature as observed for the compacted nanostructured Fe samples discussed above. For the suspended Fe particles, however, with less scope for an interfacial component existing between the nanocrystals, the component of hyperfine parameters similar to those of bulk Fe was attributed to Fe atoms within the interior of the particles, with the component of increased mean magnetic hyperfine field value compared with that of crystalline Fe being attributed to surface Fe atoms. Consistent with this interpretation, the sample of mean particle size ~10 nm exhibited a smaller contribution to the spectrum due to the component identified with surface atoms than did the sample of mean particle size ~6 nm; this reflects the decreased fraction of surface atoms for the sample of larger mean particle size. Comparison of the temperature dependence of the surface component in nanoscale Fe particles with that of the interfacial component in compacted nanostructured Fe shows, not unexpectedly, that an enhanced magnetic interaction becomes possible through the interfaces. This results in a magnetic behavior for compacted samples closer to that of a bulk behavior rather than the magnetic behavior indicative of a reduced magnetic dimensionality observed for the isolated Fe particles.

4.2. Nanostructured Alloys and Nanoglasses

As outlined in Section 3.2, the gas condensation technique has been extended to the preparation of nanostructured alloys. Two separated thermal evaporators were used by Herr *et al.*[166] to prepare consolidated binary mixtures of Fe crystals and Cu or Ag crystals of crystal sizes in the range 6–20 nm. The Cu–Fe samples containing 0.5 at.% Fe and 1 at.% Fe were found to behave in a manner similar to that of *Cu*Fe alloys cooled rapidly from the region of limited solubility[e.g.,178] or Cu–Fe samples prepared by the cosputtering of Cu and Fe onto a cooled substrate.[179] By comparison a different behavior, indicative of a different structure, was observed for samples of higher Fe concentration. In particular, the 10 K Mössbauer spectrum of Cu–30 at.% Fe (Fig. 18a) reveals magnetic hyperfine splitting with a distribution of hyperfine fields that exhibits a maximum at ~340 kOe (similar to the value for crystalline Fe) and a distribution that extends to low field values (Fig.

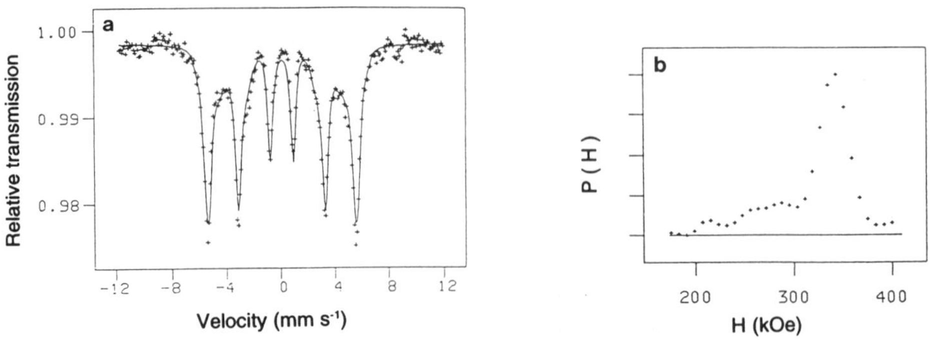

Figure 18. (a) The 10 K Mössbauer spectrum of a consolidated binary mixture of nanometer-size Cu and Fe crystals (Cu–30 at.% Fe). (b) The probability distribution of magnetic hyperfine fields for the Cu–30 at.% Fe sample as determined from the fit to the spectrum of (a). (After Herr *et al.*,[166] with permission from Pergamon Press Ltd.)

18b). The absence of an enhanced magnetic hyperfine field value above that of crystalline Fe (in comparison with nanostructured Fe, Fig. 15, which exhibits such an increase), indicated the absence of grain boundary regions between Fe crystals. On the other hand, in comparison with the behavior of thin-film and cosputtered Cu–Fe systems, it was concluded that the lower hyperfine field values (Fig. 18b) occur as the result of the formation of interphase boundaries between the Cu and Fe crystals, indicating the presence of metastable Cu–Fe alloy phases. Similarly, evidence for the formation of alloy phases of Ag–Fe over a range of concentrations was obtained from analysis of the Mössbauer spectra of a mixture of Ag and Fe crystals containing 30 at.% Fe (see Section 3.2 and Fig. 12). Given the insolubility of Ag and Fe in the solid and liquid state, as well as the extremely limited solubility of Fe in Cu even at high temperatures, the scope to prepare nanostructured alloys of the Ag–Fe and Cu–Fe systems demonstrates well the potential that nanostructured materials offer for the preparation of new materials.

A further development in the field of nanostructured materials is the preparation of nanoglasses—noncrystalline solids generated by consolidating glassy droplets of diameters $\leq$10 nm (see Section 3.3). Jing *et al.*[170] prepared a number of samples of analyzed composition $Pd_{70}Fe_3Si_{27}$ but of differing average diameter of droplets, by evaporating a $Pd_{65}Si_{35}$ alloy doped with 2 at.% ^{57}Fe at different pressures of pure He. Consolidation of these small glassy droplets by a high pressure at low temperatures—called a nanoglass—was expected to result in an atomic arrangement in the interfaces between adjacent droplets which differs from the structure in the center of the droplets. Figure 19 shows a comparison between the room-temperature Mössbauer spectrum of a consolidated $Pd_{70}Fe_3Si_{27}$ sample of mean droplet diameter ~3.6 nm and that of a $Pd_{72}Fe_{10}Si_{18}$ glass prepared by the standard melt-spinning technique. Also shown in Fig. 19 are probability distributions of the

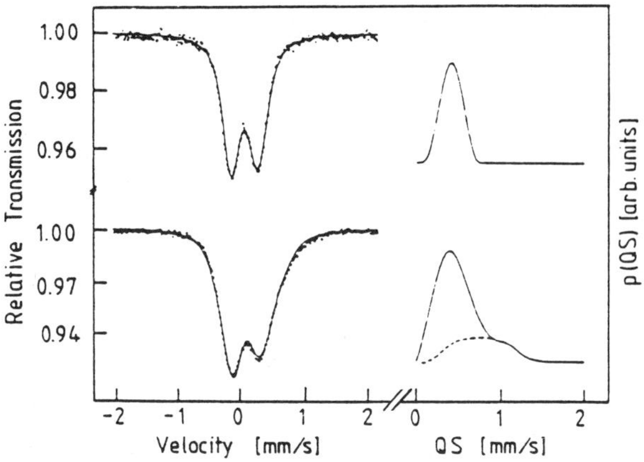

Figure 19. Room-temperature Mössbauer spectra for melt-spun $Pd_{72}Fe_{10}Si_{18}$ (upper spectrum) and a nanoglass $Pd_{70}Fe_3Si_{27}$ sample (lower spectrum) of mean droplet diameter 3.6 nm. The corresponding distributions of quadrupole splittings were derived from the fits to the data as discussed in the text. (After Jing *et al.*[170])

quadrupole splittings (QS) in these nonmagnetically split samples. The spectra were fitted assuming a linear correlation between the quadrupole splittings and the isomer shifts.[170] Several significant points can be noted:

1. The common peak in QS values at QS ~ 0.4 mm/s for both samples indicates a likely common structure for these features. For the nanoglass sample this peak is therefore linked with the core regions of the glassy droplets where the atomic arrangements are expected to be similar to those that exist in the melt-spun glass as a whole.
2. The extra component to the QS distribution of the nanoglass located around QS ~ 0.9 mm/s is considered to originate primarily from the interfaces between the consolidated glassy droplets. Chemical analysis and annealing experiments eliminated the possibilities of oxide contamination or crystalline phases causing this extra contribution to the QS distribution of the nanoglass.
3. Convincing evidence in support of the above interpretation that the extra contribution to the QS distribution of the nanoglass is due to an interfacial component is provided by the analysis embodied in Fig. 20. There the relative spectral fractions of the interfacial components (determined by fitting the QS distribution of the nanoglass to two Gaussian curves as indicated in Fig. 19) for a set of $Pd_{70}Fe_3Si_{27}$ samples of mean droplet diameters ranging from 3.6 to 7.2 nm, are plotted against the reciprocal of the droplet size. The observed increase in the spectral fraction of this component with a decrease in droplet size is consistent with the increased contribution and behavior expected of an interfacial component. The linear behavior observed in Fig. 20 further

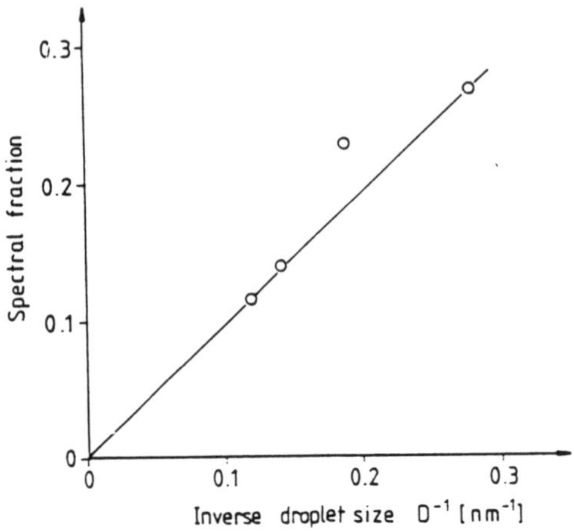

Figure 20. The relative fraction of the contribution of the interfacial components for a set of $Pd_{70}Fe_3Si_{27}$ nanoglass samples of different mean droplet diameters as a function of the reciprocal of the mean diameter of the glassy droplets. (After Jing *et al.*[170])

indicates that the thickness of the interfacial region is independent of the droplet size. For an equiaxial shape for the doublets and a constant interfacial thickness *d*, this leads to the value $d \simeq 0.4$ nm, corresponding to an interface between the droplets of about two interatomic spacings.[170]

4. This spectral analysis also leads to the result that the isomer shift of the interfacial component is increased relative to that of the inner core regions. Similar behavior was, as noted already, observed in the interfacial component of nanostructured Fe.[16] As before, the reduced *s*-electron density may reflect a reduced mass density for the interfacial region and/or electronic localization effects in the interfaces.

4.3. Nanostructured Ionic Materials

The inert-gas condensation technique has also been applied to the study of nonmetallic systems. Of particular interest was to establish whether or not the unique atomic arrangement linked with the interfacial component formed with a large concentration of grain boundaries in nanostructured metals also existed in ionic systems. On the basis of laser Raman spectroscopy measurements on nanostructured TiO_2, Melendres *et al.*[180] concluded that such an atomic arrangement did not exist. Ramasamy *et al.*[152] therefore compared powdered FeF_2 and nanostructured FeF_2 of mean crystal size 8 nm prepared by pressing powder produced using the inert-gas condensation process. FeF_2 was chosen as a convenient well-characterized model system for study. Figure 21 shows the room-temperature Mössbauer spectra for the FeF_2 in the two different sample states. Again the nanostructured material is found to be best described by a two-component fit to the spectrum (Fig. 21b). In this case, two quadrupole

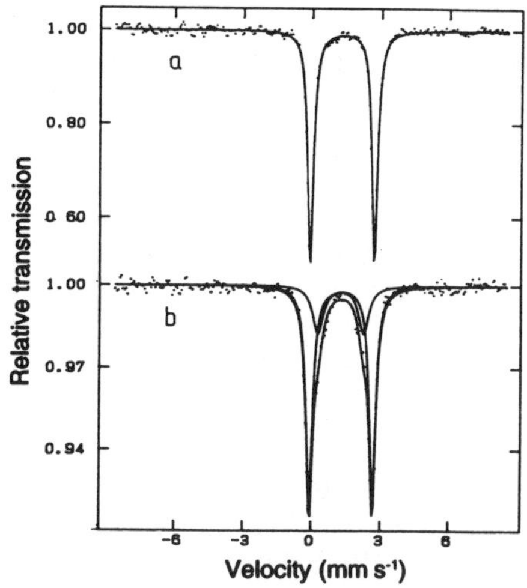

Figure 21. The room-temperature Mössbauer spectra of (a) bulk FeF_2 powder and (b) nanostructured FeF_2. The fits to the spectra are discussed in the text. (After Ramasamay *et al.*,[152] with permission from Pergamon Press Ltd.)

doublets are observed in nanostructured FeF_2: the first of fractional spectral area ~75% has hyperfine parameters similar to those of the bulk FeF_2 (Fig. 21a), with the second doublet of fractional spectral area ~25% having a linewidth about 60% greater than that of the doublet linked with the bulk FeF_2. More detailed analysis of the spectra in terms of the distributions of quadrupole splittings rather than the average quadrupole doublets shown in Fig. 21 results in a QS distribution for the nanostructured FeF_2 that is similar to the distribution of magnetic hyperfine fields observed in nanostructured Fe.

The conclusion that the additional broadened QS doublet originates from atoms in the grain boundaries was supported by a low-temperature study of the antiferromagnetic phase transition in the nanostructured FeF_2 samples.[153] Several interesting points, generally in support of the existence of the interfacial component in nanostructured FeF_2, emerged from their comprehensive comparison of the magnetic behavior of FeF_2 in the two states:

1. First, the antiferromagnetic ordering temperatures (T_N) as measured by the thermal scanning method (in which the γ-ray count rate at constant velocity changes significantly at the onset of magnetic hyperfine splitting) exhibited different characters for the two samples. T_N for the bulk sample was spread over a relatively narrow range $T_N^{bulk} \sim 75.5$–78 K in good agreement with published data whereas the antiferromagnetic transition for the nanostructured sample took place over a broad temperature range $T_N^{nano} \sim 66$–78 K. This broadening was considered to reflect a contribution from the interfacial component. A reduction in the number of near-neighbor ions and the exchange integral is expected to occur as a result of an increased average interatomic spacing in the interfacial regions, leading to the observed reduction and broadening of the transition.

2. Relaxation effects are also considered to contribute to the behavior at the magnetic transition of the nanostructured FeF_2. A spectrum at 70 K—in the middle of the transition range $T_N^{nano} \sim 66$–78 K for the nanostructured sample—revealed a quadrupole doublet in addition to the well-established magnetic splitting. This component was identified as corresponding to a contribution from the bulk component remaining in the paramagnetic state. This occurs as a result of the small crystallites of the nanostructured state exhibiting a magnetic relaxation time shorter than the Larmor precession time so that the quadrupole doublet of the paramagnetic state is observed; by comparison the larger crystallites exhibit the magnetic hyperfine splitting similar to that of the bulk FeF_2 at the same temperature.[153]
3. A comparison of the spectra of bulk and nanostructured FeF_2 at 40 K in their well-established magnetically ordered states (Fig. 22) provides further evidence in support of a contribution due to an interfacial component in the nanostructured state. First, it should be noted that in view of the large electric field gradient in FeF_2, the spectrum for the bulk sample (Fig. 22a) exhibits two extra, weak peaks (corresponding to the now weakly allowed $|\Delta m| = 2$ transitions) in addition to the familiar six-line spectrum. The overall features of this eight-line spectrum are repeated for the nanostructured sample (Fig. 22b) with additional broadening also present. This latter spectrum can, as shown

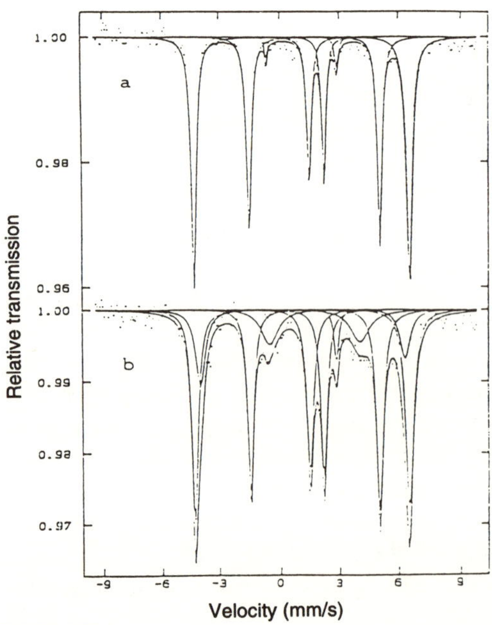

Figure 22. The 40 K Mössbauer spectra of (a) bulk FeF_2 powder and (b) nanostructured FeF_2. The fits to the spectra are discussed in the text. (After Jiang *et al.*,[153] with permission from Pergamon Press Ltd.)

by the subspectra of Fig. 22b, be further analyzed in terms of a dominant component of magnetic hyperfine parameters (mean $\bar{H}_{hf} \sim$ 309 kOe) equivalent to those of the bulk FeF_2, and a subspectrum of reduced mean hyperfine field value $\bar{H} \sim 295$ kOe, but represented by lines of increased width, reflecting a broader distribution of the hyperfine fields. In common with the spectrum for nanostructured FeF_2 in the paramagnetic state as discussed above, the spectrum of the antiferromagnetically ordered state can be well described by a component due to the grain boundary regions as well as one corresponding to the bulk state.

4. In agreement with the behavior of nanostructured Fe, a reduced Debye temperature for nanostructured FeF_2, $\theta_D^{nano} = 86$ K, was obtained relative to that for the bulk sample, $\theta_D^{bulk} = 298$ K.

4.4. Nanostructured Oxides

Compacted samples of nanoscale particles of α-Fe_2O_3 and γ-Fe_2O_3 were prepared by Jing *et al.*[181] to examine the magnetic behavior of nanostructured Fe oxides, particularly in relation to the behavior shown by fine, uncompacted particles of these oxides. The surface magnetic properties of α-Fe_2O_3 nanocrystals differ from those of the bulk.[182] Of particular interest, therefore, was the comparison between their magnetic behavior and that of the particles compacted into a solid sample so that the surfaces became interfaces with their unique atomic arrangement.

Little difference was discerned in the low-temperature (10 K) spectra of uncompacted and compacted nanocrystals of α-Fe_2O_3.[181] Both spectra could be well represented by two sextets of similar hyperfine parameters. For the uncompacted samples the two sextets ($\bar{H}^{cryst} \sim 532$ kOe, linewidth $\Gamma \sim$ 0.37 mm/s; $\bar{H}^{sur} \sim 501$ kOe, $\Gamma \sim 0.87$ mm/s) had fractional areas of ~48% for the crystalline component and ~52% for the component of increased linewidth which was associated with the behavior of the surface atoms. The compacted sample showed very similar behavior with the corresponding data being $\bar{H}^{cryst} \sim 532$ kOe, $\Gamma \sim 0.35$ mm/s and $\bar{H}^{inter} \sim 501$ kOe, $\Gamma \sim 0.93$ mm/s with a larger contribution from the latter component (identified as the interfacial component) $A^{inter} \sim 58\%$ than for the crystalline component $A^{cryst} \sim 42\%$. Significant differences between the two samples were, however, observed in their higher-temperature spectra (Fig. 23). The relative behavior of the two samples—in which the uncompacted sample at 260 K retains a dominant absorption feature near the center of the spectrum which is much reduced in the compacted sample—can be well explained in terms of superparamagnetic relaxation effects. Clearly the superparamagnetic relaxation component in the compacted sample state at 260 K is strongly suppressed compared with that observed in the uncompacted sample at the same temperature. This behavior can be explained in terms of an additional contribution to the anisotropic

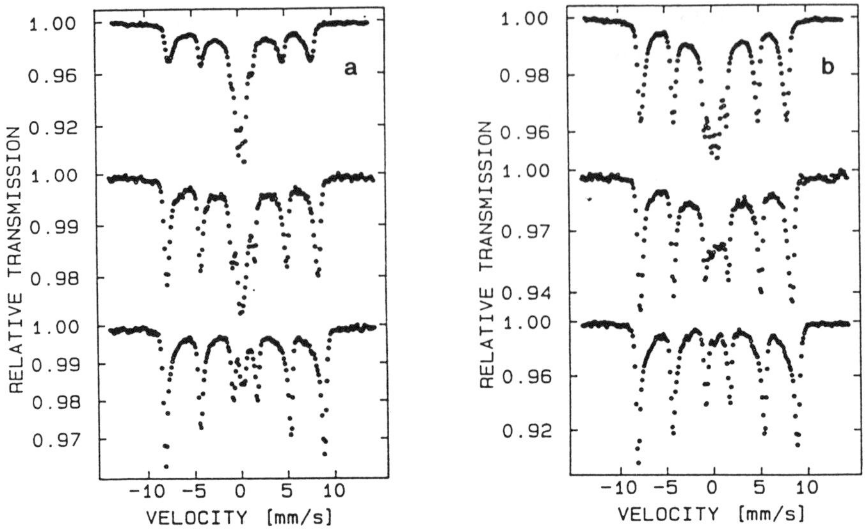

Figure 23. A comparison of the Mössbauer spectra of (a) uncompacted and (b) compacted α-Fe_2O_3 nanocrystals at various temperatures: 260 K (top), 180 K (middle), and 8 K (bottom). (After Jing *et al.*[181]).

energy due to the interface exchange energy occurring as a result of the interfaces present in the compacted sample. The total anisotropic energy of a nanocrystallite can then be represented by $E = KV + CF$ where K is the anisotropy constant, V the volume, C the interface exchange energy per unit interface area, and F the interface of the nanocrystallite. This leads to a contribution due to the interfacial exchange energy of ~30% of the total anisotropic energy for the compacted α-Fe_2O_3 sample of Fig. 23.

Whereas this exchange interaction in the interfaces was significant enough to partially suppress superparamagnetic relaxation in compacted α-Fe_2O_3, no corresponding suppression was observed in compacted α-Fe_2O_3 nanoparticles. This was explained by the fact that the dipole field interaction does not change as the ferromagnetic γ-Fe_2O_3 nanoparticles are brought closer together on compacting. Investigation of the physical properties of fine particles, particularly in relation to the behavior of the corresponding bulk material, is a major field of research in its own right (see, e.g., Refs. 183, 184 as well as references therein). Ayyub *et al.*[183] in a comprehensive investigation of Fe_2O_3 particles of different sizes (~5–70 nm) prepared by a three-component microemulsion technique, demonstrated the presence of crystal-size-induced structural transitions. α-Fe_2O_3 was found to nucleate for a particle size $\gtrsim$30 nm while γ-Fe_2O_3 was preferentially formed for $\lesssim$30 nm with amorphous Fe_2O_3 being formed at particle sizes of ~5 nm.

A further example of the suppression of superparamagnetism in nanoscale systems is provided by the investigation of Yang *et al.*[185] into Fe nanocrystals whose surface was purposely oxidized. A difference from the study of the compacted α-Fe_2O_3, however, is that after surface oxidation the powder was dispersed in evacuated silica oil, thus removing the scope for interfacial components between the particles. Nonetheless, clear evidence for suppression of superparamagnetism was observed with the effects again being explained by interface exchange anisotropy—the interface on this occasion being that between the surface oxide and the metallic cores of the nanocrystals. A Mössbauer effect study of the behavior of ultrasmall particles of Fe (~5-nm diameter) as a function of the exposure time to air has been reported recently by Linderoth *et al.*[186] Previously, Griffiths *et al.*[177] had investigated the structural and magnetic behavior of colloidal iron dispersions (~5–20 nm particles) exposed to the atmosphere using electron microscopy and magnetization measurements, respectively. The 12 K spectra of the Fe particles exposed to air for periods of up to 130 weeks showed that the oxide component of the passivated particles (as measured by the relative spectral area) increased from ~65% just after removal from the evaporation chamber to ~96% after 130 weeks' exposure.[186] Superparamagnetic relaxation was observed at room temperature for the oxide layer but not for the Fe core of the particles exposed to air for 130 weeks. The essentially unresponsive behavior of the component due to the oxide phase (other than a broadening of the lines) when subject to applied fileds of up to 4 T at 5 K indicates that the oxide layer has a noncollinear spin structure in contrast to ferromagnetic bulk oxides.[186] Given the industrial importance of ultrafine Fe particles in the fields of magnetic recording, ferrofluids and catalysis, it is clear that Mössbauer spectroscopy can continue to make a valuable contribution to improved understanding of the structural and magnetic behavior of such systems.

4.5. Nanostructured Materials: Mechanical Alloying

With the promising range of useful properties and applications that nanostructured materials offer,[11,12] a limitation in common with many of the synthesis techniques outlined in Section 2 is that they can only readily produce small quantities of material. While amounts of several hundred milligrams are more than sufficient for most research purposes, the prospect of industrial applications of nanostructured materials requires that they be produced relatively easily in large quantities. As noted already, the mechanical alloying approach to production of nanostructured materials can allow several kilograms to be produced in periods of up to 100.[96] Indeed, as reported recently by McColvin and Shaw,[187] Inco Alloys International, the world's only volume producer of mechanically alloyed materials in Ni, Fe, and Al based alloys, has (in 1990) a capacity for powder production of high-temperature alloys of 340 tonnes per year, which represent a tenfold increase since 1980. This increase in production was attained using ball-mills rather than attritors.

The fundamentals of mechanical alloying, a process that was first

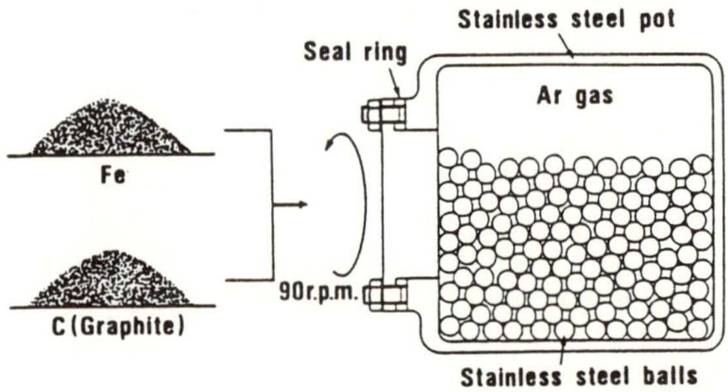

Figure 24. A schematic illustration of the application of a conventional ball-mill to mechanical alloying. (After Tanaka *et al.*[189] in their study of mechanical alloying of the Fe–C system.)

developed over 25 years ago, have recently been outlined by Benjamin.[188] The process remains a relatively straightforward one in practice, extending from the conventional (and still predominant) use of a rotating ball-mill to grind the powders using, say, stainless-steel balls in a protective atmosphere (Fig. 24[189]), to processes based on repeated rolling or pressing.[190] There is, however, a significant difference in the approach of these two techniques to the mechanically alloyed state: the former method is based on a probabilistic approach of impact, whereas the latter uses a more controlled, deterministic approach.[190,191]

An impression and overview of the state of maturity of mechanical alloying as a means of producing new phases and materials as well as nanostructured materials can be gained from the recent proceedings of the first International Symposium on Mechanical Alloying.[192] The relevance of the mechanical alloying approach to nanostructured materials is indicated by the fact that the proceedings[192] include a section (11 papers) devoted to the topic of "nano-crystalline" materials with Fe–C and Fe–C–Si,[193] Cr–Fe[194] and Al–Fe and Fe–Sn[195] being included among those systems studied by Mössbauer spectroscopy. Further examples of systems that have been prepared by mechanical alloying in recent years and that have been studied by Mössbauer spectroscopy are: Al–Fe,[95,196] Fe–V,[197] Fe and Fe–V,[198] Al–Fe, Fe–Sn, and Ag–Fe,[199] Al–Fe,[200–202] α-Fe_2O_3,[203] Fe–C,[189] Al–Fe and Ag–Fe,[204] Ag–Fe,[205] Al–Fe, Fe–Sn, Ag–Fe, Fe–C, Fe–B,[206] Fe–Cu,[207] Fe–B,[208–210] Fe–Sn,[211] Nd–Fe–B,[212] and Co–Fe–Si–B[213–215] (both without[213,214] and with [215] the aid of surface-active substances). As noted by Shingu *et al.*,[190] mechanical alloying can be used to produce a range of nonequilibrium structures: ultrafine grains (nanostructures), supersaturated solid solutions, metastable crystalline phases, and amorphous phases. Examples of these different types of structures will be encountered in the following discussion of examples of application of Mössbauer spectroscopy in the study of nanostructured materials produced by mechanical alloying.

As in the case of the application of the inert-gas condensation technique to the investigation of nanostructured alloys[166] (Section 4.2), the Ag–Fe[205] and Fe–Cu[207] systems have been chosen for study by mechanical alloying. These systems were selected by the collaborative groups at Kyoto and Osaka universities as the mutually repulsive nature of the two elements, particularly Ag–Fe, should result in distinct grain boundary features even in extremely fine grains.[205] The immiscibility of Ag and Fe and the very limited solubility of Fe in Cu and, vice versa, Cu in Fe, means that these systems represent good tests of the ability of mechanical alloying to indeed produce new phases and alloys, albeit metastable ones.

The effects of conventional low-energy ball-milling on elemental Ag–50 at.% Fe powders over extended periods using a ball-to-powder weight ratio[216] of BPR = 100 are shown by the X-ray diffraction patterns of Fig. 25.[205] The broadened peaks make it difficult to distinguish the diffraction peaks of bcc Fe from those of fcc Ag but, using Scherrer's formula, the width of the (111) Ag peak indicates a crystalline size of ~10 nm for the fcc phase after 800 h milling. High-resolution electron microscopy led to a mean size for the bcc Fe and Ag grains of ~15 nm. The room-temperature Mössbauer spectrum of the sample obtained after milling for 800 h (Fig. 26) provides evidence for alloy formation. The dominant feature of the spectrum is, of course, the contribution due to ferromagnetic bcc Fe (dashed line, Fig. 26a) but subtraction of that component leaves the subspectrum shown in Fig. 26b which Kuyama *et al.*[205] have then fitted independently. Their fitting is based on several six-line components of reduced magnetic hyperfine field values compared with ferromagnetic Fe and a central paramagnetic component. About 5% of the total number of Fe atoms (as measured by the subspectral

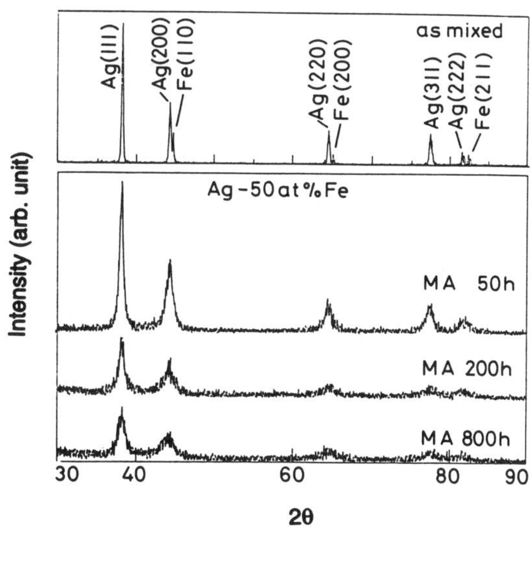

Figure 25. X-ray diffraction patterns (Cu$K\alpha$ radiation) of: the starting mixture of elemental Ag–50 at.% Fe powders (upper), and the powder mixture after mechanically alloying for the times indicated (lower). (After Kuyama *et al.*[205])

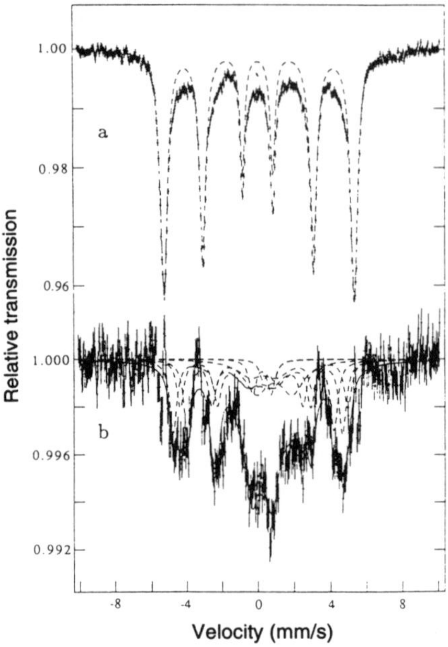

Figure 26. (a) The room-temperature Mössbauer spectrum for Ag–50 at.% Fe powders after ball-milling for 800 h. (b) The subspectrum remaining after subtracting the six-line component due to bcc Fe (dashed line in panel a). The fit to this subspectrum is discussed in the text. (After Kuyama *et al.*[205])

areas) contribute to the magnetically split subspectrum of reduced field value. By comparison the fraction of atoms that exist in grain boundaries can be estimated from $p = a/d^{1/3}r$ to yield the value $p \sim 3\%$ (d is the atomic density, r is the grain size, and the constant a is normally ≈ 1, depending on the crystal structure and grain size). The approximate agreement of these two values is taken as evidence for the formation of Ag–Fe alloy phases in the interphase boundary regions. The increased fraction of solid solution for Ag–Fe alloys obtained using the gas condensation technique[166] reflects the smaller grain sizes ~6–10 nm obtained by that technique. Similarly to Herr *et al.*,[166] the central paramagnetic absorption peak is linked with Fe atoms in the fcc Ag lattice (see also Ref.[190]) with the remaining magnetically split subspectra likely to result from metastable bcc Fe–Ag alloys of low Ag concentrations.

Mechanical alloying has also been used to extend the supersaturated solid solution of Fe in Cu[207] beyond that attained by rapid quenching from the melt or vapor deposition. Similar levels of solution of Cu in Fe were obtained for all three techniques. Transmission electron microscopy images of the mechanically alloyed powders showed that the structure was fine and homogeneous with average grain sizes in the range ~ few to 10 nm. The lattice parameter behavior shown in Fig. 27 for metastable bcc (Cu in Fe) alloys and fcc (Fe in Cu) alloys demonstrates the extension in the range of nonequilibrium phases produced by mechanical alloying over that for the phases produced by liquid quenching and sputtering. This is also brought out well by the room-

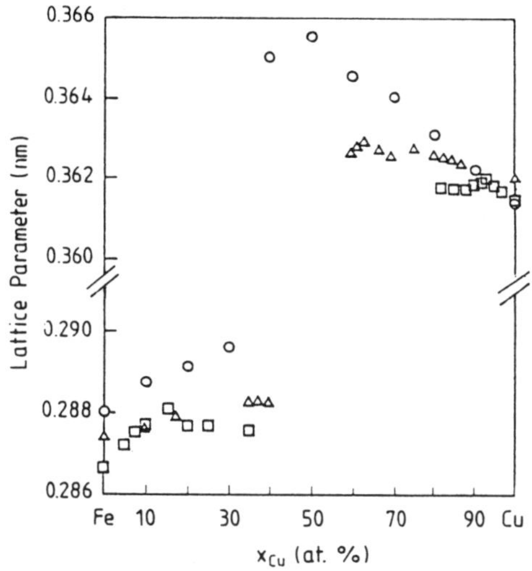

Figure 27. The lattice parameters for supersaturated solid solutions of bcc (Cu in Fe) alloys and fcc (Fe in Cu) alloys prepared by: rapid quenching from the melt[217] (squares), vapor deposition[218] (triangles), and mechanical alloying[207] (circles). (After Uenishi *et al.*[207])

temperature Mössbauer spectra for Fe–30 at.% Cu (Fig. 28) and Fe–90 at.% Cu samples (Fig. 29) ball-milled for periods of up to 400 h.[207] In the case of the Fe–30 at.% Cu mixture of powders (Fig. 28), the increased broadening of the six-line spectrum of pure bcc Fe and the decrease in the average value of the magnetic hyperfine field with increased milling time indicate increased mixing of the Fe and Cu atoms and changes in the local environment of the Fe atoms. Likewise, considering the behavior of the Cu-rich series, the gradual diminishing of the six-line spectrum of bcc Fe, and the enhancement of the central resonance feature (typical of the behavior of *Cu*Fe alloys above their magnetic ordering temperatures[178]) with increased milling time for Fe–90 at.% Cu (Fig. 29), leads to the absence of magnetic splitting and the sole absorption from the central resonance after milling for 400 h. This behavior, combined with the absence of other phases as assessed by X-ray diffraction or Mössbauer spectroscopy, is convincing evidence for the preparation of an Fe–10 at.% Cu alloy by mechanical alloying.[207]

A further example of the ability of the mechanical alloying process to produce materials in the nanostructured state is given by recent studies of the Fe–B system.[208–210] Amorphous Fe–B and related alloys exhibit attractive soft magnetic properties with many technological applications and as such have been the subject of extensive investigation.[e.g.,219] These studies[208–210] have demonstrated that mechanical alloying can, depending on composition, produce nanostructured materials and amorphous alloys. The ability to produce materials such as Fe–B and Co–Fe–Si–B[214,215] as powders is important in that it extends the potential range of technological uses of amorphous materials by allowing the amorphous powders to be consolidated into the required bulk form. The 4.2 K Mössbauer spectra of a series of ball-milled Fe–B powders of

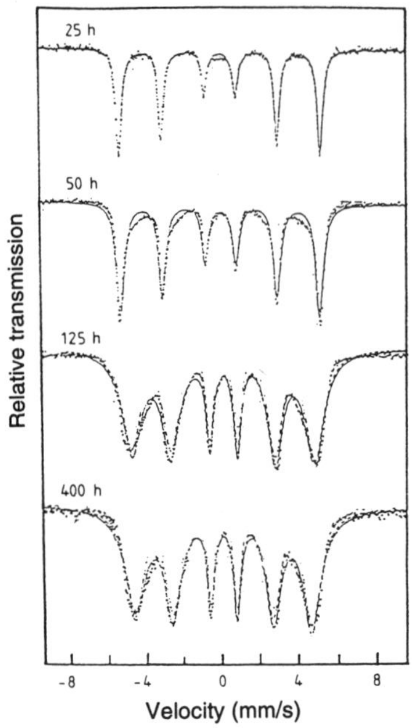

Figure 28. The room-temperature Mössbauer spectra of Fe–30 at.% Cu powders ball-milled for the times indicated. (After Uenishi *et al.*[207])

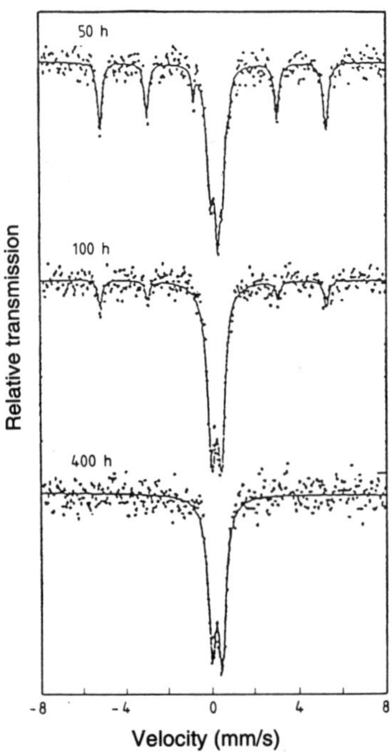

Figure 29. The room-temperature Mössbauer spectra of Fe–90 at.% Cu powders ball-milled for the times indicated. (After Uenishi *et al.*[207])

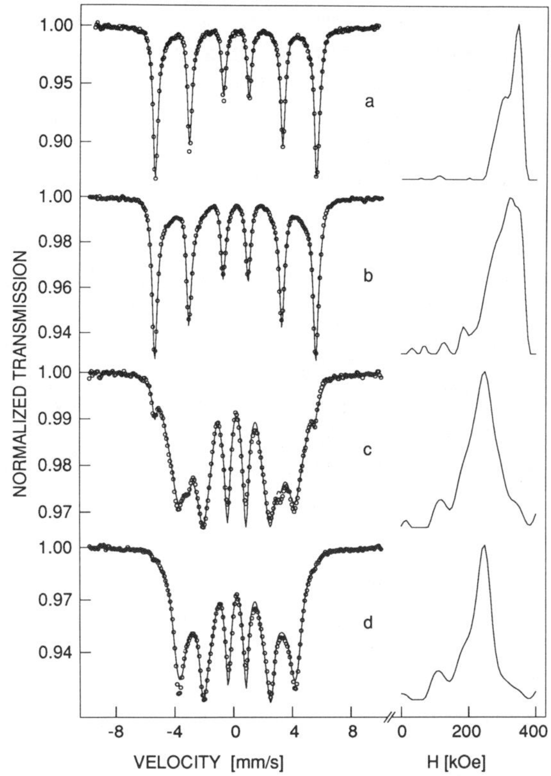

Figure 30. 4.2 K Mössbauer spectra of Fe–B powders of the following nominal starting compositions and subjected to low-energy ball-milling[99] for the times indicated: (a) $Fe_{80}B_{20}$ after 400 h, (b) $Fe_{66}B_{34}$ after 400 h, (c) $Fe_{50}B_{50}$ after 250 h, and (d) $Fe_{40}B_{60}$ after 250 h. The fits to the spectra were obtained using the program of Le Caer and Dubois[220] with the corresponding probability distribution of magnetic hyperfine fields for the interfacial or amorphous components also shown. (After Jing *et al.*[209])

nominal starting compositions $Fe_{80}B_{20}$, $Fe_{66}B_{34}$, $Fe_{50}B_{50}$, and $Fe_{40}B_{60}$ are shown in Fig. 30.[209] On the basis of these spectra and their corresponding X-ray diffraction patterns, a distinction can be made between the predominant nanostructured nature of the two Fe-rich samples (grain size $\gtrsim$ 5 nm) and the amorphouslike state of the samples with high B content ($\gtrsim$50 at.% B). The spectra of the $Fe_{80}B_{20}$ and $Fe_{66}B_{34}$ samples are similar to those of nanostructured Fe[16] in that they consist of the sharp six-line subspectrum of bcc Fe superimposed on a broader six-line subspectrum, with the combination of the subspectra giving rise to the observed asymmetrical spectra. A difference with the spectrum of nanostructured Fe, however, is that the distributed component in nanostructured Fe–B has a smaller mean value of the magnetic hyperfine field than that due to the interfacial component in nanostructured Fe. This reduction in hyperfine field can be explained as the formation of a "nanoalloy" in the interfacial component with atomic alloying occurring only in the interfacial region between Fe and B nanocrystallites while more-or-less chemically pure cores of Fe and B are maintained inside the crystallites. The distributions of magnetic hyperfine fields associated with the interfacial component (and the small fraction of amorphous phase also, on the basis of differential scanning calorimetry, considered to be present) in nanostructured $Fe_{80}B_{20}$ and $Fe_{66}B_{34}$ are shown in Fig. 30a and 30b.

Separation of the room-temperature Mössbauer spectrum of the ball-milled $Fe_{66}B_{34}$ into its two main components—crystalline and interfacial—is shown in Fig. 31a. The behavior of the interfacial component in the ball-milled $Fe_{66}B_{34}$ sample was further examined by annealing it in vacuum at temperatures in the range 250–350°C for 48-h periods.[210] The main results noted were: a decrease in the value of the mean hyperfine field of the interfacial component from the as-milled state of ~279 kOe to ~221 kOe on annealing the sample at 300°C, and an increase in the fractional spectral area of the interfacial

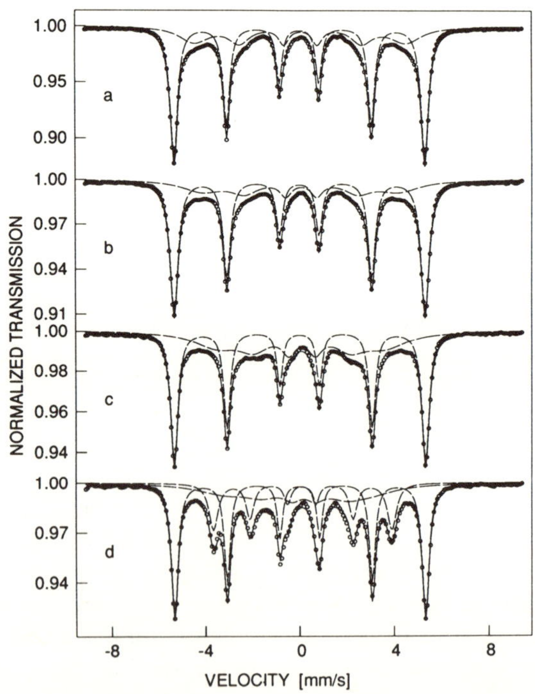

Figure 31. Room-temperature Mössbauer spectra of: (a) Fe–B powders of nominal starting composition $Fe_{66}B_{34}$ ball-milled for 400 h, and following vacuum annealing at: (b) 250°C for 48 h, (c) 300°C for 48 h, and (d) 350°C for 48 h. The spectra were fitted to a six-line subspectrum representing bcc Fe and a broad six-line subspectrum representing the interfacial component in this nanostructured sample. The fit to spectrum (d) includes a contribution from precipitated Fe_2B. (After Jing *et al.*[210])

component from ~33% to ~42%. The decrease in hyperfine field values indicates enhanced diffusion of B atoms into the interfacial component with the influence of solid-state reaction indicated by the increased fraction of the interfacial component. Annealing at 350°C did not result in a further increase in the area of the interfacial component, but rather initiated the formation of the ultimate stable intermetallic phase Fe_2B (Fig. 31d).

Comparison of the distributions of the magnetic hyperfine fields and their mean values for the ball-milled Fe–B samples (Fig. 30) with those for amorphous Fe–B alloys produced using either liquid quenching or vapor quenching techniques led to several interesting facets to the mechanical alloying process:[209]

1. Based on the mean hyperfine values, the overall B compositions of the mechanically alloyed samples were found to be considerably below those of the nominal starting compositions. This indicates that a significant fraction of the B remains in elemental form and that B atoms may have diffused into the balls and cell material during the milling process.
2. The presence of significant fractions of unreacted B was confirmed by vacuum annealing the ball-milled samples at 800°C for 0.5 h. The room-temperature Mössbauer spectra of these samples agreed with X-ray diffraction measurements on annealed samples in that the samples had reacted to give the stable phases: bcc Fe, Fe_2B, and FeB. Based on the fractions of these phases present in the annealed (and reacted) samples as estimated from their subspectral areas, the composition of the samples were now found, on the whole, to be in good agreement with the nominal starting compositions of the ball-milled powder mixtures.
3. Whereas smooth Gaussian-type distributions of magnetic hyperfine fields are obtained for amorphous Fe–B produced by rapid quenching,[221] the amorphouslike $Fe_{50}B_{50}$ and $Fe_{40}B_{60}$ ball-milled samples (Fig. 30c,d) exhibited non-Gaussian distributions with low-field peaks and pronounced shoulders on the main peaks. These effects were considered to indicate inhomogeneities in the samples and to demonstrate the degree of difficulty in obtaining homogeneous material directly from a mixture of elemental powders by a solid-state reaction due to mechanical alloying.

Nonetheless, it is clear that within certain limitations, mechanical alloying offers a relatively straightforward means of obtaining materials in the nanostructured and amorphous states. Indeed, the reproducibility of these overall effects is shown by the good agreement between the above work of Jing *et al.*[209] and the area of overlap with Okumura *et al.*[208] in their similar investigation of the mechanical alloying of $Fe_{100-x}B_x$ ($10 \le x \le 90$).

An interesting adaptation of mechanical alloying is the use of repeated

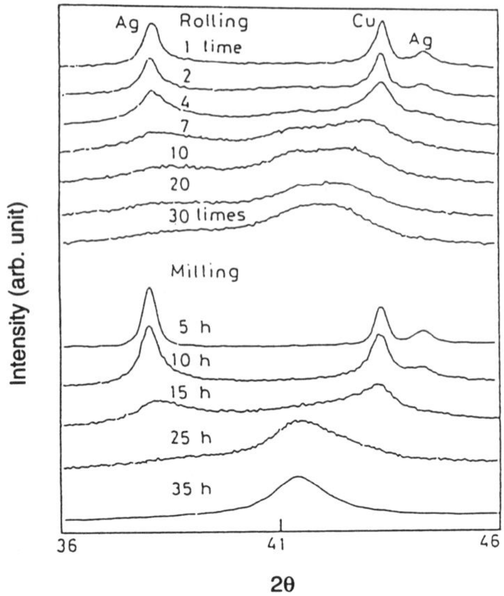

Figure 32. X-ray diffraction patterns (Cu$K\alpha$ radiation) of mixtures of Ag–70 at.% Cu powders subjected to either repeated rolling or extended ball-milling. (Reprinted with permission from Shingu *et al.*[190a])

rolling and/or pressing as an alternative to ball-milling to create new phases and materials.[190,191,204] In essence, grain size reduction can be obtained by repeated rolling and folding.[190] The parallel with ball-milling can be seen in the case of different metallic grains being sandwiched between two colliding balls thereby bringing about a reduction in grain size.[204] An impression of the equivalence of these two approaches in forming supersaturated solid solutions can be gained from Fig. 32, which shows the effects of repeated rolling and extended ball-milling on Ag–70 at.% Cu.[190] The ready formation of a supersaturated solid solution using both methods is shown by the merging of the two (111) peaks of the Ag and Cu fcc lattices either by increased milling time (to ~25 h) or by repeated rolling (~30 times).

The equivalence of the two approaches in producing similar structures and phases is also demonstrated by the room-temperature Mössbauer spectra for Al–24.4 at.% Fe powders that were mechanically alloyed by both rolling and milling.[195,204] The close similarity of the spectra for a sample rolled 10 times (Fig. 33a) and one ball-milled for 72 h (Fig. 33d) is immediately clear with both spectra well fitted by a ferromagnetic sextet and a central paramagnetic doublet. Again, on the basis of the similarities for their Mössbauer spectra, the structure and state of the sample rolled 30 times (Fig. 33c) is considered to approach that of the sample milled for 180 h (Fig. 33e). The spectra are not entirely equivalent, however, with the ferromagnetic component of the rolled samples exhibiting broader lines than the milled samples. This was taken to

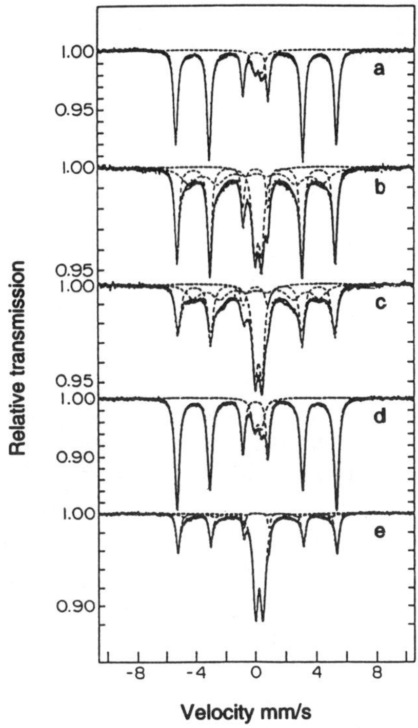

Figure 33. Room-temperatureMössbauer spectra for mechanically alloyed Al–24.4 at.% Fe powders prepared as follows: (a) rolled 10 times, (b) rolled 20 times, (c) rolled 30 times, (d) ball-milled 72 h, and (e) ball-milled 180 h. The fits to the spectra are discussed in the text. (After Nasu *et al.*[204])

indicate that larger amounts of Al atoms were present in the Fe in the rolled samples than in the milled samples.[204] The overall similarity in the phases and structures resulting from the rolling and milling approaches to mechanical alloying suggests that the process of kneading powders by an impact between colliding balls is similar to the process of thickness reduction by cold rolling.

As noted already, an essential difference between the milling and rolling approaches to mechanical alloying is that the former is a probabilistic technique whereas the latter can be described as a deterministic technique. Ball-milling and related techniques depend, among other factors, on the movements and impacts of balls and powders in a confined or partly controlled geometry[99] (see, e.g., Refs.[222–224] for discussions of: the geometry of collision events and mechanics and physics of mechanical alloying,[222] models for impacts of milling balls,[223] and energy transfer in a planetary mill as governed by the hit of the balls against the vial walls[224]). By comparison, repeated rolling offers scope for a well-defined geometry with a more controlled approach to, for example, grain size reduction. An advantage that the repeated rolling method appears to have compared with ball-milling is the reduced likelihood of contamination of the starting materials. Also the processing times are reduced considerably. On the other hand, increased handling is required for the preparation of successive packets of material for repeated rolling. Nonetheless, it is clear that both approaches to mechanical alloying complement each other well as ways of producing new phases and nanostructured materials of high quality in relatively large quantities.

4.6. Ultrafine Amorphous Particles

As noted already, the preparation of amorphous materials in powder form (particularly amorphous alloy particles) continues to attract strong interest (for a recent review see Linderoth[225]). This is due mainly to the advantages that particles have over, say, amorphous ribbons and films for practical applications. Besides their ready use in ferrofluids and catalysis, they can, as already noted, be compacted into convenient shapes for bulk applications. The ultrafine particles, which generally have sizes in the range ~5–500 nm, while not typical of the nanostructured state as exemplified by the systems considered in Sections 4.1 and 4.2, nonetheless do exhibit properties and behavior similar to those of nanostructured materials. Also of prime concern in the study of ultrafine amorphous particles is a comparison of their behavior with that of the bulk material.[225] The topic of the properties and behavior of small particles is a large and well established one (see, e.g., Pankhurst and Pollard[184] for a review of the use of applied magnetic fields in Mössbauer spectroscopy studies of fine magnetic powders). Examples of recent studies in this field that employ Mössbauer spectroscopy are given below. They serve as a basis for comparison between the properties and behavior of nanostructured materials as assessed by Mössbauer spectroscopy and outlined in this chapter, and those of ultrafine amorphous particles: finely dispersed $Fe_{82}B_{18}$ amorphous grains in an Al_2O_3 matrix,[226] amorphous Fe_2O_3 particles,[183] $Fe_{64}Cr_{16}B_{20}$ particles,[227] ultrafine Fe–B powders,[228–230] ultrafine Fe–(Ni,Co)–B powders,[231] ultrafine Fe–Ni–B powders,[232,233] transition metal–boron alloy particles,[234] amorphous iron–carbon particles (ferrofluids),[235] ultrafine amorphous Fe–B alloy particles.[236,237]

5. CONCLUDING REMARKS

The systematic study of nanostructured materials, although only established for 12 years or so,[7] is already approaching maturity as a scientific discipline. This is demonstrated, for example, by recent reviews of the structural and physical properties of nanostructured materials[13–15] and the establishment of a new series of international conferences[238] and an international journal[239] devoted to this theme. This reflects both the interesting combination of the novel physical properties of such materials and their potential technological applications (see, e.g., Refs. 11, 12) and the interest shown by condensed matter physicists and materials scientists in their aim to understand the structural nature and magnetic behavior of these materials. The full range of experimental and analytical techniques available to materials scientists has already been used to investigate the new solid-state structure that exists in the interface regions between the nanometer-scale crystal grains. An ultimate aim is the creation of new materials with enhanced properties.

Following an appraisal of the synthesis and structure of nanostructured materials, the contributions that Mössbauer spectroscopy has made to delineation of structural details and the magnetic behavior of the nanostructured state have been outlined in this chapter. Emphasis has been placed on examining the wide range of high-quality materials—nanostructured elements and alloys, nanoglasses, nanostructured ionic and covalent materials and oxides—that have been obtained by consolidation of powders prepared by condensation of evaporated atoms in an inert gas atmosphere.[16,92] Nanostructured materials prepared by mechanical alloying have also been examined.

Analysis of the Mössbauer spectra of this range of nanostructured materials has led, via the behavior of the hyperfine parameters derived from the fits to the spectra (see, e.g., Ref. 240 for an overview of data-fitting methods), to the unequivocal identification of a contribution from a component due to the unique atomic arrangement that exists in the interfacial regions between the nanocrystalline grains as well as a component due to the nanocrystalline grains themselves. This is shown most clearly in the spectrum of nanostructured Fe (Fig. 15), the first such nanostructured material to be studied systematically by Mössbauer spectroscopy.[16] The unique features of the nanostructured state have been further revealed by analysis of Mössbauer spectra obtained at different temperatures, leading to the conclusion that nanostructured materials exhibit reduced values for the Debye temperature compared with their bulk polycrystalline states.[16,153] These findings agree with the enhanced specific heat observed for nanostructured Y–7 at.% Fe compared with the polycrystalline alloy (Fig. 8[157]) and are consistent with a structural state that exhibits a wide distribution of interatomic spacings.[154] Indeed the analyses and conclusions derived from the Mössbauer studies outlined in Section 4 have been shown to fit in well with, or extend, existing knowledge of the structure of nanostructured materials described in Section 3. Current Mössbauer effect measurements being carried out on nanostructured metals and alloys under pressure[148] or subject to high magnetic fields[174] are likely to clarify further the behavior of this new state of matter. Indeed evidence for the differing magnetic natures of the intergranular and crystalline components of nanostructured FeCuNbSiB alloys has been reported recently by Zemčik *et al.*,[241] who examined the temperature dependence of the various spectral components in detail. Given the continuing advances in techniques and applications of Mössbauer spectroscopy[242]—now over 35 years since the discovery of the effect by R. L. Mössbauer—application of these more sophisticated Mössbauer techniques[242,243] is also likely to enhance our understanding of the nanostructured state. Further insight into the structure of the interfaces of nanostructured materials may also be obtained by comparison with submicrometer grain structures and dislocation substructures in strained metals and alloys.[175,244]

The measured mass densities of nanostructured materials and the nearest-neighbor configurations in the interfacial components seem to suggest that we are dealing with a solid-state structure that differs from the crystalline and the glassy states. Hence, it seems conceivable that new physical effects exist in the "defect core state" which may in turn open up the way to new technologies. Nanostructured materials should therefore not just be viewed as candidates for replacing conventional materials. Indeed, increased attention should be paid to the physical properties of solid matter in the core of defects.

Relatively little is known about structural differences between nanostructured materials with different types of chemical bonding. Most insight has so far been gained by comparing the results of experiments performed, for example, on metallic systems with data obtained for ionic or covalent materials. Given that the atomic structure may depend strongly on bonding and preparation, comparisons of this kind may be misleading. In the case of nanostructured alloys (Sections 3.2 and 4.2), it is already clear that new materials with chemical compositions and/or microstructures not available by conventional means can indeed be prepared. The remarkable potential of nanostructured alloys, as indicated by the investigations that have been carried out on them (Section 3.2), remains virtually untapped.

Of increasing importance, therefore, in the continuing investigation and development of nanostructured materials is likely to be the use of techniques such as mechanical alloying[188] to provide the larger quantities of material required for industrial applications. Such capabilities have already been reached by Inco Alloys International Ltd for powders of large particles, particularly their preparation of high-temperature alloys by ball-milling.[187] Also promising in the large-scale production of nanostructured materials is the more controllable, deterministic approach to mechanical alloying of, say, repeated rolling.[190,195] Nevertheless, it is clear that even with the attractive features that nanostructured materials exhibit, technological applications will depend to a large extent on the success in scaling up production procedures economically.

Acknowledgments. We thank the secretarial staff of the University College, UNSW, and the Universität des Saarlandes for their typing skill and assistance in the preparation of the manuscript. H.G. acknowledges the support received from the University College and the Scholarship of the Volkswagen-Stiftung, Hannover (Az: I/67410), during his stay in Canberra. Aspects of the authors' work outlined here have been supported by the Deutsche Forschungsgemeinschaft (G.W. Leibniz Programm), the Fonds der Chemischen Industrie, and the Australian Research Council.

REFERENCES

1. U. Gonser, ed., *Mössbauer Spectroscopy,* Springer-Verlag, Berlin (1975).
2. P. Gütlich, R. Link, and A. Trautwein, *Mössbauer Spectroscopy and Transition Metal Chemistry,* Springer-Verlag, Berlin (1978).
3. G. K. Wertheim, V. Jaccarino, and J. H. Wernick, *Phys. Rev.* **135A,** 151 (1964).
4. U. Gonser, ed., *Application of Nuclear Techniques to the Studies of Amorphous Metals,* International Atomic Energy Agency, Vienna (1981).
5. C. L. Chien, *Hyp. Interact.* **53**, 3 (1990).
6. J. G. Stevens, *Hyp. Interact.* **71**, 1537 (1992).
7. H. Gleiter, in *Deformation of Polycrystals,* N. Hansen, A. Horsewell, T. Leffers, and H. Lilholt, eds., p. 15, Proc. Second Risø Int. Symp. Metallurgy and Materials Science, Roskilde, Denmark (1981).
8. P. H. Hess and P. H. Parker, Jr., *J. Appl. Polym. Sci.,* **10**, 1915 (1966).
9. R. V. Morris, D. G. Agresti, H. V. Lauer, Jr., J. A. Newcomb, T. D. Shelfer, and A. V. Murali, *J. Geophys. Res.* **94**, 2760 (1989).
10. J. M. Knudsen, M. B. Madsen, M. Olsen, L. Vistisen, C. B. Koch, S. Mørup, E. Kankeleit, G. Klingelhöfer, E. N. Evlanov, V. N. Khromov, L. M. Mukhin, O. F. Prilutski, B. Zubkov, G. V. Smirnov, and J. Juchniewicz, *Hyp. Interact.* **68**, 83 (1991).
11. R. Ruthen, *Sci. Am.* **262**(5), 11 (1990).
12. R. W. Cahn, *Nature* **348**, 389 (1990).
13. H. Gleiter, *Prog. Mater. Sci.* **33**, 223 (1989).
14. R. Birringer, *Mater. Sci. Eng.* **A117**, 33 (1989).
15. H. Gleiter, *J. Appl. Crystallogr.* **24**, 79 (1991).
16. U. Herr, J. Jing, R. Birringer, U. Gonser, and H. Gleiter, *Appl. Phys. Lett.* **50**, 472 (1987).
17. K. L. Merkle, J. F. Reddy, C. L. Wiley, and J. D. Smith, *Phys. Rev. Lett.* **59**, 2887 (1987).
18. R. P. Andres, R. S. Averback, W. L. Brown, L. E. Brus, W. A. Goddard, A. Kaldor, S. G. Louie, M. Moscovits, P. S. Peercy, S. J. Riley, R. W. Siegel, F. Spaepen, and Y. Wang, *J. Mater. Res.* **4**, 740 (1989).
19. D. W. Johnson, *Adv. Ceram.* **21**, 3 (1987).
20. E. Matijevic, in *Ultrastructure Processing of Ceramics, Glasses and Composites,* L. L. Hench and D. R. Ulrich, eds., Wiley, New York (1984).
21. P. Vincencini, ed., *Ceramic Powders,* Elsevier, Amsterdam (1983).
22. H. Hausner, in: *Ceramic Materials and Components for Engines,* W. Bunk and H. Hausner, eds., Deutsche Keramische Gesellschaft, Bad Honnef, Germany, p. 27 (1986).
23. H. Hausner, *Ber. Dtsch. Keram. Ges.* **55**, 194 (1978).
24. C. Hayashi, *Phys. Today* **48**, 44 (1987).
25. C. R. Veale, *Fine Powders,* p. 5, Applied Science Publ., London (1972).
26. P. Fayet and L. Woeste, *Z. Phys.* **D3**, 177 (1986).
27. R. B. Wright, J. K. Bates, and D. M. Gruen, *Inorg. Chem.* **17**, 2275 (1978).
28. H. Abe, W. Schulze, and B. Tesche, *Chem. Phys.* **47**, 95 (1980).
29. K. Sattler, J. Muehlbach, and E. Recknagel, *Phys. Rev. Lett.* **45**, 821 (1980).
30. R. S. Bowles, J. J. Kolstad, J. M. Calo, and R. P. Andres, *Surf. Sci.* **106**, 117 (1981).
31. E. Choi and R. P. Andres, in *Physics and Chemistry of Small Clusters,* P. Jena, B. K. Rao, and S. N. Khanna, eds., p. 61, Plenum Press, New York (1987).
32. R. W. Siegel and J. A. Eastman, *Mater. Res. Soc. Symp.* **132**, 3 (1989).
33. R. W. Siegel and H. Hahn, in *Current Trends in Physics of Materials,* M. Yussouff, ed., p. 403, World Scientific Publ., Singapore (1987).
34. J. Karch, R. Birringer, and H. Gleiter, *Nature* **330**, 556 (1987).
35. J. Jing, Ph.D. thesis, Universität des Saarlandes, Saarbrücken (1989).
36. R. S. Averback, H. Hahn, H. J. Hoefler, J. L. Logas, and T. C. Shen, *Mater. Res. Soc. Symp.* **153**, 3 (1989).

37. A. Yanagida, S. Yatsuya, and K. Mihama, *Jpn. J. Appl. Phys.* **26**, L25 (1987).
38. S. Iwama, K. Hayakawa, and T. Arizumi, *J. Cryst. Growth* **66**, 189 (1984).
39. J. A. Eastman, *Mater. Res. Soc. Symp.* **132**, 27 (1989).
40. Z. Li, H. Hahn, and R. W. Siegel, *Mater. Lett.* **6**, 342 (1988).
41. S. Yatsuya, T. Kamakura, K. Yamauchi, and K. Mihama, *Jpn. J. Appl. Phys.* **25**, L42 (1986).
42. H. Hahn and R. S. Averback, *Appl. Phys. Lett.* **67,** 1113 (1990).
43. H. Trapp, private communication.
44. Y. Liu, Q.-L. Zhang, F. K. Tittel, R. F. Curl, and R. E. Smalley, *J. Chem. Phys.* **85**, 7434 (1986).
45. M. D. Morse, *Chem. Rev.* **86**, 1049 (1986).
46. K. LaiHing, R. G. Wheeler, W. L. Wilson, and M. A. Duncan, *J. Chem. Phys.* **87**, 3401 (1987).
47. E. A. Rohlfing, D. M. Cox, R. Petkovic-Luton, and A. Kaldor, *J. Phys. Chem.* **88**, 6227 (1984).
48. K. LaiHing, P. Y. Cheng, and M. A. Duncan, *J. Phys. Chem.* **91**, 6521 (1987).
49. G. W. Rice and R. L. Woodin, *J. Am. Ceram. Soc.* **71**, C181 (1988).
50. M. Cauchetier, O. Croix, M. Luce, M. Michon, J. Paris, and S. Tistchenko, *Ceram. Int.* **11**, 13 (1987).
51. L. M. Sheppard, *Adv. Mater. Processes* **4**, 53 (1987).
52. G. W. Rice and R. L. Woodin, in *Applications of Lasers to Industrial Chemistry,* Proc. SPIE, p. 458 (1984).
53. R. A. Fiato, G. W. Rice, and S. L. Soled, U. S. Patents 4659681, 4668647.
54. Degussa, Patent DE 762 723 (1942).
55. Degussa, Frankfurt, *Tech. Bull. Pigments* **56**, 12 (1982).
56. A. T. Liu and P. Kleinschmit, in *Novel Ceramic Fabrication Processes and Applications,* R. W. Davidge, ed., p. 1, Institute of Ceramics, Shelton (1986).
57. E. Wagner and H. Brunner, *Angew. Chem.* **72**, 744 (1960).
58. M. Carey Lea, *Am. J. Sci.* **37**, 479 (1989).
59. J. Turkevich, P. C. Stevenson, and J. Hillier, *Discuss. Faraday Soc.* **11**, 55 (1951).
60. R. M. Wilenzick, D. C. Russell, R. H. Morriss, and S. W. Marshall, *J. Chem. Phys.* **47**, 533 (1967).
61. H. Hausner, *Ber. Dtsch. Keram. Ges.* **55**, 194 (1978).
62. A. Henglein and R. J. Tasuch-Treml, *J. Colloid Interface Sci.* **80**, 84 (1981).
63. M. Mostafavi, J. L. Marignier, J. Amblard, and J. Belloni, *Radiat. Phys. Chem.* **34**, 605 (1989).
64. S. Mosseri, A. Henglien, and E. Janata, *J. Phys. Chem.* **93**, 6791 (1989).
65. A. Henglein, *Chem. Phys. Lett.* **154**, 473 (1989).
66. S. T. Lin, M. T. Frankling, and K. J. Klabunde, *Langmuir* **2**, 259 (1986).
67. A. E. Berkowitz and J. L. Walter, *J. Mater. Res.* **2**, 277 (1987).
68. F. Traeger and G. Pulitz, *Metal Clusters,* p. 112, Springer-Verlag, Berlin (1986).
69. J. Bourdin *Growth and Properties of Metal Clusters,* Elsevier, Amsterdam (1980).
70. S. Sugano, Y. Nishina, and S. Ohnishi, *Microclusters,* p. 126, Springer-Verlag, Berlin (1987).
71. R. Rossetti, J. L. Ellison, H. M. Gibson, and L. E. Brus, *J. Chem. Phys.* **80**, 4464 (1984).
72. M. Meyer, C. Walberg, K. Kurihara, and J. H. Fendler, *J. Chem. Commun.* **90**, 279 (1984).
73. Y. Wang, A. Suna, W. Mahler, and R. Kasowski, *J. Chem. Phys.* **87**, 7315 (1987).
74. N. F. Borrelli, D. W. Hall, H. J. Holland, and D. W. Smith, *J. Appl. Phys.* **61**, 5399 (1987).
75. P. J. Anderson and P. L. Morgan, *Trans. Faraday Soc.* **60**, 930 (1964).
76. A. W. Searcy, *J. Am. Ceram. Soc.* **70**, 155 (1983).
77. M. Kim, U. Dahmen, and A. W. Searcy, *J. Am. Ceram. Soc.* **70**, 164 (1987).
78. G. W. Kriechbaum and P. Kleinschmit, *Adv. Mater.* **10**, 330 (1989).
79. A. Fojtik, H. Weller, U. Koch, and A. Henglein, *Ber. Bunsenges. Phys. Chem.* **88**, 969 (1984).

80. M. L. Steigerwald, A. P. Alivisatos, J. M. Gibson, T. D. Harris, R. Kortan, A. J. Muller, A. M. Thayer, T. M. Duncan, D. C. Douglass, and L. E. Brus, *J. Am. Chem. Soc.* **110**, 3046 (1988).
81. R. H. Holm, *Acc. Chem. Res.* **10**, 427 (1977).
82. B. K. Teo, K. Keating, and Y.-H. Kao, *J. Am. Chem. Soc.* **109**, 3494 (1987).
83. Y. Wang and N. Herron, *J. Phys. Chem.* **91**, 257 (1987).
84. S. Umemura, in *Final Report of the Ultrafine Particles Project, Research Development Corporation of Japan,* 5-2, Nagatacho 2-chome, Chiyoda-ku, Tokyo 100, Japan (1986).
85. H. Watanabe, in *The Physics and Fabrication of Microstructures and Microdevices,* M. J. Kelly and C. Weisbuch, eds., p. 158, Springer-Verlag, Berlin (1986).
86. T. Inoshita and H. Watanabe, *Optoelectron. Devices Technol.* **1**, 33 (1986).
87. T. Inoshita and H. Watanabe, in *Microclusters,* S. Sugano, Y. Nishida, and S. Ohnishi, eds., p. 281, Springer-Verlag, Berlin (1987).
88. S. Kashu, E. Fuchita, T. Manabe, and T. Hayashi, *Jpn. J. Appl. Phys.* **23**, L910 (1984).
89. T. Takagi, *Z. Phys.* **D3**, 272 (1986).
90. I. Yamada, C. J. Palmstron, E. Kennedy, J. W. Mayer, H. Inokawa, and T. Takagi, *Mater. Res. Soc. Symp. Proc.* **37**, 401 (1984).
91. T. Ina, M. Hanai, H. Ito, and Y. Minowa, *Proc. 5th Int. Conf. Ion and Plasma Assisted Techniques,* p. 16, CEP Consultants Ltd., Edinburgh (1985).
92. R. Birringer, Ph.D. thesis, Universität des Saarlandes, Saarbrücken (1985).
93. Y. Kawamura, M. Takagi, and M. Akai, *Mater. Sci. Eng.* **98**, 449 (1988).
94. E. Hellstern, H. J. Fecht, Z. Fu, and W. L. Johnson, *J. Appl. Phys.* **65**, 305 (1989).
95. P. H. Shingu, B. Huang, S. R. Nishitani, and S. Nasu, *Jpn. Inst. Met. Trans. Suppl.* **29**, 3 (1988).
96. W. Schlump and H. Grewe, in *Proc. DGM Conf. New Materials by Mechanical Alloying Techniques,* E. Arzt and L. Schultz, eds., p. 307, Calw-Hirsau, Oct. 1988, DGM Oberursel, Germany (1992).
97. M. J. Luton, C. S. Jayanath, M. M. Disko, S. Matras, and J. Vallone, *Mater. Res. Soc Symp. Proc.* **132**, 79 (1989).
98. E. Hellstern, H. J. Fecht, C. Garland, and W. L. Johnson, *Mater. Res. Soc. Symp. Proc.* **132**, 137 (1989).
99. A. Calka and A. P. Radlinski, *Mater. Sci. Eng.* **A134**, 1350 (1991).
100. A. P. Radlinski, A. Calka, B. W. Ninham, and W. A. Kaczmarek, *Mater. Sci. Eng.* **A134**, 1346 (1991).
101. R. Z. Valiev, Y. D. Vishnyakov, R. R. Mulyukov, and G. S. Fainstein, *Phys. Status Solidi* **A117** 549 (1990); R. Z. Valiev, Y. D. Vishnyakov, R. R. Mulyukov and G. S. Fainstein, *Mater. Sci. Eng.* **A137**, 35 (1991).
102. A. K. Lee, L. E. Sanchez-Caldera, J. H. Chun, and N. P. Suh, *Mater. Res. Soc. Symp. Proc.* **132**, 87 (1989).
103. N. P. Suh, *ASME J. Eng. Ind.* **104**, 327 (1982).
104. N. P. Suh, N. Tsuda, M. G. Moon, and N. Saka, *ASME J. Eng. Ind.* **104**, 332 (1982).
105. J. P. Riviäre, P. Brouillaud, and J. F. Dinhut, *Radiation Effects and Defects in Solids* **114,** 145 (1990).
106. S. J. Savage and F. H. Froes, *J. Met.* **36**, 20 (1984).
107. T. R. Anantharaman, Metallic Glasses: Production, Properties and Applications, *Trans. Techn.*, p. 1 (1984).
108. J. Bigot, Preparation of Metallic Glasses and its Influence on the Properties, p. 18, *Summer School on Amorphous Metals,* World Scientific Publ., Singapore, p. 18, (1986).
109. H. Bestgen, *Proc. 5th Conf. Rap. Quenched Met.*, p. 433 (1985).
110. R. Sonnenberger, H. Bestgen, and G. Dietz, *Z. Phys.* **B61**, 289 (1984).
111. R. Sonnenberger, E. Pfanner, and G. Dietz, *Z. Phys.* **B63**, 203 (1986).
112. K. Boylan, D. Ostrander, U. Erb, G. Palumbo, and K. T. Aust. *Scr. Metall.* **25**, 2711 (1991).
113. S. Veprek, Z. Iqbal, H. R. Oswald, and A. B. Webb, *J. Phys.* **C14**, 295 (1981).

114. H. J. Schalditz, *Z. Metall.* **59**, 18 (1968).
115. H. G. F. Wilsdorf, O. T. Inal, and L. E. Murr, *Z. Metall.* **69**, 701 (1978).
116. B. M. Gallois, R. Mathur, S. R. Lee, and J. Y. Yoo, *Mater. Res. Soc. Symp. Proc.* **132**, 49 (1989).
117. R. A. Roy and R. Roy, *Abstracts, Materials Research Society Annual Meeting,* p. 377, Boston (1982).
118. D. W. Hoffman, R. Roy, and S. Komarneni, *Ceram. Bull.* **62**, 375 (1983).
119. R. A. Roy and R. Roy, *Mater. Res. Bull.* **19**, 169 (1984).
120. D. W. Hoffman, R. Roy, and S. Komarneni, *J. Am. Ceram. Soc.* **67**, 468 (1984).
121. R. Roy, S. Komarneni, and D. M. Roy, in *Better Ceramics Through Chemistry,* C. J. Brinker, D. E. Clark, and D. R. Ulrich, eds., p. 347, Elsevier, North Holland, New York (1984).
122. D. W. Hoffman, R. Roy, and S. Komarneni, *Mater. Lett.* **2**, 245 (1984).
123. D. W. Hoffman, S. Komarneni, and R. Roy, *J. Mater. Sci. Lett.* **3,** 439 (1985).
124. Y. Suwa, R. Roy, and S. Komarneni, *J. Am. Ceram. Soc.* **68**, C-238 (1985).
125. R. Roy, Y. Suwa, and S. Komarneni, in *Science of Ceramic Chemical Processing,* L. L. Hench and D. R. Ulrich, eds., p. 247, Wiley, New York (1986).
126. Y. Suwa, S. Komarneni, and R. Roy, *J. Mater. Sci. Lett.* **5**, 21 (1986).
127. Y. Suwa, R. Roy, and S. Komarneni, *Mater. Sci. Eng.* **83**, 151 (1986).
128. G. Vilmin, S. Komarneni, and R. Roy, *J. Mater. Sci.* **22**, 3556 (1987).
129. R. Birringer, U. Herr, and H. Gleiter, *Jpn. Inst. Met. Trans. Suppl.* **27**, 43 (1986).
130. R. W. Siegel and H. Hahn, in *Current Trends in Physics of Materials,* M. Yussouff, ed., p. 403, World Scientific Publ., Singapore (1987).
131. C. Surayanarayana and F. H. Froes, in *Proc. Physical Chemistry of Powder Metals Production and Processing,* W. Murray and D. G. V. Robertson, eds., p. 269, TMS Publ., Warrandale, PA (1990).
132. P. Marquardt and H. Gleiter, *Verh. Dtsch. Phys. Ges.* **15**, 328 (1980).
133. W. Wunderlich, Y. Ishida, and R. Maurer, *Scr. Metall.* **24**, 403 (1990).
134. W. Wang, H. Fuchs, J. Ying, and H. Gleiter, *Ultramicroscopy* **42–44**, 594 (1992).
135. M. Klingel, Ph.D. thesis, Universität des Saarlandes, Saarbrücken (1992).
136. E. Jorra, H. Franz, J. Peisl, G. Wallner, W. Petry, R. Birringer, H. Gleiter, and T. Haubold, *Philos. Mag.* **B60**, 159 (1989).
137. J. E. Epperson, R. W. Siegel, J. W. White, T. E. Klippert, A. Narayanasamy, J. A. Eastman, and F. Trouw, *Mater. Res. Soc. Symp. Proc.* **132**, 15 (1989).
138. R. Kirchheim, T. Muetschele, W. Kieninger, H. Gleiter, R. Birringer, and T. D. Koblé, *Mater. Sci. Eng.* **99**, 457 (1988).
139. T. Muetschele and R. Kirchheim, *Scr. Metall.* **21**, 135 (1987).
140. H. Hahn, H. J. Hoefler, and R. S. Averback, *Defect and Diffusion Forum* **66–69**, 549 (1989).
141. J. Horvath, R. Birringer, and H. Gleiter, *Solid State Commun.* **62**, 319 (1987).
142. S. Schumacher, R. Birringer, R. Strauss, and H. Gleiter, *Acta Metall.* **37**, 2485 (1989).
143. H. J. Hoefler, Ph.D. thesis, Universität des Saarlandes, Saarbrücken (1991).
144. H. Gleiter, *Solid State Phys. (b)* **172**, 41 (1992).
145. A. P. Sutton, *Philos. Mag.* **63**, 793 (1991).
146. S. R. Phillipot, D. Wolf, and S. Yip, *MRS Bull.* **12**, 38 (1990).
147. D. Wolf, *Acta Metall.* **37**, 1983 (1990).
148. T. Limbach, S. Trapp, H. Gleiter, X. Zhao, and S. J. Campbell, (to be published).
149. R. Würschum, W. Greiner, G. Soyez, and H. E. Schaefer, in *Positron Annihilation: Proceedings of the 9th International Conference* (Aug. 26–30, 1991, Szombathely, Hungary), Zs. Kajcsos and Cs. Steles, eds., Zurich, Trans. Tech. 1992 (Materials Science Forum, pp. 105–110).
150. T. Haubold, R. Birringer, B. Lengeler, and H. Gleiter, *Phys. Lett.* **135**, 461 (1989).
151. T. Haubold, W. Krauss, and H. Gleiter, *Philos. Mag.* **63**, 245 (1991).
152. S. Ramasamy, J. Jiang, H. Gleiter, R. Birringer, and U. Gonser, *Solid state Commun.* **74**, 851 (1990).

153. J. Jiang, S. Ramasamy, R. Birringer, U. Gonser, and H. Gleiter, *Solid State Commun.* **80**, 525 (1991).
154. M. Wagner, *Phys. Rev.* **B45**, 635 (1992).
155. J. Rupp and R. Birringer, *Phys. Rev.* **B36**, 7888 (1987).
156. E. Hellstern, H. J. Fecht, Z. Fu, and W. L. Johnson, *J. Appl. Phys.* **65**, 305 (1989).
157. H. Klein, Diploma thesis, Universität des Saarlandes, Saarbrücken (1992).
158. S. L. Sass and J. Fitzsimmons, *Phys. Rev. Lett.* **61**, 2237 (1988).
159. H. J. Klam, H. Hahn, and H. Gleiter, *Acta Metall.* **35**, 2101 (1987).
160. H. E. Schaefer, R. Würschum, and R. Birringer, *Phys. Rev.* **B38**, 9549 (1988).
161. J. Weissmueller, R. Birringer, and H. Gleiter, *Proc. TMS Annual Meeting,* New Orleans, 1991, p. 273, TMS, Warrandale, PA. (1992).
162. X. Zhu, R. Birringer, U. Herr, and H. Gleiter, *Phys. Rev.* **B35**, 9085 (1987).
163. M. R. Fitzsimmons, J. A. Eastman, M. Mueller-Stach, and G. Wallner, *Phys. Rev. B.* **44**, 2452 (1991).
164. M. S. Duesbury and G. Y. Richardson, *Solid State Mater. Sci.* **17**, 1 (1991).
165. A. Tschoepe, Ph.D. thesis, Universität des Saarlandes, Saarbrücken (1992).
166. U. Herr, J. Jing, U. Gonser, and H. Gleiter, *Solid State Commun.* **76**, 197 (1990).
167. G. Herzer and H. Warlimont, *J. Nanostruct. Mater.* **1,** 263 (1992).
168. J. Karch, Ph.D. thesis, Universität des Saarlandes, Saarbrücken (1991).
169. J. Weissmueller, R. Birringer, and H. Gleiter, *Phys. Lett. A* **145**, 130 (1990).
170. J. Jing, A. Krämer, R. Birringer, H. Gleiter, and U. Gonser, *J. Non-Cryst. Solids* **113**, 167 (1989).
171. U. Gonser, *Hyp. Interact.* **68**, 71 (1991).
172. A. Krämer, J. Jing, and U. Gonser, *Hyp. Interact.* **54**, 591 (19901).
173. B. Hu, X. Zhang, Y. Fang, and X. Yang, *Nucl. Instrum. Methods Phys. Res. B* **76,** 50 (1993).
174. J. Chadwick, R. J. Pollard, S. J. Campbell, and H. Gleiter, (in preparation).
175. Z. Valiev, R. R. Mulyukov, and V. V. Ovchinnikov, *Philos. Mag. Lett.* **62**, 253 (1990).
176. J. Jing, X. Yang, Y. Hsia, and U. Gonser, *Surf. Sci.* **233**, 351 (1990).
177. C. H. Griffiths, M. P. O'Horo, and T. W. Smith, *J. Appl. Phys.* **50**, 7108 (1979).
178. S. J. Campbell, P. E. Clark, and P. R. Liddell, *J. Phys. F* **2**, L114 (1972); S. J. Campbell and P. E. Clark, *J. Phys. F.* **4**, 1073 (1974).
179. K. Sumiyama and Y. Nakamura, in *Rapidly Quenched Metals,* S. Steeb and H. Warlimont, eds., p. 859, Elsevier, Amsterdam (1985).
180. C. A. Melendres, A. Narayanasamy, V. A. Maroni, and R. W. Siegel, *J. Mater. Res.* **4**, 1246 (1989).
181. J. Jing, F. Zhao, X. Yang, and U. Gonser, *Hyp. Interact.* **54**, 571 (1990).
182. A. M. van der Kraan, *Phys. Status Solidi A* **18**, 215 (1973).
183. P. Ayyub, M. Maltani, M. Barma, V. R. Palkar, and R. Vijayaraghavan, *J. Phys. C* **21**, 2229 (1988).
184. Q. A. Pankhurst and R. J. Pollard, in *Mössbauer Spectroscopy Applied to Magnetism and Materials Science,* Vol. 1, G. J. Long and F. Grandjean, eds., Plenum Press, New York, (1993).
185. X. Yang, J. Jing, H. Engelmann, and U. Gonser, *Nuovo Cimento* **12D**, 35 (1990).
186. S. Linderoth, M. D. Bentzon, and S. Mørup, *Nucl. Instrum. Methods Phys. Res. B* **76,** 173 (1993).
187. G. M. McColvin and M. J. Shaw, *Mater. Sci. Forum* **88–90**, 235 (1992).
188. J. S. Benjamin, *Mater. Sci. Forum* **88–90**, 1 (1992).
189. T. Tanaka, S. Nasu, K. N. Ishihara and P. H. Shingu, *J. Less-Common Met.* **171**, 237 (1991).
190. P. H. Shingu, K. N. Ishihara, K. Uenishi, J. Kuyama, B. Huang, and S. Nasu, in *Solid State Powder Processing,* A. H. Claver and J. J. deBarbadillo, eds., p. 21, The Minerals, Metals and Materials Society, Warrendale, PA, USA, (1990); P. H. Shingu, K. N. Ishihara, and J. Kuyama, *Proc. 34th Jpn. Congr. Mater. Res.*, p. 19, *Society of Materials Science,* Kyoto (1991).

191. P. H. Shingu, in *Proc. Conf. on Metastable Phase Formation in the Solid State,* Irsee, Germany (1992).
192. Proc. Int. Symp. on Mechanical Alloying, Kyoto, Japan, May, 1991, *Mater. Sci. Forum* **88–90** (1992).
193. T. Tanaka, S. Nasu, K. Nakagawa, K. N. Ishihara, and P. H. Shingu, *Mater. Sci. Forum* **88–90**, 269 (1992).
194. H. Kuwano, H. Ouyang, and B. Fultz, *Mater. Sci. Forum* **88–90**, 561 (1992).
195. S. Nasu, S. Imaoka, S. Morimoto, H. Tanimoto, B. Huang, T. Tanaka, J. Kuyama, K. N. Ishihara, and P. H. Shingu, *Mater. Sci. Forum* **88–90**, 569 (1992).
196. P. H. Shingu, B. Huang, J. Kuyama, K. N. Ishihara, and S. Nasu, in *DGM Conference, New Materials by Mechanical Alloying Techniques* (1988).
197. B. Fultz, G. Le Caër, and P. Matteazzi, *J. Mater. Res.* **4**, 1450 (1989).
198. L. Lanotte and P. Mateazzi, *J. Magn. Magn. Mater.* **88**, 58 (1990).
199. S. Nasu, P. H. Shingu, K. N. Ishihara, and F. E. Fujita, *Hyp. Interact.* **55**, 1043 (1990).
200. W. H. Wang, K. Q. Xiao, Y. D. Dong, Y. Z. He, and G. M. Wang, *J. Non-Cryst. Solids* **124**, 82 (1990).
201. G. M. Wang, D. Zhang, W. H. Wang, and Y. D. Dong, *J. Magn. Magn. Mater.* **97**, 73 (1991).
202. Y. D. Dong, W. H. Wang, L. Liu, K. Q. Xiao, S. H. Tong, and Y. Z. He, *Mater. Sci. Eng.* **A134**, 867 (1991).
203. P. Mateazzi and G. Le Caër, *Hyp. Interact.* **68**, 177 (1991).
204. S. Nasu, S. Morimoto, H. Tanimoto, B. Huang, T. Tanaka, J. Kuyama, K. N. Ishihara, and P. H. Shingu, *Hyp. Interact.* **67**, 681 (1991).
205. J. Kuyama, H. Inui, S. Imaoko, S. Nasu, K. N. Ishihara, and P. H. Shingu, *Jpn. J. Appl. Phys.* **30**, L854 (1991).
206. S. Nasu and P. H. Shingu, in *Ordering and Disordering in Alloys,* A. R. Yavari, ed., p. 402, Elsevier, Amsterdam (1992).
207. K. Uenishi, K. F. Kobayashi, S. Nasu, H. Hatano, K. N. Ishihara, and P. H. Shingu, *Z. Metallk.* **83**, 132 (1992).
208. H. Okumura, K. N. Ishihara, P. H. Shingu, H. S. Park, and S. Nasu, *J. Mater. Sci.* **27**, 153 (1992).
209. J. Jing, A. Calka, and S. J. Campbell, *J. Phys. Condens. Matter* **3**, 7413 (1991).
210. J. Jing, S. J. Campbell, A. Calka, and A. V. J. Edge, *Hyp. Interact.* **69**, 475 (1991).
211. G. Le Caër, P. Matteazzi, and B. Fultz, *J. Mater. Res.* **7**, 1387 (1992).
212. S. J. Campbell, K. D. Jayasuriya, A. Calka, J. M. Cadogan, J. Jing, and E. Wu, in *Proc. 7th Int. Symposium on Magnetic Anisotropy and Coercivity in Rare-Earth Transition Metal Alloys,* Canberra, July, 1992, p. 68, Hi-Perm Laboratory, Research Centre for Advanced Mineral and Materials Processing, The University of Western Australia (printed Scott Four Colour Print, Perth, WA).
213. A. Calka, A. P. Pogany, R. A. Shanks, and H. Engelmann, *Mater. Sci. Eng.* **A128**, 107 (1990).
214. S. J. Campbell, K. D. Jayasuriya, A. Calka, and J. Jing, *Nucl. Instrum. Methods Phys. Res. B* **76,** 81 (1993).
215. K. D. Jayasuriya, S. J. Campbell, A. Calka, and J. Jing, *Nucl. Instrum. Methods Phys. Res. B* **76,** 85 (1993).
216. C. Suryanarayana, G.-H. Chen, and F. H. Froes, *Scr. Metall. Mater.* **26**, 1727 (1992).
217. P. H. Duwez, R. H. Willens, and W. Klement, Jr., *J. Appl. Phys.* **31**, 113 (1960).
218. K. Sumiyama, T. Yoshitake, and Y. Nakamuara, *Acta Metall.* **33**, 1791 (1985).
219. *Amorphous Metallic Alloys,* F. E. Luborsky, ed., Butterworths, London (1983).
220. G. Le Caër and J. M. Dubois, *J. Phys. E* **12**, 1082 (1982).
221. C. L. Chien and K. M. Unruh, *Phys. Rev. B* **25**, 5790 (1982).
222. D.R. Maurice and T. H. Courtney, *Metall. Trans.* **21A**, 289 (1990).
223. H. Hashimoto and R. Watanabe, *Mater. Sci. Forum* **88–90**, 89 (1992).

224. M. Magini, *Mater. Sci. Forum* **88–90**, 121 (1992).
225. S. Linderoth, *Hyp. Interact.* **68**, 107 (1991).
226. J.-R. Cui, J. L. Dormann, C. Sella, and P. Renaudin, *IEEE Trans. Magn.* **18**, 1589 (1982).
227. C. Djega-Mariadassou, J. L. Dormann, L. Bessais, P. Renaudin, E. Agostinelli, and D. Fiorani, *Hyp. Interac.* **55**, 933 (1990).
228. J. Jiang, F. Zhao, P. Gao, I. Dézsi, and U. Gonser, *Hyp. Interact.* **55**, 981 (1990).
229. M. Ghafari, J. Saida, and Y. Nakamura, *Hyp. Interact.* **69**, 595 (1991).
230. F.-S. Li, D.-S. Xue, M.-Z. Mao, and Y.-D. Zhang, *Hyp. Interact.* **69**, 651 (1991).
231. M. Ghafari, J. Saida, R. A. Brand, and Y. Nakamura, *Hyp. Interact.* **69**, 599 (1991).
232. Y. Xu, X. Yang, X. Wu, C. Ye, and H. Jiang, *Hyp. Interact.* **69,** 681 (1991).
233. Z. Hu, J. Shen, L. Zhang, Y. Chen, and Y. Hsia, *Hyp. Interact.* **69**, 693 (1991).
234. S. Linderoth and S. Mørup, *J. Appl. Phys.* **69**, 5256 (1991).
235. P. V. Hendriksen, S. Mørup, and S. Linderoth, *Hyp. Interact.* **70**, 1079 (1992).
236. S. Linderoth, *J. Magn. Magn. Mater.* **104–107**, 128 (1992).
237. S. Linderoth, *J. Magn. Magn. Mater.* **104–107**, 167 (1992).
238. *First Int. Conf. on Nanostructured Materials,* Cancun, September, 1992.
239. *Nanostructured Materials,* companion journal to *Acta Metallurgica* and *Scripta, Metallurgica et Materiala,* Pergamon Press, New York.
240. S. J. Campbell and F. Aubertin, in *Mössbauer Spectroscopy Applied to Inorganic Chemistry,* G. J. Long and F. Grandjean, eds., p. 183, Plenum Press, New York (1989).
241. T. Zemčik, F. Aubertin, and U. Goner, *J. Mater. Sci.* (submitted).
242. *Nuclear Zeeman Effect and Recent Advances in Mössbauer Spectroscopy,* Proc. Int. Workshop in honor of Prof. S. S. Hanna, P. Boolchand, ed., *Hyp. Interact.* **72**, 1992.
243. *Mössbauer Spectroscopy II,* U. Gonser, ed., *Topics in Current Physics* **25**, Springer-Veralg, Berlin (1981).
244. O. Schneeweiss, *Mater. Lett.* **11**, 43 (1991).

8

Zinc-67 Mössbauer Spectroscopy

W. Potzel

1. INTRODUCTION

Since the early days of the Mössbauer effect, the main attraction of the 93.31-keV transition in ^{67}Zn has been the extremely high resolution for the determination of small changes in the γ-ray energy.[1–9] This exceptionally high sensitivity has been used for precision measurements of hyperfine interactions and in gravitational redshift experiments. In recent years, several new developments have occurred. From the point of view of the instrumentation, a new type of spectrometer has been built that allows the application of external pressures up to $\sim$10 GPa in the temperature range from 1.5 to 100 K. In addition, it has been demonstrated that the ^{67}Zn Mössbauer resonance is also a highly favorable tool for studying lattice dynamic effects.[10–12] As a consequence, this resonance has been applied to investigate the volume and temperature dependence of hyperfine interactions and lattice dynamics in a variety of new systems including Zn metal and Zn alloys, semiconducting and insulating compounds, in particular Zn chalcogenides as well as normal and inverse spinels. Fortunately, this experimental activity also stimulated theorists' interest and theoretical calculations have become available in recent years on ZnF_2, Zn chalcogenides, Zn metal, and alloys. We will describe most of these new developments. Since emphasis will be put on intrinsic properties of materials, we will not mention experiments on phase modulation of recoilless γ radiation,[13,14] and more recent measurements of the gravitational redshift.[8] The early history of ^{67}Zn Mössbauer spectroscopy and a large part of the results published up to 1982 have been covered by two previous review articles.[15,16]

W. Potzel • Department of Physics, Technical University of Munich, D-85747 Garching, Germany.

Mössbauer Spectroscopy Applied to Magnetism and Materials Science, Volume 1, edited by Gary J. Long and Fernande Grandjean. Plenum Press, New York, 1993.

2. EXPERIMENTAL DEVELOPMENTS

Several nuclear parameters together with the decay scheme are summarized in Fig. 1. The 93.31-keV Mössbauer state in ^{67}Zn can be populated from ^{67}Ga, $T_{1/2} = 78.26$ h, and from ^{67}Cu, $T_{1/2} = 61.9$ h. Due to the relatively long half-life of 9.1 μsec, the minimum observable width (FWHM) in a Mössbauer experiment is $2\Gamma_0 = 0.31\ \mu$m/s $\triangleq$ 23 kHz. Since the transition energy is relatively large, the recoil-free fraction f is small: $f \approx 2\%$ at 0 K for a

Source Production:

^{68}Zn (p, 2n) ^{67}Ga

28 MeV protons (Cyclotron) on natural (19% ^{68}Zn) Zn compound (i.e.: ZnO)

^{65}Cu (α, 2p) ^{67}Cu ⟷ ^{65}Cu (α, 2n) ^{67}Ga

46 MeV alphas (Cyclotron) on natural (31% ^{65}Cu) Cu compound (i.e.: Cu–metal)

46 MeV alphas (Cyclotron) on natural (31% ^{65}Cu) Cu compound (i.e.: Cu–metal)

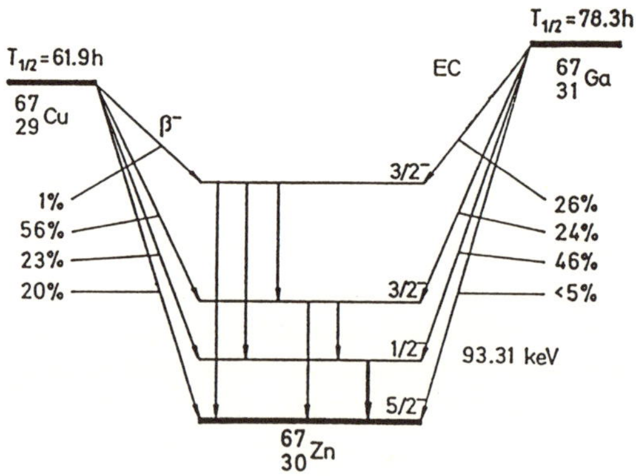

Spins and magnetic moments:	$I_e^\pi = \frac{1}{2}^-$	$\mu_e = 0.58\,\mu_N$
	$I_g^\pi = \frac{5}{2}^-$	$\mu_g = 0.8754\,\mu_N$
Multipolarity:	E2	
Quadrupole moment (groundstate):	0.15 b	
Half-life of the 1/2⁻-state:	9.1 μs	
Conversion coefficient:	0.89	
Min. observable width (FWHM):	$9.6 \cdot 10^{-11}\ eV = 23\ kHz = 0.31\ \mu m/s$	
Relative resolution:	$1.0 \cdot 10^{-15}$	
Max. resonance cross section:	$4.96 \cdot 10^{-20}$ cm^2 ($\sim$ 2% of ^{57}Fe)	
Recoil-free fraction:	$\approx$ 1.8% for $T \to 0$ and $\Theta = 300$ K	
	$\approx$ 0.03% for $T \to 0$ and $\Theta = 150$ K	

Figure 1. Basic features of the 93.31-keV Mössbauer transition in ^{67}Zn and main methods of source production.

Debye temperature $\Theta_D = 310$ K. As a consequence, ^{67}Zn Mössbauer experiments usually have to be performed at liquid-He temperatures. In favorable cases the temperature regime can be extended up to ~100 K. Since the natural abundance of ^{67}Zn is only 4.1%, absorbers have to be prepared from enriched material. This makes ^{67}Zn Mössbauer experiments very costly.

2.1. Sources of High Specific Activity

In most of the ^{67}Zn experiments, the decay from ^{67}Ga is used. The advantage is that the ^{67}Ga activity can be produced *in situ* of appropriate source matrices which contain Cu or Zn by the application of nuclear reactions in a cyclotron. Two source matrices proved to be highly successful: Cu metal disks and single crystals of ZnO.

The disks of Cu metal, 8 mm in diameter and 0.5 mm thick, are irradiated by 46-MeV α particles, the main reaction being ^{65}Cu(α, 2n)^{67}Ga. With the ZnO single crystals (dimensions $6 \times 6 \times 1$ mm), the reaction ^{68}Zn(p, 2n)^{67}Ga with 28-MeV protons is more favorable even though the natural abundance of ^{68}Zn is only 18.8%. If the targets are immersed in cooling water during the irradiation, the particle current can be as high as 25 μA. Thus, sources with ^{67}Ga activity in excess of 50 mCi can be produced with irradiation times of ~24 h. After allowing the short-lived activity to decay, the Cu sources can be used without annealing. The ZnO single crystals have to be annealed at 700°C in an oxygen atmosphere for 7 h and thereafter slowly (~50°C/h) cooled to room temperature. For many applications the single crystals are oriented with the hexagonal c axis parallel to $\mathbf{k}_\gamma$, the direction of observation of the 93.31-keV γ rays. This has the advantage that the emission spectrum of the source consists of two lines only (instead of three, as expected from the quadrupole splitting of the spin-5/2 ground state) with relative intensities of 2:1. The splitting of the two lines has been accurately determined by frequency modulation techniques[5–7] and can easily be taken into account in the data analysis. The ZnO sources exhibit almost the natural linewidth of 0.155 μm/s.

Other materials containing ^{68}Zn in natural abundance can be used as source matrices in a fashion similar to ZnO, [e.g., Zn metal (see Section 3.1) or spinels (see Section 4.3)]. The annealing procedure, however, may be different each time and has to be adjusted accordingly.

Because of the fcc structure of Cu, the Cu matrix provides a single-line source. Its width is still about three times the natural linewidth. The center shift between the ZnO and Cu sources is $-(110.7 \pm 0.1)$ μm/s. For semiconducting and insulating compounds ZnO sources are favorable, whereas for metallic systems Cu sources are more appropriate.

2.2. Spectrometer for Ambient Pressure

The small linewidth requires Doppler velocities in the micrometer per second range. Conventional double-loudspeaker drives are not accurate

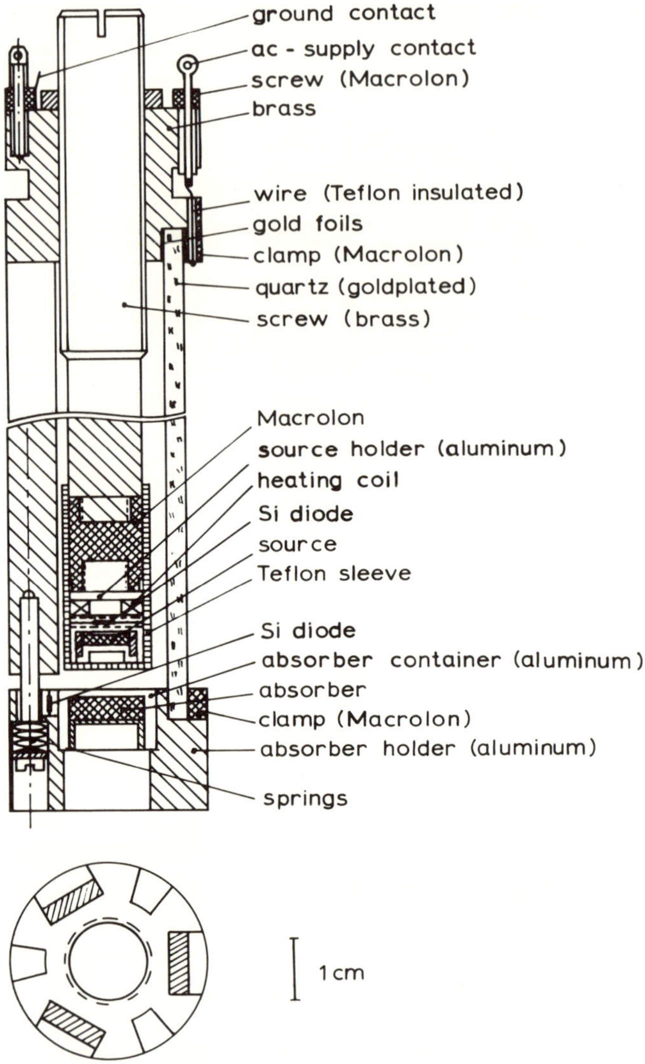

Figure 2. Design of the piezoelectric Mössbauer drive. The geometrical arrangement of the three quartz crystals is shown at the bottom. The source can be heated to elevated temperatures. (Taken from Ref. 18.)

enough. Therefore, drive systems have been developed that use the inverse piezoelectric effect in quartz or in piezoelectric ceramics.[15,16] Figures 2 and 3 show the construction of a quartz drive that allows heating the source independently of the absorber.

Three quartz crystals (only one is shown in the upper part of Fig. 2) are arranged in the shape of a regular triangle (see lower part of Fig. 2). The

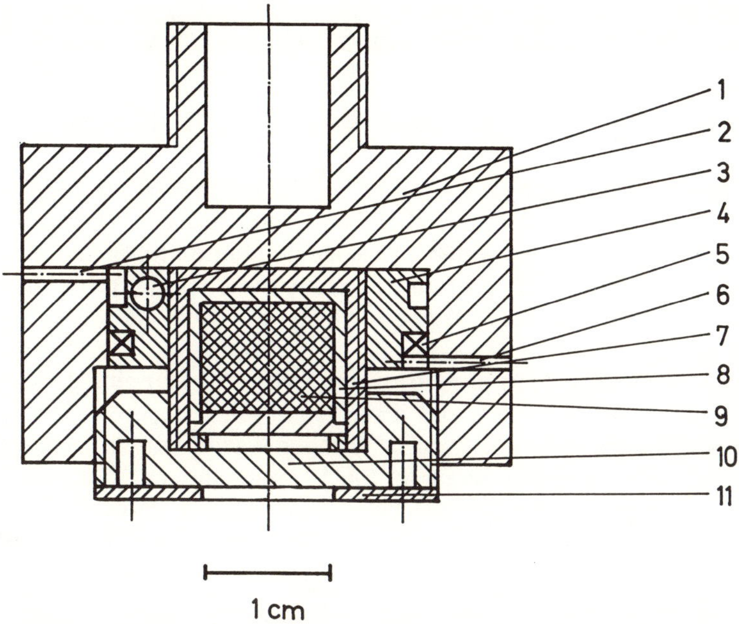

Figure 3. Holder for heating the absorber. 1, Makrolon holder, 2, hole for electrical leads (Si diode); 3, Si diode; 4, aluminum ring, 5, heating coil; 6, hole for electrical leads (heating coil); 7 and 8, absorber containers; 9, ZnF_2 powder absorber, 10, Makrolon cover; 11, Ta collimator. (Taken from Ref. 121.)

quartz crystals are $\times$-cut and are 100 mm long (y direction), 10 mm wide (z direction), and 3 mm thick (x direction). Usually, a sinusoidal voltage is applied to the (y, z) faces which are gold plated. The piezoelectric elongations and contractions of interest occur in the y direction.[17,18] With sinusoidal voltage profiles, velocities up to $\sim$300 μm/s have been scanned. The system has also been used for observing the natural width of ^{67}Zn resonance lines.[19]

The temperature of the source can be increased by a small heating coil wrapped around the aluminum source holder. The temperature is measured by a calibrated Si diode of type N-446 or BAY46 placed immediately above the source and maintained at the desired temperature by a controller. To avoid excess heat flow, the source holder is thermally insulated from the brass screw by a Makrolon connector and shielded by a Teflon sleeve. A second N-446 Si diode that is located at the aluminum absorber holder monitors the temperature of the absorber.

If it is necessary to heat the absorber, the holder shown in Fig. 3 can be screwed into the aluminum holder of the drive (see Fig. 2). The assembly is surrounded by a Makrolon container which provides good thermal insulation. The drive system containing source and Makrolon holder is suspended by soft springs inside a sealed stainless-steel container filled with He exchange gas at reduced pressure.[18] In this manner it is possible to keep the source temperature below ~18 K even at absorber temperatures of ~60 K.

2.3. Spectrometer for External Pressures up to 10 GPa

^{67}Zn Mössbauer spectroscopy puts stringent conditions on a high-pressure apparatus: in addition to the special drive system, the small f-factors require the measurements to be performed at cryogenic temperatures and demand the use of absorbers of large dimensions. To meet these special requirements, the "clamp technique" has been applied which is based on the principle of "opposed anvils".[20]

Figure 4 shows the high-pressure clamp with the Doppler drive mounted on top.[21] The clamp is made from a hardened Cu–Be alloy, Beryvac 200,[22] which stays sufficiently elastic even at cryogenic temperatures. The clamp is 205 mm long with an outer diameter of 105 mm and an inner diameter of 70 mm. As described in Ref. 23, the pressure is applied to the sample at room temperature by means of a hydraulic press. Thereafter the clamp can be cooled

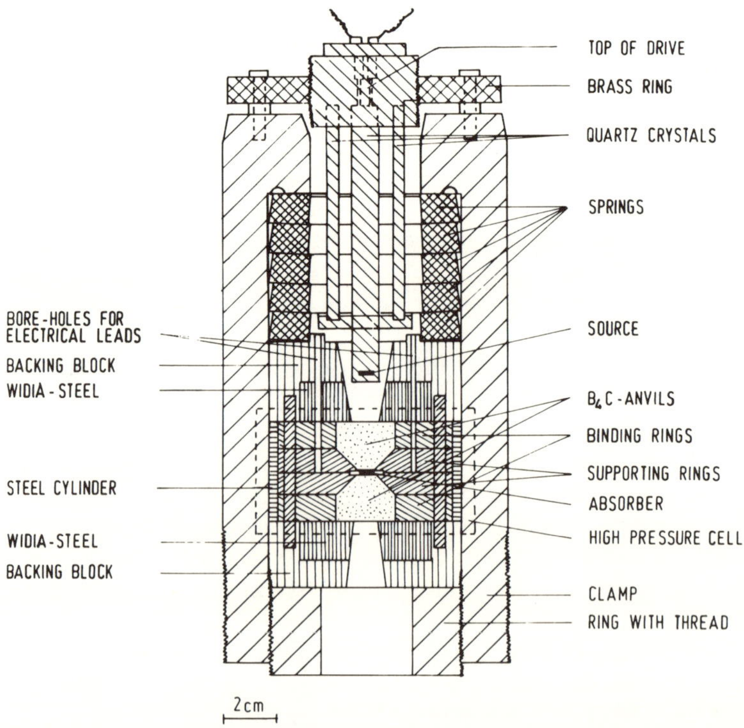

Figure 4. High-pressure clamp with piezoelectric quartz drive mounted on top. The innermost part indicated by broken lines is the high-pressure cell (see Fig. 5). (Taken from Ref. 21.)

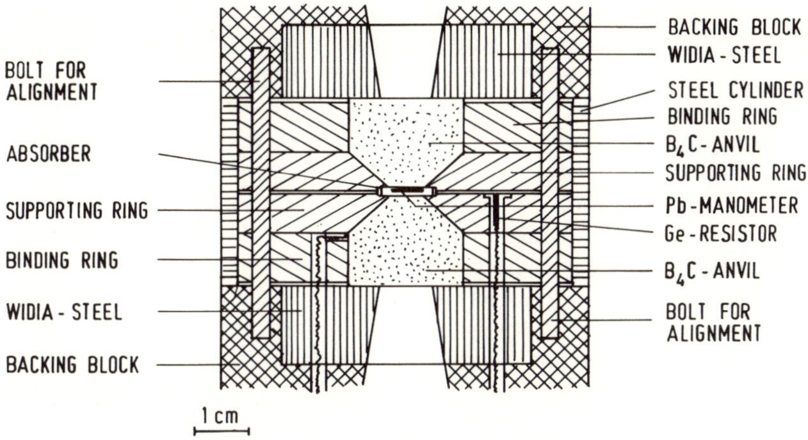

Figure 5. High-pressure cell indicated by broken lines in Fig. 4. (Taken from Ref. 21.)

to cryogenic temperatures to perform the experiments. The force is stored in five steel springs. The springs also compensate for the differences in thermal expansion between the clamp and the components of the high-pressure cell when cooling to cryogenic temperatures.

Figure 5 displays the high-pressure cell in greater detail. The two symmetrical anvils made from sintered B_4C apply the pressure to the absorber material.[24] The absorber is embedded in talcum powder, which is used as the pressure-transmitting medium to obtain quasihydrostatic conditions. The large dimensions of the absorber (~7 mm in diameter, ~2 mm high) require a sandwich gasket,[21,25] which is depicted in Fig. 6. Four tantalum rings each 0.1 mm thick, are sandwiched between 0.35-mm-thick pyrophyllite rings. Thus, the inner friction forces in the sandwich gasket are increased. This is necessary

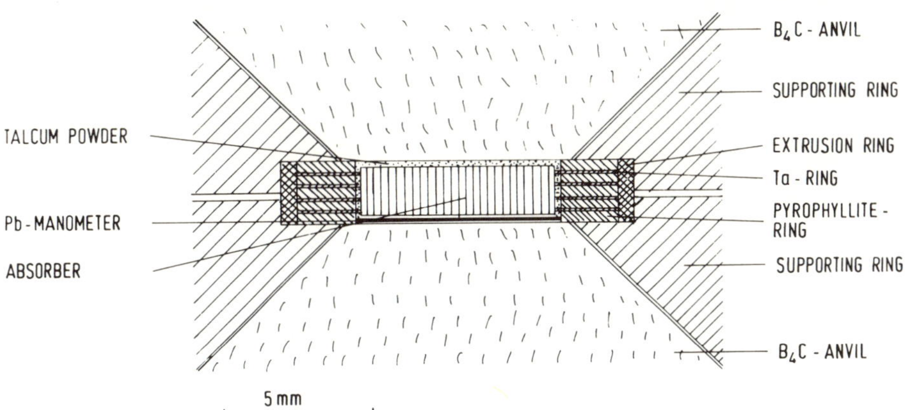

Figure 6. Details of the sandwich gasket. (Taken from Ref. 21.)

for such a large sample to allow the pressure to decrease mainly within the gasket. In this way, quasihydrostatic pressure over the volume of the sample is achieved. The gasket is centered and further supported by an extrusion ring made of a Cu–Be alloy, Beryvac 200, which in turn is backed by the two supporting rings. The gasket serves in addition as a highly effective collimator for the 93.31-keV γ radiation. The sintered B_4C anvils can withstand pressures up to ~15 GPa. They are, however, very brittle and have to be stabilized by supporting and binding rings made from steel, US-ultra 2. The force is transmitted to the anvils via the backing blocks which are enforced by tungsten carbide (Widia) inserts.[26] Even at a maximal force of $\sim 1 \times 10^6$ N, the Widia inserts show only minor deformations which are caused by the B_4C anvils. Except for the gasket and the extrusion ring, all parts of the high-pressure cell can be reused in further experiments with different absorbers. Binding and supporting rings of the high-pressure cell are centered by a steel cylinder and fixed in place with respect to the backing blocks by two alignment bolts.

The whole system is inserted from the top into the liquid-helium vessel of a metal cryostat. The pressure and the pressure gradient at the sample are determined *in situ* by a lead manometer,[27] which uses the known pressure dependence of the superconducting transition temperature of lead[28] as a gauge. As indicated in Figs. 5 and 6, a lead strip which lies between the absorber and a B_4C anvil is contacted by four wires to measure its resistivity. The transition temperature is measured by a calibrated Ge resistor (see Fig. 5). From the width of the transition it is possible to determine the pressure gradient across the absorber.[27]

Despite the large dimensions of the samples, quasihydrostatic pressures up to 10 GPa with gradients smaller than 10% can be obtained. If pressures greater than 10 GPa have to be applied, the B_4C anvils must be replaced by a harder material. Preliminary experiments with sintered diamond anvils look very promising.[29]

In order to perform Mössbauer experiments with the high-resolution ^{67}Zn resonance, the spectrometer has to be decoupled from mechanical vibrations of the cryostat and of the boiling liquid helium. This is accomplished by suspending the clamp by springs inside a sealed stainless-steel Dewar as indicated in Fig. 7. The drive is mounted on top of the clamp. The Dewar, which can be filled with He exchange gas at reduced pressure, is inserted from the top into the liquid-helium bath of a conventional metal cryostat. It takes about 4 h to cool the spectrometer from room temperature to 4.2 K. To change the pressure, the clamp together with the piezo-drive have to be warmed up to room temperature.

2.4. Velocity Calibration and Nuclear Pulse Counting

Contrary to conventional Mössbauer spectroscopy, the piezoelectric drive is usually operated without a feedback loop. This requires that the (sinusoidal) drive voltage must not show distortions $>10^{-3}$. Further details are given in Ref. 17.

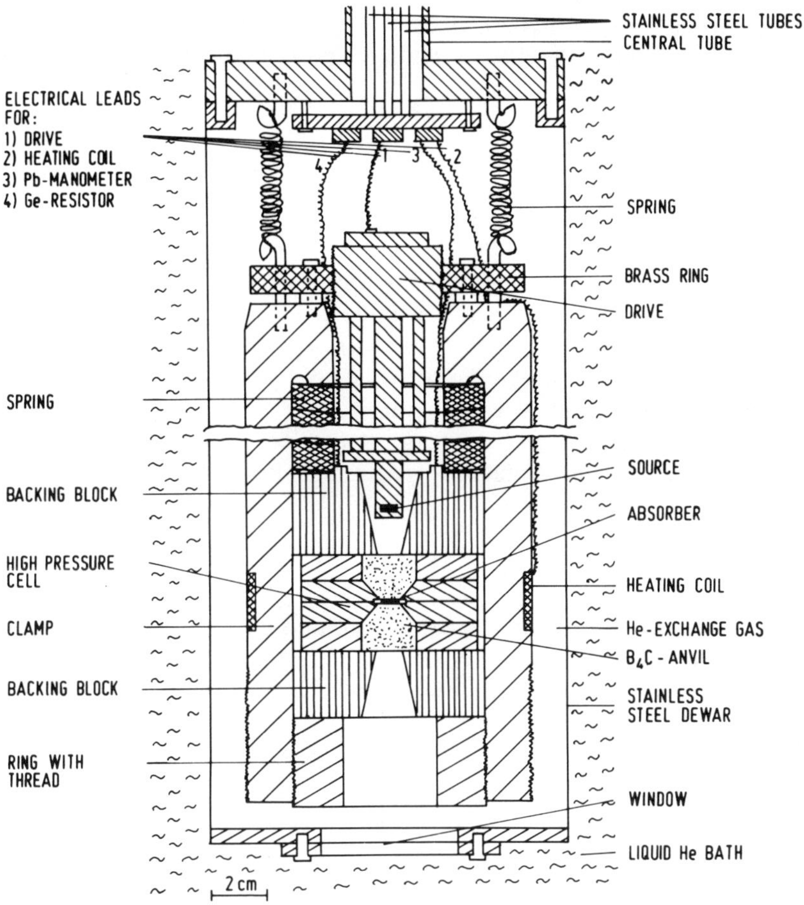

Figure 7. High-pressure clamp with piezoelectric quartz drive as mounted inside a liquid-helium cryostat. (Taken from Ref. 21.)

The spectrometer is calibrated in velocity using the quadrupole interactions in ^{67}Zn metal and in ^{67}ZnO, which have been measured absolutely by the frequency modulation technique.[5–7,30]

To detect the 93.31-keV γ rays, a NaI scintillation counter is in principle sufficient. However, for investigating the temperature or volume dependence of the Lamb–Mössbauer factor, an intrinsic Ge detector is mandatory in order to determine the signal-to-noise ratio more accurately. Such a detector should then be optimized with respect to high counting rates and still sufficient energy resolution by using a fast preamplifier–main amplifier chain. Modern fast amplifiers, which are commercially available,[31] can stand integral count rates up to $\sim 1 \times 10^6\ s^{-1}$. In the case of ^{67}Zn this corresponds to ~200,000 counts/s in the 93.31-keV window. The signal-to-noise ratio is then ~70%. The data pulses are stored in a multichannel analyzer operated in time mode. Since the drive frequency is typically 200 Hz, short channel dwell times are required.

3. METALLIC SYSTEMS

3.1 Zn Metal

Zn metal has provoked numerous theoretical[32–34] and experimental[35–37] investigations, mainly because it crystallizes in the hexagonal close-packed structure with an unusually large c/a ratio ($c/a = 1.86$), and hence many of its solid-state properties are highly anisotropic. A further consequence of the c/a ratio being far from the ideal value ($c/a = \sqrt{8/3} \approx 1.633$) is the presence of a pronounced electric field gradient (efg) at the Zn nucleus. Drastic changes of hyperfine interactions and of lattice-dynamic properties are observed when the c/a ratio is altered by pressure.

3.1.1. Hyperfine Interactions

Figure 8 shows Mössbauer spectra obtained at 4.2 K. As sources, ^{67}Ga in Zn single crystals were used with three different orientations ϑ of the c axis with respect to the direction of observation of the 93.31-keV γ rays.[18] The absorber was β'-brass, which provides a rather narrow single absorption line (see Section 3.2).

Due to the quadrupole interaction in Zn metal, the spin-$\frac{5}{2}$ ground state of ^{67}Zn splits into three sublevels (the spin-$\frac{1}{2}$ excited state remains unchanged) which give rise to the three Mössbauer lines. The relative intensities of the lines are changing with orientation. The efg tensor can easily be derived from the line separations, since the nuclear quadrupole moment Q of the $\frac{5}{2}$ ground state is known,[38] unfortunately only to an accuracy of 10%: $Q = (0.150 \pm 0.015)$ b. The asymmetry parameter η is determined from the ratio $\Delta E_1/\Delta E_2$, where ΔE_1 and ΔE_2 are the energy separations between the $\pm\frac{5}{2}$ and $\pm\frac{3}{2}$, and the $\pm\frac{3}{2}$ and $\pm\frac{1}{2}$ sublevels, respectively. $\Delta E_1/\Delta E_2 = 2$, if $\eta = 0$. The experiment[18] gives $e^2qQ/h = (12.30 \pm 0.08)$ MHz independent of temperature from 4.2 up to 47 K. Higher temperatures are not easily accessible with ^{67}Zn Mössbauer spectroscopy (see Section 3.1.2). From time-differential perturbed angular distribution (TDPAD) measurements, however, it is known that the efg at ^{67}Zn in Zn metal remains constant up to ~170 K.[39] The asymmetry parameter is found to be $\eta = 0$ in the same temperature range.

The results of modern theoretical calculations[32–34] for the efg in Zn metal are generally in good agreement with experimental values in the low-temperature regime. In particular, this is true for modern band-structure calculations of the efgs for all hcp elementary metals, including zinc, at ambient pressure. They show that the efg is mainly determined by the

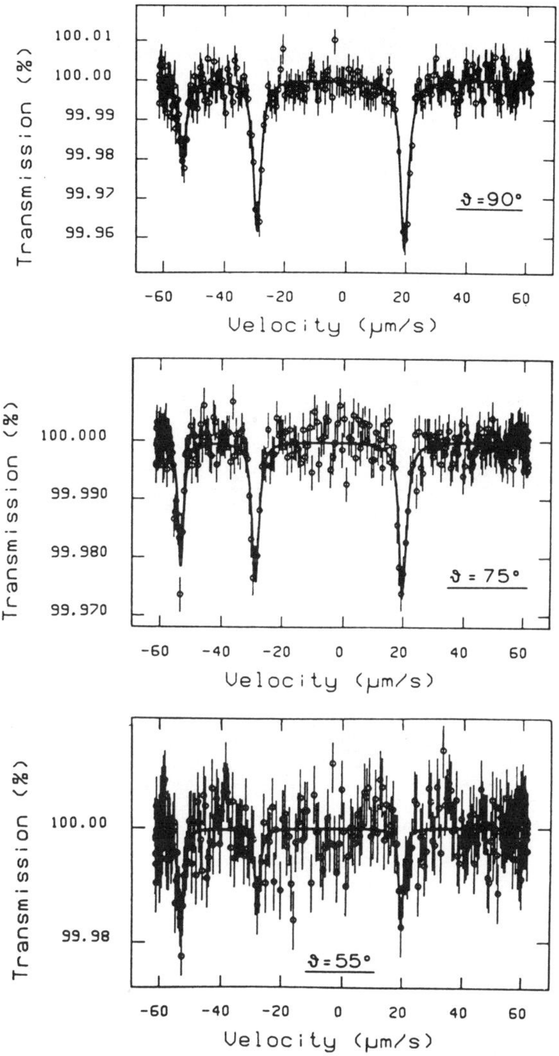

Figure 8. Mössbauer absorption spectra obtained at 4.2 K. Sources are ^{67}Ga in Zn metal single crystals with three different orientations ϑ of the c axis with respect to the direction of observation of the 93.31-keV γ rays. The absorber is β'-brass. (Taken from Ref. 18.)

nonspherical distribution of the valence electrons, in particular the p electrons. However, the understanding of the temperature dependence of efgs is still poor.[18] Here, the experimental results call for considerable refinements of the theoretical approaches.

Figure 9 displays Mössbauer absorption spectra of zinc metal recorded at 4.2 K and various pressures.[37] The source is ^{67}Ga in Cu kept also at 4.2 K. The

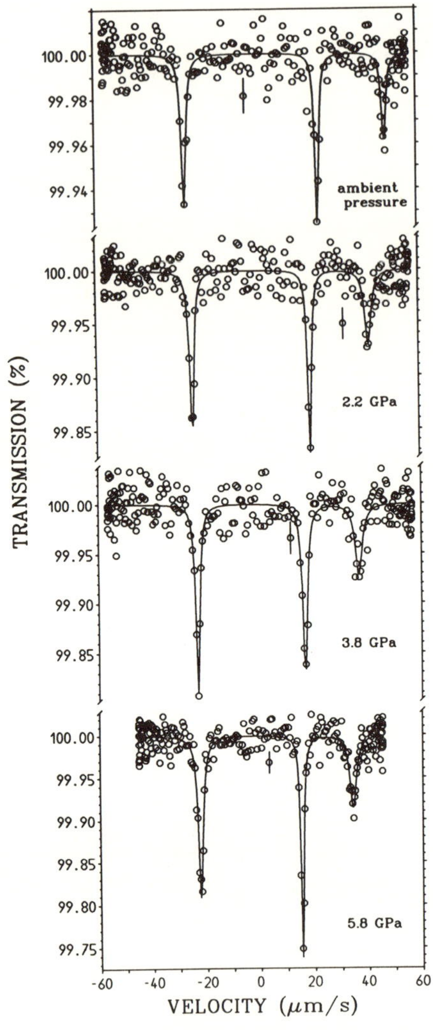

Figure 9. Mössbauer absorption spectra of ^{67}Zn metal recorded at 4.2 K and various pressures. The source is ^{67}Ga in Cu also at 4.2 K. (Taken from Ref. 37.)

results are summarized in Table 1. Figure 10 shows the pressure dependence of the main component of the efg. It decreases linearly in the whole pressure range investigated:

$$q_{\mathrm{efg}} = [(3.433 + 0.006) - (0.186 \pm 0.003)P] \times 10^{17}\ \mathrm{V/cm^2}$$

where P is in GPa. Full axial symmetry ($\eta = 0$) has been observed at all pressures.[37] In terms of reduced volume the linear relation gives

$$q_{\mathrm{efg}} = [(-11.22 \pm 0.23) + (14.65 \pm 0.23)V/V_0] \times 10^{17}\ \mathrm{V/cm^2}$$

Table 1. Zinc Metal: Pressure Dependence of the Lamb–Mössbauer Factor (f_{LMF}), of the Main Component of the Electric Field Gradient Tensor (q_{efg}), and of the Center Shift (S_C) Relative to the ^{67}Ga*Cu* Source

Pressure (GPa)	f_{LMF} (%)	$q_{efg}(\times 10^{17}\ V/cm^2)$	$S_C(\mu m/s)$
ambient pressure	0.90 ± 0.10	3.42 ± 0.01	14.95 ± 0.05
2.2 ± 0.2	1.62 ± 0.15	3.04 ± 0.01	12.14 ± 0.08
3.8 ± 0.3	2.10 ± 0.20	2.75 ± 0.01	10.21 ± 0.09
4.5 ± 0.5	2.45 ± 0.18	2.60 ± 0.01	9.09 ± 0.05
5.2 ± 0.2	—	2.37 ± 0.03	7.62 ± 0.19
5.9 ± 0.2	2.89 ± 0.60	2.26 ± 0.05	6.63 ± 0.31
6.5 ± 0.1	—	2.25 ± 0.02	7.09 ± 0.12
6.7 ± 0.3	—	2.21 ± 0.04	6.11 ± 0.25
8.5 ± 0.3	—	1.93 ± 0.09	3.69 ± 0.41

The volume dependence of the efg at constant temperature T can be deduced from the pressure coefficients[40]:

$$\left(\frac{\Delta \ln(q_{efg})}{\Delta \ln P}\right)_T = \left(\frac{\Delta \ln(q_{efg})}{\Delta \ln(c/a)}\right)_{V,T}\left(\frac{\Delta \ln(c/a)}{\Delta \ln P}\right)_T + \left(\frac{\Delta \ln(q_{efg})}{\Delta \ln V}\right)_{c/a,T}\left(\frac{\Delta \ln V}{\Delta \ln P}\right)_T \quad (1)$$

The Mössbauer result is

$$[\Delta \ln(q_{efg})/\Delta \ln P]_T = -(5.4 \pm 0.1) \times 10^{-6}$$

The compressibility data[41,42] give

$$[\Delta \ln(c/a)/\Delta \ln P]_T = -0.94 \times 10^{-6}$$

and

$$[\Delta \ln V]/\Delta \ln P]_T = -1.42 \times 10^{-6}$$

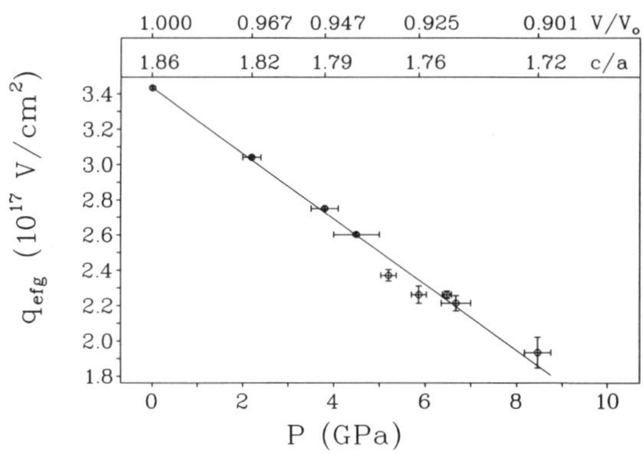

Figure 10. Pressure dependence of the electric field gradient at the ^{67}Zn nucleus in Zn metal at 4.2 K. The straight line is a least-squares fit to the data points.

In order to determine the remaining terms

$$[\Delta \ln(q_{\text{efg}})/\Delta \ln(c/a)]_{V,T}$$

and

$$[\Delta \ln(q_{\text{efg}})/\Delta \ln V]_{c/a,T}$$

separately, high-pressure measurements on single crystals would have to be performed. All theoretical models,[33,34] however, predict that the efg should vanish close to the ideal c/a ratio, $c/a = 1.633$. If a linear c/a dependence is assumed,

$$[\Delta \ln(q_{\text{efg}})/\Delta \ln(c/a)]_{V,T} = 8.3$$

Using this estimate one obtains

$$[\Delta \ln(q_{\text{efg}})/\Delta \ln V]_{c/a,T} = -(1.7 \pm 0.1)$$

The change of the efg with pressure is caused by two counteracting effects: decreasing the c/a ratio reduces the efg, whereas decreasing the volume increases the efg. The former process by far outweighs the latter.

The Mössbauer results on Zn metal can be compared with those obtained by time-differential perturbed angular correlation (TDPAC) measurements, where ^{111}Cd, as an impurity, was diffused into Zn metal.[40] Although the TDPAC experiments were performed at room temperature and only to pressures up to ~3 GPa, the results are in good agreement. The TDPAC data gave

$$[\Delta \ln(q_{\text{efg}})/\Delta \ln P]_T = -(5.6 \pm 0.4) \times 10^{-6}$$

and

$$[\Delta \ln(q_{\text{efg}})/\Delta \ln V]_{c/a,T} = -(2.6 \pm 0.5)$$

where again a linear decrease of the efg with the c/a ratio was assumed.

The question whether the efg vanishes close to the ideal c/a ratio could be answered by experiments at pressures of ~20 GPa, where c/a is already close to 1.633.[41,42] Such experiments have not as yet been possible.

At ambient pressure the center shift S_C with respect to ^{67}Ga*Cu* is $+(14.95 \pm 0.05)\ \mu$m/s. The experimentally observed variation of the center shift with pressure at constant temperature T can be expressed in terms of a volume change using compressibility data for Zn metal.[41,42] The volume dependence can be written

$$\left(\frac{\partial S_C}{\partial V}\right)_T = \left(\frac{\partial S_{\text{SOD}}}{\partial V}\right)_T + \left(\frac{\partial S}{\partial V}\right)_T \tag{2}$$

The first term describes the variation of the second-order Doppler shift (SOD).[43] The second term represents the explicit dependence of the isomer shift S due to changes of the electron density when the volume of the unit cell is reduced.

The determination of S_{SOD} is a serious problem in ^{67}Zn spectroscopy, because in general S and S_{SOD} are of about the same magnitude. In the particular case of Zn metal, one needs to know the phonon frequency distributions at various pressures (volumes of the unit cell). Unfortunately, such distributions are not available. However, an estimate of S_{SOD} can be obtained within the Debye model, if the volume dependence of the Lamb–Mössbauer factor (LMF) is determined.[37] The pressure and volume dependence of the LMF at 4.2 K are plotted in Fig. 11. Thc straight-line fits give the results:

$$f_{LMF} = [(25.1 \pm 3.0) - (24.2 \pm 3.0)]V/V_0](\%)$$

$$f_{LMF} = [(0.896 \pm 0.096) + (0.332 \pm 0.043)P](\%)$$

The LMF increases linearly from $f_{LMF} \approx 0.9\%$ at ambient pressure to $f_{LMF} \approx 2.9\%$ at 5.9 GPa (see Table 1). This corresponds to an increase of the effective Debye temperature from $\Theta_{eff} = 258$ K to $\Theta_{eff} = 343$ K from which the change ΔS_{SOD} of the SOD can be derived. For the ^{67}Zn resonance, the

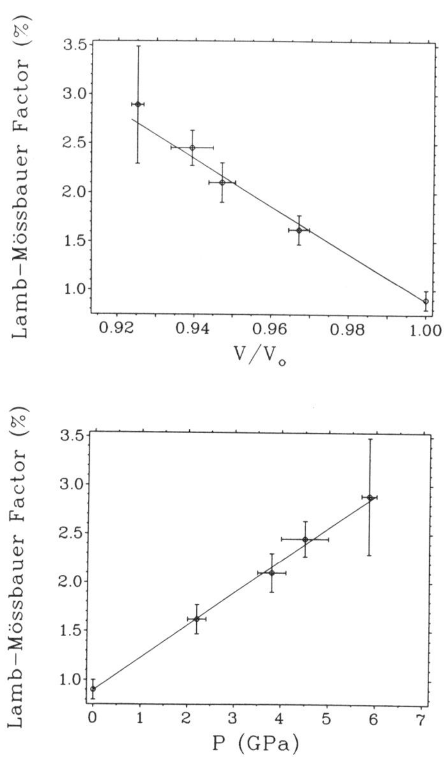

Figure 11. Pressure and volume dependence of the Lamb–Mössbauer factor of Zn metal. The straight lines are least-squares fits to the data points.

equation is:

$$\Delta S_{\mathrm{SOD}} = 0.231[\mu\mathrm{m/(s\,K)}] \times (\Theta_1 - \Theta_2) \tag{3}$$

where Θ_1 and Θ_2 are effective Debye temperatures (in K) of the same sample at two different pressures, or of two different samples, at the same temperature.[43]

The results are shown in Fig. 12 where the changes of S_{C} and S_{SOD} are plotted versus reduced volume. The straight lines are given by the equations

$$\Delta S_{\mathrm{C}} = -(99.3 \pm 1.4)\left(1 - \frac{V}{V_0}\right)\mu\mathrm{m/s}$$

$$\Delta S_{\mathrm{SOD}} = -(261 \pm 23)\left(1 - \frac{V}{V_0}\right)\mu\mathrm{m/s}$$

The difference $\Delta S_{\mathrm{C}} - \Delta S_{\mathrm{SOD}}$ describes the change of isomer shift ΔS due to an increase of the s electron density $\rho(0)$ at the ^{67}Zn nucleus when the volume is reduced. This increase of $\rho(0)$ is mainly caused by the conduction electrons. If the conduction electrons are characterized by a $4s4p$ electronic configuration, the $4s$ density ρ_{4s} is $\sim 7.5a_0^{-3}$.[44] Applying a simple scaling assumption for the change of ρ_{4s}, namely, $\Delta\rho_{4s}/\rho_{4s} = 1 - V/V_0$, and using $\Delta\langle r^2\rangle = (17.8 \pm 3.9) \times 10^{-3}\,\mathrm{fm}^2$,[45] a threefold larger change of isomer shift ΔS is predicted as derived above from the difference $\Delta S_{\mathrm{C}} - \Delta S_{\mathrm{SOD}}$. Most likely the $4s4p$ configuration overestimates the s density of the conduction electrons. In addition, the simple scaling assumption is expected[46] to give slightly too high changes of $\Delta\rho$. A description of the conduction electrons by an sp^3 hybridization would indeed agree much better with the observed change in $\Delta\rho$. Still, it should again be emphasized that the Debye model for such an anisotropic system as Zn metal is only an approximation. The model may be sufficient, because the SOD is independent of the anisotropy of the crystal lattice.

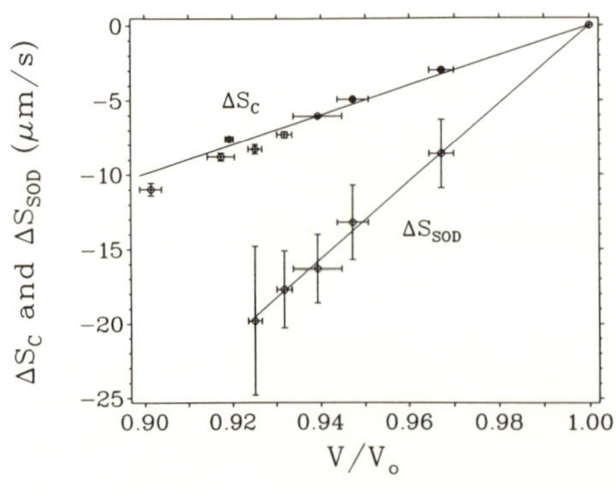

Figure 12. Change of the center shift, ΔS_{C}, and of the second-order Doppler shift, ΔS_{SOD}, in Zn metal with reduced volume of the unit cell. The straight lines are least-squares fits to the data points.

3.1.2. Lattice-Dynamic Effects

3.1.2.1. Basic Aspects. The Mössbauer effect is well suited for investigating lattice dynamics mainly by virtue of the Lamb–Mössbauer factor (f-factor) and the second-order Doppler shift (SOD).

The f-factor can be written as:

$$f = \exp\{-\mathbf{k}^2\langle x^2\rangle_{\mathbf{k}}\} \tag{4}$$

where $\mathbf{k}$ is the wave vector of the γ radiation and $\langle x^2\rangle_{\mathbf{k}}$ is the mean-square atomic displacement in the direction of $\mathbf{k}$. In general, f is anisotropic. In the case of axial symmetry, as e.g. for Zn metal, one obtains;

$$f(\vartheta) = \exp\{-\mathbf{k}^2[\langle x^2\rangle_{\perp} + (\langle x^2\rangle_{\parallel} - \langle x^2\rangle_{\perp}) \cdot \cos^2\vartheta]\} \tag{5}$$

or

$$\ln[f(\vartheta)] = -\mathbf{k}^2\langle x^2\rangle_{\perp} - \mathbf{k}^2(\langle x^2\rangle_{\parallel} - \langle x^2\rangle_{\perp}) \cdot \cos^2\vartheta \tag{6}$$

with the mean-square displacements parallel and perpendicular to the c axis, ϑ being the angle between $\mathbf{k}$ and the c axis.

By measuring $f(\vartheta)$ at two, or more, different angles at various temperatures, $\langle x^2\rangle_{\perp}$ and $\langle x^2\rangle_{\parallel}$ as well as their temperature dependences can be obtained. Often the notation is used:

$$f_{\perp} = \exp\{-\mathbf{k}^2\langle x^2\rangle_{\perp}\} \quad \text{and} \quad f_{\parallel} = \exp\{-\mathbf{k}^2\langle x^2\rangle_{\parallel}\} \tag{7}$$

According to the theory of lattice dynamics, for monoatomic cubic systems, where $\langle x^2\rangle_{\perp} = \langle x^2\rangle_{\parallel} = \langle x^2\rangle$, the mean-square atomic displacement can be calculated by integration over the phonon frequency spectrum $Z(\omega)$:

$$\langle x^2\rangle = \text{const} \cdot \int_0^{\omega_{\max}} d\omega \cdot Z(\omega) \cdot W_1(\omega) \tag{8}$$

where $\omega_{\max}$ is the maximum phonon frequency and the "weighting factor" W_1 is given by

$$W_1(\omega) = \omega^{-1} \cdot [\exp(\hbar\omega/k_B T) + 1] \cdot [\exp(\hbar\omega/k_B T) - 1]^{-1} \tag{9}$$

where k_B is the Boltzmann constant.

In the more general case of a monoatomic *anisotropic* system, $\langle x^2\rangle$ of Eq. (8) represents an average over all directions in space, i.e., $\langle x^2\rangle$ is proportional to the sum of the values $\langle x^2\rangle_i$, where i denotes the main directions of the anisotropy tensor.[29,47] For example, in the special case of axial symmetry $\langle x^2\rangle = (2\langle x^2\rangle_{\perp} + \langle x^2\rangle_{\parallel})/3$. Unfortunately, in general for anisotropic systems, which are investigated as powder samples, $\langle x^2\rangle$ cannot be measured directly. For a powder sample and no texture being present, only the recoilfree fraction averaged over all directions in space can be determined experimentally:

$$\langle f\rangle_{\text{pow}} = (1/4\pi) \int_0^{\pi} \int_0^{2\pi} f(\vartheta, \varphi) \sin\vartheta \, d\vartheta \, d\varphi \tag{10}$$

From $\langle f \rangle_{\text{pow}} = \exp\{-k^2 \langle x^2 \rangle_{\text{pow}}\}$ a value $\langle x^2 \rangle_{\text{pow}}$ can formally be derived. It can be shown[29] that always $\langle x^2 \rangle_{\text{pow}} \leq \langle x^2 \rangle$ of Eq. (8), if no texture is present in the sample. For this reason, knowledge of $Z(\omega)$ is usually not sufficient when comparing experimental data on powder samples of anisotropic systems with Eq. (8). In the case of Zn metal, for example, the phonon frequency spectrum $Z(\omega)$ is known and will be discussed in Section 3.1.2.2. Still, in order to compare with Eq. (8), one has to have additional information on the anisotropy of the mean-square displacement, e.g., via Goldanskii–Karyagin effect[36] or from single-crystal experiments.[18]

At low temperatures, where most of the ^{67}Zn measurements have been performed, and if $(\hbar\omega/k_B T) \gg 1$,

$$W_1(\omega) \propto \omega^{-1} \tag{11}$$

i.e., the f-factor weights $Z(\omega)$ with ω^{-1} and for this reason is sensitive to the low-frequency part of $Z(\omega)$.

In the general case of multiatomic systems, the theoretical evaluation of the Lamb–Mössbauer factor is much more intricate because the individual contribution of each mode to $\langle x^2 \rangle$ has to be calculated. We will return to this eigenvalue problem of the "dynamic matrix" in Sections 4.1.5 and 4.2.3, where lattice-dynamic models for the zinc chalcogenides and for ZnF_2, respectively, will be considered.

The second-order Doppler shift in velocity units is given by

$$S_{\text{SOD}} = -\langle v^2 \rangle / (2c) \tag{12}$$

with the mean-square velocity $\langle v^2 \rangle$ of the nucleus in the lattice. Again, S_{SOD} can be measured at various temperatures to derive the temperature dependence of $\langle v^2 \rangle$. For monoatomic systems

$$\langle v^2 \rangle = \text{const}' \cdot \int_0^{\omega_{\max}} d\omega \cdot Z(\omega) \cdot W_2(\omega) \tag{13}$$

with

$$W_2(\omega) = \omega \cdot [\exp(\hbar\omega/k_B T) + 1] \cdot [\exp\hbar\omega/k_B T) - 1]^{-1} \tag{14}$$

At low temperatures, $W_2(\omega) \propto \omega$, i.e., the second-order Doppler shift weights $Z(\omega)$ with ω^1.

Also the theoretical calculation of the second-order Doppler shift is much more involved for multiatomic systems. This question will be addressed for the zinc chalcogenides and for ZnF_2 in Sections 4.1.5 and 4.2.3, respectively.

Analogously, the specific heat C_V may be written as

$$C_V = \text{const}'' \cdot \int_0^{\omega_{\max}} d\omega \cdot Z(\omega) \cdot W_3(\omega) \tag{15}$$

with

$$W_3(\omega) = (\omega/T)^2 \cdot \exp(\hbar\omega/k_B T) \cdot [\exp(\hbar\omega/k_B T] - 1]^{-2} \tag{16}$$

At low temperature, $W_3(\omega) \propto (\omega/T)^2 \cdot [\exp(\hbar\omega/k_B T)]^{-1}$.

The weighting factors W_1, W_2, and W_3 are drastically different. Thus, measurements of f-factors and S_{SOD} in monoatomic systems provide critical tests for theories describing the phonon frequency spectrum. In the case of C_V this also holds for multiatomic lattices.

3.1.2.2. Anisotropy in Zn Metal. The anisotropy of the LMF at ambient pressure is very large. Measurements on Zn single-crystal sources with different orientations ϑ between the c axis and the direction of observation of the γ rays (see Fig. 8) show that the ratio $(f_\perp/f_\parallel)$ is ~25 at 4.2 K and rises to ~2200 at 47 K.[18] Table 2 summarizes the LMFs and mean-square atomic displacements. As a consequence of the small value of $f_\parallel$, measurements with crystal orientations of ϑ much below $\vartheta = 60°$ are very difficult.

The large anisotropy is clearly reflected in the phonon frequency spectrum,[48] which is displayed in Fig. 13. The large peak at $\omega \approx 12$ THz mainly originates from lattice vibrations parallel to the c axis. The temperature variations of $f_\perp$ and $f_\parallel$ are well described by calculations based on the modified axially symmetric model (MAS).[48–50] This can be seen from Fig. 14, where the corresponding mean-square atomic displacements are plotted versus temperature. For comparison, the results obtained in an extended Debye model (EDM) are also indicated. The EDM is characterized by two Debye temperatures, $\Theta_\perp = 267$ K and $\Theta_\parallel = 157$ K which are derived from the values $f_\perp$ and $f_\parallel$ at 4.2 K, respectively. Using these characteristic temperatures, the agreement of the calculated values for $f_\perp$ and $f_\parallel$, the SOD, and the specific heat with the experimental data is very good.[18,51] This suggests that an EDM gives reasonable predictions for other highly anisotropic cases, e.g., nearly two-dimensional systems.

A slightly more elaborate model,[52] where zinc metal is visualized as a layered structural crystal, involves phonon frequency distributions which

Table 2. Lamb–Mössbauer Factors $f_\perp$ and $f_\parallel$, Perpendicular and Along the c Axis of Zn Metal, Respectively, and Corresponding Mean-Square Atomic Displacements at Various Temperatures

Lamb–Mössbauer factors		Mean-square atomic displacements (Å^2)		Temperature (K)
$f_\perp$	$f_\parallel$	$\langle x^2\rangle_\perp$	$\langle x^2\rangle_\parallel$	
$(1.07^{+0.13}_{-0.12}) \times 10^{-2}$	$(4.3^{+8.8}_{-3.0}) \times 10^{-4}$	0.00203 ± 0.00005	0.0035 ± 0.0005	4.2
$(0.80^{+0.18}_{-0.15}) \times 10^{-2}$	$(9.8^{+53}_{-8.0}) \times 10^{-4}$	0.00216 ± 0.00009	0.0031 ± 0.0008	20.8
$(0.40^{+0.10}_{-0.05}) \times 10^{-2}$	$(1.8^{+67}_{-1.7}) \times 10^{-6}$	0.00247 ± 0.00010	0.0059 ± 0.0013	47

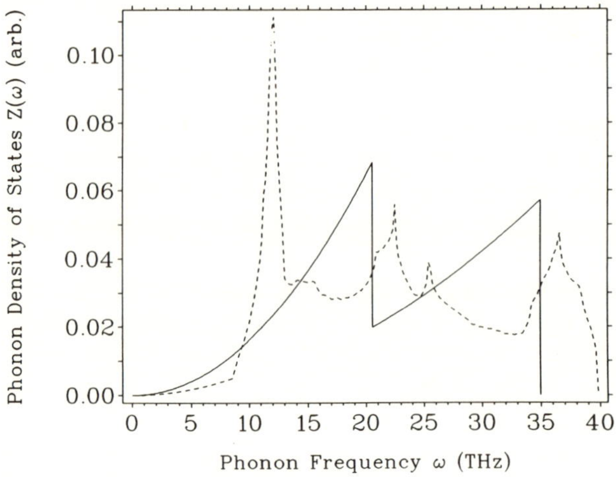

Figure 13. Phonon frequency distribution of Zn metal. Dashed line represents results of the MAS model based on neutron scattering data at room temperature. Solid line is the sum of two Debye-type distributions characterized by $\Theta_{\perp} = 267$ K and $\Theta_{\parallel} = 157$ K. The relative weight of the two distributions is 2:1.

consist of quadratic and linear terms for both the *c* direction and perpendicular to it. This model, which incorporates the presence of planar modes, leads, however, to *four* characteristic temperatures as adjustable parameters. This model is able to describe very well the Mössbauer results and the specific heat data on Zn metal.

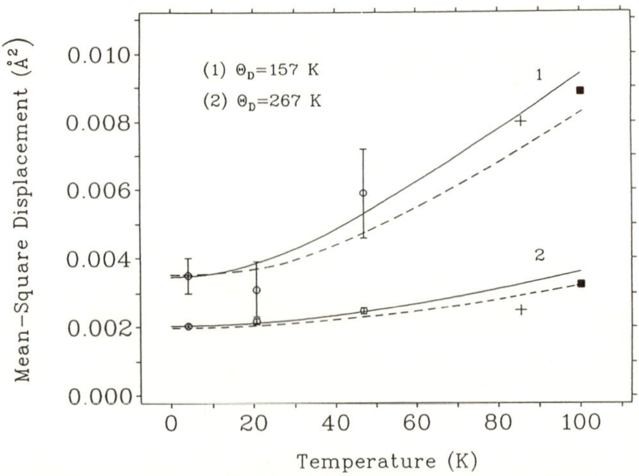

Figure 14. Temperature dependence of the mean-square atomic displacements in Zn metal, parallel and perpendicular to the *c* axis. Circles: ^{67}Zn Mössbauer effect. Solid squares and crosses: results deduced from X-ray experiments.[18] Solid line: fit to data points by extended Debye model of Fig. 13. Dashed line: results of calculations based on MAS model.

As mentioned above, the LMF drastically increases at high pressures. A more detailed discussion[37] shows that this increase is not just due to a reduction of anisotropy but is mainly a consequence of a drastic change of the phonon frequency spectrum at reduced volume. This pressure-dependent change in the phonon frequency distribution function should affect the electron–phonon interaction and hence superconductivity. On the basis of the EDM model described above, the effect of crystal anisotropy and of high pressure on superconductivity in zinc metal has been calculated.[53] Crystal anisotropy turns out to be highly significant. With increasing pressure the superconductivity critical temperature, T_c, is predicted to decrease; at 5.8 GPa the value of $T_c(\approx 0.12\,\text{K})$ is only about one-seventh of that at ambient pressure, where $T_c = (835 \pm 20)\,\text{mK}$.[11] Unfortunately, experimentally it is very difficult to determine the changes of the anisotropy of the LMF in zinc metal at high pressures, since, for this purpose, pressures of several gigapascals would have to be applied to single crystals. Modern cluster calculations, which can give reliable values for force constants between atoms, will be able to clarify these questions and, in particular, will help to solve the problem of second-order Doppler shift when the volume of the unit cell is reduced as mentioned in Section 3.1.1.

3.2. Cu–Zn Alloys

3.2.1. Hyperfine Interactions

The Cu–Zn alloy system, commonly known as brass, has been of enormous technological importance. In addition, the rich phase diagram[54–56] which is shown in Fig. 15 has drawn the attention of theorists, who tried to relate the phase transitions to electronic properties of the alloy.[57–63] Hume-Rothery first pointed out that phase stability properties may be driven by the average concentration of conduction electrons, or electron-to-atom ratio, and formulated his famous rules. Brass has been the representative of a class of alloys today known as Hume-Rothery alloys[57,58,64,65] or "electron phases". However, modern more advanced theoretical band structure calculations[57,58] could find only little justification for the Hume-Rothery rules. In this connection the question of short-range order in the α phase arises. To prove the existence of short-range order experimentally is not an easy task. X-ray diffraction does not work well, because the atomic form factors of Cu and Zn are nearly identical. ^{67}Zn Mössbauer investigations, however, have clearly demonstrated that in the α phase, which forms the fcc structure, short-range order is present.[66,67] The ability of α-brass, with composition around Cu_3Zn, to form long-period superstructures was later suggested from measurements of the diffuse scattering intensity in neutron diffraction experiments[68] and has also been revealed recently by first-principle phase stability calculations.[69] In addition, from the results of dilatometry and electron microscopy it has been

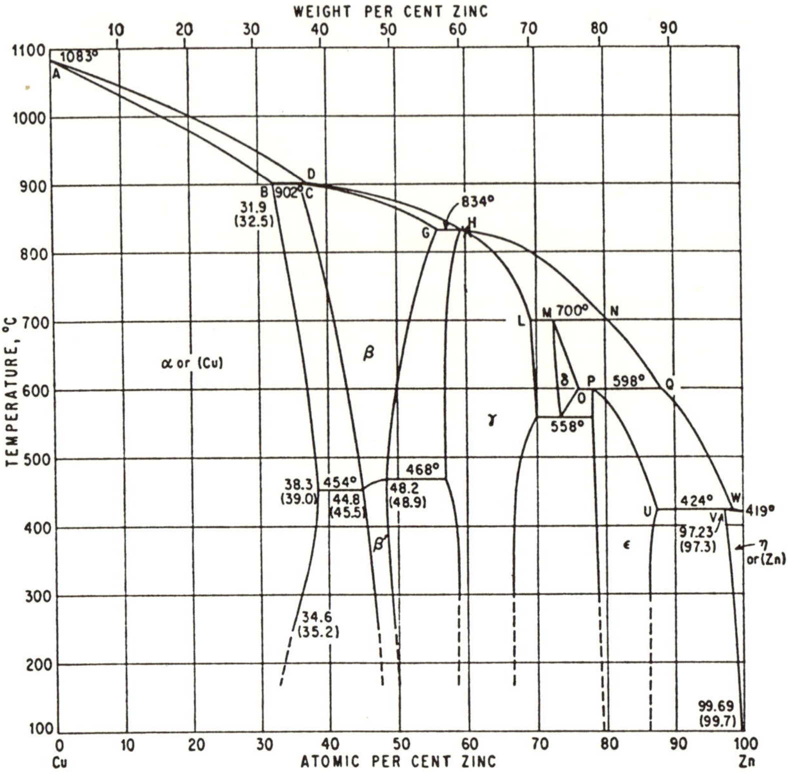

Figure 15. Phase diagram of the Cu–Zn system. (Taken from Refs. 54 and 55.)

Table 3. Center Shifts, S_C, Relative to the ^{67}Ga*Cu* Source, for Cu–Zn Alloys with Various Concentrations of Zn, c_{Zn}, at Ambient Pressure

c_{Zn} (at.%Zn)	Crystallographic phase	S_C (μm/s)
15.9	α phase (fcc)	-6.48 ± 0.15
		0.00 ± 0.10
		$+1.73 \pm 0.16$
		$+7.20 \pm 0.25$
49.2	β phase (CsCl)	-6.15 ± 0.02
63.0	γ phase ($D8_2$)	$+0.6 \pm 0.2$ (inner tetrahedron)
		-36.1 ± 0.2 (cubo-octahedron)
80.5	ε phase (hcp)	-11.6 ± 0.4
		-6.9 ± 0.2
		$+2.8 \pm 0.2$
		$+15.1 \pm 0.5$
100.0	Zn metal (hcp)	$+14.95 \pm 0.05$

suggested that the phase diagram (see Fig. 15) might have to be revised. In particular, in the low-temperature range, the region of the β and β' phases might have to be extended considerably towards lower Zn concentration at the expense of the α phase.[70] ^{67}Zn Mössbauer spectroscopy has been able to contribute significantly to solve these problems by investigating all major phases. Table 3 gives a summary of the values for the center shifts.

Figures 16 and 17 display Mössbauer spectra of α- and β'-brass absorbers. The spectra of Fig. 16 cover the range between 4.3 and 24.6 at.%Zn of the pure α phase. Figure 17 shows two spectra belonging to the mixed α and β' phases and one spectrum of the pure β' phase. As indicated by the bar diagram, the Mössbauer spectrum of 4.3 at.%Zn in Cu consists of a four-line

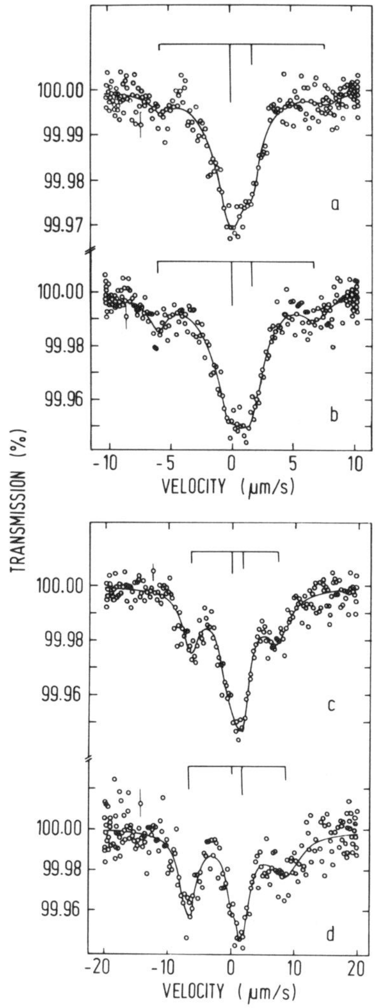

Figure 16. Mössbauer transmission spectra obtained at 4.2 K of brass alloy absorbers with Zn concentrations of 4.3 (a), 10.0 (b), 15.9 (c), and 24.6 at.% (d). The solid lines are least-squares fits to the four-line patterns indicated by the bar diagrams. (Taken from Ref. 66.)

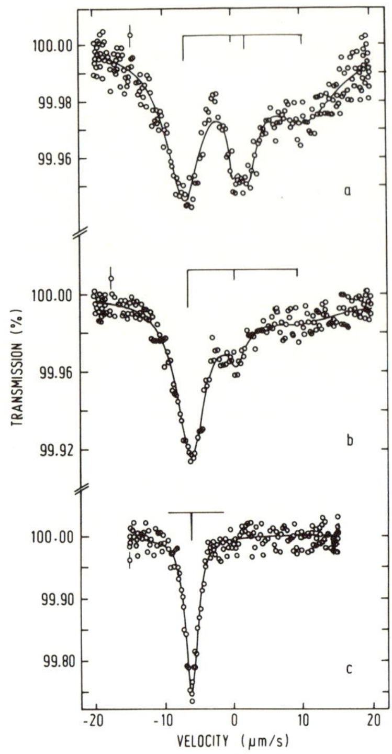

Figure 17. Mössbauer transmission spectra obtained at 4.2 K of brass alloy absorbers with Zn concentrations of 38.1 (a), 41.9 (b), and 48.0 at.% (c). The solid lines are least-squares fits to the multiline patterns indicated by the bar diagrams. (Taken from Ref. 66.)

pattern. When the Zn concentration c_{Zn} is increased, these four lines are more pronounced, but surprisingly no additional absorption lines appear and the line positions remain virtually unchanged even at values of c_{Zn} as high as 42%. Since the relative intensities widely change with c_{Zn}, the absorption lines do not originate from quadrupole interaction.[66,71,72] The peak at $\sim -6\,\mu m/s$ is observed in all spectra of Figs. 16 and 17. Its position remains unchanged, its relative intensity, however, increases continuously with c_{Zn} until it forms the single resonance line of β'-brass (see Fig. 17c). In addition, high-pressure ^{67}Zn Mössbauer experiments were carried out on two alloys, containing 15.9 at.% (α phase) and 49.2 at.%Zn (β' phase).[67] Surprisingly, the center shift, S_C, of the line at $\sim -6\,\mu m/s$ decreases by the same amount as S_C of the β' phase when the volume of the unit cell is reduced (see below). These observations could be interpreted that already at rather low Zn concentrations, β'-brass is formed and that the phase diagram indeed might need revision. This, however, is not the case. ^{67}Zn Mössbauer measurements performed up to 60 K.[72] on the two alloys (see Section 3.2.2) clearly demonstrate that the Lamb–Mössbauer factor and the second-order Doppler shift decrease more rapidly for β'-brass than for the line at $\sim -6\,\mu m/s$ in α-brass, i.e., there is no β'-brass present below the phase boundary shown in Fig. 15.

Which Cu–Zn configurations then cause the four absorption lines in

α-brass? Up to now this problem has only been partially solved. The absorption line at 0 μm/s, i.e., unshifted with respect to the emission line of the source, is to be identified with a configuration of one Zn atom surrounded by 12 nearest Cu neighbors, i.e., the same Cu–Zn configuration as in the source.[66,67,72] The absorption lines at positive velocities of ~1.8 and ~7.5 μm/s are caused by Cu–Zn configurations with more than one Zn atom in the nearest-neighbor shell.[66,67] For these configurations it is expected from theoretical calculations[59,62] that the s electron density $\rho(0)$ at the Zn nucleus is increased. These theoretical calculations show that s electron transfer from Zn to Cu is important. More Zn atoms in the nearest-neighbor shell reduce this transfer effect per Zn atom and thus $\rho(0)$ in such configurations is increased. The exact Cu–Zn configurations, however, are still unknown.

The absorption line at ~−6 μm/s is due to a highly stable Cu–Zn configuration which exhibits stronger chemical bonds than the other configurations and gives rise to a line at a negative Doppler velocity because of the second-order Doppler shift.[72] According to first-principle phase stability calculations,[69] the most probable candidate is the Cu_3Zn configuration. Investigations of lattice-dynamic effects, in particular Lamb–Mössbauer factor and second-order Doppler shift, at temperatures up to 60 K and high-pressure measurements on a sample with c_{Zn} = 15.9 at.% give strong support for this interpretation[67,72] (see Section 3.2.2).

At Zn concentrations of ~48 at.% the β phase is established which—below 725 K—forms an ordered alloy, β'-brass, with CsCl structure, i.e., each Zn atom is surrounded by eight Cu atoms. As expected for a cubic structure, the Mössbauer spectrum exhibits a single, rather narrow, Lorentzian line (see Fig. 17c). Most surprisingly, the line position is shifted to a negative velocity of −6.1 μm/s with respect to the source ^{67}Zn in fcc Cu, where each ^{67}Zn atom is surrounded by 12 Cu atoms in the nearest-neighbor shell. From the point of view of the Hume-Rothery rules, one would expect a positive shift due to the increase of conduction electrons per atom. Since, as will be discussed below, the phonon frequency spectrum of the source is harder than that for β'-brass, i.e., the effective Debye temperature of the source is higher than that for β'-brass, a negative shift cannot be caused by the second-order Doppler shift either.[72] Numerically, the negative shift corresponds to a rather large decrease $\Delta\rho(0)$ in contact density which amounts to 16% of the total change in $\rho(0)$ between Zn metal and ZnO! The origin of the low contact density in the CsCl structure of β'-brass is not understood at present. It has been partially attributed to an increased shielding effect of the rather narrow $3d$ band of Zn in β'-brass relative to α-brass.[11,73,74] Modern theoretical calculations[69,75,76] are called for clarifying the origin of this large decrease in $\rho(0)$.

Three alloys with zinc concentrations of 15.9, 41.9, and 48.0 at.% have been studied at 4.2 K and pressures up to 6.2 GPa.[67] The data show that the short-range order leading to the four different Cu–Zn configurations remains unaffected when the volume is reduced. There is also no indication of a phase transition from β' to α phase or vice versa. Such a transformation might have been expected on the basis of the Hume–Rothery rules and the interplay between Fermi surface of the conduction electrons and Brillouin zone boundary at a crystallographic transition.[77]

When pressure is applied, all absorption lines move toward more negative Doppler velocities. This decrease of the center shifts, S_C, is due to the second-order Doppler effect, SOD, which is partially compensated by the increase of isomer shift, S, i.e., by the increase of $\rho(0)$ when the conduction electrons are compressed. As already mentioned, in α-brass, one of the four (Cu–Zn) configurations, which corresponds to the line at $\sim -6\,\mu$m/s, surprisingly exhibits a center shift, and its change with pressure, which closely resembles S_C of the β' phase. Still, as will be shown in Section 3.2.2, this (Cu–Zn) configuration does *not* belong to the β'-brass.

In addition, β'-brass exhibits unexpected features at high pressures. Already at rather modest pressures of ~1 GPa, the single absorption line splits into a doublet with unequal intensities. The line with lower intensity and more negative center shift S_C corresponds to a new high-pressure phase which is most probably caused by a partially completed low-temperature martensitic transition.[67] The change of S_C with pressure is slightly larger for the new phase than for β'-brass. Since the effective Debye temperature is higher for the martensitic phase[78] than for the β' phase, the more negative value of S_C for the former mainly originates from an increased second-order Doppler shift.

The increase of $\rho(0)$ with pressure for the β' phase as well as for the high-pressure phase is a factor of ~1.5 smaller than for the α phase. This is astonishing, since, according to the Hume-Rothery rules and a simple band-structure model, the number of conduction electrons per atom is higher in β'- than in α-brass. To clarify these questions, modern theoretical calculations are of utmost importance.

The Mössbauer spectrum of γ-brass, containing 62.5 at.%Zn (see Table 3 and Fig. 15), which crystallizes in a complex structure of $D8_2$ type with 52 atoms in the unit cell, demonstrates the existence of two Zn sites with highly different surroundings.[11,79] The cubo-octahedral (CO) site exhibits a center shift of $-36.1\,\mu$m/s, which is far more negative than that of the inner tetrahedron (IT) with $+0.6\,\mu$m/s. In addition, at the IT site an efg of $eq = -0.6 \times 10^{17}$ V/cm^2 with asymmetry parameter $\eta = 0$ is observed. These results prove that the bonding at both sites is highly different. Furthermore, the spectrum shows approximately equal intensities for both Zn sites, whereas from the crystallographic structure a threefold higher intensity for the CO site is expected. This discrepancy is again a result of the different bonding at the two sites: the shift for the CO site is close to that for ZnTe and shows the importance of covalent bonding, whereas for the IT site, metallic bonding is dominant.

The ε-phase (see Table 3) is the only case where the relative line intensities are in accordance with a binomial distribution.[11,79] The widths of the absorption lines are relatively large, which points to the presence of an efg at the Zn sites. This is indeed expected in an hcp lattice with $c/a \approx 1.56$.

For larger c_{Zn}, another phase transition from the ε to the η phase (Zn metal) is observed. It is not clear if this transition is driven by conduction electrons because both phases belong to the hcp structure. The c/a ratio, however, is vastly different and makes Zn metal a highly interesting system as described in more detail in Section 3.1.

3.2.2. Lattice-Dynamic Effects

At present, two alloys with $c_{Zn} = 15.9$ at.% (α-brass) and $c_{Zn} = 49.2$ at.% (β'-brass) have been investigated up to temperatures of 60 K. The change of center shift, S_C with increasing temperature is plotted in Fig. 18 for both alloys. The change of S_C with temperature is nearly the same for all four Cu–Zn configurations of the α phase.[72]

The experimentally observed temperature variation of S_C can be written[80,81]

$$(\partial S_C/\partial T)_P = (\partial S_{SOD}/\partial T)_P + (\partial S/\partial T)_V + (\partial S/\partial \ln V)_T(\partial \ln V/\partial T)_P \quad (17)$$

The first term represents the second-order Doppler shift. The second term describes the explicit temperature dependence of the isomer shift S at constant

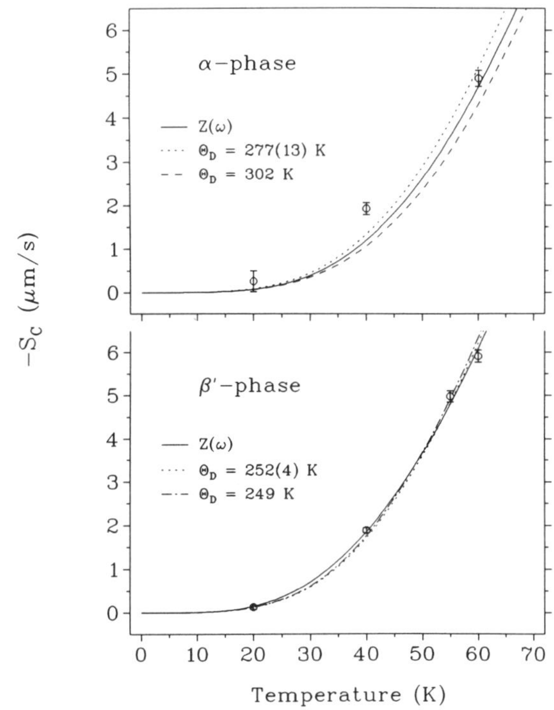

Figure 18. Temperature dependence of the center shift, S_C, of α-brass (15.9 at.%Zn) and β'-brass (49.2 at% Zn). Solid lines: calculation of the second-order Doppler shift using the phonon frequency distributions, $Z_\alpha(\omega)$ and $Z_{\beta'}(\omega)$, respectively, derived from inelastic neutron scattering data. Dotted lines: best fits to the Debye model. Dashed lines: Debye model with $\Theta_D = 302$ K for α-brass derived from specific-heat data and with $\Theta_D = 249$ K for β'-brass derived from the f-factor measurements. (Taken from Ref. 72.)

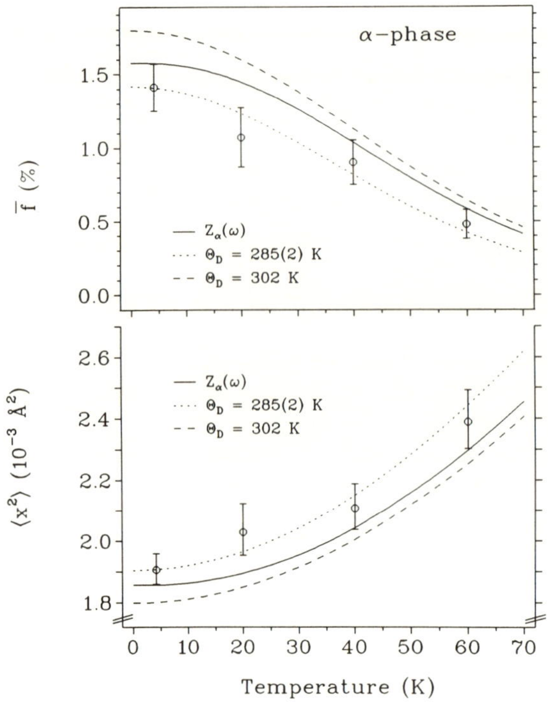

Figure 19. Temperature dependence of the average Lamb–Mössbauer factor, $\bar{f}$, and of the corresponding average mean-square atomic displacement of ^{67}Zn in α-brass with 15.9 at.% Zn. Solid lines: predictions obtained from $Z_\alpha(\omega)$. Dotted lines: best fit to the Debye model. The dashed lines correspond to a Debye temperature $\Theta_D = 302$ K derived from specific-heat data of α-brass. (Taken from Ref. 72.)

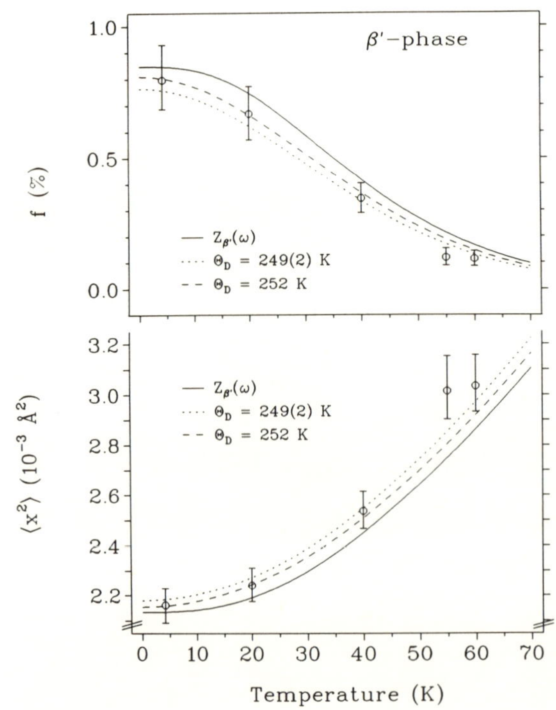

Figure 20. Temperature dependence of the Lamb–Mössbauer factor, f, and of the corresponding mean-square atomic displacement of ^{67}Zn in β'-brass with 49.2 at% Zn. Solid lines: predictions obtained from $Z_{\beta'}(\omega)$. Dotted lines: best fit to the Debye model. The dashed lines correspond to a Debye temperature $\Theta_D = 252$ K derived from the data on the second-order Doppler shift. (Taken from Ref. 72.)

volume due to changes of the electron density caused by the electron–phonon interaction upon the electronic states of the conduction electrons.[82] The third term accounts for the volume dependence of S caused by thermal expansion of the lattice. The last two terms are expected to change very little with temperature. For a detailed discussion see Ref. 18.

The temperature dependence of the LMF is depicted in Figs. 19 and 20. The LMF is derived from the total area under the absorption lines after correction for nonresonant background radiation. The LMF of the ^{67}Ga*Cu* source, $f_s = (2.0 \pm 0.2) \times 10^{-2}$, was determined earlier using the lattice-dynamic properties of ^{67}ZnO which are described in Sections 4.1.1 and 4.1.2. The temperature dependence of LMF is similar for all four Cu–Zn configurations in the α phase.[72] In Fig. 19 the temperature dependence of the average LMF ($\bar{f}$) is plotted.

For pure Cu metal[83] as well as for the alloys Cu_3Zn (α phase)[84] and $Cu_{0.53}Zn_{0.47}$ (β' phase),[85] the phonon frequency distributions $\tilde{Z}(\omega)$ have been determined earlier on the basis of phonon dispersion relations derived from inelastic neutron scattering data. In the case of Cu_3Zn, model calculations were performed which included interactions up to the sixth nearest-neighbor shell and allowed to determine 18 force constants between the atoms of the lattice. For the alloy $Cu_{0.53}Zn_{0.47}$ it was sufficient to take into account interactions up to the fourth nearest-neighbor shell involving 8 force constants. The phonon frequency distributions of both alloys have to be modified slightly in order to represent the chemical composition of the brass systems discussed here. This can be accomplished via the Lindemann formula[84,86] and results in a multiplication of the frequency axis of $\tilde{Z}(\omega)$ by factors of 1.0166 and 0.9926 to obtain $Z_\alpha(\omega)$ and $Z_{\beta'}(\omega)$, respectively. Figure 21 depicts these modified distributions together with $Z_{Cu}(\omega)$ for pure Cu metal.

Using $Z_{Cu}(\omega)$, we calculate for the LMF of pure Cu metal, $f_{Cu} = 2.06\%$ which is in good agreement with the value for the source ^{67}Zn*Cu*, $f_s = (2.0 \pm 0.2)\%$ mentioned above. Also, the Mössbauer results on the lattice-dynamic effects of both alloys can very well be described by the corresponding distributions $Z_\alpha(\omega)$ and $Z_{\beta'}(\omega)$. This is indicated by the solid lines in Figs. 18–20. In particular, the experimental data on the center shift (see Fig. 18) are in good agreement with the predicted second-order Doppler shift. Thus, the center shift in both alloys is mainly due to the SOD; the explicit temperature dependence of the isomer shift is negligibly small. In addition, for the α phase, experimental specific-heat data are available.[87,88] They are shown in Fig. 22. The overall agreement of the calculation using the phonon frequency distribution $Z_{sp}(\omega)$ with the experimental data is satisfactory. Still, as indicated in Fig. 22, the experimental data systematically tend toward lower values. This discrepancy can be attributed to the presence of short-range order effects. These are not taken into account properly neither by $Z(\omega)$ nor by the Lindemann approximation which assume a homogeneous alloy with binomial distribution of Zn in Cu.

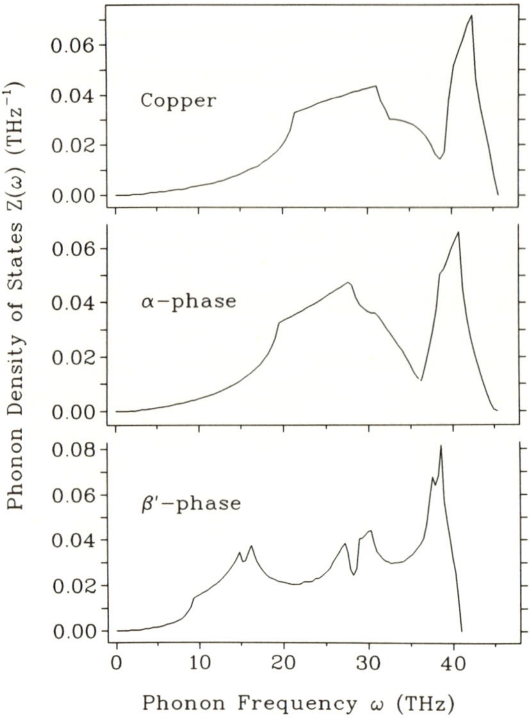

Figure 21. Phonon frequency distributions for Cu metal, $Z_{\mathrm{Cu}}(\omega)$, and the α and β' alloys, $Z_{\alpha}(\omega)$, $Z_{\beta'}(\omega)$, of Figs. 18–20. These distributions were obtained from model calculations based on inelastic neutron scattering data. (Taken from Ref. 72.)

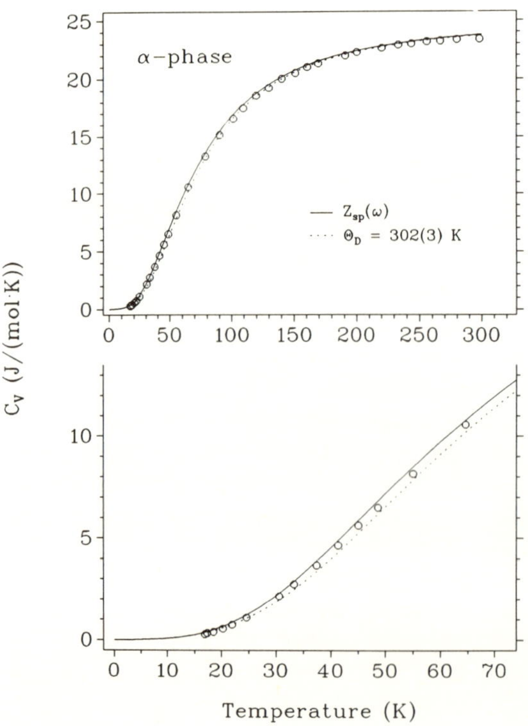

Figure 22. Specific heat at constant volume, C_V, for α-brass containing 19.77 at.%Zn. The error bars for C_V are smaller than the symbols used for the data points. Solid line: calculation using the phonon frequency distribution $Z_{\mathrm{sp}}(\omega)$ derived from inelastic neutron scattering data and slightly modified to represent the chemical composition of this alloy. Dotted line: best fit to the Debye model with $\Theta_D = (302 \pm 3)$ K. The bottom panel gives an expanded view of the temperature range up to 70 K. (Taken from Ref. 72.)

The Mössbauer data provide stringent tests for the phonon frequency distributions $Z_{\alpha}(\omega)$ and $Z_{\beta'}(\omega)$ and show that the model calculations mentioned above are well suited to predict lattice-dynamic effects. For comparison, in Figs. 18–20 and 22 are also shown the results obtained by fitting the Debye model to the data. Although the effective Debye temperatures as determined from LMF, SOD, and specific-heat data differ somewhat within each alloy, this simple model describes all lattice-dynamic effects surprisingly well. This is probably due to the highly symmetric (cubic) crystallographic structures of both alloys and the rather small difference in masses between Cu and Zn.

Finally, we would like to stress another result from the high-pressure experiments[67] on the brass systems mentioned in Section 3.2.1. In α- as well as in β'-brass, the LMF increases by almost a factor of two when the highest external pressure of 6.2 GPa is reached. Together with data on bulk moduli, such an increase in LMF can be translated into a Grüneisen parameter, γ, with a value of $\gamma \approx 4$ for both alloys.[37,67] Such a γ value appears too large by a factor of ~2. A similar situation has also been found for Zn metal.[37] This discrepancy is not understood at present. Again, modern cluster calculations should be able to clarify these questions.

4. SEMICONDUCTORS AND INSULATORS

4.1. Chalcogenides

The *sp*-bonded binary compound semiconductors of groups II and VI of the periodic table are an important class of technological materials. At ambient pressure, most of them crystallize in either the hexagonal wurtzite or the cubic zinc-blende structure or both. Each cation is tetrahedrally coordinated by four anions and vice versa, a coordination which is typical of sp^3 bonding. Under large external pressure, these compounds tend to transform to structures with higher coordination, e.g., from wurtzite or zinc blende to rock salt structure.

In recent years, a number of theoretical models have been developed to gain insight into structural properties and crystal stability.[89–91] For zinc chalcogenides and for ZnO in particular, also band structure[92,93] and cluster calculations[76] have been carried out to describe electronic transport as well as hyperfine properties.

4.1.1. ZnO at Ambient Pressure

The compound studied most thoroughly by ^{67}Zn Mössbauer spectroscopy is ZnO.[13,17,71,94,95] Figure 23 displays Mössbauer absorption spectra recorded at source temperatures between 4.2 and 77.3 K with a ZnO single crystal source oriented with its c axis parallel to the direction of observation of the 93.31-keV

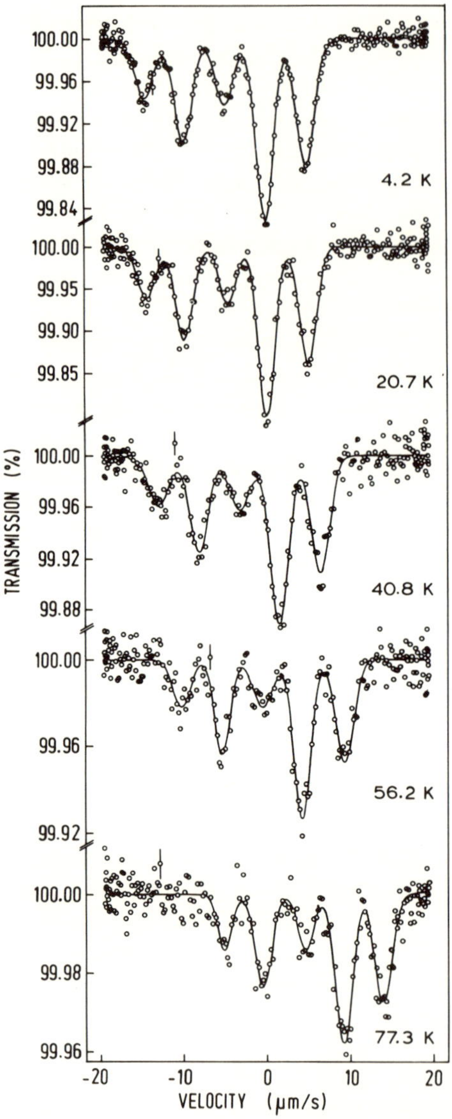

Figure 23. Mössbauer absorption spectra recorded at source temperatures between 4.2 and 77.3 K. The source is ^{67}Ga in ZnO single crystal with the *c* axis parallel to the direction of observation of the 93.31-keV γ rays. The absorber is enriched ^{67}ZnO powder. (Taken from Ref. 71.)

γ rays ($\vartheta = 0$). The absorber is an enriched powder. Due to the hexagonal structure of ZnO, an efg is present at the Zn nucleus both in the source as well as in the absorber. With relative intensities of the emission lines in the source of 0:2:1 and three absorption transitions of equal intensities in the powder absorber, the transmission spectrum exhibits five peaks with relative intensities of 1:2:1:3:2. Experiments carried out with different cuts ($\vartheta = 45°$) of the ZnO single crystal clearly showed that the recoil-free fraction is nearly isotropic resulting in effective Debye temperatures of $\Theta_{\perp} = 319$ K and $\Theta_{\parallel} = 315$ K.[71] This result is in sharp contrast with the highly anisotropic behaviour of Zn metal (see Section 3.1.2). In hexagonal ZnO, each Zn atom is surrounded by four oxygen atoms arranged in a slightly distorted tetrahedron. The nearly isotropic LMF indicates that the chemical bonds to the four oxygens are almost equal. This is further corroborated by the relatively small quadrupole interaction of $e^2qQ/h = +(2.401 \pm 0.004)$ MHz which corresponds to an efg $V_{zz} = +(6.68 \pm 0.01) \times 10^{16}\ \mathrm{V/cm^2}$. The changc of the efg with temperature up to 77.3 K is smaller than 1%. The asymmetry parameter η can be different from zero but is usually small at ambient pressure. This problem will be discussed below.

Most interesting is the behavior of the center shift, S_C. A change of S_C with temperature is clearly visible in Fig. 23. It has increased by $+(9.01 \pm 0.03)\ \mu$m/s between 4.2 and 77.3 K. As in the case of Cu–Zn alloys (see Section 3.2.2), the temperature variation of S_C can be written

$$(\partial S_C/\partial T)_P = (\partial S_{SOD}/\partial T)_P + (\partial S/\partial T)_V + (\partial S/\partial \ln V)_T(\partial \ln V/\partial T)_P \quad (18)$$

The meaning of the symbols is described in Section 3.2.2. From high-pressure Mössbauer data,[94,95] it is found that the contribution of the third term is negligibly small. Thus, to a very good approximation, the center shift is due to the second-order Doppler shift, S_{SOD}, and the explicit temperature dependence of the isomer shift, S_{ET}.

The separation of S_{SOD} from S_C can be accomplished by using specific-heat data.[71] It turns out that S_{ET} is an essential contribution to S_C already at cryogenic temperatures. The change of S_{ET} is such that the s electron density, $\rho(0)$, at the Zn nucleus is *reduced* with increasing temperature. As discussed in detail in Ref. 71, dynamic charge transfer appears to play the important role due to the partially covalent chemical bond in ZnO. The dynamic phonon-induced electron transfer from partially occupied 4s and 4p orbitals of zinc to the neighboring oxygen atoms is mediated by the orbital–lattice interaction. Such an electron transfer indeed decreases $\rho(0)$. From the measured value of ~3.5 μm/s (at 77.3 K) a charge transfer which corresponds to a fraction of 0.01 of a 4s electron can be estimated.

The temperature dependence of the dynamic charge transfer was predicted[96] to cause S_{ET} to vary according to

$$S_{ET} = KT^4\Phi(T) \quad (19)$$

where K is a constant containing the electronic matrix elements. In Ref. 96 the Debye approximation was used with

$$\Phi(T) = \int_0^{\Theta/T} z^3(e^z - 1)^{-1}\, dz \quad (20)$$

where Θ is an effective Debye temperature.

The experiment gives $K = (3.0 \pm 1.0) \times 10^{-8}\ \mu\mathrm{m\,s^{-1}\,K^{-4}}$ and $\Theta =$

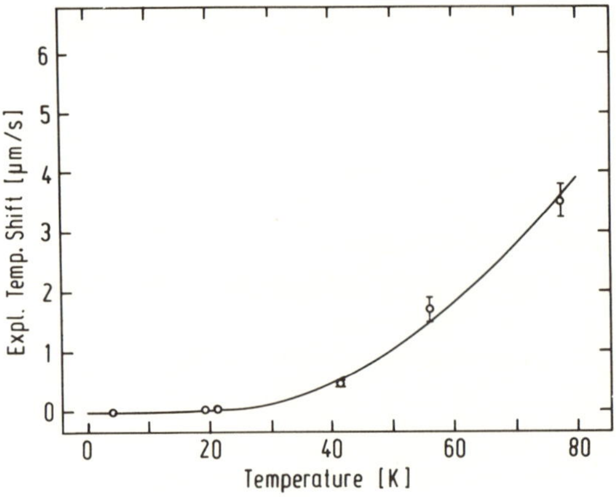

Figure 24. Explicit temperature dependence of the isomer shift, S_{ET}. The solid line is a fit to the data using the equation $S_{ET} = KT^4\Phi(T)$. At low temperatures a T^4 dependence is observed. (Taken from Ref. 71.)

(275 ± 3) K. The solid line in Fig. 24 is calculated with these parameters. The agreement with the experimental data is very good. In addition, the value of Θ agrees perfectly with the Debye temperature, Θ_D^{sp}, derived from specific-heat data in this temperature range. At low temperatures, $\Phi(T)$ is temperature independent to a good approximation. Indeed, if $(\Theta/T) > 7$, the deviation of $\Phi(T)$ from $\Phi(0) = \pi^4/15$ is less than 5%. Therefore,

$$S_{ET} = K(\pi^4/15)T^4 \tag{21}$$

Although the change of the s electron density at the ^{67}Zn nucleus is extremely small, the high-resolution spectroscopy of the 93.31-keV transition makes it possible to determine phonon-induced charge transfer with high precision.

4.1.2. ZnO at High Pressure

4.1.2.1. Experimental Results. A phase transition from the hcp (wurtzite) to the fcc (NaCl) structure is observed with increasing pressure.[95] Figure 25 shows the normalized peak areas of the $[110]_H$ reflection, representative of the hexagonal phase, and of the $[200]_C$ peak of the cubic rock salt phase. Increasing pressure, the phase change starts at ~6.5 GPa and is completed only at ~10 GPa. Releasing pressure, the rock salt structure is maintained to ~2.5 GPa and is transformed back to the wurtzite structure at ~0.7 GPa. This is the first time that this phase change in ZnO has been observed without using a catalyst stabilizing the rock salt structure.[95,97]

At room temperature the following values were obtained for the lattice constants and bulk moduli of ZnO in the wurtzite phase: $a_0 = (3.254 \pm 0.008)$ Å; $c_0 = (5.220 \pm 0.011)$ Å; $\delta(c/a)/\delta P = -(0.0007 \pm 0.0003)\ \text{GPa}^{-1}$; $B_0 = (204 \pm 9)$ GPa; for the rock salt phase: $a_0 = (4.270 \pm 0.009)$ Å; $B_0 = (254 \pm 7)$ GPa. In both cases B_0' turned out to be negligibly small at pressures up to 10.5 GPa. The phase transformation is accompanied by a drastic

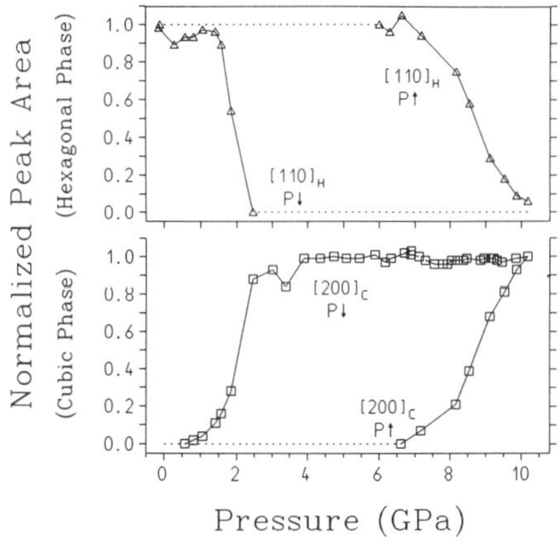

Figure 25. Phase transition of ZnO from hexagonal wurtzite to cubic rock salt structure as determined by X-ray diffraction from the normalized peak area of the $[110]_H$ (wurtzite) and $[200]_C$ (rock salt) reflections. Note the large hysteresis of the phase transition when increasing ($P\uparrow$) and decreasing ($P\downarrow$) pressure. (Taken from Ref. 95.)

reduction in volume per ZnO molecule of ~19%. The bulk modulus B_0 of the cubic phase is ~25% larger than in the second modification.

Figure 26 displays Mössbauer transmission spectra recorded at 4.2 K and ambient pressure as well as at pressures of 0.7 and 4.4 GPa. At pressures below the phase transition the spectra consist of five absorption lines because of the quadrupole interaction in both, source and absorber.[71] However, already at low pressures, smaller than 0.7 GPa, where ^{67}ZnO is still in the wurtzite phase, the Mössbauer spectrum changes markedly. This is mainly due to the sharp increase of the asymmetry parameter η of the efg tensor. Also, the main component of the efg tensor increases with reduced volume, V,

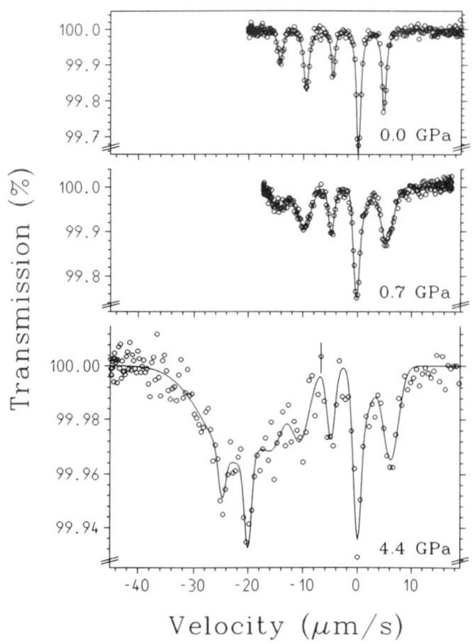

Figure 26. Mössbauer transmission spectra of ^{67}ZnO powder at 4.2 K and ambient pressure, at 0.7 GPa, and at 4.4 GPa after releasing the pressure from ~8 GPa. The source is ^{67}Ga*ZnO* single crystal at 4.2 K with the c axis parallel to the direction of observation of γ rays. The spectrum obtained at 4.4 GPa shows a superposition of the absorption lines of the wurtzite and rock salt phases of ^{67}ZnO. Note the restricted velocity range. (Taken from Ref. 95.)

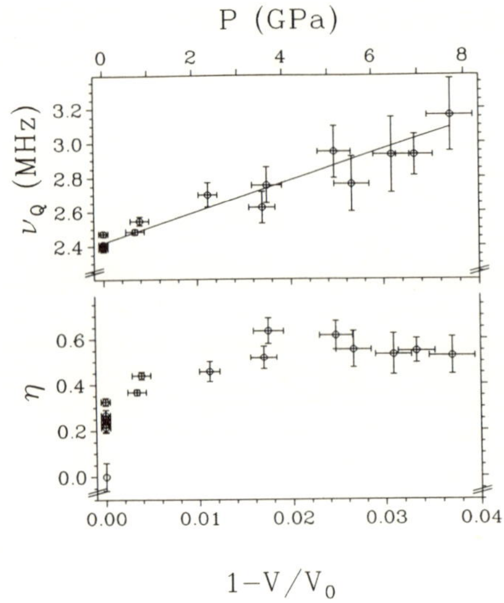

Figure 27. Quadrupole frequency ν_Q of the main component of the electric field gradient tensor and asymmetry parameter η of ^{67}ZnO (wurtzite) versus reduced volume of the unit cell.

according to

$$\nu_Q = e^2qQ/h = +[(2.418 \pm 0.008) + (18.2 \pm 1.9)(1 - V/V_0)]\ \text{MHz}$$

where V_0 is the volume of the unit cell at ambient pressure. The results on ν_Q and η are summarized in Fig. 27. The measurements clearly point out that the wurtzite structure is highly unstable even against small nonaxial lattice distortions. Small stresses within the ZnO crystallites themselves or originating from lattice contractions, when cooling down to liquid-helium temperatures, are sufficient to cause $\eta \neq 0$.[94,95] This instability has various consequences. For example, it prevents the precise determination of the gravitational redshift of the 93.31-keV γ rays in the gravitational field of the Earth when using ZnO as absorber material. The gravitational redshift experiment has been described in detail in Ref. 8 and will not be considered in the present review, where the physics of intrinsic properties of materials is emphasized. Other examples are the theoretical calculations for ZnO, where deviations of η from $\eta = 0$ have to be taken into account (see Section 4.1.2.2).

The Lamb–Mössbauer factor, f_{hcp}, of the hcp phase increases linearly with reduced volume:

$$\Delta f_{\text{hcp}}/\Delta(V/V_0) = -(34 \pm 13)\%$$

The center shift slightly increases:

$$S_C = [(0.014 \pm 0.010) + (17.3 \pm 2.4)(1 - V/V_0)]\ \mu\text{m/s}$$

The change of S_C at constant temperature can be written as

$$\left(\frac{\Delta S_C}{\Delta V}\right)_T = \left(\frac{\Delta S_{\text{SOD}}}{\Delta V}\right)_T + \left(\frac{\Delta S}{\Delta V}\right)_T \tag{22}$$

Within the Debye model, the ΔS_{SOD} can be estimated from the change of the recoil-free fraction, f_{hcp}. The change of isomer shift ΔS can then be calculated: $\Delta S/\Delta(V/V_0) = -(374 \pm 88)\,\mu\text{m/s}$. This corresponds to an *increase* of *s* electron density, $\rho(0)$, at the ^{67}Zn nucleus with reduced volume,[45] of $\Delta\rho/\Delta(V/V_0) = -(10.7 \pm 3.4)\,e/a_0{}^3$. The experimental results strongly emphasize that in the wurtzite structure a reduction of the unit cell volume of ^{67}ZnO induces an increase of covalency of the Zn–O bond and thus leads to an increase of $\rho(0)$.

At pressures between ~6.5 and ~10 GPa the phase transition is incomplete and a superposition of the absorption lines of both structures is observed, as shown in the lowest spectrum of Fig. 26. The ZnO sample was pressurized to ~8 GPA, the Mössbauer spectrum was recorded after releasing the pressure to ~4.4 GPa. The fit represents the five-line absorption pattern, centered at $0\,\mu\text{m/s}$, of the wurtzite phase and the spectrum of the rock salt phase, centered at $-20\,\mu\text{m/s}$, which consists of two lines because of the quadrupole interaction in the ^{67}Ga*ZnO* single-crystal source. In addition, a rather broad unresolved absorption dip has to be taken into account. Because it is also centered at $-20\,\mu\text{m/s}$, it reflects ^{67}ZnO grains in the rock salt phase, which, however, are heavily distorted. This observation is supported by X-ray investigations: the diffraction lines of the rock salt phase are slightly broadened, in particular within the pressure range in which the crystallographic transition takes place.[95]

The S_C of the cubic phase is by $20\,\mu\text{m/s}$ more negative than S_C of the wurtzite structure. This is surprising, because the phase transition is accompanied by a drastic decrease of ~19% of the unit cell volume. A decrease usually favors covalent bonding and thus should lead to a larger *s* electron density at the ^{67}Zn nucleus. In the case of ZnO, however, the volume collapse is mainly due to the increase in coordination number, from 4 to 6, of nearest-neighbor (nn) atoms. The nn distance in the rock salt structure is in fact larger by 7% than in the wurtzite phase. Thus, covalency of the Zn–O bond is reduced and its ionic character is larger. In terms of *s* electron density at the ^{67}Zn nucleus, the fcc (NaCl) modification exhibits a lower value of $\rho(0)$, extrapolated to ambient pressure, than the hcp (wurtzite) structure.[94] The difference is $\Delta\rho(0) = -0.14e/a_0{}^3$. This decrease of $\rho(0)$ corresponds to ~1/50 of a 4*s* electron. When pressure is applied, $\rho(0)$ rises considerably in the hcp phase, as mentioned above, whereas only a very small increase, if any, can be observed in the fcc structure. To detect such minute variations requires the extreme sensitivity of the 93.31-keV Mössbauer resonance.

4.1.2.2. Theoretical Calculations of Hyperfine Interactions. From a theorist's point of view, it is an enormous challenge to calculate tiny changes in $\rho(0)$ from first principles, within the theoretical models presently available. Older methods of computations are not sufficient. For example, a theoretical calculation[98] based on an ionic model which, in addition, took overlap effects into account, gave $\Delta\rho(0)/\Delta(V/V_0) = -(28.7 \pm 6.2)e/a_0{}^3$ in clear disagreement with the experimental result given in Section 4.1.2.1. It has been attempted, however, to meet this challenge by performing self-consistent, nonrelativistic Hartree–Fock calculations on a finite cluster of atoms with zinc at the center.[75,76] The clusters $(ZnO_4)^{-6}$ and $(ZnO_6)^{-10}$ for the wurtzite and cubic phases, respectively, were chosen. To take into account the Madelung potential of the infinite solid, the clusters were embedded in an array of point charges. This Hartree–Fock cluster procedure includes covalency effects in a first-principles manner and will be addressed in more detail in Section 4.1.4. It is the first time that theoretical calculations have been applied to investigate such small changes of $\rho(0)$.

The calculations show that indeed covalency of the 4*s* and 4*p* electrons in the Zn–O bond is most important in both modifications. At ambient pressure, a quadrupole coupling constant in the hcp phase, e^2qQ/h, of $+(2.57 \pm 0.07)$ MHz was obtained,[76] in good agreement with the experimental value of $+(2.401 \pm 0.004)$ MHz. A different cluster calculation has been given in Ref. 99. The approximations used will be described in more detail in Section 4.2.3, in the discussion of ZnF_2. The agreement of the latter calculation with experimental data is only within ~40%.

The results of both types of cluster calculations emphasize that the main component of the efg is dominated by an unbalanced occupation of the 4*p* orbitals of Zn in the valence states. Furthermore, the efg turns out to react very sensitively to details of the crystallographic structure. Unfortunately, the distortions of the lattice, which lead to $\eta \neq 0$ when pressure is applied, are not precisely known. For this reason, it was not possible to calculate for the wurtzite phase the changes of e^2qQ and of $\rho(0)$ with reduced unit cell volume.

As already mentioned, at the phase transition the covalency of the Zn–O bond is reduced leading to a larger ionic character. The cluster calculations, however, seem to overestimate $\Delta\rho(0)$ at the phase transition. They give $\Delta\rho(0)_{th} \sim -1.02e/a_0^3$, whereas experimentally a value of $\Delta\rho(0)_{ex} \sim -0.14e/a_0^3$ is derived. The origin of this discrepancy is unclear. It should be emphasized again that experimentally the separation of isomer shift and second-order Doppler shift is a difficult task[71] and a large part of the discrepancy might be connected just to this problem.

Within the fcc (NaCl) structure, covalency increases when the unit cell volume is reduced leading to an increase of the 4*s* electrons at the zinc atom and therefore to a higher *s* electron density at the ^{67}Zn nucleus. Both calculations and experiment demonstrate that this increase of $\rho(0)$ is relatively small, in agreement with the larger ionic character of the Zn–O bond in the fcc phase relative to the hcp structure.

The calculations show another very important aspect. Although the 4*s* electrons play the essential role, their contribution to $\rho(0)$ is partially counteracted by the inner 1*s*, 2*s*, and 3*s* electrons. This is most surprising, since these closed shells are embedded deep inside the electronic core. Still, their influence amounts to ~35% of the 4*s* electron contribution and cannot be neglected. Although relativistic calculations might change all of these numbers somewhat, the method of "freezing" inner core electrons which makes computations easier and faster is questionable when changes of *s* electron density have to be calculated.

4.1.3 ZnS, ZnSe, and ZnTe

Figure 28 displays transmission spectra of ZnS, ZnSe and ZnTe with cubic zinc-blende structure. The source is ^{67}Ga*Cu*. The results are summarized in Table 4. As expected for cubic systems, a single absorption line is recorded. Surprisingly, this is also the case for the wurtzite modification of ZnS: no quadrupole interaction is observed. In fact, the Mössbauer parameters for both ZnS structures are closely similar.[17] This is in agreement with the observation that the hexagonal and the cubic modifications of ZnS differ only by the stacking sequence of the Zn and S layer-pairs and the nearest-neighbor bond lengths Zn–S are nearly equal in the two structures.[100] Looking only at the nearest-neighbor atoms, it is not possible to tell whether the crystal is in the cubic or hexagonal phase. Thus, a difference in the hyperfine interactions, like a possible quadrupole interaction, can only originate from the second-nearest-neighbor or even higher-order shells.

Within the zinc chalcogenides, the center shift S_C (with respect to a

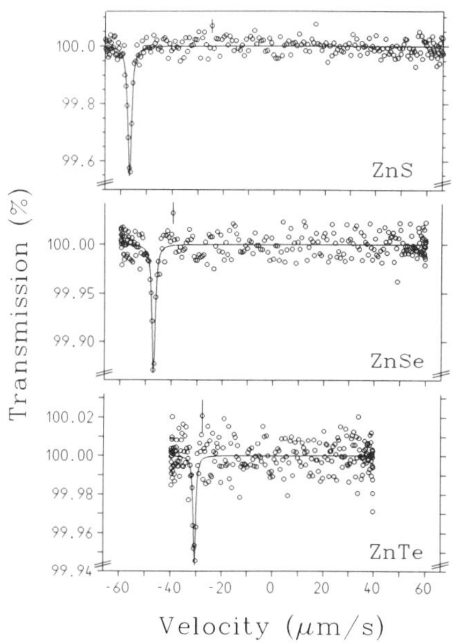

Figure 28. Mössbauer transmission spectra of ^{67}ZnS, ^{67}ZnSe, and ^{67}ZnTe recorded at 4.2 K. The source is ^{67}Ga in Cu metal kept at 4.2 K.

Table 4. Mössbauer Parameters of Zn Chalcogenides and ZnF_2[a]

Compound	S_C (μm/s)	Linewidth (μm/s)	e^2qQ (MHz)	η
^{67}ZnTe	−30.7 ± 0.1	1.1 ± 0.2	—	—
^{67}ZnSe	−47.0 ± 0.1	1.4 ± 0.1	—	—
$^{67}ZnS_{hcp}$	−57.5 ± 0.2	2.6 ± 0.5	—	—
$^{67}ZnS_{cub}$	−56.5 ± 0.1	1.6 ± 0.1	—	—
$^{67}ZnO_{hcp}$	−110.7 ± 0.1	1.8 ± 0.2	+2.401 ± 0.004	small
$^{67}ZnO_{NaCl}$	−130.7 ± 1.1	1.5 ± 0.3	—	—
$^{67}ZnF_2$	−174.1 ± 0.3	2.2 ± 0.2	+8.20 ± 0.06	0.15 ± 0.02

[a]The center shifts S_C are given relative to the ^{67}Ga*Cu* source. The linewidth is the full width at half-maximum. The quadrupole coupling constant is denoted by e^2qQ, η is the asymmetry parameter.

^{67}Ga*Cu* source) changes drastically from $-(110.7 \pm 0.1)$ μm/s for ZnO to $-(30.7 \pm 0.1)$ μm/s for ZnTe. Again, the problem arises to separate S_{SOD} from S_C in order to derive the isomer shift, i.e., changes of $\rho(0)$. The procedure of calculating S_{SOD} from the Debye model[17] might lead to wrong results, because the Debye model only poorly approximates the experimentally determined phonon frequency distributions of the Zn chalcogenides. In addition, the ratio of the masses of Zn and the ligands changes from ~4 for ZnO to ~0.5 for ZnTe. As a consequence, the eigenvectors of the dynamic matrix are important and lead to a different weighting of the different branches of the phonon dispersion relation. This problem will be discussed in Section 4.1.5. We will first describe theoretical calculations of *s* electron densities. These calculations will then enable us to address the second-order Doppler shift.

4.1.4. Theoretical Calculations of s-Electron Densities

The change of *s* electron density $\rho(0)$ at the ^{67}Zn site within the chalcogenides has been calculated by two groups[75,76,92,101] using very different approaches. In Ref. 92 the relativistic first-principles linear muffin-tin orbital (LMTO) method with the atomic sphere approximation (ASA) has been employed. The LMTO method allows the crystal potential and charge distributions to be iterated until self-consistency is reached. In ASA the crystal unit cell is approximated by spheres around the atomic sites. For simplicity it was further assumed that ZnO—which, as mentioned before, crystallizes in slightly distorted wurtzite structure—also possesses sphalerite structure, at the same density as natural ZnO. In addition, in all cases the atomic cores were treated as frozen, i.e., their contribution to $\rho(0)$ is a constant.

A rather different method has been used in Refs. 75, 76 and 101. Nonrelativistic self-consistent Hartree–Fock calculations have been carried out on a finite cluster of atoms with Zn at the center. The Madelung potential of the infinite solid has been included by embedding the cluster in an array of point charges with the same symmetry as the crystal. Called the Hartree–Fock cluster procedure, this approach includes covalency effects in a first-principles manner. The core was not frozen, i.e., all electrons inside the cluster have been taken into account. The major drawback, however, is that the cluster which can be computed in this way, has to be rather small. For this reason, sometimes the core is treated as frozen and only the electronic states in the valence region are iterated to self-consistency in a now considerably larger cluster.[99] Usually a correction factor which depends on the atomic number of the central ion is applied to adjust for relativistic effects.

Table 5 summarizes the changes $\Delta\rho(0)$ obtained by the LMTO–ASA and the cluster (core not frozen) methods. A relativistic correction factor[44,99] of 1.36 was used in the cluster approach to take into account relativistic effects. As can be seen from Table 5, the agreement between both methods is very good. This makes us confident that both methods are able to calculate $\Delta\rho(0)$ in zinc chalcogenides correctly. Both calculations demonstrate that covalency effects in Zn chalcogenides lead to sizable changes in $\rho(0)$.

With the experimental value[45] of $\Delta\langle r^2\rangle = +(17.8 \pm 3.9) \times 10^{-3}\,\text{fm}^2$, we are now able to calculate the isomer shifts, S, and deduce the S_{SOD} using the experimentally determined S_{C}. Figure 29 shows a plot of the changes in S_{C} versus $\Delta\rho(0)$ as derived from cluster calculations. The data are taken from Table 5. The rather large errors connected to the values for S and S_{SOD} are mainly due to the uncertainty of $\Delta\langle r^2\rangle$. For comparison, the values for $S^{\text{D}}_{\text{SOD}}$, which result from the Debye model, are also included in Table 5. In the case of

Table 5. Comparison of Results for Change of s Electron Densities, $\Delta\rho(0)$, from LMTO and Cluster Calculations[a]

Compound	$\Delta\rho(0)$ (e/a_0^3) LMTO	Cluster	S_{C} (μm/s)	S (μm/s)	S_{SOD} (μm/s)	$S^{\text{D}}_{\text{SOD}}$ (μm/s)	Debye temp.[b] (K)
ZnTe	+2.4 ± 0.2	+2.3[c]	+80.0 ± 0.1	+80.6 ± 17.7	−0.6 ± 17.7	+23.6 ± 2.6	215 ± 10
ZnSe	+1.8 ± 0.2	+1.6[c]	+63.7 ± 0.1	+55.1 ± 12.1	+8.6 ± 12.1	+19.6 ± 2.6	232 ± 10
ZnS_{cub}	+1.3 ± 0.2	+1.3[c]	+54.2 ± 0.1	+46.4 ± 10.2	+7.8 ± 10.2	+16.2 ± 2.6	247 ± 10
ZnO_{hcp}	0	0	0	0	0	0	317 ± 5
ZnO_{NaCl}	—	−1.0[c]	−20.0 ± 1.0	−35.5 ± 7.8	+15.6 ± 7.9	−15.0 ± 9.3	382 ± 40
ZnF_2	—	−1.7[c]	−63.4 ± 0.3	−60.7 ± 13.3	−2.7 ± 13.3	+8.7 ± 1.3	280 ± 5
		−2.1[d]	−63.4 ± 0.3	−71.5 ± 15.7	+8.1 ± 15.7	+8.7 ± 1.3	280 ± 5

[a] With $\Delta\langle r^2\rangle = +17.8\,\text{fm}^2$ the values for the isomer shifts, S, relative to ZnO_{hcp} are obtained. The experimental center shift, also relative to ZnO_{hcp}, is denoted by S_{C}. The change of the second-order Doppler shift (SOD) is calculated from $S_{\text{SOD}} = S_{\text{C}} - S$. The change of SOD when using the Debye model is given by $S^{\text{D}}_{\text{SOD}}$. In the last column the values for the Debye temperatures are summarized.
[b] Determined from the recoil-free fraction.
[c] Ref. 101.
[d] Ref. 99.

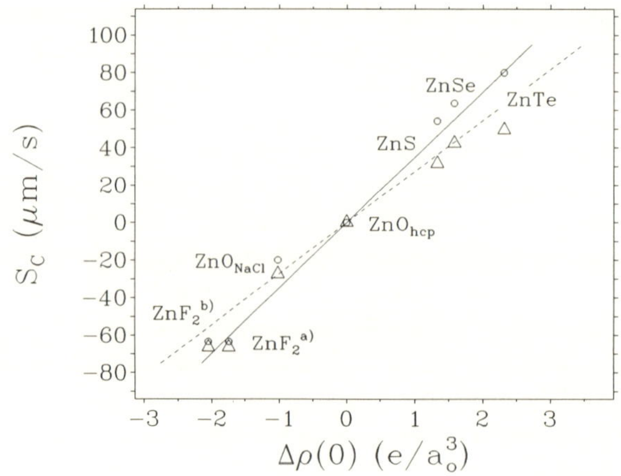

Figure 29. Plot of center shifts, S_C (○), relative to ZnO wurtzite phase, for various zinc compounds versus the change of s electron density, $\Delta\rho(0)$, as derived from cluster calculations. For ZnF_2 the results of two different cluster calculations are indicated: a, Ref. 101, b, Ref. 99. The solid line is obtained with $\Delta\langle r^2\rangle = +17.8 \times 10^{-3}\,\text{fm}^2$. The deviation of the shift of a particular compound from this line gives the correction for the second-order Doppler shift, S_{SOD}. The dashed line is obtained with $\Delta\langle r^2\rangle = +13.9 \times 10^{-3}\,\text{fm}^2$. If the center shifts are corrected by the second-order Doppler shift S_{SOD}^{cal}, as calculated from force constant models (see Table 6), the isomer shift values denoted by the symbol Δ are obtained. The agreement with $\Delta\langle r^2\rangle = +13.9 \times 10^{-3}\,\text{fm}^2$ is quite satisfactory.

the ^{67}Zn resonance, $S_{SOD}^{D} = 0.231[\mu\text{m}/(\text{s K})] \times [\Theta_A - \Theta_B)$, where Θ_A and Θ_B are the effective Debye temperatures as determined, for example, from the LMFs for the compounds A and B, respectively, at the same temperature.[43] Table 5 clearly shows that the Debye model provides a reasonable estimate of the second-order Doppler shift for ZnS, ZnSe, and ZnTe, which all crystallize with cubic zinc blende structure, only if the lower limit for $\Delta\langle r^2\rangle = 13.9 \times 10^{-3}\,\text{fm}^2$ is accepted. Otherwise, no correction at all for the second-order Doppler shift appears to be a better approximation. We will come back to this problem in Section 4.1.5.

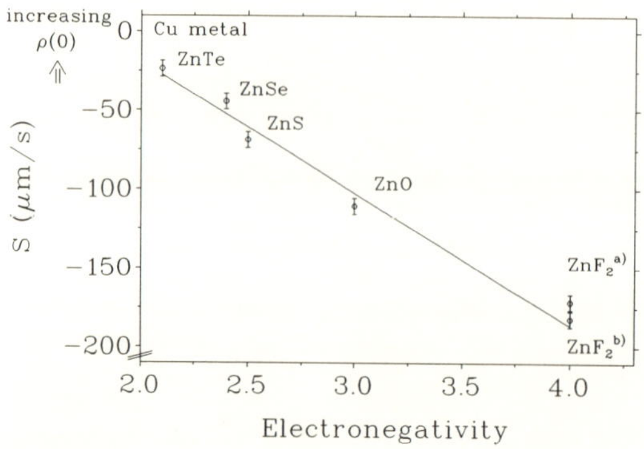

Figure 30. Plot of isomer shifts, S, versus Pauling electronegativity of the ligands for divalent Zn compounds. The shifts are given relative to the ^{67}Ga*Cu* source. For ZnF_2, two different results are indicated: a, Ref. 101; b, Ref. 99.

In Fig. 30 the isomer shifts, S, of the zinc chalcogenides are plotted versus the Pauling electronegativity, EN, of the ligands bonded to the divalent Zn atom.[102] Also, ZnF_2, which will be discussed in Section 4.2, has been included. A linear correlation is found: $S = [-(83.75 \pm 0.34)\,\text{EN} + (148.59 \pm 0.97)]\ \mu\text{m/s}$. The chemical bond in divalent zinc involves the $4s$, $4p$, and $4d(3d)$ electrons, whose density at the zinc site decreases[17] when going from ZnTe, with the ligand of lowest electronegativity, to ZnSe, ZnS, ZnO, and finally to ZnF_2 with the ligand of highest electronegativity. In compounds with ligands of larger electronegativity, the s electron density, $\rho(0)$, decreases because the direct contribution to $\rho(0)$ of the $4s$ electrons by far outweighs the shielding effect of $4p$ and $4d(3d)$ electrons.[103] As will be discussed in more detail in Section 4.2, even in ZnF_2, the chemical bond is not completely ionic. The change in isomer shift corresponds to a decrease, $\Delta\rho(0)$, of $-2.3e/a_0^3$ between ZnTe and ZnO and a further decrease, $\Delta\rho(0)$, of $-1.7e/a_0^3$ between ZnO and ZnF_2. Covalency is important and determines also the efg in ZnF_2. In addition, high-pressure measurements on ZnO (see Section 4.1.2) and on ZnS[104] clearly support the picture described above and show that the covalent part of the chemical bond determines the changes of $\rho(0)$.

4.1.5. Model Calculations of Lattice-Dynamic Effects

The theoretical calculations of LMF and S_{SOD} are an enormous challenge. Very few results have been published. We have tried to calculate[105] these lattice-dynamic quantities for zinc chalcogenides and for ZnF_2 using relatively simple force constant models.

In the harmonic approximation, the potential energy of the lattice is written[106]

$$V = \tfrac{1}{2} \sum_{ll',\kappa\kappa'} \sum_{\alpha\beta} u_\alpha\binom{l}{\kappa} \cdot \Phi_{\alpha\beta}\binom{ll'}{\kappa\kappa'} \cdot u_\beta\binom{l'}{\kappa'}$$

where l and l' label primitive cells of the lattice, κ and κ' are the basis indices within the lattice cells l and l', respectively, $u_\alpha\binom{l}{\kappa}$ is the component α of the displacement of the atom denoted by l and κ, and $\Phi_{\alpha\beta}\binom{ll'}{\kappa\kappa'}$ are the general second-order expansion coefficients of the potential energy, usually referred to as the *force constants.*[106]

Using plane waves to solve the equation of motion leads to an eigenvalue problem of the dynamic matrix[105,106] which yields for each eigenvalue j a relation between the eigenfrequencies $\omega_j(\mathbf{q})$ and the wave vector $\mathbf{q}$. If n is the size of the basis, $j = 1, 2, \ldots, 3n$. The solution of this eigenvalue problem also gives the eigenvectors, which describe the polarization of the modes.

The LMF can be written[107]

$$f = \exp\left[-\frac{\hbar}{m_\kappa N} \sum_{\mathbf{q}_j} \frac{|\mathbf{k} \cdot \mathbf{e}(\kappa\,|{}^{\mathbf{q}}_{j})|^2}{\omega_j(\mathbf{q})} \left(n_j(\mathbf{q}) + \tfrac{1}{2}\right)\right] \tag{23}$$

The second-order Doppler shift can be generalized in a similar fashion. In velocity units,

$$S_{\mathrm{SOD}} = -\frac{\hbar}{2m_\kappa cN}\sum_{\mathbf{q}j} \omega_j(\mathbf{q}) \cdot |\mathbf{e}(\kappa\,|^{\mathbf{q}}_{j})|^2 \cdot (n_j(\mathbf{q}) + \tfrac{1}{2}) \tag{24}$$

where $\mathbf{k}$ is the wave vector of the γ ray, c is the vacuum velocity of light, κ labels the ^{67}Zn site within the unit cell, and m_κ is the mass of the ^{67}Zn atom. The eigenvectors $\mathbf{e}(\kappa\,|^{\mathbf{q}}_{j})$ indicate the polarization of the mode $(\mathbf{q}, j)$ associated with the eigenfrequency $\omega_j(\mathbf{q})$. N is the number of unit cells of the finite crystal.

The phonon occupation number at temperature T is given by

$$n_j(\mathbf{q}) = \left(\exp\left(\frac{\hbar\omega_j(\mathbf{q})}{k_{\mathrm{B}}T}\right) - 1\right)^{-1} \tag{25}$$

For the case of monoatomic systems ($n = 1$), Eqs. (23)–(25) lead to the familiar formulas for f and S_{SOD} described in Section 3.1.2.1. In general, however, Eqs. (23) and (24) involve the eigenvectors as well as the eigenfrequencies.

To calculate f and S_{SOD} according to Eqs. (23)–(25), a force constant model is usually employed. Among the different models available in the literature, two relatively simple models have been investigated.[105] Both models are suitable for ionic systems which are partially covalent.

The *rigid ion model* (RIM)[108] describes the interaction between two ions in the crystal by two contributions: a short-range interaction due to the covalent bonding and a long-range interaction due to the Coulomb forces between the ions. The short-range interaction usually extends to the next and second nearest neighbors. For the Coulomb interaction, the ions are assumed to be point charges with no internal structure, i.e., "rigid ions."

The basic assumptions of the *shell model* (SM)[109,110] are the same as in the RIM, except that the point charges of the ions are replaced by polarizable ions consisting of a massless shell and a heavy core. Both shell and core are charged and the intraionic shell–core interaction is described by an isotropic harmonic force. In addition, shell and core interact with the shells and cores of neighboring ions via the Coulomb interaction and short-range interactions.

In both models, the short-range interaction potential may be expressed as a function of the bond angles, leading to bond bending forces, and of bond lengths, giving rise to bond stretching forces.[111] In these so-called "valence force field models", the number of adjustable parameters is often further reduced compared with the rigid ion and shell models.

The parameters of these phenomenological models have to be determined empirically. This is usually accomplished by fitting the calculated phonon dispersion relations to inelastic neutron scattering data, critical point frequencies determined by first- or second-order Raman scattering or macroscopic quantities as elastic constants, and, in case of the shell model, the dielectric constant in the limits of zero and infinite frequencies. Unfortunately, these quantities are not sensitive to the eigenvectors of the dynamic matrix but only to the frequency spectrum, thus allowing only the determination of the eigenvalues of the dynamic matrix.[106] However, as already mentioned, the calculation of LMF and S_{SOD} from these models requires the knowledge of the eigenvectors as well as of the frequency spectrum.

Calculations on the zinc chalcogenides were performed[105] using the "valence overlap shell model."[112,113] It is a modified version of the shell model with the valence force field approximation. The polarization of overlap charges is simulated by a positive shell charge at the cation site. This model has been optimized to reproduce the results of second-order Raman scattering but is also expected to be reliable in terms of the eigenvector representation.[113,114]

The calculations[105] show that the model can well reproduce the experimental dispersion relations, specific heat data, and the phonon frequency distributions. The latter are compared in Fig. 31 with results published in Ref. 115. The agreement is quite satisfactory. The overall shape is the same for the three distributions. In particular, one outstanding feature is common to all three distributions: acoustic and optical branches are well separated. In the calculations of the LMFs, these high-frequency phonons are less important, because $Z(\omega)$ is weighted by ω^{-1}. However, for the second-order Doppler

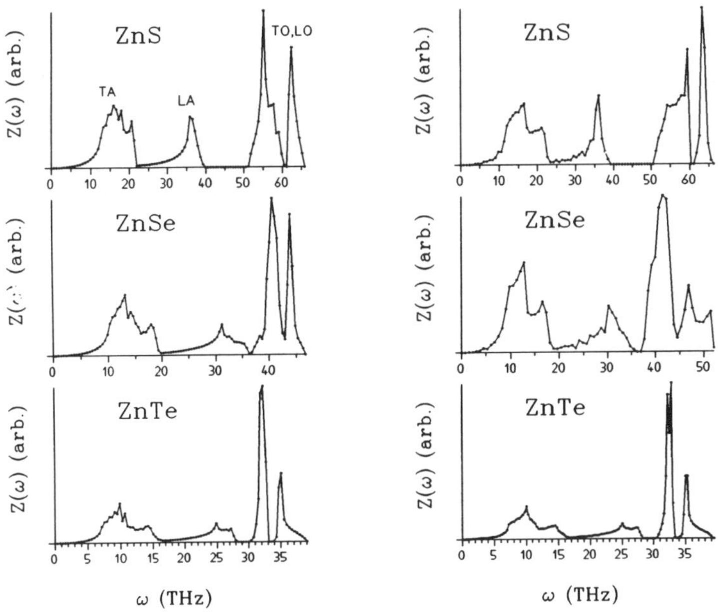

Figure 31. Phonon frequency distributions $Z(\omega)$ of ZnS, ZnSe, and ZnTe according to Ref. 115 (right) and lattice-dynamic calculations[105] (left).

Table 6. Comparison of Calculated and Experimental Results for the Lamb–Mössbauer Factor, f, and Second-Order Doppler Shift Relative to ZnO_{hcp}

Compound	$f^{cal}(\%)$	$f(\%)$	$S_{SOD}^{cal}(\mu m/s)$	$S_{SOD}(\mu m/s)^a$
ZnTe	0.26^b	0.30 ± 0.05	$+29.6^b$	-0.6 ± 17.7
ZnSe	0.60^b	0.46 ± 0.06	$+20.6^b$	$+8.6 \pm 12.1$
ZnS	0.64^b	0.74 ± 0.05	$+21.9^b$	$+7.8 \pm 10.2$
ZnO_{cub}	1.87^b	—	$+6.7^b$	—
ZnO_{hcp}	$f_x = f_y \approx f_z = 2.2^c$	2.0 ± 0.1	0^d	0
ZnF_2	$f_x = 2.98^e$	$f_x \geq 2.4 \pm 0.2$	$+2.4^e$	-2.7 ± 13.3^f
	$f_y = 1.74^e$	$f_y \geq 2.1 \pm 0.2$		$+8.1 \pm 15.7^g$
	$f_z = 1.03^e$	$f_z \geq 1.3 \pm 0.2$		

[a]Values taken from Table 5.
[b]Force constant model of Ref. 113. In this reference also for ZnO sphalerite structure was assumed.
[c]Force constant model of Ref. 116.
[d]Using the force constant model of Ref. 116, the calculated value of ZnO_{hcp} is $-89.5\ \mu m/s$. The values given for S_{SOD}^{cal} are relative to this number.
[e]Force constant model of Ref. 127.
[f]Ref. 101.
[g]Ref. 99.

shift, the weighting factor is ω^1 (see Section 3.1.2.1) and high-frequency phonons are essential.

The results for f and S_{SOD} are summarized in Table 6 for comparison with experimental data. The LMF can well be reproduced by the model calculations. For ZnO two different force constant models were used, one that assumed zinc blende structure also for ZnO[112,113] and a second model[116] that is based on the wurtzite structure. The results of both models are in good agreement. Thus, there appears to be very little change of the second-order Doppler shift when a system crystallizes in the sphalerite and wurtzite structures. Another example is ZnS. All Mössbauer parameters of ZnS in both structures are very similar.[17] In particular, the center shifts agree within $\sim 1.0\ \mu m/s$. It is extremely unlikely that changes in S and S_{SOD} would just compensate each other to such a precise degree. Therefore, S_{SOD} appears to be the same for both structures. The second-order Doppler shift is probably more strongly influenced by the coordination number of Zn which is four for the wurtzite and sphalerite structures. ZnO, however, exhibits a phase transition at high pressures from hcp wurtzite to NaCl structure, where the coordination number of Zn is increased from four to six. Here the second-order Doppler shift changes by $\sim +16\ \mu m/s$ (see Table 5).

The values for S_{SOD}^{cal}, however, are less satisfying. In general, they are far too large. In fact, the model calculations give very similar values as the Debye model (see Tables 5 and 6). Again, one can argue that the upper limits of the experimental values of S_{SOD} are more appropriate. This then means that $\Delta\langle r^2 \rangle = +13.9 \times 10^{-3}\ fm^2$ as indicated by the dashed line in Fig. 29. Indeed, except for ZnTe, the isomer shifts $S = S_C - S_{SOD}^{cal}$ of all chalcogenides fall very close to the dashed line. Also, the value $\Delta\rho(0) = -2.1e/a_0^3$ calculated[99] for

ZnF_2 fits rather nicely, whereas the result[101] of $\Delta\rho(0) = -1.7e/a_0^3$ is considerably too small. At present this problem remains unsolved. However, we would like to stress, in addition, that the results of the lattice-dynamic calculations are strongly model dependent. If force constant models are used that are less appropriate in terms of the eigenvector representation, the LMFs, e.g., can be off by a factor of three and the S_{SOD} can vary by $\pm 10\,\mu m/s$.[105] Thus, reliable force constant models are of utmost importance for this otherwise fine method.

4.2. ZnF_2

This compound crystallizes with tetragonal rutile structure, where each Zn is surrounded by a distorted fluorine octahedron.[117,118] Several physical properties show anisotropic behavior, e.g., the coefficients of thermal expansion at low temperatures, the axial compressibilities, and the dielectric constants.[119]

4.2.1. Hyperfine Interactions

Figure 32 shows the Mössbauer transmission spectrum at 4.2 K of a $^{67}ZnF_2$ powder absorber.[120,121] The quadrupole splitting in ZnF_2 causes three widely separated hyperfine lines. The splitting in the ZnO source results in two lines separated by $\sim 4.8\,\mu m/s$ with intensity ratio 2:1. Thus, the observed spectrum consists of three pairs of lines, each pair reflecting the shape of the emission spectrum of the ZnO source.

From the observed quadrupole frequency, $\nu_Q = eV_{zz}Q/h = +(8.1 \pm 0.1)$ MHz the main component of the efg tensor $V_{zz} = +(2.25 \pm 0.23) \times 10^{17}$ V/cm^2 is derived, where $Q = (0.150 \pm 0.015)$ b was used. The asymmetry

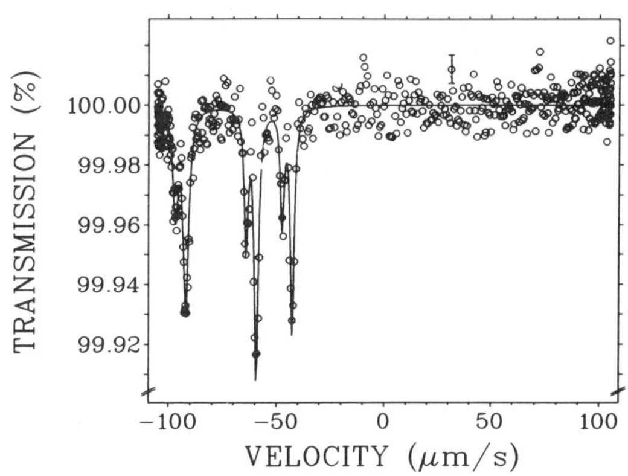

Figure 32. Mössbauer transmission spectrum of a $^{67}ZnF_2$ powder absorber measured with a ^{67}Ga in ZnO single-crystal source, both at 4.2 K. (Taken from Ref. 121.)

parameter $\eta = (0.15 \pm 0.02)$. Theoretical cluster calculations (see Section 4.2.3) predict the y axis of the efg tensor to point toward the two nearest fluorine atoms at the apexes of the distorted octahedron, the x axis being parallel, the z axis, with the main component, perpendicular to the crystallographic c axis. The x axis (z axis) is bisecting the shorter (longer) edges of the basal plane of the distorted octahedron.

Contrary to expectation, the efg tensor in ZnF_2 disagrees with the predictions of a simple point charge model,[122] which gives a value too large for η and even the wrong sign for V_{zz}. This discrepancy has been discussed thoroughly in Ref. 120. Further insight into these questions provides detailed theoretical calculations[75,99] which we will describe in Section 4.2.3.

The center shift S_C of ZnF_2 relative to ZnO is $-(63.44 \pm 0.32)\ \mu m/s$. Despite the anisotropic behavior of several physical properties, the temperature behavior, up to 55.3 K, of the Lamb–Mössbauer factor, of the second-order Doppler shift, as well as the specific-heat data in the low-temperature limit (12 K) are well described by the same effective Debye temperature.[121] Due to the different Debye temperatures of ZnO, $\Theta = 317$ K, and ZnF_2, $\Theta = 280$ K, the second-order Doppler effect gives $S_{SOD} = (8.7 \pm 1.3)\ \mu m/s$ which would result in an isomer shift of $S = -(72.1 \pm 1.3)\ \mu m/s$. However, as pointed out earlier, the Debye model might be insufficient to derive S_{SOD}. We will address this question in connection with the change of $\rho(0)$ in Section 4.2.3.

4.2.2. Lattice-Dynamic Effects

For an isotropic lattice and random orientation of the crystallites in a powder absorber, the relative areas under the three pairs of absorption lines (see Fig. 32) are expected to be equal. For ZnF_2, however, the relative intensity of the innermost pair at $v \approx -40\ \mu m/s$ is ~20% smaller than those of the other two pairs. This has mainly been attributed[121] to the Goldanskii–Karyagin effect[123] originating from an anisotropic LMF. All data obtained with a powder absorber can very well be explained by introducing an anisotropic but axially symmetric LMF. The symmetry axis ($f_{\parallel}$) is taken identical to the direction of the main component of the efg tensor discussed in Sections 4.2.1 and 4.2.3. In this model, $f_{\parallel} = 0.51^{+0.24}_{-0.16}\%$ and $f_{\perp} = 1.74^{+0.25}_{-0.21}\%$, resulting in effective Debye temperatures of $\Theta_{\parallel} = (229 \pm 6)$ K and $\Theta_{\perp} = (306 \pm 4)$ K if an extended Debye model is used.

A slightly different point of view has been taken in Ref. 124, which visualizes ZnF_2 as consisting of chains of distorted fluorine octahedra, the chains running parallel and perpendicular to the c axis. The dynamics of Zn is then expected to be influenced by one-dimensional modes (chain modes) and by different degrees of binding of the chains parallel to the c axis and perpendicular to it. The phonon frequency distribution of this model involves quadratic and linear terms for both the c direction and perpendicular to it.

Using the experimental Mössbauer data obtained for the ZnF_2 powder absorber,[121] four characteristic temperatures were adjusted[124] which reflect the Zn dynamics: the chains running parallel to the c axis are more loosely bound ($\Theta_1 = 100$ K) between themselves in comparison with the binding between the chains perpendicular to the c direction ($\Theta_2 = 200$ K). The intrachain vibrations parallel and perpendicular to the c axis remain the same ($\Theta_3 = \Theta_4 = 350$ K). Also, this model fits very well the data on the powder absorber.

To reveal the anisotropy of the LMF completely, experiments with ZnF_2 single-crystal sources have recently been carried out.[29] These measurements clearly demonstrate that the LMF is largest along the x direction, i.e., parallel to the crystallographic c axis, and smallest along the z axis which is perpendicular to the c axis and bisecting the longer edges of the basal plane of the octahedron. Although the y axis is pointing to the fluorine neighbors at the octahedron apexes, which have the shortest distance to zinc, the LMF along the y axis is not the largest. Specifically, $f_x \approx 2.4\%$, $f_y \approx 2.1\%$, and $f_z \approx 1.3\%$; i.e., the LMF is smallest along the z direction, where the efg is the largest, and the LMF is the largest parallel to the c axis. With these numbers the LMF for a ZnF_2 powder averaged over all directions in space is predicted[29] to be $f_{pow} \approx 1.9\%$ which is considerably larger than the actually measured value of 1.23%, indicating that the powder material contains a sizeable amount of defects.

Because f_y appears to be only ~10% smaller than f_x, the anisotropic but axially symmetric models mentioned above represent reasonable approximations to describe the lattice dynamics of Zn in ZnF_2.

4.2.3. Theoretical Calculations

As mentioned in Section 4.2.1, a simple point charge model is not sufficient to reproduce the efg tensor. An ionic model is only acceptable if the polarizabilities of the zinc and fluorine ions are allowed for. The experimental results can be reproduced by taking into account even relatively small values for the dipole and quadrupole polarizabilities of the ions.[99] The main component of the efg tensor at the zinc site is predicted to point toward the two *nearest* fluorine neighbors at the apexes of the distorted octahedron. Such a model, however, suffers from too many parameters which have to be adjusted to fit the experimental data. Much more satisfying are cluster[75,76,99,101] and band structure[125] calculations. Two different nonrelativistic cluster calculations have been reported. The basic procedure of the first type[75,76,101] has already been described in Section 4.1.4. The second type[99] uses a local-density approximation for the exchange potential and a subsequent muffin-tin approximation of the total one-electron potential. An X_α-type dependence of the exchange potential on the charge distribution is introduced. Both calculations give very similar results for the efg tensor. The chemical bond in ZnF_2 is partially covalent due to a rather strong overlap of $4d(3d)$, $4s$, and $4p$(Zn) with 2*s* and 2*p* (F) wave functions. The covalent part of the chemical bond causes a charge transfer from F^- to Zn^{2+} which reduces the ionic charges. Both calculations show that the efg at the Zn site ($V_{zz}^{\text{calc}} \approx +2.4 \times 10^{17}\,\text{V/cm}^2$, $\eta^{\text{calc}} \approx 0.2$) originates from 4*p* and 4*d*(3*d*) electrons. The agreement with the experimental results is very good. Compared with the results of the point charge models, these calculations also predict a different orientation of the efg tensor. As already mentioned, surprisingly not the *z* axis but rather the *y* axis is pointing toward the fluorine atoms at the octahedron apexes, the *x* axis (*z* axis) being parallel (perpendicular) to the crystallographic *c* axis is bisecting the shorter (longer) edges of the basal plane of the octahedron.

The band structure calculation[125] employs the full-potential linearized augmented plane wave (LAPW) method based on the local density approximation.[32,126] A fully relativistic description for the core states, but a scalar relativistic description, neglecting spin–orbit coupling, for the valence states is used. The efg is calculated directly from the resulting self-consistent charge densities. Also, this calculation predicts the orientation just mentioned of the efg tensor and reproduces the main component at the zinc site very well. The result is $V_{zz} = +2.28 \times 10^{17}\,\text{V/cm}^2$. For the asymmetry parameter, however, $\eta = 0.05$ is obtained as compared with the experimental value of $\eta = 0.15$.

Unfortunately, charge densities at the zinc nucleus have not been calculated with this method.

In order to check the predictions for the orientations of the efg tensor, experiments have been carried out using ZnF_2 single crystals with various orientations as sources.[29] Unfortunately, the crystallographic structure is such that neighboring chains of fluorine octahedra along the *c* axis are rotated by 90°, making it impossible to observe orientations along the *y* (or *z*) axis separately: In relation to the crystallographic axes, the *y* axis in one-half of the octahedra corresponds to the *z* axis in the other half. The measurements clearly show that the main component of the efg tensor is oriented perpendicular to the crystallographic *c* axis which, unfortunately, is not sufficient to distinguish the predictions of the point charge model from those of the cluster and band structure calculations.

Concerning the change in $\rho(0)$ between ZnF_2 and ZnO (wurtzite), the agreement between both cluster calculations is less satisfactory. Ref. 101 predicts that $\rho(0)$ in ZnF_2 is reduced relative to ZnO by $\Delta\rho(0) \approx -1.7e/a_0^3$, whereas Ref. 99 obtains $\Delta\rho(0) \approx -2.1e/a_0^3$. Using $\Delta\langle r^2 \rangle = +17.8 \times 10^{-3}\,\text{fm}^2$, the latter value gives an isomer shift $S \approx -73\,\mu\text{m/s}$ and the S_{SOD} estimated from the Debye model (see Section 4.2.1 and Table 5) would just be correct. With $\Delta\rho(0) = -1.7e/a_0^3$, however, one obtains $S \approx -61\,\mu\text{m/s}$ and a correction for S_{SOD} of *opposite* sign would be required. The discrepancy in $\Delta\rho(0)$ is not understood at present. The relatively small size of the clusters does not appear to be a possible cause.[99] Still, both calculations show that by far the larger contribution to $\Delta\rho(0)$ originates from the 4*s* orbitals, whereas the 1*s*, 2*s*, and approximately also the 3*s* electrons remain almost unaffected. Even though S_C is much more negative relative to ZnO, the covalent part of the chemical bond in ZnF_2 is important for $\rho(0)$ and essentially determines the efg tensor.

In Fig. 29 the results of all cluster calculations for $\Delta\rho(0)$ with respect to the ZnO wurtzite phase are plotted versus the experimental center shifts S_C. Deviations of the shifts from the straight line, which is obtained with the known value for $\Delta\langle r^2\rangle$, give the corrections for S_{SOD}. Lattice-dynamic theories face the enormous challenge of validating these predictions for S_{SOD} and the LMFs for the various compounds of different crystallographic structures. First results of lattice-dynamic calculations for S_{SOD}^{cal}, discussed in Section 4.1.5, are also shown in Fig. 29. They favor $\Delta\langle r^2\rangle = +13.9 \times 10^{-3}\,fm^2$.

Model calculations of the type described in Section 4.1.5 were also carried out[105] for ZnF_2 using the force constant model described in Ref. 127. The results are included in Table 6. The anisotropy of the LMF is pronounced, and the agreement with experimental data is reasonable considering the relatively low symmetry of the crystallographic structure. The calculated value for the second-order Doppler shift is $S_{SOD}^{cal} = +2.4\,\mu m/s$, which is rather close to the value of $\sim 8.7\,\mu m/s$ obtained from the Debye model (see Table 5). The small value for S_{SOD}^{cal} favors the electron density calculation of Ref. 99, a conclusion that was reached in Section 4.1.5.

4.3. Oxide Spinels

The oxide spinels comprise a large group of compounds with enormous technological importance.[128] They are isostructural with the mineral spinel, $MgAl_2O_4$, and crystallize in an fcc structure, space group *Fd*3*m*, with eight molecules in the unit cell.[129]

Basically, there are two types of spinels, the normal and the inverse spinels. The normal spinels have the general formula $(A)[B_2]O_4$, where A and B are divalent and trivalent ions, respectively. Typical examples are $(Zn)[Al_2]O_4$ and $(Zn)[Fe_2]O_4$. The A site is tetrahedrally, the B site octahedrally coordinated by oxygens. For an ideally close-packed nondistorted structure, the oxygen parameter is $u = 3/8 = 0.375$. Most spinels, however, have $u \neq 0.375$, which leads to a trigonally distorted octahedron at the B site, whereas the tetrahedron at the A site remains undistorted.

The inverse spinels have the general formula $(X)[YX]O_4$, where often, but not always, X and Y are divalent and tetravelent ions, respectively. Examples are $(Zn)[SnZn]O_4$ and $(Zn)[TiZn]O_4$. However, there also exist mixtures of the normal and inverse structures. In addition, more than two ions can be involved.

Results from ^{67}Zn Mössbauer experiments are only available on the Zn-containing spinels mentioned above.[130] Sources that have been used include ^{67}Ga in ZnO single crystals, ^{67}Ga in Cu metal, and ^{67}Ga in sintered

Table 7. Summary of ^{67}Zn Mössbauer Parameters, the Center Shift, S_C, the Linewidth (FWHM), Γ, and the Main Component, V_{zz}, of the Electric Field Gradient Tensor for Various Normal and Inverse Spinels[a]

	Compound	S_C(μm/s)	Γ(μm/s)	V_z(×10^{17} V/cm^2)
Normal spinels				
(^{67}Zn)[Al_2]O_4	(A site)	−13.19 ± 0.01	2.1 ± 0.1	—
	(B site)	−26.43 ± 0.05	2.9 ± 0.2	−3.14 ± 0.31
(^{67}Zn)[Fe_2]O_4	(A site)	−5.72 ± 0.15	3.5 ± 0.7	—
Inverse spinels				
(^{67}Zn)[Sn^{67}Zn]O_4	(site 1)	−14.9 ± 0.1		−2.17 ± 0.22
	(site 2)	−20.9 ± 0.1	5 ± 1	+2.06 ± 0.20
	(site 3)	−22.7 ± 0.1		+1.51 ± 0.15
(^{67}Zn)[Ti^{67}Zn]O_4	(site 1)	+2.4 ± 0.1		+0.91 ± 0.09
	(site 2)	−11.1 ± 0.1	4 ± 1	−0.95 ± 0.10
	(site 3)	−20.7 ± 0.1		−0.78 ± 0.08
	(site 4)	−49.7 ± 0.1		+0.72 ± 0.07

[a] Asymmetry parameter $\eta = 0$ for all compounds listed. The center shift is given relative to ZnO (wurtzite). To obtain the center shift relative to the ^{67}Ga*Cu* source, 110.7 μm/s has to be subtracted from the values for S_C.

powder of the spinel (Zn)[Al_2]O_4. All absorbers were enriched (91.5%) in ^{67}Zn. The results are summarized in Table 7.

4.3.1. Normal Spinels

Figure 33 shows the Mössbauer spectrum of (^{67}Zn)[Al_2]O_4 obtained with a ZnO single-crystal source. As expected, the ^{67}Zn at the A site exhibits a single absorption transition. The splitting observed in Fig. 33 originates from the

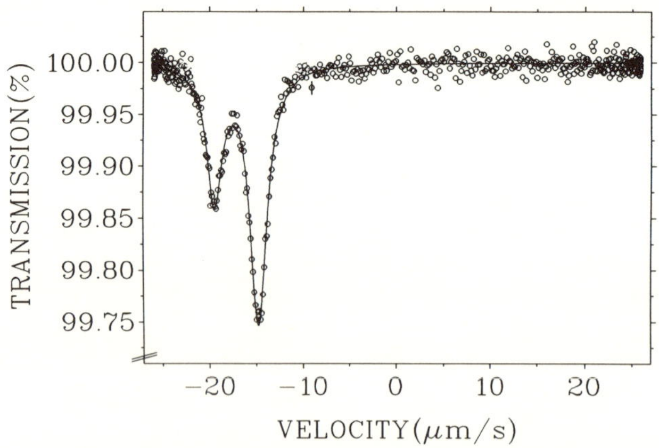

Figure 33. Mössbauer transmission spectrum of a (^{67}Zn)[Al_2]O_4 spinel absorber obtained with a ^{67}Ga in ZnO single-crystal source, both at 4.2 K. The splitting into two lines is due to the quadrupole interaction in the source. (Taken from Ref. 130.)

^{67}Ga*ZnO* source which exhibits two emission lines, separated by 4.785 μm/s, the intensity ratio being 2:1 (see Section 4.1.1). The recoil-free fraction of the spinel is ~3.8% and is the largest ever observed for a zinc compound or a zinc-containing metallic system. The effective Debye temperature is (370 ± 30) K.

Although the tetrahedral site is expected to be of cubic symmetry, the absorption lines are slightly broadened which may be caused by a small quadrupole interaction due to lattice distortions of next-nearest neighbors. This observation is supported by previous NMR results.[131,132] Nevertheless, the line broadening is rather small and (^{67}Zn)[Al_2]O_4 is an excellent single-line absorber.

In order to obtain microscopic information on the B site, a source or emission experiment was performed. A pellet of $ZnAl_2O_4$ powder was irradiated with 28-MeV protons (see Section 2.1). This source was measured against the single-line absorber $^{67}ZnAl_2O_4$ described above. Figure 34 exhibits the Mössbauer spectrum. It shows a triplet of approximately equal intensity of the lines and a rather broad absorption peak which consists of two lines with center shifts of ~−6 and ~0 μm/s. The latter originates from ^{67}Zn atoms at the A site whereas the former may be caused by ^{67}Zn atoms located at interstitial sites, also with cubic symmetry.

The triplet is due to ^{67}Zn ions at B sites. ^{67}Ga atoms produced by *p*-bombardment are expected to readily replace Al, since Ga favors the trivalent state, as does Al. After the decay to ^{67}Zn, the triplet in the Mössbauer spectrum is caused by the large quadrupole interaction at the trigonally distorted octahedral site. From the known quadrupole moment $Q = +(0.015)$ b[38] of the 5/2 ground state, a negative efg, V_{zz}, of $-(3.14 \pm 0.31) \times 10^{17}$ V/cm^2 is deduced, and the asymmetry parameter η is found to be

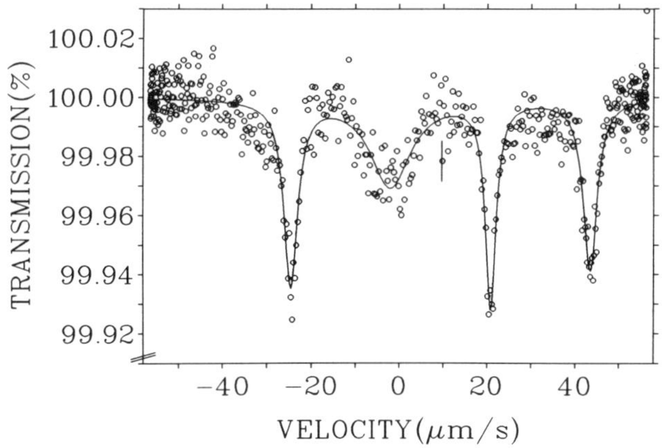

Figure 34. Mössbauer emission spectrum: Source: ^{67}Ga in (Zn)[Al_2]O_4, absorber: (^{67}Zn)[Al_2]O_4, both at 4.2 K. (Taken from Ref. 130.)

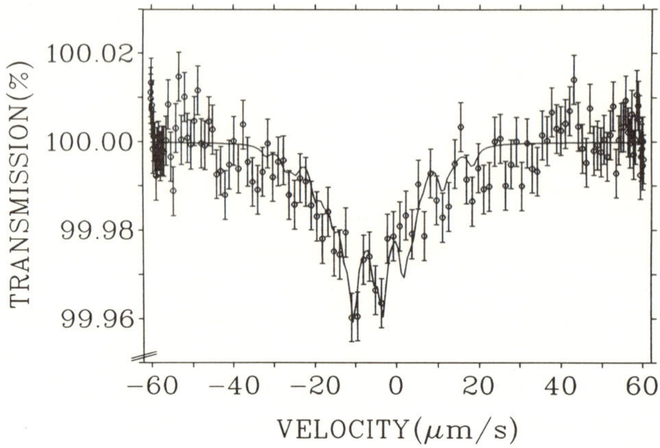

Figure 35. Mössbauer transmission spectrum of a (^{67}Zn)[Fe_2]O_4 spinel absorber recorded with a ^{67}Ga in ZnO single-crystal source, both at 10 K. The solid line is a fit indicating an effective magnetic field of 0.2 T.

zero. $|V_{zz}|$ is much larger than $V_{zz} = \pm 0.92 \times 10^{17}$ V/cm^2 derived for ^{27}Al from NMR results[131,132] (see Section 4.3.3) but comparable to $V_{zz} = +2.68 \times 10^{17}$ V/cm^2 obtained from ^{57}Fe Mössbauer measurements,[132] with ^{57}Fe at the B site and using a quadrupole moment of 0.28 b for ^{57}Fe. Since the quadrupole moment for ^{57}Fe is probably smaller by a factor of two or so, the value for V_{zz} might be larger correspondingly. Simple point charge calculations[122] are not able to reproduce the large value of $|V_{zz}|$ for ^{67}Zn at the B site. Therefore, covalent contributions to the bonding at the B site as well as lattice relaxation, slight distortion due to ^{67}Zn at the Al site, appear to be most important. Both effects have to be investigated by more advanced theoretical calculations, in particular by cluster calculations (see Section 4.3.3). As indicated in Table 7, the center shift of the B site is 13.24 μm/s more negative than the A site center shift.

The spinel (^{67}Zn)[Fe_2]O_4 orders antiferromagnetically at ~10 K.[128] Due to superexchange via the oxygen atoms, a transferred hyperfine field is expected at the zinc site. The Mössbauer spectrum at 11 K is similar to the absorption pattern of (^{67}Zn)[Al_2]O_4, shown in Fig. 33. The tetrahedral (A) site, which ^{67}Zn occupies, exhibits cubic point symmetry as expected from the crystallographic structure. The center shift is 7.47 μm/s more positive than the A site center shift of the Al spinel. At ~10 K, onset of magnetic order is detected as can be noticed in Fig. 35. For the first time, a transferred hyperfine field has been observed at the ^{67}Zn atom. At 10 K, this field is estimated to be 0.2 T and at 4.2 K about 1 T.

4.3.2. Inverse Spinels

In completely inverse spinels, zinc occupies the A and B sites in equal proportions. From the data on normal spinels, (^{67}Zn)[Al_2]O_4, and using a

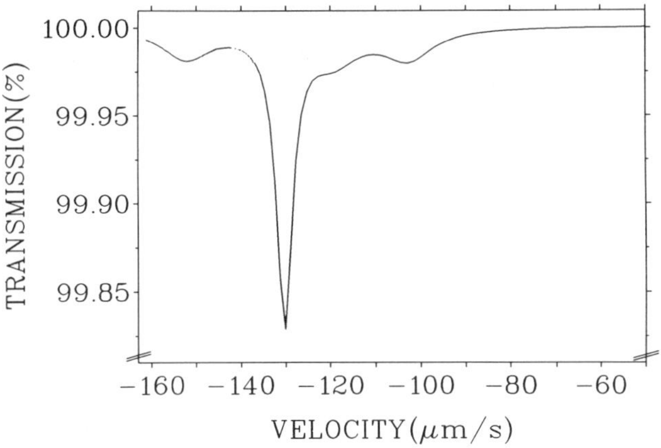

Figure 36. Simulated spectrum for an inverse spinel assuming that Zn occupies A and B sites in equal proportions and that quadrupole interaction exists at the B site only.

single-line ^{67}Ga*Cu* source, a spectrum as shown in Fig. 36 is expected. It exhibits a superposition of a single-line spectrum, corresponding to the A site, and a triplet, corresponding to the B site, with a center shift of 5 μm/s relative to the position of the single line. The relative intensities of the triplet lines are equal. The total intensities of the triplet and the singlet are the same. However, such a spectrum is not what is observed.[130] Figures 37 and 38 show Mössbauer transmission spectra of the two inverse spinels, (^{67}Zn)[Sn^{67}Zn]O_4 and (^{67}Zn)[Ti^{67}Zn]O_4. The source is ^{67}Ga*Cu* which exhibits a single emission line. If a fit on the basis of Fig. 36 is performed, the intensity of the single-line spectrum of the A site is much too low relative to the intensity of the quadrupolar split spectrum of the B site. This would indicate site exchanges of Zn and Sn, and of Zn and Ti. Neutron scattering data, however, prove that

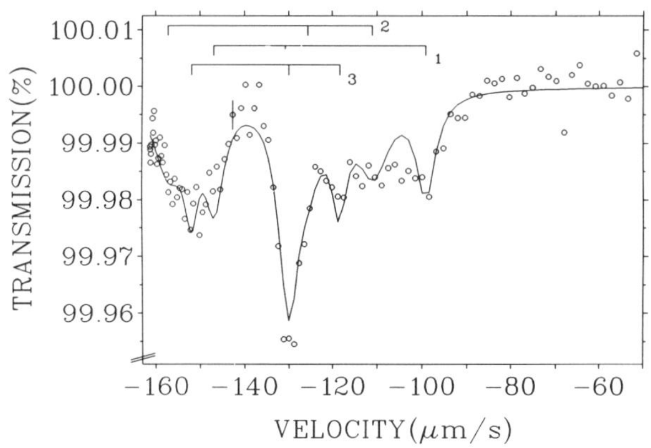

Figure 37. Mössbauer transmission spectrum of a (^{67}Zn)[Sn^{67}Zn]O_4 inverse spinel absorber with a ^{67}Ga in Cu source, both at 4.2 K. (Taken from Ref. 130.)

this is not the case; both spinels are completely inverse. The least-squares fit of Fig. 37 indicates that only three different ^{67}Zn sites are present in $^{67}Zn_2SnO_4$ with electric quadrupole interaction being observed at all of them. The relative intensities corresponding to the three sites are 3:3:2. Two sites exhibit positive efgs, one site a negative efg. The results are given in Table 7. Surprisingly, the efgs at the A and B sites are of about the same magnitude, $|V_{zz}| \approx 2 \times 10^{17}$ V/cm^2; $\eta = 0$, i.e., also the A site is distorted! In addition, the center shifts are very close at ~ -130 μm/s relative to ^{67}Ga*Cu*, or at ~ -20 μm/s relative to ZnO (wurtzite). Unfortunately, up to now it was not possible to attribute the A and B sites to the corresponding subspectra. At both sites the efgs and the center shifts are sharply defined. The width of the absorption lines is ~ 5 μm/s, which is astonishingly small. These observations strongly indicate that Zn and Sn are *not* distributed binomially on the B sites; on the contrary, short-range order rather leads to mainly two different (Zn, Sn) configurations at the octahedral sites.

The absorption spectrum of $^{67}Zn_2TiO_4$ (see Fig. 38) can well be fitted by four different ^{67}Zn sites. All four sites show efgs which are smaller in magnitude than for $^{67}Zn_2SnO_4$. The results are summarized in Table 7. Also for the Ti compound, the efgs and the center shifts are distinctly pronounced, leading to the conclusion that Zn and Ti are not distributed binomially on the B sites but that again short-range order effects are predominant in this case. As in $^{67}Zn_2SnO_4$, the point symmetry at the A site is not cubic, because an efg is observed. The absolute values of the four efgs cover the range between 0.72×10^{17} and 0.95×10^{17} V/cm^2. Contrary to the Sn spinel, the center shifts of the four subspectra are highly different, the most negative and most positive ones being ~ -160 and ~ -108 μm/s relative to ^{67}Ga*Cu*, i.e., ~ -50 and $\sim +2$ μm/s relative to ZnO (wurtzite), respectively.

Clearly, a point charge model is not sufficient to describe these results on

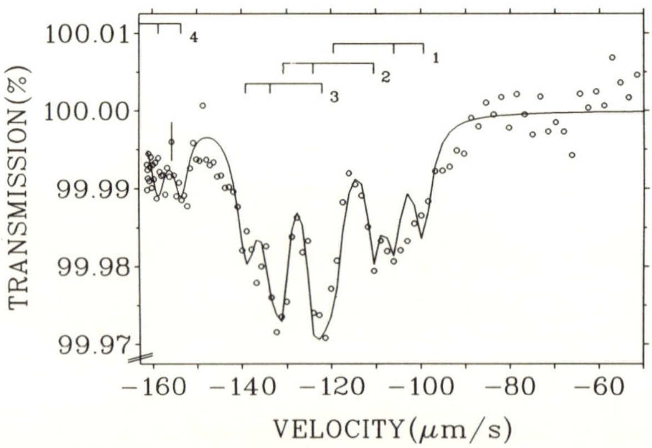

Figure 38. Mössbauer transmission spectrum of a (^{67}Zn)[Ti^{67}Zn]O_4 inverse spinel absorber with a ^{67}Ga in Cu source, both at 4.2 K. (Taken from Ref. 130.)

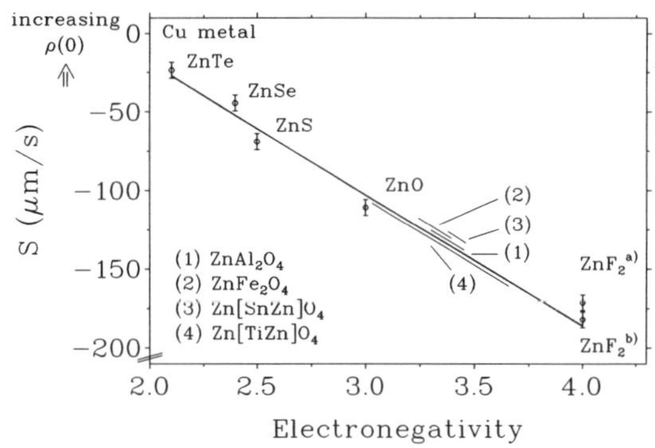

Figure 39. Isomer shifts, relative to ^{67}Ga in Cu, versus Pauling electronegativity of the ligands in compounds containing divalent ^{67}Zn (see Fig. 30). Also indicated are the shifts observed in spinels. The large range of shifts for the four ^{67}Zn sites in the Ti spinel may originate from covalent *d* electrons of Ti involved in the chemical bond.

the inverse spinels. Covalent contributions to the chemical bond are most important. This can easily be seen from Fig. 39, where, neglecting second-order Doppler shift, the ranges of center shifts are marked in a plot of isomer shifts versus electronegativity as described earlier (see Section 4.1.3). The shifts observed for the four different sites in the $^{67}Zn_2TiO_4$ spinel cover the total range between ZnO and ZnF_2, whereas for the tin spinel, the shifts are very close. The reason for the different behavior of the titanium spinel is not clear. However, one might argue that contrary to tin, 3*d* electrons of titanium are playing an essential role. Due to the shielding effect of *d* electrons, the *s* electron density at the ^{67}Zn nucleus can be reduced very effectively.[44] The participation of *d* electrons of titanium in the chemical bond might have implications also concerning superconductivity, which has been observed in some titanium oxide spinels.[133] Band structure as well as cluster calculations will be able to provide deeper insight into these questions.

4.3.3. Theoretical Calculations

Modern theoretical calculations for ^{67}Zn in oxide spinels are just beginning to appear. First results, however, look very promising. Recently, HF cluster calculations[134] of the type described in Sections 4.1.2.2, 4.1.4, and 4.2.3 have been performed on $(Zn)[Al_2]O_4$. To obtain the efg for Al at the B site, a cluster $(AlO_6)^{-9}$ was used. For the case where Zn substitutes for Al the HF calculations were carried out on a $(ZnO_6)^{-10}$ cluster embedded in the same crystallographic point charge lattice used with the $(AlO_6)^{-9}$ cluster. Lattice relaxation due to distortion of the lattice around the Zn impurity has not been taken into account. The following results were obtained: $V_{zz} = -0.97 \times 10^{17}\,\mathrm{V/cm^2}$, $\eta = 0$, for Al at the B site with the orientation of V_{zz} along the trigonal (1, 1, 1) axis. This is in good agreement with the experimental value[132] of $V_{zz} = \pm 0.92 \times 10^{17}\,\mathrm{V/cm^2}$ taking into account the quadrupole moment $Q(^{27}\mathrm{Al}) = 0.165\,\mathrm{b}$.[135] The efg at the Al site is negative due to the large contribution from the Al(3*p*) valence orbitals which is greater in magnitude than the positive contribution to V_{zz} from the core Al(2p) orbitals. These results show the importance of covalency effects in the calculation of V_{zz} in this spinel.

When Zn substitutes for Al, an efg of $V_{zz} = -2.28 \times 10^{17}$ V/cm^2, $\eta = 0$, at the Zn site is obtained. This agrees within ~30% with the experimental value[130] of $V_{zz} = -3.14 \times 10^{17}$ V/cm^2. The relative closeness indicates that the lattice is not very much distorted around the Zn impurity. Furthermore, because of the difference in formal ionic charge between Zn^{2+} and Al^{3+} one expects a charged defect. However, the Mulliken analysis gives a charge of +1.91 for the Zn impurity and a charge of +2.18 for Al in the pure spinel. This is a hint that the relative size and charge differences between Zn and Al at the Al site in the lattice are substantially reduced from those for the free ions Zn^{2+} and Al^{3+}.

For the (^{67}Zn)[Al_2]O_4 spinel the s electron density $\rho(0)$ at the ^{67}Zn nucleus at the A site has been calculated. Relative to ZnO (wurtzite) $\rho(0)$ has decreased by $-0.9e/a_0^3 \triangleq -31.3\ \mu$m/s. Thus, a second-order Doppler shift of $S_{SOD} = +18.1\ \mu$m/s is expected. Unfortunately, for a detailed discussion no lattice-dynamic calculations are available.

The magnetic hyperfine field at the Fe site in (Zn)[Fe_2]O_4 was determined[136] to be 51.5 T, when extrapolated to 0 K. Covalency effects which reduce the local moment[129] of Fe from 5.0 to 4.2μ_B are also convincingly established in the cluster calculation of this compound. Refinements of such calculations are clearly necessary. The understanding of the transferred hyperfine field, from iron to zinc, is of enormous importance.

The calculations will also help to clarify the cause for the difference in center shifts of $-13.2\ \mu$m/s between the A and B sites in the aluminum spinel. If second-order Doppler shift corrections were negligible, this difference would indicate that $\rho(0)$ is smaller at the B site than at the A site. In the similar case[137] of ^{57}Fe diluted in $MgAl_2O_4$, ^{57}Fe Mössbauer measurements show that ^{57}Fe is more covalently bound at the B site than at the A site. Therefore, in the case of ^{67}Zn, the second-order Doppler shift probably plays an important role.

A further example where modern calculations will bring more physical insight is the participant of d electrons of titanium in the chemical bond of the zinc–titanium spinel and their possible implications in superconductivity.

5. HIGH-T_C SUPERCONDUCTORS

^{67}Zn Mössbauer spectroscopy has been used to determine the ^{67}Zn quadrupole interaction constants and asymmetry parameters of the efg tensors at the Cu sites in the $YBa_2Cu_3O_7$ and $YBa_2Cu_3O_6$ systems.[138] In both cases emission or source experiments were carried out at 4.2 K with enriched ^{67}ZnS as absorber. The 93.31-keV level was excited from ^{67}Cu to make sure that the ^{67}Zn Mössbauer nuclei are formed at the Cu lattice sites after the decay of

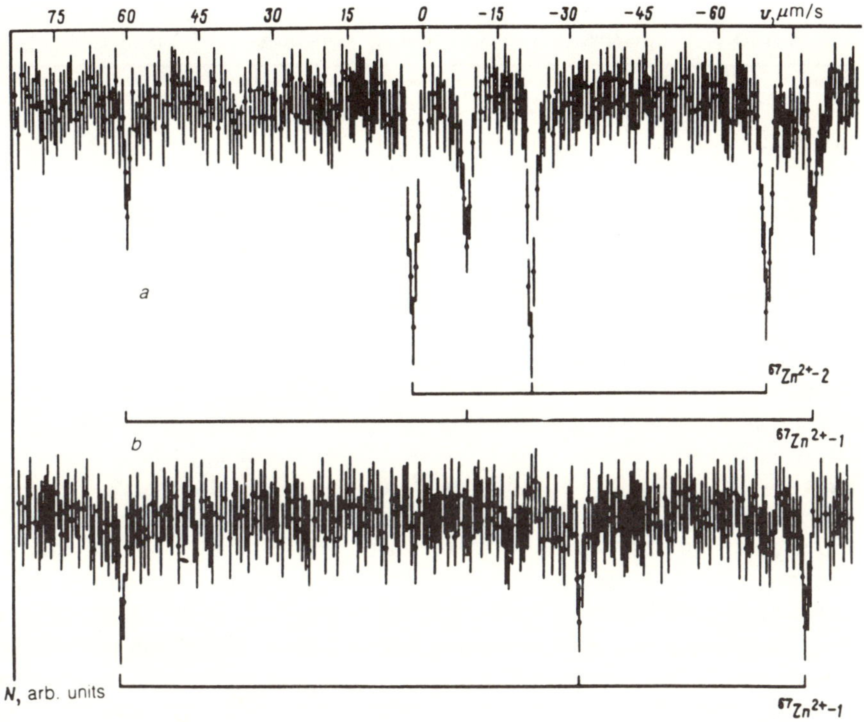

Figure 40. Mössbauer emission spectra obtained at 4.2 K. Sources: ^{67}Cu in $YBa_2Cu_3O_7$ (a) and ^{67}Cu in $YBa_2Cu_3O_6$ (b). Absorber: ^{67}ZnS. ($^{67}Zn^{2+}$-1) and ($^{67}Zn^{2+}$-2) denote ^{67}Zn centers at Cu-1 and Cu-2 sites, respectively. (Taken from Ref. 138.)

^{67}Cu. Figure 40 shows emission Mössbauer spectra of both compounds. In $YBa_2Cu_3O_7$ the Cu atoms occupy two structurally nonequivalent lattice sites with a 1:2 population ratio.[139] Accordingly, two quadrupole triplets are seen in Fig. 40a with an intensity ratio of ~1:2. The less intense spectrum has been attributed to ^{67}Zn on Cu-1 sites (^{67}Zn-1), and the more intense spectrum to ^{67}Zn on Cu-2 sites (^{67}Zn-2). The system $YBa_2Cu_3O_6$ shows only one triplet (see Fig. 40b) which has been attributed to ^{67}Zn on Cu-1 sites. Because the Cu-2 sublattice in $YBa_2Cu_3O_6$ is expected to order antiferromagnetically below 450 K,[140] the Mössbauer spectrum of ^{67}Zn on Cu-2 sites is probably spread out over a large velocity interval and not seen in Fig. 40b.

Table 8 summarizes the results on the efg tensors. In $YBa_2Cu_3O_7$ the ratio of the main components of the efg tensors for (^{67}Zn-1) and (^{67}Zn-2) is ~1.87. The asymmetry parameters are very different too: at (^{67}Zn-1), $\eta \sim 0.95$, whereas at (^{67}Zn-2), $\eta \leq 0.2$. For (^{67}Zn-1) in $YBa_2Cu_3O_6$, the sign of the main component of the efg tensor is negative, whereas it is positive for (^{67}Zn-1) in $YBa_2Cu_3O_7$. The asymmetry parameter is small, ≤0.2, in $YBa_2Cu_5O_6$, and large, 0.95, in $YBa_2Cu_3O_7$. The transition from one compound to the other is also accompanied by changes of the efg tensor at the Ba site.[138]

Table 8. Properties of ^{67}Zn Centers in $YBa_2Cu_2O_7$ and $YBa_2Cu_3O_6$

Compound	Center	e^2qQ (MHz)	η	V_{zz} ($\times 10^{17}$ V/cm^2)	$S_C{}^a$ (μm/s)	$S_C{}^b$ (μm/s)
$YBa_2Cu_3O_7$	^{67}Zn-1	+20.1 ± 0.5	0.95 ± 0.05	+5.5 ± 0.5	0	−49
	^{67}Zn-2	+11.8 ± 0.5	≤0.2	+3.3 ± 0.3	+21 ± 3	−28 ± 3
$YBa_2Cu_3O_6$	^{67}Zn-1	−23.5 ± 0.5	≤0.2	−6.5 ± 0.6	+7 ± 2	−42 ± 2

[a]Center shift relative to (^{67}Zn-1) in $YBa_2Cu_3O_7$.
[b]Center shift relative to ^{67}Ga*Cu*.

Unfortunately, no center shifts are reported in Ref. 138. However, from Fig. 40 the center shifts can be estimated and are also given relative to (^{67}Zn-1) in $YBa_2Cu_3O_7$ and to ^{67}Ga*Cu* in Table 8. It appears that (^{67}Zn-2) is shifted by ~21 μm/s toward higher *s* electron density $\rho(0)$ at the ^{67}Zn nucleus as compared with (^{67}Zn-1) in $YBa_2Cu_3O_7$, and (^{67}Zn-1) in $YBa_2Cu_3O_6$ is shifted toward higher $\rho(0)$ by ~7 μm/s as compared with (^{67}Zn-1) in $YBa_2Bu_3O_7$. If the second-order Doppler shift can be neglected, these center shifts indicate that covalency is highest at (^{67}Zn-2) and lowest at (^{67}Zn-1) in $YBa_2Cu_3O_7$ with (^{67}Zn-1) in $YBa_2Cu_3O_6$ at an intermediate position. All shifts observed are more positive than those for ZnS. The (^{67}Zn-2) site in $YBa_2Cu_3O_7$ exhibits a shift which is even more positive than the shift for ZnTe. Conduction electrons might well play an essential role. It is a challenging task for theorists to determine the origin of these shifts as well as of the efg tensors in these compounds and to clarify the role of covalency effects on the center shifts and on the efg tensors at the various Zn sites.

From an experimental point of view, using ^{67}Cu to prepare sources in superconductors is not an easy task, because this activity cannot be prepared *in situ*, within the superconductor, by particle bombardment in a cyclotron. Experiments employing ^{67}Cu, however, are of considerable importance because even a small concentration of a few percent zinc drastically reduces the transition temperature.[141] The experiments described here provide, for the first time, information on the Y–Ba–Cu system for the limit of extremely low zinc concentration. An important step now will be to increase the zinc concentration and monitor the ^{67}Zn Mössbauer spectra. Also, lattice dynamic investigations are highly desirable in this connection.

6. SYNTHESIS ON LATTICE-DYNAMIC EFFECTS

6.1. Second-Order Doppler Shift and Debye Model

The examples of zinc systems given in the previous sections demonstrate that the ^{67}Zn Mössbauer resonance is a sensitive tool for investigating lattice-dynamic effects via the Lamb–Mössbauer factor and the second-order Doppler shift. In the low-temperature regime, the temperature variation of S_{SOD} is usually dominant relative to the explicit temperature-dependent isomer shift and large relative to the experimental linewidth and hence easy to measure. A serious problem arises when the isomer shifts, S, between different compounds or of the same system at different pressures or temperatures have to be determined. Often S is masked by S_{SOD} and it is necessary to measure or calculate one of them independently. In Section 4.1.4 an example is given where S has been calculated theoretically. Usually, however, such calculations are not available. The determination of S_{SOD} requires not only a model for force constants of the system based on phonon dispersion relations or Raman scattering results but in addition the knowledge of the eigenvectors of the dynamic matrix. Most models are optimized to reproduce the phonon dispersion relations but only little information on the actual eigenvectors is given. Because of this unsatisfying situation the experimenter takes refuge in simpler models, in particular the Debye and even the Einstein models. The systematics on zinc systems investigated by the ^{67}Zn resonance give some empirical hints on when the Debye model is sufficient, in cases where the measuring temperature $T \leq \Theta/4$. The Debye model may provide a good approximation if the effective Debye temperatures, Θ, as determined from the LMF extrapolated to 0 K, from the temperature dependence of the LMF, from the temperature dependence of S_{SOD}, and from the specific heat, all fall within a narrow range, say within $\pm 0.1 \times \Theta$.

Favorable cases are monoatomic systems with high crystallographic symmetry. In systems with two kinds of atoms, nearly equal atomic masses, as in the brass alloys, are a good assurance, but for example ZnO[71] only barely fulfills the criterion mentioned above. However, even nearly equal masses and high crystallographic symmetry are by no means sufficient conditions as the example of the zinc chalcogenide ZnSe demonstrates. With the chalcogenides, the criterion is violated, in particular by the specific heat data.[105] As mentioned in Section 4.1.5, the main characteristic is the phonon frequency distribution, $Z(\omega)$, which exhibits distinct separations of optical and acoustic phonon branches, as shown in Fig. 31. Fortunately, for zinc chalcogenides and ZnF_2, force constant models are available which make lattice-dynamic calculations for the Lamb–Mössbauer factor, the second-order Doppler shift, and the specific heat possible. As demonstrated in Section 4.1.5 and Table 6, the results of such calculations for the Lamb–Mössbauer factor are in good agreement with experiment if care is taken that only such force constant models are employed which are reliable in terms of eigenvector representation. Surprisingly, in all cases studied (see Tables 5 and 6), the values for S^{cal}_{SOD} calculated from these models are in reasonable agreement with the estimates S^{D}_{SOD} derived from the Debye approximation. Thus, it turns out that the Debye model is sufficient for the Lamb–Mössbauer factor and for the second-order Doppler shift. But even at rather low temperatures, the Debye model fails to explain the specific-heat data. They are, however, convincingly reproduced by the force constant models.[105] It appears that for the systems mentioned in Table 6, specific heat is discriminating more sharply against insufficient models than the Lamb–Mössbauer factor and second-order Doppler shift.

As emphasized by Fig. 29, the lattice-dynamic calculations based on force constant models as described in Section 4.1.5 favor $\Delta\langle r^2\rangle = +13.9 \times 10^{-3}\,\mathrm{fm}^2$, the lower limit of the value for $\Delta\langle r^2\rangle$ given in Ref. 45.

6.2. Phase Transitions

All zinc chalcogenides exhibit a high-pressure phase transition from zinc blende or wurtzite to the rock salt structure. ZnO is an example, which has been discussed in Section 4.1.2. There are several theoretical models which describe the structural properties and crystal stability of these *sp*-bonded solids.[89–91,142] Some calculations[89] predict a drastic softening of the transverse optical phonons across the phase transition. For the zinc chalcogenides, the drop in these phonon frequencies is expected to be a factor of ~10. The high sensitivity of the ^{67}Zn Mössbauer resonance can certainly play a decisive role in the investigation of the transition and test of these predictions.

7. SUMMARY

In many respects, high-resolution ^{67}Zn Mössbauer spectroscopy acts like conventional Mössbauer spectroscopy under a magnifying glass or sometimes even under a microscope. In fact, the range of Doppler velocities of a typical spectrum is fully contained within the natural linewidth of ^{57}Fe. The measuring technique has been considerably refined over the last 10 years, in particular, experiments can now be carried out at high external pressures, and the temperature range has been extended up to ~100 K.

A large number of new material properties has been studied which is indicative of the broad range of research fields to which ^{67}Zn Mössbauer spectroscopy has been applied. The formation of short-range order in the α phase of brass is exciting, not only for metallurgists but also for theorists who try to investigate the stability of Hume-Rothery phases. A new high-pressure phase has been discovered in β'-brass.

^{67}Zn Mössbauer spectroscopy allows determination of tiny changes in energy of the 93.31-keV γ rays as in gravitational redshift experiments and in precision measurements of hyperfine interactions. In addition, however, the interpretation of the spectra requires investigation of lattice dynamics. One reason is that S_{SOD} is not negligible as is often the case with conventional Mössbauer resonances. To derive the isomer shift, S, from experiment, it is crucial to know S_{SOD} and *vice versa*. Another reason is that the resonance in ^{67}Zn is highly sensitive to lattice-dynamic effects, to S_{SOD}, and to changes of

LMF. The anisotropy of LMF in zinc metal and in ZnF_2 has been discussed in more detail. As a consequence, to deepen the physical understanding of the experimental results, the experimenter needs support from two directions, from theorists whose research field is hyperfine interaction and from theorists who are devoted to lattice dynamics. The zinc chalcogenides and ZnF_2 have been given as examples in this respect because for these compounds cluster calculations have become available to reproduce the parameters of hyperfine interactions and force constant model calculations have been performed to derive the LMF and S_{SOD}.

First results on oxide spinels and high-T_c superconductors have been described. Many problems like the origin of superconductivity and the mechanisms of transfer of magnetic fields, as for example in $(Zn)[Fe_2]O_4$, are still open. This is for the future and ^{67}Zn Mössbauer spectroscopy can contribute significantly by applying its powerful microscope.

ACKNOWLEDGMENTS. This work would not have been possible without the help of many scientists. We thank Dr. P. Blaha, Technische Universität Wien, Vienna, Drs. G. Dietzmann and K. Melzer, Universität Leipzig, Prof. Dr. J. Gal and I. Halevy, Ben-Gurion University and Nuclear Research Center Negev, Beer- Sheva, Israel, Prof. Dr. U. Gonser, Universität des Saarlandes, Saarbrücken, Dr. A. Martin, Friedrich-Schiller-Universität Jena, Jena, and Prof. Dr. G. M. Kalvius, H. Karzel, M. Köfferlein, M. Peter, W. Schiessl, and M. Steiner, Technische Universität München, Garching, for many helpful discussions. We are very grateful to Prof. T. P. Das and D. W. Mitchell, State University of New York at Albany, and Dr. S. Nagel for making results of cluster calculations available to us before publication. We gratefully acknowledge the continuous support and the excellent cooperation of the cyclotron group at the Kernforschungszentrum Karlsruhe, and thank especially Dr. H. Schweickert, K. Assmus, and W. Maier. This work has been funded by the German Federal Minister for Research and Technology (Bundesminister für Forschung und Technologie) under Contract No. KA2TUM and by funds of the Kernforschungszentrum Karlsruhe.

REFERENCES

1. R. V. Pound and G. A. Rebka, Jr., *Phys. Rev. Lett.* **3**, 439 (1959).
2. I. Y. Barif, M. I. Podgoretskii, and F. L. Shapiro, *Zh. Eksp. Teor. Fiz.* **38**, 301 (1960).
3. R. V. Pound and G. A. Rebka, Jr., *Phys. Rev. Lett.* **4**, 337 (1960).
4. R. V. Pound and J. L. Snider, *Phys. Rev. B* **140**, 788 (1965).
5. G. J. Perlow, W. Potzel, R. M. Kash, and H. de Waard, *J. Phys. (Paris)* **35**, C6-197 (1974).
6. W. Potzel, T. Obenhuber, A. Forster, and G. M. Kalvius, *Hyp. Interact.* **12**, 135 (1982).
7. P. Helistö, E. Ikonen, T. Katila, W. Potzel, and K. Riski, *Phys. Rev. B* **30**, 2345 (1984).

8. W. Potzel, C. Schäfer, M. Steiner, H. Karzel, W. Schiessl, M. Peter, G. M. Kalvius, T. Katila, E. Ikonen, P. Helistö, J. Hietaniemi, and K. Riski, *Hyp. Interact.* **72**, 214 (1992).
9. T. Katila and K. Riski, *Phys. Lett.* **83A**, 51 (1981).
10. W. Potzel, U. Närger, T. Obenhuber, J. Zänkert, W. Adlassnig, and G. M. Kalvius, *Phys. Lett.* **98A**, 295 (1983).
11. T. Obenhuber, W. Adlassnig, J. Zänkert, U. Närger, W. Potzel, and G. M. Kalvius, *Hyp. Interact.* **33**, 69 (1987).
12. W. Potzel, *Hyp. Interact.* **71**, 1515 (1992).
13. E. Ikonen, J. Hietaniemi, and T. Katila, *Phys. Rev. B* **38**, 6380 (1988).
14. E. Ikonen, P. Helistö, J. Hietaniemi, and T. Katila, *Phys. Rev. Lett.* **60**, 643 (1988) and references therein.
15. W. T. Vetterling, *Adv. Chem. Ser.* **194**, 329 (1981).
16. T. Katila and K. Riski, *Hyp. Interact.* **13**, 119 (1983).
17. A. Forster, W. Potzel, and G. M. Kalvius, *Z. Phys. B* **37**, 209 (1980).
18. W. Potzel, W. Adlassnig, U. Närger, T. Obenhuber, K. Riski, and G. M. Kalvius, *Phys. Rev. B* **30**, 4980 (1984).
19. W. Potzel, A. Forster, and G. M. Kalvius, *J. Phys. (Paris)* **37**, C6-691 (1976).
20. P. W. Bridgman, *J. Appl. Phys.* **12**, 461 (1941).
21. W. Adlassnig, W. Potzel, J. Moser, C. Schäfer, M. Steiner, and G. M. Kalvius, *Nucl. Instrum. Methods A* **277**, 485 (1989).
22. Deutsche Beryllium GmbH, D-6370 Oberursel/Ts, Germany.
23. U. F. Klein, G. Wortmann, and G. M. Kalvius, *J. Magn. Magn. Mater.* **3**, 50 (1976).
24. Elektroschmelzwerk Kempten GmbH, D-8000 München 33, Germany.
25. J. Moser, J. Gal, W. Potzel, G. Wortmann, G. M. Kalvius, B. D. Dunlap, D. J. Lam, and J. C. Spirlet, *Physica* **102B**, 199 (1980).
26. Krupp Widia GmbH, D-4300 Essen 1.
27. J. Willer and J. Moser, *J. Phys. E.* **12**, 886 (1979).
28. A. Eichler and J. Wittig, *Z. Angew. Phys.* **25**, 319 (1968).
29. M. Steiner, Ph.D. thesis, Technische Universität München, D-8046 Garching, Germany (1993).
30. T. Obenhuber, A. Forster, W. Potzel, and G. M. Kalvius, *Nucl. Instrum. Methods* **214**, 361 (1983).
31. W. Potzel and N. Halder, *Nucl. Instrum. Methods* **226**, 418 (1984), and N. Halder, Elektronik GmbH, Seeleiten 1, D-8110 Seehausen, Germany.
32. P. Blaha, K. Schwarz, and P. H. Dederichs, *Phys. Rev. B* **37**, 2792 (1988) and references therein.
33. T. P. Das and P. C. Schmidt, *Z. Naturforsch.* **41a**, 47 (1986) and references therein.
34. P. Blaha, *Hyp. Interact.* **60**, 773 (1990).
35. R. V. Pound and G. A. Rebka, Jr., *Phys. Rev. Lett.* **4**, 397 (1960).
36. W. Potzel, A. Forster, and G. M. Kalvius, *Phys. Lett.* **67A**, 421 (1978), and *J. Phys. (Paris) Colloq.* **40**, C2–29 (1979).
37. W. Potzel, W. Adlassnig, J. Moser, C. Schäfer, M. Steiner, and G. M. Kalvius, *Phys. Rev. B* **39**, 8236 (1989).
38. N. S. Laulainen and M. N. McDermott, *Phys. Rev.* **177**, 1606 (1969).
39. J. Christiansen, P. Heubes, R. Keitel, W. Klinger, W. Loeffler, W. Sandner, and W. Witthuhn, *Z. Phys. B* **24**, 177 (1976).
40. J. A. da Jornada and F. C. Zawislak, *Phys. Rev.* **20**, 2617 (1979).
41. D. B. McWhan, *J. Appl. Phys.* **36**, 664 (1965).
42. O. Schulte, A. Nikolaenko, and W. B. Holzapfel, *High Press. Res.* **6**, 169 (1991).
43. D. L. Williamson, in *Mössbauer Isomer Shifts*, G. K. Shenoy and F. E. Wagner, eds., p. 331, North-Holland, Amsterdam (1978).
44. I. M. Band and V. I. Fomichev, *At. Data Nucl. Data Tables* **23**, 296 (1979).
45. F. Buheitel, W. Potzel, and D. C. Aumann, *Hyp. Interact.* **47–48**, 606 (1989).

46. L. D. Roberts, D. O. Patterson, J. O. Thomson, and R. P. Levey, *Phys. Rev.* **179**, 656 (1969).
47. R. M. Housley and F. Hess, *Phys. Rev.* **146**, 517 (1966).
48. L. J. Raubenheimer and G. Gilat, *Phys. Rev.* **157**, 586 (1967).
49. W. T. Vetterling and D. Candela, *Phys. Rev. B* **27**, 5394 (1983).
50. R. E. DeWames, T. Wolfram, and G. W. Lehman, *Phys. Rev.* **138**, A717 (1965).
51. R. V. Pound, L. Niesen, H. de Waard, and G.-L. Zhang, *Phys. Rev. B* **29**, 6086 (1984).
52. S. P. Tewari and P. Silotia, *J. Phys. Condens. Matter* **1**, 5165 (1989).
53. S. P. Tewari and P. K. Gumber, *J. Phys. Condens. Matter* **1**, 6307 (1989) and *Physica C* **160**, 545 (1989).
54. M. Hansen and K. Anderko, in *Constitution of Binary Alloys,* p. 650, McGraw–Hill, New York (1958).
55. R. P. Elliott, in *Constitution of Binary Alloys,* 1st Suppl., p. 390, McGraw–Hill, New York (1967).
56. C. J. Smithells, in *Metals Reference Book,* 4th ed., Vol. II, Butterworths, London (1967).
57. V. Heine and D. Weaire, in *Solid State Physics,* Vol. 24, p. 249, Academic Press, New York (1970).
58. J. Hafner and V. Heine, *J. Phys. F* **13**, 2479 (1983).
59. A. Bansil, H. Ehrenreich, L. Schwartz, and R. E. Watson, *Phys. Rev. B* **9**, 445 (1974).
60. R. Zeller and P. H. Dederichs, *Phys. Rev. Lett.* **42**, 1713 (1979).
61. R. Prasad, S. C. Papadopoulos, and A. Bansil, *Phys. Rev. B* **23**, 2607 (1981).
62. R. Prasad and A. Bansil, *Phys. Rev. Lett.* **48**, 113 (1982).
63. G. M. Stocks, M. Boring, D. M. Nicholson, F. J. Pinski, D. D. Johnson, J. S. Faulkner, and B. L. Gyorffy, in *Noble Metal Alloys.* Conference Proceedings of The Metallurgical Society of AIME, T. B. Massalski, W. B. Pearson, L. H. Bennett, and Y. A. Chang, eds., p. 27, Febr. 24–28, 1985.
64. W. Hume-Rothery, *J. Instrum. Methods* **35**, 307 (1926).
65. T. B. Massalski and U. Mizutani, *Prog. Mater. Sci.* **22**, 151 (1978).
66. T. Obenhuber, W. Adlassnig, U. Närger, J. Zänkert, W. Potzel, and G. M. Kalvius, *Europhys. Lett.* **3**, 989 (1987).
67. W. Adlassnig, W. Potzel, J. Moser, W. Schiessl, U. Potzel, C. Schäfer, M. Steiner, H. Karzel, M. Peter, and G. M. Kalvius, *Phys. Rev. B* **40**, 7469 (1989) and references therein.
68. L. Reinhard, B. Schönfeld, G. Kostorz, and W. Bührer, *Phys. Rev. B* **41**, 1727 (1990).
69. P. E. A. Turchi, M. Sluiter, F. J. Pinskii, D. D. Johnson, D. M. Nicholson, G. M. Stocks, and J. B. Staunton, *Phys. Rev. Lett.* **67**, 1779 (1991).
70. E. Hornbogen, *Z. Metallkde.* **78**, 689 (1987).
71. C. Schäfer, W. Potzel, W. Adlassnig, P. Pöttig, E. Ikonen, and G. M. Kalvius, *Phys. Rev. B* **37**, 7247 (1988).
72. M. Peter, W. Potzel, M. Steiner, C. Schäfer, H. Karzel, W. Schiessl, G. M. Kalvius, and U. Gonser, *Hyp. Interact.* **69**, 463 (1991), and *Phys. Rev. B* **47**, 753 (1993).
73. G. K. Wertheim, M. Campagna, and S. Hüfner, *Phys. Condens. Matter.* **18**, 133 (1974).
74. T. Obenhuber, W. Adlassnig, J. Zänkert, U. Närger, W. Potzel, and G. M. Kalvius, *Hyp. Interact.* **28**, 1033 (1986).
75. H. H. Klauss, N. Sahoo, P. C. Kelires, T. P. Das, W. Potzel, G. M. Kalvius, M. Frank, and W. Kreische, *Hyp. Interact* **60**, 853 (1990).
76. D. W. Mitchell, S. B. Sulaiman, N. Sahoo, T. P. Das, W. Potzel, and G. M. Kalvius, *Phys. Rev. B* **44**, 6728 (1991).
77. N. F. Mott and H. Jones, in *The Theory of the Properties of Metals and Alloys,* Oxford University Press, London (1936).
78. M. Ahlers, *Prog. Mater. Sci.* **30**, 135 (1986).
79. W. Potzel, in *Proc. XXth Winter School on Physics,* p. 294, Zakopane, Poland (1985).
80. R. V. Pound, G. B. Benedek, and R. Drever, *Phys. Rev. Lett.* **7**, 405 (1961).
81. G. M. Rothberg, S. Guimard, and N. Benczer-Koller, *Phys. Rev. B* **1**, 136 (1970).

82. R. V. Kasowski and L. M. Falicov, *Phys. Rev. Lett.* **22**, 1001 (1969).
83. R. M. Nicklow, G. Gilat, H. G. Smith, L. J. Raubenheimer, and M. K. Wilkinson, *Phys. Rev.* **164**, 922 (1967).
84. E. D. Hallmann and B. N. Brockhouse, *Can. J. Phys.* **47**, 1117 (1969).
85. G. Dolling and G. Gilat, in *Inelastic Scattering of Neutrons,* Vol. 1, p. 343, IAEA, Vienna (1965).
86. J. A. Reissland, *The Physics of Phonons,* Wiley, New York (1973).
87. Y. S. Touloukian and E. H. Buyco, *Thermophysical Properties of Matter,* Vol. 4, p. 346, IFI/Plenum Press, New York/Washington, D.C. (1970).
88. M. Peter, Diploma thesis, Technische Universität München, D-8046 Garching, Germany (1990).
89. J. A. Majewski and P. Vogl, *Phys. Rev. Lett.* **57**, 1366 (1986); *Phys. Rev. B* **35**, 9666 (1987); *Acta Phys. Pol. A* **73**, 341 (1988).
90. M. Causa, R. Dovesi, C. Pisani, and C. Roetti, *Phys. Rev. B* **33**, 1308 (1986).
91. K. J. Chang and M. L. Cohen, *Phys. Rev. B* **30**, 4774 (1984).
92. A. Svane and E. Antoncik, *Phys. Rev. B* **33**, 7462 (1986).
93. J. E. Jaffe, R. Pandey, and A. B. Kunz, *Phys. Rev. B* **43**, 14030 (1991).
94. W. Potzel, *Hyp. Interact.* **40**, 171 (1988), **71**, 1515 (1992).
95. H. Karzel, W. Potzel, C. Schäfer, M. Steiner, J. Moser, W. Schiessl, M. Peter, G. M. Kalvius, D. W. Mitchell, S. B. Sulaiman, N. Sahoo, and T. P. Das, *Hyp. Interact.* **70**, 1067 (1992).
96. K. N. Shrivastava, *Hyp. Interact.* **24–26**, 817 (1985).
97. C. H. Bates, W. B. White, and R. Roy, *Science* **137**, 993 (1962).
98. K. N. Shrivastava, *Phys. Rev. B* **13**, 2782 (1976).
99. S. Nagel, *Z. Phys. B* to be published, and private communication.
100. L. G. Berry and B. Mason, in *Mineralogy,* p. 308, Freeman, San Francisco (1959).
101. D. W. Mitchell, S. B. Sulaiman, N. Sahoo, T. P. Das, W. Potzel, and G. M. Kalvius, *Phys. Rev. B* to be published, and private communication.
102. L. Pauling, *The Nature of the Chemical Bond,* Cornell University Press, Ithaca, N.Y. 1960.
103. A. J. Freeman and D. E. Ellis, in *Mössbauer Isomer Shifts,* G. K. Shenoy and F. E. Wagner, eds., p. 111, North-Holland, Amsterdam (1978).
104. W. Potzel, A. Forster, J. Moser, and G. M. Kalvius, *Phys. Lett.* **88A**, 307 (1982).
105. M. Köfferlein, Diploma thesis, Technische Universität München, D-8046 Garching, Germany (1993).
106. G. Venkataraman, I. A. Feldkamp, and V. C. Sahni, in *Dynamics of the Perfect Crystal,* MIT Press, Cambridge, Mass. (1975).
107. P. Brüesch, in *Phonons in Theory and Experiments II,* Springer, Berlin (1986).
108. M. Born and K. Huang, in *Dynamical Theory of Crystal Lattices,* Oxford University Press, London (1954).
109. W. Chochran, R. A. Cowley, G. Dolling, and M. M. Elcombe, *Proc. R. Soc. London Ser. A* **293**, 433 (1966).
110. J. G. Traylor, H. G. Smith, R. M. Nicklow, and M. K. Wilkinson, *Phys. Rev. B* **3**, 3457 (1971).
111. D. L. Price, J. M. Rowe, and R. M. Nicklow, *Phys. Rev. B* **3**, 1268 (1971).
112. H. Bilz, M. Buchanan, K. Fischer, R. Haberkorn, and U. Schröder, *Solid State Commun.* **16**, 1023 (1975).
113. K. Kunc and H. Bilz, *Solid State Commun.* **19**, 1027 (1976).
114. D. N. Talwar, M. Vandevyver, K. Kunc, and M. Zigone, *Phys. Rev. B* **24**, 741 (1981).
115. H. Bilz and W. Kress, in *Phonon Dispersion Relations in Insulators,* Springer, Berlin (1979).
116. K. Thoma, B. Dorner, G. Duesing, and W. Wegener, *Solid State Commun.* **15**, 1111 (1974).
117. W. H. Baur and A. A. Khan, *Acta Crystallogr. B* **27**, 2133 (1971).
118. W. H. Baur, *Acta Crystallogr. B* **32**, 2200 (1976).
119. J. K. Vassiliou, *J. Appl. Phys.* **59**, 1125 (1986).

120. W. Potzel and G. M. Kalvius, *Phys. Lett.* **110A**, 165 (1985).
121. M. Steiner, W. Potzel, C. Schäfer, W. Adlassnig, M. Peter, H. Karzel, and G. M. Kalvius, *Phys. Rev. B* **41**, 1750 (1990).
122. F. W. de Wette, *Phys. Rev.* **123**, 103 (1961).
123. S. V. Karyagin, *Fiz. Tverd. Tela (Leningrad)* **8**, 1739 (1966) [*Sov. Phys. Solid State* **8**, 1387 (1966)], **9**, 2514 (1967) [*Sov. Phys. Solid State* **9**, 1979 (1968)].
124. S. P. Tewari and P. Silotia, *Phys. Rev. B* **45**, 7471 (1992).
125. P. Sorantin, Ph.D. thesis, Technische Universität Wien, Vienna, Austria (1990).
126. P. Blana, K. Schwarz, and P. Herzig, *Phys. Rev. Lett.* **54**, 1192 (1985).
127. C. Benoit and J. Giordano, *J. Phys. C.* **21**, 5209 (1988).
128. R. E. Vandenberghe and E. De Grave, in *Mössbauer Spectroscopy Applied to Inorganic Chemistry*, G. J. Long and F. Grandjean, eds., Vol. 3, p. 59, Plenum Press, New York (1989).
129. U. König, *Techn. Mitt. Krupp Forsch.-Ber.* **30**, 1 (1972).
130. W. Schiessl, W. Potzel, H. Karzel, C. Schäfer, M. Steiner, M. Peter, G. M. Kalvius, I. Halevy, J. Gal, W. Schäfer, and G. Will, *Hyp. Interact.* **68**, 161 (1991).
131. E. Brun and S. Hafner, *Z. Kristallogr.* **117**, 37 (1962).
132. P. Rüegsegger, F. Waldner, and E. Brun, *Helv. Phys. Acta* **43**, 799 (1970).
133. D. C. Johnston, H. Prakash, W. H. Zachariasen, and R. Viswanathan, *Mater. Res. Bull.* **8**, 777 (1973).
134. D. W. Mitchell, S. B. Sulaiman, N. Sahoo, T. P. Das, W. Potzel, G. M. Kalvius, W. Schiessl, H. Karzel, and M. Steiner, *Hyp. Interact.* (in press).
135. J. E. Rodgers, R. Roy, and T. P. Das, *Phys. Rev.* **A14**, 543 (1976).
136. U. König, E. F. Bertaut, and G. Chol, *Solid State Commun.* **8**, 759 (1970).
137. D. L. Dickson and K. K. P. Srivastava, *J. Phys. Chem. Solids* **37**, 447 (1976).
138. P. P. Seregin, F. S. Nasredinov, V. F. Masterov, and G. T. Daribaeva, *Pis'ma Zh. Eksp. Teor. Fiz.* **51**, 593 (1990) [*JETP Lett.* **51**, 676 (1990)].
139. J. M. Tarascon, P. Barboux, L. H. Greene, B. G. Bagley, G. W. Will, Y. LePage, and W. R. McKinnon, *Physica C* **153–155**, 566 (1988).
140. P. Burlet, C. Vettier, M. J. G. M. Jurgens, J. Y. Henry, J. Rossat-Mignod, H. Noel, M. Potel, P. Gougeon, and J. C. Levet, *Physica C* **153–155**, 1115 (1988).
141. B. Jayaram, S. K. Agarwal, C. V. Narasimha Rao, and A. V. Narlikov, *Phys. Rev. B* **38**, 2903 (1988).
142. D. G. Pettifor and R. Podloucky, *Phys. Rev. Lett.* **53**, 1080 (1984), and *J. Phys. C* **19**, 315 (1986).

9

Mössbauer Spectroscopy of New Materials Containing Gadolinium

Gordon Czjzek

1. INTRODUCTION

The first observation of the Mössbauer effect with 86.5-keV γ rays from ^{155}Gd nuclei was reported as early as 1962,[1] only a few years after Mössbauer's discovery of recoil-free emission and resonance absorption of γ rays by nuclei embedded in a solid.[2] The further history of Gd Mössbauer spectroscopy is an illuminating example for the interplay of methodological developments on one side and of the evolving interest in rare-earth compounds, in their properties and their potential for technical applications on the other side.

For a long time, interest in rare-earth compounds was focused primarily on aspects of fundamental character, with few exceptions of technical applications. About 30 years ago, the attention shifted increasingly toward the exploitation of the outstanding magnetic properties of rare-earth compounds for the fabrication of materials for permanent magnets. Another impetus came quite surprisingly through the discovery of oxide compounds with spectacularly high transition temperatures T_c to the superconducting state[3] because rare earths are essential constituents of many of these compounds. Both for permanent-magnet materials and for high-T_c compounds, the possibility of obtaining detailed information about structural aspects and about electronic and magnetic properties from local probes at rare-earth sites by Mössbauer spectroscopy is of considerable interest.

For some years, the experimental development of Gd Mössbauer spectroscopy was characterized by the systematic exploitation of the suitability of all

Gordon Czjzek • Kernforschungszentrum Karlsruhe, INFP, W-7500 Karlsruhe 1, Germany.

Mössbauer Spectroscopy Applied to Magnetism and Materials Science, Volume 1, edited by Gary J. Long and Fernande Grandjean. Plenum Press, New York, 1993.

Table 1. Mössbauer Transitions in Gadolinium Isotopes

A	a (%)	Parent isotope/$t_{1/2}$	$(I^{\pi})_g$	E_{γ} (keV)	$(I^{\pi})_e$	$t_{1/2}$ (ns)	σ_0 $(10^{-23}\,m^2)$	Γ_{nat} (mm/s)
154	2.18	^{154}Eu/8.5 years	0^+	123.1	2^+	1.17	3.7	0.950
155	14.80	^{155}Eu/4.96 years; CE	$\frac{3}{2}^-$	60.0	$\frac{5}{2}^+$	0.19	1.0	12.0
				86.5	$\frac{5}{2}^+$	6.34	3.47	0.250
				105.3	$\frac{3}{2}^+$	1.18	1.76	1.10
156	20.47	^{156}Eu/15 days; nc; CE	0^+	89.0	2^+	2.22	3.15	0.693
157	15.65	^{157}Eu/15 h; CE	$\frac{3}{2}^-$	54.5	$\frac{5}{2}^-$	0.13	0.96	19.3
				64.0	$\frac{5}{2}^+$	460	4.55	0.0046
158	24.84	^{158}Tb/150 years; nc; CE	0^+	79.5	2^+	2.53	2.76	0.680
160	21.86	^{160}Eu/0.8 min; CE	0^+	75.3	2^+	2.7	2.6	0.673

A: atomic mass number.
a: isotopic abundance.
Parent isotope: the isotope with the longest $t_{1/2}$ is given; nc (CE) indicates that excited states have been populated by neutron capture (Coulomb excitation).
$(I^{\pi})_{g(e)}$: spin I and parity π of the nuclear ground (excited) state.
$\sigma_0 = \frac{2\pi c^2\hbar^2}{E_{\gamma}^2}\frac{1+2I_e}{1+2I_g}\frac{1}{1+\alpha_T}$: cross section for resonance absorption of γ rays of energy E_{γ}; α_T is the total conversion coefficient.
$\Gamma_{nat} = \frac{\hbar \ln 2}{t_{1/2}}\frac{c}{E_{\gamma}}$: natural linewidth. The smallest observable linewidth, in the limit of vanishing absorber thickness, is $2\Gamma_{nat}$.

transitions in the six Gd isotopes having low-energy excited states (Table 1) for Mössbauer spectroscopy. These studies and their results will be summarized briefly in Section 2.

At present, only the 86.5-keV transition in ^{155}Gd is employed extensively in studies of solids. It yields Mössbauer spectra of modest to good resolution, and the preparation of single-line sources is not forbiddingly difficult, as demonstrated for the first time in the pioneering work of Prowse *et al.*[4] Consequently, investigations with this transition will be treated in the remainder of this chapter. In Section 3, nuclear parameters and relevant aspects of hyperfine interactions at ^{155}Gd nuclei, and in Section 4, experimental techniques and methods for analyzing the Mössbauer spectra will be described. For ^{155}Gd, as well as for all other Mössbauer transitions in Gd nuclides, only resonance absorption spectra have been observed. No resonance scattering and no attempts at conversion electron Mössbauer spectroscopy have been reported in the literature. Results obtained for metallic alloys and compounds are the subject of Section 5, with emphasis on materials of interest for permanent magnets. Section 6 is devoted to a report on studies of Gd compounds with predominantly ionic or covalent bonds (oxides and sulfides). Investigations of superconductors, in particular of superconducting cuprates and related compounds, will be treated in Section 7.

Earlier reviews of Gd Mössbauer spectroscopy were contained in reports on rare-earth Mössbauer spectroscopy in general,[5,6] and on isomer shifts in rare-earth compounds.[7] Although many years have gone by since publication of these reviews, their contents are by no means obsolete and will not be duplicated here. Rather, they will be referred to whenever appropriate.

The two nuclides ^{155}Gd and ^{157}Gd are also useful probes for nuclear magnetic and quadrupole resonance spectroscopy (NMR, NQR). Information obtained by these techniques to a large extent overlaps with that derived from Mössbauer spectroscopy. There are, however, complementary aspects, and a combination of results obtained by the different spectroscopies may increase by manyfold the amount of information derived from a single technique. Furthermore, results of nuclear resonance spectroscopy frequently can serve as a valuable guideline for investigations by Mössbauer spectroscopy. Concerning the topics of this chapter, a recent review of NMR and NQR studies of intermetallic compounds containing rare earths[8] is of interest. For high-T_c cuprates, these techniques have provided extremely valuable information, mainly through nuclear resonance spectroscopy with Cu isotopes and with ^{17}O. A comprehensive review of this work has not yet been published, but overviews of important aspects were given recently by Kumagai[9] and by Takigawa.[10]

Finally, it should be stressed that results derived from Mössbauer spectroscopy, NMR, and NQR have a strictly local character. Implications concerning spatial correlations—for example, on the type of magnetic order—are practically impossible unless additional information on the material studied is available. Generally, the combination of (magnetic) neutron diffraction data with results of local nuclear-probe studies can give a truly complete picture. Measurements of bulk properties such as magnetic susceptibility or specific heat frequently serve well to provide complementary information needed for the interpretation of results obtained by nuclear spectroscopies.

2. *MÖSSBAUER SPECTROSCOPY WITH ^{154}Gd, ^{156}Gd, ^{157}Gd, ^{158}Gd, AND ^{160}Gd*

The first experiments leading to the observation of the Mössbauer effect with ^{156}Gd and ^{158}Gd employed the population of the excited nuclear levels by neutron capture in ^{155}Gd and ^{157}Gd, respectively.[11,12] The large capture cross sections of these nuclei (of order 10^5 b) were favorable for the application of the neutron-capture technique. Initially, oxide targets led to very broad, unresolved spectra.[11] Even with an improved single-line target, the resonance lines were rather broad due to the short half lives of the excited nuclear states (Table 1). Nevertheless, magnetic hyperfine fields, quadrupole splittings, and isomer shifts could be deduced from the spectra.[12] Isomer shift values resulting from these studies, supplemented by results for ^{154}Gd (obtained with a $^{154}EuF_3$ source), were used to establish values for the ratios of $\Delta\langle r^2\rangle_N$, the changes of the nuclear mean-square radii, of the isotopes ^{154}Gd, ^{155}Gd, ^{156}Gd, and ^{158}Gd.[13]

Another unconventional technique to populate excited nuclear states, by Coulomb excitation in a target bombarded with a beam of charged nuclei (protons, α particles) or ions (O^{n+}), accelerated to an energy of several megaelectron volts, was employed by a group at John Hopkins University.[14,15] Mössbauer spectra were obtained for ^{155}Gd (60-keV transition), ^{156}Gd, ^{158}Gd, ^{160}Gd, and other rare-earth nuclei (^{164}Dy, ^{168}Er). From spectra obtained with oxide targets and absorbers, estimates for the ratios between quadrupole moments of the first excited states in ^{156}Gd, ^{158}Gd, and ^{160}Gd and the ground-state quadrupole moment of ^{155}Gd were derived.

The Coulomb-excitation technique was later extended by Russell *et al.*[16] Very thin targets were bombarded with ^{35}Cl ions, and the recoil energy of the excited nuclei (^{158}Gd and ^{160}Gd) was used to implant them into an Al catcher foil. In this way, a truly single-line source was obtained. From the difference between center shifts for $GdAl_2$, $GdFe_2$, and Gd–metal absorbers, estimates for $\Delta\langle r^2\rangle_N$ in ^{158}Gd and ^{160}Gd were derived.

The technique of Coulomb excitation recoil implantation Mössbauer effect (CRIME) appeared very promising in several aspects:

1. It offers the possibility of observing Mössbauer spectra with low-energy nuclear states that are not populated in any radioactive decay.
2. Single-line sources can be obtained by implantation into nonmagnetic metals of cubic crystal structure.
3. The recoiling ions are expected to produce a large number of defects in the host lattice before they come to rest, and consequently, the Mössbauer spectra should show effects of nearby defects.

The information actually derived from these experiments, however, was not nearly as rich as expected. It hardly justified the strenuous efforts involved. Thus, the technique was more or less given up after the demonstration of its feasibility. Only recently, considerable improvement in experimental techniques has brought about remarkable results on the structure and dynamics of defects,[17] so far solely for CRIME with ^{57}Fe. Extensions to experiments with other nuclei are conceivable.

Mössbauer spectroscopy with ^{156}Gd is also possible with conventional sources. Sources for Mössbauer spectroscopy with ^{155}Gd are usually produced by neutron capture in ^{154}Sm, initially yielding ^{155}Sm which decays under β^- emission ($t_{1/2} = 24$ min) to ^{155}Eu ($t_{1/2} = 5$ years). However, ^{155}Eu has a very large neutron-capture cross section ($\sigma_{nc} = 1.4 \times 10^4$ b). Thus, during the activation of the source material in a reactor, strong activity of ^{156}Eu ($t_{1/2} = 15$ days) is produced. In the decay of ^{156}Eu, the 89-keV state of ^{156}Gd is populated, and for a limited time these sources are useful for Mössbauer spectroscopy with ^{156}Gd. Initially, the intensity of the 86.5-keV γ rays of ^{155}Gd is less than 5% of that in the 89-keV line. This technique has been used, for example, by Henning[18] in an investigation aiming at the determination of $\Delta\langle r^2\rangle_N$ in ^{155}Gd and ^{156}Gd, and by Armon *et al.*[19] who have determined magnetic and quadrupole moments of excited states in the two Gd isotopes.

Although the resolution of Mössbauer spectra obtained for ^{156}Gd is inferior to that of Mössbauer spectra with 86.5-keV γ rays of ^{155}Gd, spectra of ^{156}Gd may be advantageous in obtaining certain information. The asymmetry caused by the Goldanskii–Karyagin effect,[20] for example, is much weaker in spectra of ^{155}Gd than in those of ^{156}Gd.[21,22]

In ^{157}Gd, two transitions can be used for Mössbauer spectroscopy (Table 1). The first excited state at 54.5 keV has a very short half-life, and an enormous linewidth, larger than 70 mm/s, was found in the Mössbauer spectrum,[23] much larger than any hyperfine splitting to be expected. In contrast, the second excited state at 64.0 keV has by far the longest half-life, $t_{1/2} = 460$ ns, of all Mössbauer states in the rare-earth region. Hence, γ rays emitted in the decay of ^{157}Gd nuclei from this state should in principle give rise to Mössbauer spectra with an extraordinary resolution as the natural linewidth is only about 10^{-9} eV. Unfortunately, this implies a very high sensitivity to defects in sources which, combined with the short half-life of the parent nucleus ^{157}Eu ($t_{1/2} = 15$ h), makes Mössbauer spectroscopy with this transition a very difficult project. Consequently, only a few attempts to perform Mössbauer spectroscopy with the 64-keV γ rays of ^{157}Gd were undertaken. The first observation of the effect was reported by Prange and Kienle.[24] The parent nuclei ^{157}Eu were produced by the photonuclear reaction ^{158}Gd(γ, p)^{157}Eu in a Gd_2O_3 target with high-energy γ rays generated in an electron linear accelerator. After extraction from the target, ^{157}Eu was precipitated with Gd_2O_3, and a Mössbauer spectrum of this radiation source with a Gd_2O_3 absorber was measured. It consisted of three resonance lines which were broadened about five times compared with the theoretical width for the given absorber thickness. With improved experimental approaches—in particular with transfer of ^{157}Eu into EuF_2, a compound with cubic symmetry at Eu sites—Mössbauer spectra of several absorbers were obtained.[25,26] Line broadening was severe for the EuF_2 sources as well, and the information deduced from the spectra was mainly values of the magnetic and quadrupole moments of the excited state and of $\Delta\langle r^2\rangle_N$. Since then, no further attempts to observe Mössbauer spectra with this transition have been reported.

In ^{155}Gd, three transitions to low-energy excited states (60.0, 86.5, 105.3 keV) can be employed for Mössbauer spectroscopy. The short half-life of the 60-keV state ($t_{1/2} = 0.19$ ns) leads to a very broad resonance line (Γ_{obs}: 29 mm/s,[15] 25 mm/s,[21] and 36 mm/s[23]). No difference between the linewidths obtained with a single-line $Gd_{0.15}Y_{0.85}Al_2$ absorber and a $GdFe_2$ absorber was found,[21] although the magnetic hyperfine field at Gd nuclei in $GdFe_2$ has a large value. This observation demonstrates that the transition is not suited for solid-state investigations. For transitions to the third excited state at 105.3 keV, the resolution is somewhat better,[21,27] but it is definitely inferior to that achievable with 86.5-keV γ-rays. Thus, from Mössbauer spectra with 105.3-keV γ rays, information on the excited nuclear state has been extracted,[21,27] but for solid-state investigations, the 86.5-keV transition has been used exclusively.

3. NUCLEAR PARAMETERS AND HYPERFINE INTERACTIONS FOR ^{155}Gd

3.1. Nuclear Spins and Moments

The nuclear level scheme of ^{155}Gd is shown in Fig. 1. Values of nuclear g factors and of quadrupole moments of the nuclear states involved in the 86.5-keV transition are listed in Table 2. The magnetic moment of the nuclear state (s) is given by $\mu_{(s)} = g_{(s)} \cdot \mu_N \cdot I_{(s)}$ where $\mu_N = 5.0505 \times 10^{-27}$ J/T = 3.1524×10^{-8} eV/T is the nuclear magneton. Ground state moments of ^{155}Gd and ^{157}Gd were determined by atomic beam resonance.[28] Results for magnetic moments were of high accuracy. However, the value of the quadrupole moment, $Q_0(^{155}\text{Gd}) = (1.59 \pm 0.16)$ b, has a large uncertainty. A determination of the quadrupole moments with muonic X rays[29] yielded $Q_0(^{155}\text{Gd}) = (1.30 \pm 0.02)$ b and $Q_0(^{157}\text{Gd}) = (1.36 \pm 0.02)$ b. Within error limits, the ratio between these two values is in agreement with the result $Q_0(^{157}\text{Gd})/Q_0(^{155}\text{Gd}) = 1.06534 \pm 0.00003$ reported by Unsworth.[28]

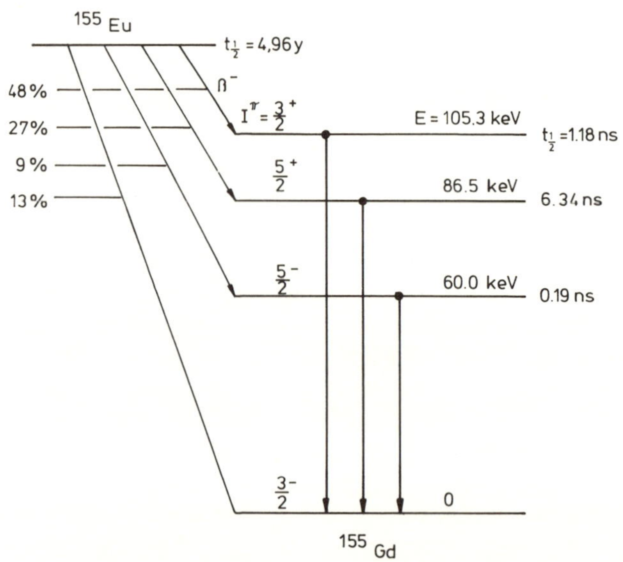

Figure 1. Nuclear level scheme of ^{155}Gd.

Table 2. g-Factors and Quadrupole Moments of the Nuclear States of ^{155}Gd Involved in the 86.5-keV Mössbauer Transition

	Ground state	86.5-keV state
I^{π}	$\frac{3}{2}^-$	$\frac{5}{2}^+$
g-factor	-0.1723 ± 0.0003	-0.2127 ± 0.0014
Quadrupole moment (b)	1.30 ± 0.02	0.113 ± 0.008

The moments of the 86.5-keV state were derived from Mössbauer spectra. The values now generally accepted are those determined by Armon *et al.*[21] Due to the small quadrupole moment of the 86.5-keV state, the quadrupole splitting of this level is not resolved, even in compounds with very large field gradients. This has rendered the determination of the moment quite difficult. Armon *et al.*[21] could not unambiguously deduce the sign. A positive value was preferred on the basis of theoretical considerations. This was later confirmed experimentally[30] by a spectrum of $Gd_2Ti_2O_7$ in an applied field of 5 T.

The parities of the ground state and of the 86.5-keV state have opposite sign. Thus, the γ radiation emitted has electric dipole (E1) character. For E1 radiation, the Lorentzian shape of resonance absorption lines is modified by a dispersive term:

$$\sigma(x) = \sigma_0 \frac{1 + 2\xi x/\Gamma}{1 + 4(x/\Gamma)^2} \tag{1}$$

with $x = E - E_0$.

The origin of the dispersive term is an interference between the following processes:

- Nuclear resonance absorption of γ rays followed by deexcitation via internal conversion and emission of an atomic electron.
- Photoelectric absorption of γ radiation leading to ejection of an atomic electron

Observation of a dispersive term in the Mössbauer line of ^{181}Ta was reported by Sauer *et al.*[31] The interpretation was given independently by Trammel and Hannon[32] and by Kagan *et al.*[33]

An approximate expression for the factor ξ is[32]

$$\xi = \varepsilon \frac{E_\gamma}{\hbar c} \sqrt{\frac{\alpha_T \sigma_e}{6\pi}} \tag{2}$$

where α_T is the total conversion factor (for this transition, $\alpha_T = 0.41$ was measured[34]), σ_e is the cross section for photoelectric absorption, and ε is a correction factor with absolute value less than 1.

In a study of interference effects for several E1 transitions, Henning *et al.*[35] found for the 86.5-keV transition in ^{155}Gd the value $\xi = 0.025 \pm 0.005$.

For an accurate determination of the value of ξ, sample purity is an essential condition. Even small traces of impurities introducing slightly asymmetric components in the Mössbauer spectra influence the value of ξ resulting from fits to the spectra. From our own experience, the best value is $\xi = 0.023 \pm 0.001$, in good agreement with the result reported by Henning *et al.*[35]

3.2. Isomer Shift

Extensive experimental work by Cashion *et al.*[36] provided the basis for the establishment of an isomer shift scale for the 86.5-keV transition of ^{155}Gd. Implications and the comparison with scales for other Mössbauer transitions of rare-earth nuclei were discussed by Cashion *et al.*[37,38] and summarized by Bauminger *et al.*[7]

The trends in the experimental data clearly show that the change of the mean-square nuclear radius $\Delta\langle r^2\rangle_N$ is negative. Numerical values quoted for $\Delta\langle r^2\rangle_N$ vary from[39] $-5.6 \times 10^{-3}\,\mathrm{fm}^2$ to[38] $-8.8 \times 10^{-3}\,\mathrm{fm}^2$. For negative $\Delta\langle r^2\rangle_N$, increasing electronic density at Gd nuclei leads to shifts toward negative velocities (with the standard convention that positive velocity implies increasing transition energy in the absorber).

The compound GdF_3, in which bonding presumably is closest to purely ionic, has been recommended[40] as the reference material for the isomer shift scale of ^{155}Gd. However, considerable nonaxial quadrupole splitting of the spectrum for this compound[41] prohibits the determination of the center shift with high accuracy. Hence, now most authors refer shifts to the single-line source, ^{155}Eu:*Pd*. The shift of a GdF_3 absorber with respect to this source material amounts to (0.684 ± 0.009) mm/s.[40]

From Mössbauer spectra of Gd^{3+} (configuration $4f^7$) and Gd^{2+} (configuration $4f^7 5d^1$) in CaF_2, an estimate for the shift due to the presence of a $5d$ electron—$\delta_{IS}(4f^7 5d^1) - \delta_{IS}(4f^7) = (0.35 \pm 0.09)$ mm/s—has been derived.[42] Similarly, $6p$ electrons are expected to reduce the electron density at Gd nuclei whereas $6s$ electrons, having nonzero density at the nuclei themselves, increase the density.

The very large range of shifts observed in metallic Gd compounds (Fig. 2), from about 0 to nearly 0.9 mm/s, reflects a considerable variation of conduction electron character, with the largest $6s$ contributions for the smallest values of δ_{IS}, strong $5d$ and/or $6p$ predominance for the largest δ_{IS}.

In studies of the concentration dependence of isomer shifts, δ_{IS}, in mixed alloy systems, variations due to a changing character of conduction and/or valence electrons are of primary interest. Variations of δ_{IS} due to changes in the atomic volume, which are of minor interest, can be accounted for, at least approximately, if the variation of shifts with pressure has been observed for a material which is related to the system to be studied. No experiments under pressure have been reported for ^{155}Gd or any other Gd Mössbauer nucleus. However, pressure-dependent shifts have been studied for ^{151}Eu and for

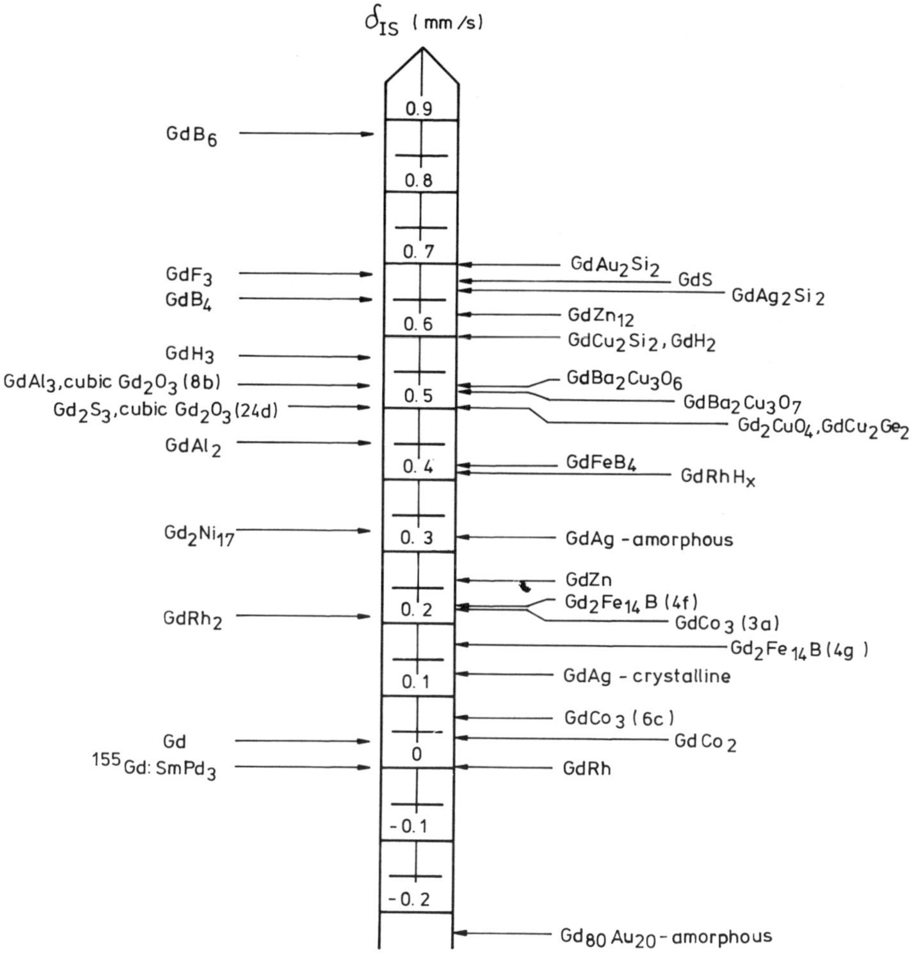

Figure 2. Isomer-shift scale for the 86.5-keV transition of ^{155}Gd. On the left side of the scale, some data reported by Bauminger *et al.*[7] are indicated for comparison with the more recent results shown on the right side of the scale. For compounds with inequivalent Gd sites, shifts for different sites are labeled in parentheses in Wyckoff's notation.

^{153}Eu.[43] No significant differences between the volume dependence of shifts for compounds with Eu^{2+} which has the same $4f^7$ configuration as Gd^{3+}, and for compounds with Eu^{3+} which has the same charge state as Gd^{3+} have been found. Thus, a conversion of results obtained for Eu can be considered as a reasonable estimate for volume-dependent shifts in ^{155}Gd. The results reported[43] for Eu yield the estimate $0.3\ \text{mm/s} \leq (\partial\delta_{IS}/\partial \ln V) \leq 0.6\ \text{mm/s}$ for ^{155}Gd. The main contribution to the uncertainty stems from the uncertainty of the values of $\Delta\langle r^2\rangle_N$.

Second-order Doppler shifts δ_{SOD} which in some cases contribute significantly to observed shifts,[44] mainly in light nuclei with small changes of the nuclear radius, can be neglected in the case of ^{155}Gd. According to calculations by Gettys and Stevens,[45] the difference between δ_{SOD} for absorbers with $\Theta_D = 100$ K and $\Theta_D = 400$ K, the extreme limits of Θ_D to be expected, amounts to $\delta_{SOD}(\Theta_D = 100\text{ K}) - \delta_{SOD}(\Theta_D = 400\text{ K}) = 0.03$ mm/s. Compared with the wide ranges of shifts due to changes of electronic densities at ^{155}Gd nuclei, this is a minor correction, indeed. Also, the variation of δ_{SOD} with temperature is negligible for the range of temperatures accessible to Mössbauer spectroscopy with ^{155}Gd.

For the derivation of precise shift values from the spectra, the dispersive term due to interference must be accounted for correctly. It introduces an asymmetry in the resonance lines which yields an artificial shift if a fit with symmetrical Lorentz lines is attempted. One must be cautious, in particular, when comparing shift data reported by different groups. A comparison leads to correct implications only if in all cases the dispersive term was taken into account.

3.3. Quadrupole Interaction

At sites of lower than cubic point symmetry, the interaction between the electric field gradient V_{zz}(EFG) and the nuclear quadrupole moment splits the nuclear energy levels into a series of Kramers doublets. The formalism and the generally accepted notations and conventions as presented and discussed by Dunlap[46] are used throughout this chapter. Mössbauer spectra of ^{155}Gd consist of five lines in the case of axial symmetry, that is, if $\eta = (V_{xx} - V_{yy})/V_{zz} = 0$. In the case of orthorhombic or lower symmetry, the spectra consist of six lines. Actually, however, due to the small quadrupole moment of the 86.5-keV state, only the ground-state splitting S_Q is resolved, where

$$S_Q = \tfrac{1}{2}eQ_0\,|V_{zz}|\sqrt{1 + \tfrac{1}{3}\eta^2} \tag{3}$$

Consequently, the spectra have the appearance of doublets. Neither the sign of V_{zz} nor the value of η can be deduced from these spectra.

The origin of the EFG is the electrostatic potential due to the distribution $\rho(\mathbf{r})$ of all charges in the surroundings of the probe nucleus. The formal expression is given by the volume integral

$$V_{zz} = \int \rho(\mathbf{r})\frac{3\cos^2\theta - 1}{r^3}\,d\tau \tag{4}$$

where the probe nucleus is at the origin of the coordinate system and θ is the polar angle with respect to the z axis. From measurements of the quadrupole splitting, information about the symmetry and other aspects of the distribution of electronic and ionic changes in the solid investigated can be deduced. Of particular interest are frequently changes observed in series of closely related compounds with variation of constituents or in alloy systems with varying concentration. For the evaluation of experimental results, a decomposition into contributions due to charges at greater distance (the lattice EFG V_{zz}^{latt}) and due to local charges (V_{zz}^{loc}, the EFG caused by valence or conduction electrons near the probe site) is generally advantageous:

$$V_{zz} = (1 - R) \cdot V_{zz}^{\text{loc}} + (1 - \gamma_\infty) \cdot V_{zz}^{\text{latt}} \tag{5}$$

The Sternheimer factors $(1 - R)$ and $(1 - \gamma_\infty)$ account for the attenuation or amplification of V_{zz} resulting from the deformation of closed shells induced by V_{zz}^{loc} and by V_{zz}^{latt}, respectively. The decomposition as expressed in Eq. (5) must be considered as an approximation. In particular, the expression for the local contribution in this form is hardly justified. Strictly, for this term, a factor $(1 - \gamma(r))$ should be inserted in the integrand in Eq. (4) where $\gamma(r)$ is a function which varies rapidly with r in the region where core and valence charges overlap.[47]

For rare earths with nonzero orbital angular momentum, the dominant contribution to the EFG is that of the $4f$ shell, a local contribution. For the S-state ion Gd^{3+}, this contribution is negligible, and V_{zz}^{loc} is due to nonspherical local charge distributions of conduction electrons or of electrons involved in covalent bonds, mostly of $5d$ and/or $6p$ character. The evaluation of these local contributions to V_{zz} requires rather elaborate calculations.[48–51]

The lattice contribution to the EFG is frequently calculated assuming the charge distribution at all ions to have spherical symmetry, the point-charge model. The integral in Eq. (4) is then replaced by a sum over ionic sites, each associated with an appropriate charge. Barton and Cashion[52] have shown that the quadrupole splitting measured in several ionic Gd compounds is linearly related to the splitting calculated with the point-charge model (Fig. 3). The slope yields the experimental estimate for the Sternheimer factor: $(1 - \gamma_\infty) = 98 \pm 9$. This value is larger than results of theoretical calculations[47,53] which yield values for γ_∞ in the range -70 to -80 for trivalent rare-earth ions. Of the compounds studied by Barton and Cashion, only the result for $GdVO_3$ deviates considerably from the linear correlation between experimental and calculated splittings. The authors ascribe this to a high degree of covalent bonding in this compound and hence to an important local contribution to V_{zz}.

Note that the value for the Sternheimer factor quoted here as deduced from the work of Barton and Cashion deviates somewhat from the value given in their original publication.[52] This occurs for two reasons:

- Barton and Cashion have used the old result for the quadrupole moment of the ^{155}Gd ground state, $Q_0 = 1.59$ b,[28] whereas the value quoted here is derived with $Q_0 = 1.30$ b.[29]
- Barton and Cashion have allowed for a constant term in the linear relation between experimental and calculated values of S_Q. As they

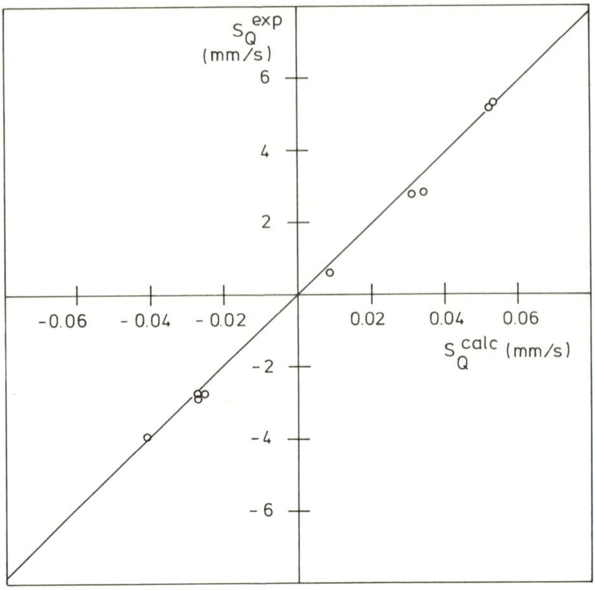

Figure 3. Comparison of experimental quadrupole splittings in several ionic compounds with results of point-charge calculations. Adapted from Ref. 52 with small modifications as discussed in the text.

discuss in their paper,[52] no physical reason for this constant term can be recognized, and consequently it was not included in the present reevaluation.

The results for V_{zz} and for η from ^{155}Gd Mössbauer spectra can provide estimates for the quadrupolar terms B_2^0 and B_2^2 in the crystal field acting on the 4f electrons of rare earths with nonzero orbital angular momentum.[54] In compounds with crystal structure of low symmetry, these terms frequently play a decisive role for the magnetocrystalline anisotropy.

In analogy with the nuclear quadrupole interaction, crystal field terms are products of a multipole moment of the 4f state and a corresponding potential originating from the distribution of charges in the crystal lattice. Thus, the quadrupolar terms can be expressed as

$$B_2^m = Q_{4f} \cdot A_2^m \tag{6}$$

where $Q_{4f} = \alpha_J \cdot \langle r^2 \rangle_{4f}$ is the quadrupole moment of the 4f state. The definition of the Stevens factor α_J is given in Ref. 54 along with tables of numerical values of α_J and of $\langle r^2 \rangle_{4f}$ for all rare earths. The definition of A_2^0 is identical to that of V_{zz} in Eq. (4). Although a decomposition similar to that of the EFG [Eq. (5)] usually is not made, the remarks following Eq. (5) apply to A_2^0 as well. Since the 4f charge density extends over a much larger range than the nuclear charge density, reaching beyond the range of core charges, shielding effects, expressed by the shielding factor $(1 - \sigma_2)$, are substantially weaker than those modifying the nuclear quadrupole interaction. Values of σ_2, derived from calculations[55] or from experimental results,[56] lie in the range between 0.4 and 0.7.

The conversion from V_{zz} to A_2^0 is given by

$$A_2^0 = -\frac{1}{4}\frac{(1 - \sigma_2)}{(1 - \gamma_\infty)} V_{zz} \tag{7}$$

The term A_2^2 is directly related to the asymmetry parameter η:

$$A_2^2 = \eta \cdot A_2^0 \tag{8}$$

Thus, A_2^2 is zero along with η for a site of axial symmetry.

For several compounds, independent determinations of both V_{zz} and B_2^0 led to values for the ratio $(1 - \gamma_\infty)/(1 - \sigma_2)$.[55,57] These values scatter over a very large range, from 120 to above 300. Most results, however, cluster between 200 and 250. The value of this ratio seems to be essentially constant for all rare earths.

This approach can yield reliable estimates for the crystal field parameters only when the lattice contribution is dominant. When local contributions are important, Eq. (7) may even give the wrong sign for A_2^0.[49] Thus, Eq. (7) should be applied with extreme caution to results obtained for metallic compounds and for solids with strongly covalent bonding. Nevertheless, for many metallic and covalently bonded solids, the sign of A_2^0 derived from the quadrupole splitting of ^{155}Gd Mössbauer spectra was found to agree with that inferred from other properties,[58,59] occasionally even the value agrees well with that deduced from other experimental data.[60,61]

3.4. Magnetic Hyperfine Interaction

When ^{155}Gd nuclei in a site of cubic point symmetry experience a magnetic field, the Mössbauer spectrum is split into 12 lines. Because of the near equality of the two g factors involved, there are three groups of 4 lines each. Transitions with $\Delta M_I = 0$ give rise to the central group, and the two outer groups arise from transitions with $\Delta M_I = \pm 1$. This structure of the spectrum facilitates the detection of deviations from an isotropic distribution of moment directions in the absorber. In case of isotropy, the intensities of the three groups are equal. If there is a preference for moment directions parallel (perpendicular) to the direction of observation, the intensity of the central group decreases (increases) relative to the intensity of the outer groups. This is illustrated in Fig. 4 by spectra of ferrimagnetic $GdFe_2$ and of antiferromagnetic $GdBe_{13}$ in an applied field.

In the general case, when both magnetic and quadrupole interactions must be taken into account, the hyperfine Hamiltonian has the simplest form when

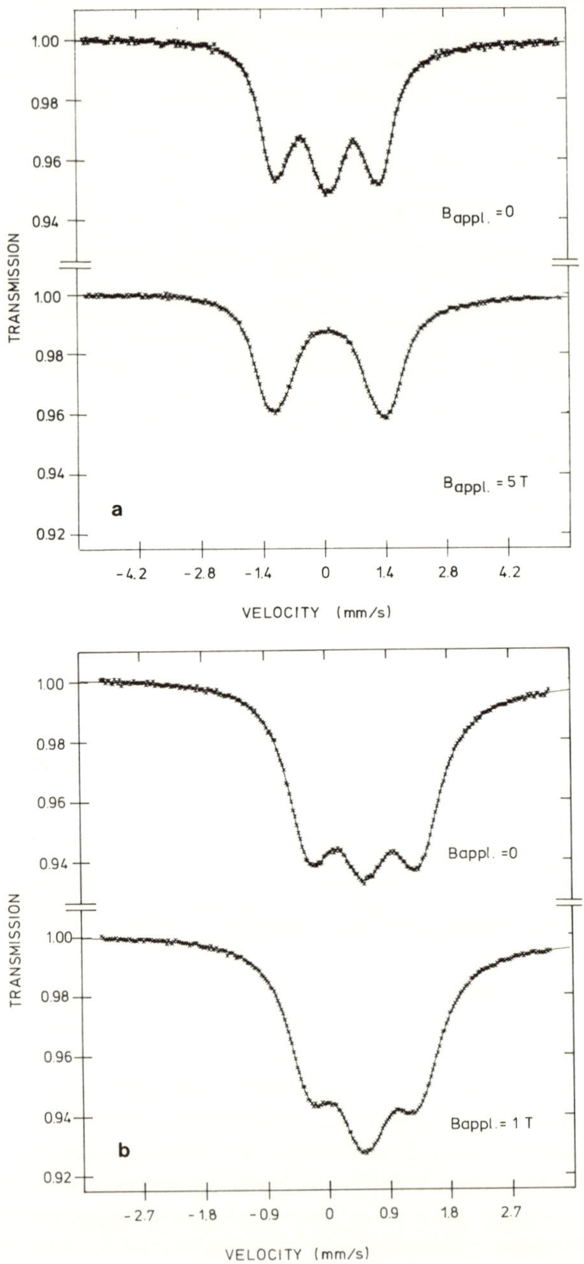

Figure 4. Mössbauer spectra of ^{155}Gd in (a) $GdFe_2$ and (b) $GdBe_{13}$, measured at 4.2 K in zero field and in an applied field parallel to the direction of observation of strength as indicated. In the ferrimagnet $GdFe_2$, the moments are aligned parallel to the field. In the antiferromagnet $GdBe_{13}$, in a field above the spin-flop field (about 0.5 T), the moments are approximately perpendicular to the applied field.

expressed in the coordinate system defined by the principal axes of the EFG tensor:

$$\mathcal{H}_{\rm hf} = \Delta_{\rm m}[\hat{I}_z \cos\theta + \tfrac{1}{2}(\hat{I}_+ e^{-i\phi} + \hat{I}_- e^{i\phi})\sin\theta] + \frac{\Delta_{\rm Q}}{I(2I-1)}[3\hat{I}_z^2 - \hat{I}^2 + \tfrac{1}{2}\eta(\hat{I}_+^2 + \hat{I}_-^2)] \tag{9}$$

with

$$\Delta_{\rm m} = g \cdot \mu_{\rm N} \cdot B_{\rm hf} \cdot c/E_\gamma$$

and

$$\Delta_{\rm Q} = \tfrac{1}{4} \cdot e \cdot Q \cdot V_{zz} \cdot c/E_\gamma$$

The splitting parameters $\Delta_{\rm m}$ and $\Delta_{\rm Q}$ are given in units corresponding to the Doppler velocity which is most convenient for fitting the spectra. For the 86.5-keV transition in ^{155}Gd, the conversion of the ground-state parameters $\Delta_{\rm m}^{(0)}$ and $\Delta_{\rm Q}^{(0)}$ to field quantities is:

$$B_{\rm hf}/[{\rm T}] = 53.16 \cdot \Delta_{\rm m}^{(0)}/[{\rm mm\,s^{-1}}]$$

$$V_{zz}/[10^{21}{\rm V \cdot m^{-2}}] = 8.87 \cdot \Delta_{\rm Q}^{(0)}/[{\rm mm \cdot s^{-1}}]$$

In most cases, the hyperfine Hamiltonian, Eq. (9), must be diagonalized numerically since $\Delta_{\rm m}^{(0)}$ and $\Delta_{\rm Q}^{(0)}$ have comparable values. Approximations frequently used in the analysis of Mössbauer spectra of ^{57}Fe where in nearly all cases $|\Delta_{\rm Q}| \ll |\Delta_{\rm m}|$ are not applicable. With the computational power of modern computers, this is no problem, and it brings the great advantage that the spectra are sensitive to all parameters involved. The analysis leads to results for $B_{\rm hf}$, for V_{zz} (including the sign), for η, and for the polar angles θ and ϕ, that is, for the direction of the hyperfine field with respect to the principal axes of the EFG. For the combination of nuclear spins $I_0 = \frac{3}{2}$, $I_{\rm ex} = \frac{5}{2}$, complete information on all parameters involved can be derived from powder spectra, in contrast to the situation for ^{57}Fe, where from powder spectra unambiguous values can be obtained only for $B_{\rm hf}$, for V_{zz}, and for $\delta_{\rm IS}$.[62]

In metallic solids, three contributions to the hyperfine field at Gd nuclei are usually distinguished[8]:

$$B_{\rm hf} = B_{\rm CP} + B_{\rm S} + B_{\rm N} \tag{10}$$

The core-polarization field $B_{\rm CP}$ is due to exchange interactions between the local 4f electrons and core electrons. Its value,[63,64] $B_{\rm CP} = (-34 \pm 2)$T, is generally assumed not to depend on the environment of Gd atoms in solids. The self-polarization field $B_{\rm S}$ is ascribed to polarization of conduction electrons by the local Gd moment, and $B_{\rm N}$ is the transferred field due to conduction electrons polarized by magnetic atoms in the surroundings. Many detailed studies of alloy systems aiming at the determination of the different contributions to $B_{\rm hf}$ have been performed, mainly by NMR.[8]

The sign of the hyperfine field is defined with respect to the direction of the local $4f$ moment in the atomic shell of the probe nucleus. In ferromagnetic materials, the sign is easily determined by comparison of the magnetic splitting in zero field and in an applied field. The hyperfine field is positive (negative) if the splitting increases (decreases) with the applied field. For ferrimagnetic materials the same technique can be applied, provided the field available is sufficiently strong to overcome the anisotropy field. The outcome then depends on the sublattice which has the larger magnetic moment. In $GdFe_2$, for example, the Gd moment outweighs the Fe moments, and the behavior is the same as in ferromagnets. In $Gd_2Fe_{14}B$, on the other hand, the Fe moments dominate. In this case, a decrease of the magnetic splitting in an applied field implies a positive hyperfine field.[58]

4. EXPERIMENTAL TECHNIQUES AND DATA ANALYSIS

Splittings observed in Mössbauer spectra of ^{155}Gd have in no case yet exceeded velocities of ± 8 mm/s. This is a comfortable velocity range, and any conventional Mössbauer spectrometer,[65,66] either with constant-acceleration drive or with sine-wave drive, can be used.

However, because of the high energy of the γ rays. Mössbauer spectroscopy with ^{155}Gd is restricted to low temperatures. Both the source and the absorber must be in a He cryostat. The cryostat design should be chosen such that the source temperature stays constant (best at liquid-He temperature) as the absorber temperature is varied. The highest absorber temperature accessible depends somewhat on the source strength and on the inclination of the experimenter concerning the time he or she is willing to spend for the acquisition of a spectrum. In any case, the recoilless fraction decreases very rapidly above 100 K, and absorber temperatures above 100 K usually are considered only if it is of particular interest to take spectra in this temperature range (an example is the high-T_c superconductor $GdBa_2Cu_3O_7$ with $T_c \sim$ 95 K).

The γ spectra are best recorded with a Ge detector which clearly resolves all lines emitted by the source (Fig. 5). It is possible to work with a NaI scintillator which does not resolve the 86.5-keV and the 105.3-keV lines. In this case, a Pb foil of thickness d in the range $0.5\,\text{mm} < d \leq 1.0\,\text{mm}$ can be used as filter. The K edge of Pb is at 88 keV, and the attenuation length λ_e ($I = I_0 \cdot e^{-d/\lambda_e}$) for γ rays of energy 105.3 keV is $\lambda_e^{(105.3)} = 0.19$ mm, whereas for 86.5-keV γ rays it is $\lambda_e^{(86.5)} = 0.67$ mm. Thus, a sufficiently thick Pb foil effectively suppresses the 105.3-keV contamination in the Mössbauer spectrum.[67]

Topics to be discussed in more detail in this section concern sources, absorbers, and the analysis of the Mössbauer spectra and of the recoilless fraction $f(T)$.

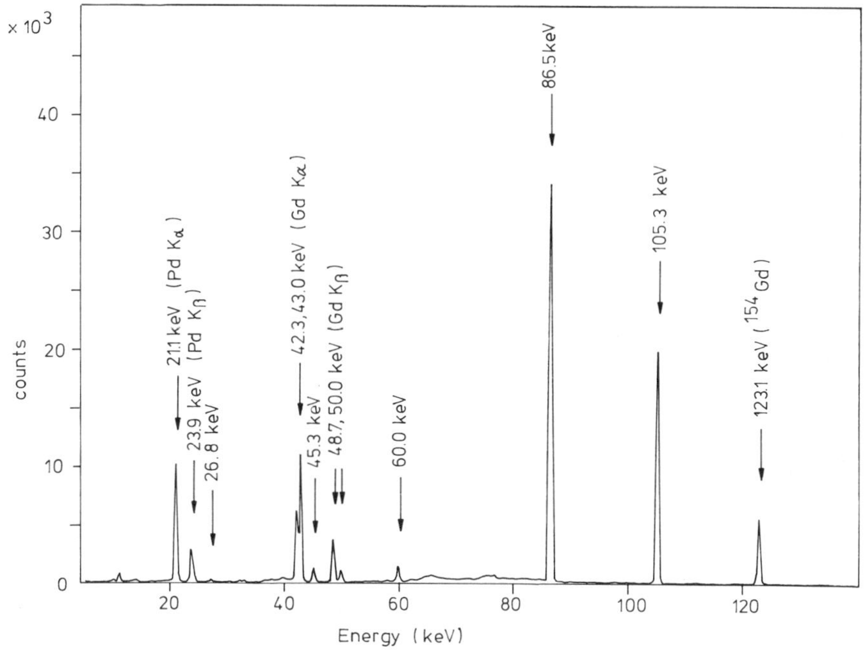

Figure 5. γ-ray spectrum of a ^{155}Eu:*Pd* source measured with an intrinsic Ge detector. Lines with energies 26.8, 45.3, 60.0, 86.5, and 105.3 keV originate from ^{155}Gd nuclei.

4.1. Sources and Absorbers

In the first experiments, intending primarily the demonstration of the existence of the Mössbauer effect in ^{155}Gd, oxide was used as source material.[1] This is definitely not a single-line source because of the presence of two inequivalent Gd sites and of strong field gradients at both sites. For subsequent studies, several source materials were tried: SmH_2, maintained at 20 K, above the magnetic ordering temperature,[68] is also a source emitting a quadrupole-split spectrum. Similarly, an alloy of Sm in Al, held at 10 K, did not emit a truly single-line spectrum.[41] The first single-line source was $SmRh_2$ which had to be maintained at 40 K, above the Curie temperature ($T_C = 22$ K).[21]

The final breakthrough came with the proposal[4] to use either a dilute solution of ^{155}Eu diffused into Pd or the compound $^{154}SmPd_3$ which after neutron irradiation is a useful source for Mössbauer spectroscopy with ^{156}Gd for a short while as described in Section 2. In the long run, it is a very efficient source for Mössbauer spectroscopy with ^{155}Gd. These ^{155}Gd sources emit a single line of width $\Gamma_S = (0.35 \pm 0.01)$ mm/s, slightly broader than the natural width, $\Gamma_{nat} = 0.25$ mm/s. Furthermore, no splitting or additional broadening was found at low temperature down to 1.5 K.

The abundance of ^{155}Gd in natural Gd is 14.8%. In most cases, this is sufficient to prepare absorbers with natural Gd. Only for materials with low Gd concentrations, especially in combination with other heavy elements, is enriched ^{155}Gd needed. Because of the high energy of the γ rays, it is highly advisable to optimize the absorber thickness using the approach described by Long *et al.*[69]

4.2. Analysis of the Spectra and Results

For the high γ-ray energy of 86.5 keV, electronic absorption is relatively weak, and the electronic attenuation length λ_e is relatively large. Consequently, the optimized absorber thickness can be rather large.[69] In most cases, the values of the effective thickness[70] T_a considerably exceeds the value $T_a = 1$. Then, the Lorentz approximation (LA), mostly used in the analysis of Mössbauer spectra, is not applicable. It is necessary to fit the spectra with the transmission integral (TI).[70] Cook and Cashion[71] gave a drastic example for the failure of the Lorentz approximation. Examples that appear less severe at first sight are shown in Fig. 6. The overall features of the spectra are fairly well reproduced by the Lorentz approximation, although a close look reveals deficiencies in details, and the χ^2 values for the Lorentz-approximation fits significantly surpass those for the transmission-integral fits (Table 3). Most severe, however, are the differences between several parameter values derived from the two approaches to describe the spectra (emphasized in Table 3). This demonstrates that results deduced from Lorentz-approximation fits to Mössbauer spectra of ^{155}Gd must be considered with great caution.

Line broadening due to slow relaxation processes has hardly ever been observed in Mössbauer spectra of ^{155}Gd. Apparently, only in two cases broadening ascribed to slow relaxation has been reported: a ^{155}Eu:*Pd* source below 1 K,[72] and $GdBa_2Cu_3O_7$ near the Néel temperature $T_N = 2.25$ K.[73,74] In the first case, the reduced relaxation rate probably is a consequence of the extremely low concentration of rare-earth moments in the nonmagnetic Pd matrix. In the second case, the slowing down of relaxation presumably is due to the nearly two-dimensional character of magnetic interactions in $GdBa_2Cu_3O_7$.

The predominance of fast relaxation processes in Gd compounds is in contrast to observations reported for many compounds of other rare earths even with high concentrations of magnetic atoms. As discussed by Gubbens *et al.*, the lack of a direct transition path between states of opposite moments resulting from crystal-field splitting frequently is a cause for slowing down of relaxation rates.[75] This mechanism is excluded in the case of Gd since crystal-field splittings are negligible.

In alloys and in other mixed compounds in which certain sites are occupied at random by two or more different chemical species, the magnetic hyperfine field, the electric field gradient, and the isomer shift generally vary from site to site, depending on the varying local environment. The differences

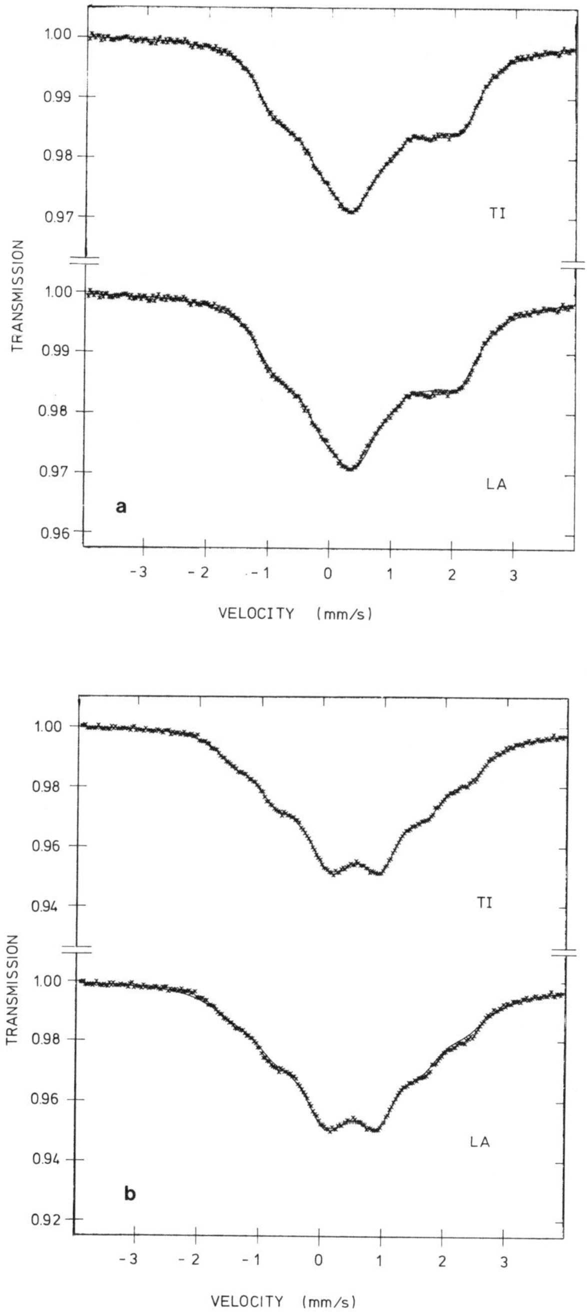

Figure 6. Mössbauer spectra of (a) $GdBa_2Cu_3O_7$ and (b) Gd_2CuO_4. Comparison of fits with the transmission integral (TI) and with the Lorentz approximation (LA).

Table 3. Comparison of Results Deduced from Mössbauer Spectra by Fits with the Transmission Integral (TI) and with the Lorentz Approximation (LA)[a]

	$GdBa_2Cu_3O_7$		Gd_2CuO_4	
	TI	LA	TI	LA
δ_{IS} (mm/s)	0.530(2)	0.537(3)	0.500(2)	0.508(3)
$\Delta_Q^{(0)}$ (mm/s)	**−0.672(2)**	**−0.701(3)**	**−0.644(2)**	**−0.629(3)**
η	**0.55(1)**	**0.37(2)**	0.74(14)	0.84(24)
$\Delta_m^{(0)}$ (mm/s)	0.550(2)	0.551(3)	**0.550(2)**	**0.576(3)**
θ (deg.)	0	0	46(1)	42(2)
ϕ (deg.)	—	—	**70(2)**	**90(1)**
χ^2	0.85	1.51	1.10	2.57

[a]Standard derivations derived in the fit calculations are given in parentheses. Unit is in each case the last decimal of the parameter value displayed.

between hyperfine interactions in different sites usually are too small to give rise to clearly separated subspectra. Rather, the variations result in a broadening of resonance lines. Variation of the isomer shift δ_{IS} broadens all lines to the same degree; variations of V_{zz} and of B_{hf} affect the width of different lines in a different way. An approximate treatment of these broadening effects in terms of broadened Lorentz lines (in the absorption cross section which appears in the exponential in the transmission integral) which involves no assumptions concerning the shapes of the distributions of hyperfine-interaction parameters except for the existence of second and higher moments has been proposed by Czjzek and Berger.[76] This approach is easily generalized to the case of complex spectra of ^{155}Gd when both magnetic and quadrupole interactions have to be considered. In this general case, the spectra consist of 24 lines. For a line centered at the Doppler velocity $v_l(\Delta_m, \Delta_Q, \delta_{IS})$, the linewidth is

$$\Gamma_l^2 = \Gamma_0^2 + 8\Bigg[\left(\frac{\partial v_l}{\partial \Delta_m}\right)^2 \cdot \sigma^2(\Delta_m) + \left(\frac{\partial v_l}{\partial \Delta_Q}\right)^2 \cdot \sigma^2(\Delta_Q) + 2\alpha_{MQ}\frac{\partial v_l}{\partial \Delta_M}\cdot\frac{\partial v_l}{\partial \Delta_Q}\cdot \sigma(\Delta_M)\cdot\sigma(\Delta_Q) + 2\alpha_{M\delta}\frac{\partial v_l}{\partial \Delta_m}\cdot\sigma(\Delta_m)\cdot\sigma(\delta_{IS}) + 2\alpha_{Q\delta}\frac{\partial v_l}{\partial \Delta_Q}\cdot\sigma(\Delta_Q)\cdot\sigma(\delta_{IS})\Bigg] \tag{11}$$

where $\sigma(\Delta) = \sqrt{\langle\Delta^2\rangle - \langle\Delta\rangle^2}$ is the root-mean-square width of the distribution of the parameter Δ. The coefficients α are the coefficients of correlations between the distributions of pairs of parameters indicated by the pairs of subscripts. Γ_0 is assumed to include uniform broadening due to a distribution of isomer shifts: $\Gamma_0^2 = \Gamma_{nat}^2 + 8\sigma^2(\delta_{IS})$. The relative intensities of individual lines have to be renormalized to preserve the relative areas of the lines at the values appropriate for the unbroadened situation: $I_l = I_l^{(0)} \cdot \Gamma_0/\Gamma_l$.

Special care is necessary in the analysis of polarized absorbers,[77] either if a magnetic field is applied as in the case of the spectra shown in Fig. 4 or for single-crystalline absorbers.[58,78] The two situations require somewhat different treatments. A detailed description of the analysis of Mössbauer spectra obtained with γ rays transmitted through polarized absorbers has been given by Gonser and Fischer.[79]

Most results of interest, such as the magnetic hyperfine field and its direction, the electric field gradient, etc., are derived directly from the fits to the spectra. There are no general recipes for the description of the variation of these parameters with temperature or with concentration in mixed systems, for example. The situation is different for the temperature dependence of the recoilless fraction $f(T)$. Under the assumption of harmonic lattice vibrations which is justified for the vast majority of solids, $f(T)$ is in full generality expressed in terms of the frequency spectrum $g_M(\nu)$ of the vibrations of the atom M containing the probe nucleus.[44,80] For M = Gd:

$$f(T) = \exp\left\{ -\frac{E_\gamma^2}{hM_{Gd}c^2} \int_0^\infty [\tfrac{1}{2} + (e^{\beta h\nu} - 1)^{-1}]\nu^{-1} g_{Gd}(\nu)\, d\nu \right\} \tag{12}$$

In this equation, M_{Gd} is the mass of the ^{155}Gd atom, $\beta = 1/(k_B T)$, and $g_{Gd}(\nu)$ is normalized to 1, $\int_0^\infty g_{Gd}(\nu)\, d\nu = 1$.

Frequently, experimental results for $f(T)$ are analyzed under the assumption of a Debye spectrum for $g_M(\nu)$:

$$g_M(\nu) = g_D(\nu) = \frac{3\nu^2}{\nu_D^3} \Theta(\nu_D - \nu) \tag{13}$$

where $\Theta(x) = 0$ for $x < 0$, $\Theta(x) = 1$ for $x \geq 0$.

This is an approximation which is justified only for monoatomic lattices. For more complex structures, only the spectrum of acoustic vibrations can be described by the Debye approximation. Optical modes usually give rise to sharp structures in $g_M(\nu)$ which are not accounted for in the Debye approximation. In spite of the limited temperature range accessible to Mössbauer spectroscopy with ^{155}Gd, the temperature dependence of the recoilless fraction is quite sensitive to details of the frequency spectrum $g_{Gd}(\nu)$, and experimental results cannot be reproduced by the Debye approximation in most cases.[78,81] The agreement is greatly improved when an optical mode is accounted for by an Einstein term $g_E(\nu) = \delta(\nu - \nu_E)$ in the frequency spectrum:

$$g_{Gd}(\nu) = w_D g_D(\nu) + w_E g_E(\nu) \tag{14}$$

with the empirical values for the weight factors[44,82] $w_D = M_{Gd}/M_{UC}$, $w_E = 1 - \omega_D$ where M_{UC} is the mass of the *primitive* unit cell. When information on the vibrational spectrum derived from inelastic neutron scattering was available, the results deduced from this analysis of $f(T)$ were found to be in very good agreement with the neutron data.[81,83]

5. METALLIC COMPOUNDS AND ALLOYS

5.1. Crystalline and Amorphous Alloys

Studies of hyperfine interactions in Gd metal by measurements of neutron-resonance cross-sections,[84] by Mössbauer spectroscopy,[85] by perturbed angular correlations (PAC),[86] and by NMR[87] have yielded widely differing results for the values of the magnetic hyperfine field and of the EFG. A careful investigation of a Gd single crystal, of Gd powders obtained from different sources, and of dilute solutions of Tm and Tb in Gd by ^{155}Gd Mössbauer spectroscopy has revealed that hyperfine interactions in Gd are highly sensitive to the presence of small amounts of impurities.[88] The best values, measured for the Gd single crystal, are: $B_{hf} = (-37.2 \pm 0.4)$ T, $V_{zz} = (3.35 \pm 0.09) \times 10^{21}$ V m^{-2}. For the angle θ between the field direction and the EFG z axis (which coincides with the crystallographic c axis) $\theta = (28 \pm 2)°$ was found, in good agreement with the value deduced from neutron diffraction.[89] In the powder samples and in Tb- and Tm-doped materials the values of both the magnetic hyperfine field and θ vary strongly. From their results, the authors infer a considerable anisotropy of the magnetic hyperfine field in Gd.[88]

Although binary rare-earth alloys present a rich variety of interesting magnetic phenomena involving moment orientations, phase transitions, and multicritical points,[90] very few alloy studies by Mössbauer spectroscopy with ^{155}Gd have been carried out. Mössbauer spectra were recorded in the alloy systems Gd–Y, Gd–Sc,[30,91] and Gd–Lu[92] over wide concentration ranges. In addition, hyperfine fields both at Gd and at Lu nuclei in Gd–Lu alloys were measured by PAC.[86] From Mössbauer spectra obtained for dilute alloys of 3% Gd in Sc, Y, and Lu at 30 K, results for the quadrupole splitting were derived.[88]

In the limit of vanishing Gd concentration, the transferred field B_N in Eq. (10) goes to zero in alloys of Gd with nonmagnetic metals. Then, the hyperfine field B_{hf} is completely determined by the core-polarization field B_{CP} (which is expected to be constant for Gd in all solids when the Gd moment is saturated) and by the self-polarization term B_S. Usually, B_S is assumed to be independent of alloy concentration. Tomala has pointed out that the difference between B_{hf} in dilute *Sc*Gd compared with the other two alloys proves this assumption not to be justified in all cases.[30] Assuming $B_{CP} = -34$ T, the experimental data for Gd in Sc yield the estimate $B_S \approx 30$ T, nearly two times the value derived for Gd in Y and in Lu, $B_S \approx 16$ T. The difference may result from spin fluctuations. They affect the linear term in the low-temperature specific heat for Sc more strongly than for the other two metals.[93] Furthermore, an anomalous enhancement of the Gd moment in dilute *Sc*Gd alloys has been observed.[94]

Mössbauer spectra of nuclei in amorphous materials can yield valuable information about various aspects of magnetism in glasses.[95] In addition, EFGs arising from the charge distribution in the environment of the probe nuclei, carry information about the atomic-scale structure which is complementary to radial pair distributions, derived from neutron or X-ray diffraction patterns.[96] There has been a long-standing debate about the question of whether the atomic structure is best described by the dense random packing model (DRP),[97] or whether the structure is better described as consisting of locally ordered structural units, such as capped trigonal prisms.[98] The information contained in radial pair distributions is not sufficient to discriminate between these views. However, these models differ considerably with regard to the symmetry of the local environment of individual atoms, an aspect which is reflected in the distribution of the parameters V_{zz} and η, that is, in the distribution of EFGs experienced by nuclear probes in amorphous solids. An analytic expression could be derived for the distribution function appropriate for a structure corresponding to the DRP model, in particular.[99] This distribution function describes very well Mössbauer spectra of ^{155}Gd in amorphous Gd–Ni alloys.[99] Spectra obtained for amorphous alloys of Gd and Eu with Au, in contrast, do not correspond to the prediction for the DRP model.[100] In these alloys, there seems to be a stronger tendency to short-range order than in the Gd–Ni alloys.

Remarkable differences between the magnetic properties of crystalline and of amorphous alloys $Gd_xY_{50-x}Ag_{50}$ ($10 \leq x \leq 50$) have been found.[101] Crystalline alloys order antiferromagnetically for all Gd concentrations with a noncollinear arrangement of Gd moments induced by negative biquadratic exchange interactions. In amorphous alloys, the magnetic order is ferromagnetic for large Gd concentrations $x \geq 40$. For lower Gd concentrations, $x \leq 30$, properties typical for magnetic cluster glasses are observed. A comparison of the magnetic hyperfine splitting observed in a dilute amorphous alloy ($x = 10$) in zero field and in an applied field of 6 T shows an unusual increase of the hyperfine field in the presence of the applied field. This result indicates that the frozen Gd moments in the spin-glass state are substantially smaller than $7\mu_B$.

5.2. Intermetallic Compounds

Several overviews over the large manifold of intermetallic compounds of rare earths containing extensive compilations of structural and magnetic data have been published.[102–105]

Results obtained in two studies of Gd compounds with magnetic *3d* transition metals by Mössbauer spectroscopy[106,107] have been tabulated in the review by Taneja and Kimball.[6] Here, qualitative trends resulting from this work will be discussed. In the analysis of hyperfine fields in these compounds, contributions to the transferred hyperfine field B_N in Eq. (10) are due to conduction electron polarization by $3d$ moments (B_N^{3d}) and by $4f$ moments (B_N^{4f}): $B_N = B_N^{3d} + B_N^{4f}$. In compounds with Fe, a strong positive term B_N^{Fe} is dominant, leading to positive hyperfine fields at Gd nuclei. Similarly, B_N^{Co} is positive, but its value is smaller than B_N^{Fe}. It nearly cancels the negative contribution of $B_{CP} + B_S$, and Gd hyperfine fields in compounds with Co mostly have rather small values. Because of the strong magnetic anisotropy of Gd–Co compounds, the sign of the net hyperfine field cannot be deduced from Mössbauer spectra in an applied field. For $GdCo_5$, a negative sign of B_{hf} at Gd was deduced from the concentration dependence of B_{hf} in a series of compounds $Gd(Co_{1-x}Cu_x)_5$.[108] These data also confirm the positive sign of B_N^{Co}. In the same study, a transferred field due to Gd moments, $B_N^{Gd} = 7$ T was inferred from the concentration dependence of B_{hf} in $(Gd_{1-x}La_x)Co_5$. This value seems rather large for B_N^{Gd}. Generally, it is assumed that B_N^{Gd} is practically negligible in compounds of this type. Possibly, the expansion of the lattice resulting from the substitution of La for Gd contributes to the observed variation of B_{hf} with La concentration, since transferred hyperfine fields depend considerably on interatomic distance.[109] However, results for the pseudobinary series[110] $Gd_{1-x}Y_xCO_2$ and for[111] $Gd_{1-x}Y_xFe_2$ also indicate transferred hyperfine fields due to Gd moments of similar strength, although replacement of Gd by Y hardly modifies the lattice parameters.

In intermetallic compounds with rare earths, Ni has at most a very small magnetic moment. Thus, B_N^{Ni} is negligible, and the Gd hyperfine field is negative in all Gd–Ni compounds.

In the compounds GdT_3, Gd atoms occupy two inequivalent sites. In one site, the local symmetry and the transition-metal coordination are similar to the Gd site in the cubic Laves-phase compounds GdT_2, whereas the other site in GdT_3 is similar to the Gd site in GdT_5. These similarities are almost perfectly reflected in the values of B_{hf} and of δ_{IS} for the two sites in GdT_3, being nearly equal to the values for the corresponding site in GdT_2 or in GdT_5, respectively. This shows that influences of the *local* environment dominate B_{hf} and δ_{IS} in these Gd–transition metal compounds.

Ternary compounds of composition GdT_nAl_{12-n} (T = Cr, Mn, Fe, Cu, Rh; n = 4, 5, 6), with the tetragonal $ThMn_{12}$ structure, were the subject of a series of systematic investigations by X-ray diffraction, magnetization measurements, and Mössbauer spectroscopy.[112–116] Of the three inequivalent Mn sites in the $ThMn_{12}$ structure, two are preferentially occupied by transition-metal atoms. Only a small fraction of the transition-metal atoms occupies the third site. Among the transition metals considered in these studies, only Fe was found to carry a magnetic moment. Results of ^{155}Gd Mössbauer spectroscopy are summarized in Table 4. Obviously, the concentration of the

Table 4. Results Derived from ^{155}Gd Mössbauer Spectra for Compounds GdT_nAl_{12-n} Where T Is a Transition Metal[112–116a,b]

T	$\lvert B_{hf}\rvert$ (T)			θ(deg.)		
	$n = 4$	$n = 5$	$n = 6$	$n = 4$	$n = 5$	$n = 6$
Cr	17.2(4)	—	16.8(4)	80(7)	—	64(4)/52(4)
Mn	10.2(5)	—	20.1(5)	46(6)	—	83(5)
Fe	18.8(5)	27.7(6)	20.9(2)	44(2)	40(5)	43(4)
Cu	16.6(4)	—	24.6(2)	60(6)	—	48(3)/61(3)
Rh	—	—	18.6(4)	—	—	60(2)
	$\Delta_Q^{(0)}$ (mm/s)			δ_{IS}(mm/s)		
Cr	1.08(2)	—	±0.20(3)	0.44(5)	—	0.41(2)
Mn	0.78(1)	—	0.68(1)	0.43(5)	—	0.46(3)
Fe	−0.85(2)	−0.40(3)	−0.39(2)	0.4(1)	0.38(5)	0.38(2)
Cu	−0.30(1)	—	±0.21(2)	0.4(1)	—	0.53(2)
Rh	—	—	0.95(2)	—	—	0.10(3)

[a]Error estimates are given in parentheses, in units of the last figure of the parameter value quoted.
[b]For tbe compounds with Cr and Cu, $n = 6$, the sign of $\Delta_Q^{(0)}$ could not be determined. For positive and negative $\Delta_Q^{(0)}$, different values for the angle θ between the fourfold c axis and the direction of $\mathbf{B}_{hf}$ are obtained, as indicated.

transition metal, n, has at most a weak influence on the magnetic hyperfine field and on the isomer shift. Only the quadrupole splitting shows considerable variation with n. The direction of $\mathbf{B}_{hf}$ is neither parallel nor perpendicular to the tetragonal c axis. In compounds with Fe, for one of the Fe sites, the values of θ deduced from ^{57}Fe Mössbauer spectra agree with those found for Gd, in accordance with antiferromagnetic Gd–Fe coupling deduced from an analysis of magnetization data. It appears somewhat surprising that the direction of Gd moments in compounds with nonmagnetic transition metals is quite similar to that in compounds with Fe.

Intensive efforts were devoted to the analysis and interpretation of isomer shifts in intermetallic Gd compounds. In the two pseudobinary alloy systems $GdCu_{1-x}Ga_x$ and $GdAg_{1-x}In_x$ with CsCl structure, the isomer shift was found to increase at the same rate with increasing concentration of the trivalent elements In or Ga.[117] Furthermore, the isomer shift in GdZn is nearly the same as in $GdCu_{0.5}Ga_{0.5}$. The decreasing electron density at Gd nuclei with increasing occupation of conduction electron bands points to a predominant $5d$ character of the additional conduction electrons at Gd sites.

A model, originally proposed by Miedema and van der Woude[118] for the interpretation of ^{197}Au isomer shifts in metallic compounds and alloys, was applied to the analysis of isomer shifts measured for ^{155}Gd in a large number of binary Gd compounds,[119,120] and later extended to isomer shifts in ternary compounds.[121]

According to this model, the primary causes for isomer shifts in binary compounds Gd_xM_y are[119]

1. The electronegativity difference $\phi^*_M - \phi^*_{Gd}$ between the atoms in the compound, leading to charge transfer to the atom of higher electronegativity ϕ^*.
2. The difference in electron density at the boundaries between Wigner–Seitz cells belonging to dissimilar atoms, $n^M_{WS} - n^{Gd}_{WS}$, leading to an intraatomic transfer between *s* and *d* electrons counteracting any density mismatch at the boundary.
3. In compounds of Gd with elements M which are not transition metals, an additive term R' accounts for hybridization of Gd 5*d* bands with *s, p* bands of M

The largest shift for a given combination Gd–M, for compounds of very low Gd content in which Gd is surrounded only by M atoms, is

$$\delta^{max}_{IS} = P'(\phi^*_M - \phi^*_{Gd}) + Q'(n^M_{WS} - n^{Gd}_{WS})/n^{Gd}_{WS} + R' \tag{15}$$

The values of the constants $P' = 0.27$ mm/s and $Q' = -0.11$ mm/s are derived empirically from a comparison of shifts calculated according to Eq. (15) with experimental data for several series of compounds with transition metals.[119] For compounds with higher Gd content, the maximum shift according to Eq. (15) must be multiplied by the coordination fraction f^{Gd}_M. A prescription to calculate values for f^{Gd}_M in terms of atomic concentrations and volumes has been given by Miedema *et al.*[122,120] In a refined version of the model,[120] volume effects which are particularly important in the highly contracted Laves-phase compounds GdT_2 with 3*d* transition metals are taken into account as volume-dependent modifications of the factors P' and Q'.

In terms of this model, the results for the pseudobinary alloy systems Gd(Cu, Ga) and Gd(Ag, In) quoted above are primarily a consequence of the appearance of the hybridization term R' associated with the introduction of the nontransition elements Ga or In.

The overall agreement between isomer shifts calculated according to this model—in particular in the version which accounts for volume effects—and an enormous number of experimental data is quite impressive, indeed. Yet two minor deficiencies ought to be mentioned:

1. Differences between inequivalent sites in a given compound are not accounted for. In view of the results for the two sites in the compounds GdT_3 mentioned above, this is not quite satisfactory. The problem could be overcome by definition of *local* coordination fractions $f^{Gd,site}_M$ which account for the site-specific coordination.
2. Effects of site symmetry are not considered. Such effects are expected for the following reason: Energy levels of 6*s* electrons are insensitive to the local symmetry, but 5*d* (and 6*p*) levels are progressively split as the

symmetry is lowered. As only the lowest-energy levels of the 5*d* manifold are occupied, lowering of the symmetry results in a transfer from 6*s* to 5*d* states. This could appropriately be accounted for by symmetry-dependent modifications of $n_{\mathrm{WS}}^{\mathrm{Gd}}$.

5.3. Hydrides

Metal hydrides have a great potential as energy carriers, for energy storage, and for many other applications.[123] Thus, studies of their properties are of some importance. However, because of the modest resolution for the ^{155}Gd Mössbauer line and the limitation of Mössbauer spectroscopy to low temperatures, ^{155}Gd is not a suitable probe for investigations of some phenomena of general interest, such as the diffusion of hydrogen and its dependence on temperature. Valuable information on hydrogen diffusion were derived from ^{57}Fe Mössbauer spectra[124] and from ^{181}Ta Mössbauer spectra in combination with other experimental techniques.[125] Hence, only a limited number of ^{155}Gd Mössbauer spectroscopy studies of hydrides have been reported.

Values of hyperfine interaction parameters of ^{155}Gd in GdH_2, in GdH_3, and in the alloys $Sc_{1-x}Gd_xH_2$ have been determined by Cashion *et al.*,[36] however, these authors did not comment on their results. Similar results were reported by Lyle and co-workers[126] whose primary goal was the extraction of information on the charge state of hydrogen. The isomer shifts, in both hydrides close to that in GdF_3, favor a negative hydrogen charge. Supporting evidence could not be derived from the quadrupole splitting which is zero in the cubic compound GdH_2, and in GdH_3 the sign of V_{zz} could not be determined.

In a source experiment, producing the parent nucleus ^{155}Eu by neutron capture in $^{154}SmH_2$, Wortmann *et al.*[127] observed a quadrupole splitting at 20 K, above $T_N = 13 \pm 2$ K, in agreement with earlier data,[68] although SmH_2 has a cubic structure. Wortmann *et al.* ascribe the quadrupole splitting to the ejection of one hydrogen from the eightfold cubic H-coordination shell of the Gd sites.

In several ternary hydrides $GdTH_x$ (T = Rh, Pd, Cu, Ag, Au), drastic increases of magnetic hyperfine fields and of isomer shifts relative to the hydrogen-free compounds were reported.[128] The changes of B_{hf} presumably result from reduced positive or even negative contributions B_S and B_N due to the transfer of Gd 6*s* electrons to hydrogen, evidenced by the positive isomer shifts. Results for [129] $GdCu_2H_x$ as well as for[130] $GdRh_2H_x$ and $GdRu_2H_x$ are quite similar.

The hydride $LaNi_5H_x$ which can absorb up to 6.7 hydrogen atoms per formula unit has properties which make it of direct interest for technical applications.[123] Structural and magnetic effects due to hydrogen absorption in $Gd_{0.1}La_{0.9}Ni_5H_x$ ($x = 0$, 4.6, 6.7) and in $Gd_{0.1}La_{0.9}Co_5H_x$ ($x = 0$, 1.8, 3.0, 3.8,

4.2) were studied by ^{155}Gd Mössbauer spectroscopy.[131] In the nonmagnetic nickel compounds, single-component spectra were obtained for $x = 0$ and $x = 6.7$. For intermediate concentrations, the spectra are a superposition of the components found for the unhydrided and the maximally hydrided compounds. Analogously, in the cobalt series, three single-phase compounds, for $x = 0$, 3.2, and 4.2, were found, and superpositions of the corresponding components at intermediate hydrogen concentrations. In both systems, loading with hydrogen drastically reduces the quadrupole splitting. In compounds with Co, the magnetic hyperfine field increases considerably in the hydride with $x = 3.2(|B_{hf}(x = 3.2)| = (36.6 \pm 0.4)\mathrm{T}$, compared with ~10 T for $x = 0$). For the highest H content ($x = 4.2$), the field is strongly reduced to about 5 T.

5.4. Ternary Silicides and Germanides

Ternary silicides with the general formula RT_2Si_2 where R stands for a rare earth and T for a late transition metal (beginning with Cr, Ru, and Os for the 3*d*, 4*d*, and 5*d* transition-metal series, respectively) are one of the most intensively studied groups of compounds. They allow an extraordinarily wide range of combinations between rare earths and transition metals, exhibiting many interesting magnetic phenomena.[132] Nearly all of these compounds crystallize in the body-centered tetragonal $ThCr_2Si_2$ structure. There are a few exceptions, notably the silicides with Pt, RPt_2Si_2 whose primitive tetragonal $CaBe_2Ge_2$ structure is closely related to the $ThCr_2Si_2$ structure.

In several of these compounds with Ce, Eu, and Yb, an instability of the 4*f* valence is indicated by unusual properties. The best known example is $CeCu_2Si_2$, the first heavy-fermion compound in which a transition to superconductivity was discovered.[133] For the evaluation of results which are characteristic for the unstable valence, a comparison with data for compounds with stable 4*f* valence is of some importance. Besides compounds with the nonmagnetic rare earths Y, La, and Lu, those with Gd are useful for this purpose because of the magnetic isotropy of the $4f,^8$ $S_{7/2}$ ground state.

Results derived from ^{155}Gd Mössbauer spectra for ternary Gd silicides were reported by several groups.[134–138,59] Data obtained by different groups are in good mutual agreement. Studies of germanides with the $ThCr_2Si_2$ structure have so far been restricted to compounds with 3*d* transition metals.[139,140]

In the Gd compounds with Mn, the Mn moments order antiferromagnetically below about 450 K and 365 K, the Gd moments order at much lower temperatures, 65 K and 96 K in the silicide and in the germanide, respectively.[132] In compounds with Ru and Rh, anomalies in the temperature and field dependence of the magnetization were interpreted as an indication for itinerant magnetism of the transition-metal *d* electrons.[136,141] All other transition metals are nonmagnetic both in the silicides and in the germanides. The absence of a magnetic moment at Fe in compounds RFe_2Si_2 was confirmed by ^{57}Fe Mössbauer spectra.[135,59]

In most silicides and germanides, the magnetic hyperfine field has a unique value at all Gd nuclei, pointing in a direction perpendicular—or nearly perpendicular—to the fourfold *c* axis. Solely in $GdCr_2Si_2$, the field direction is parallel to the *c* axis.[138] Mössbauer spectra of ^{155}Gd in $GdRu_2Si_2$ and in $GdOs_2Si_2$ are more complex, indicating a helical moment arrangement.[138,59]

Low-temperature magnetization curves revealed the occurrence of metamagnetic phase transitions in $GdRu_2Si_2$, $GdRh_2Si_2$, $GdOs_2Si_2$, and $GdPt_2Si_2$.[59] For $GdPt_2Si_2$, Gignoux *et al.* observed the same phenomena.[142] The metamagnetic transition in $GdRu_2Si_2$ may have implications for the interpretation of a metamagnetic transition in $CeRu_2Si_2$ which has been associated with the heavy-fermion character of this compound.[143] In Gd, the stability of the $4f^7$ configuration is beyond any doubt. Thus, the similarity of the metamagnetic transitions in the two compounds raises questions concerning the interpretation of the transition in the case of $CeRu_2Si_2$.

The variations of isomer shifts and quadrupole splittings with the atomic

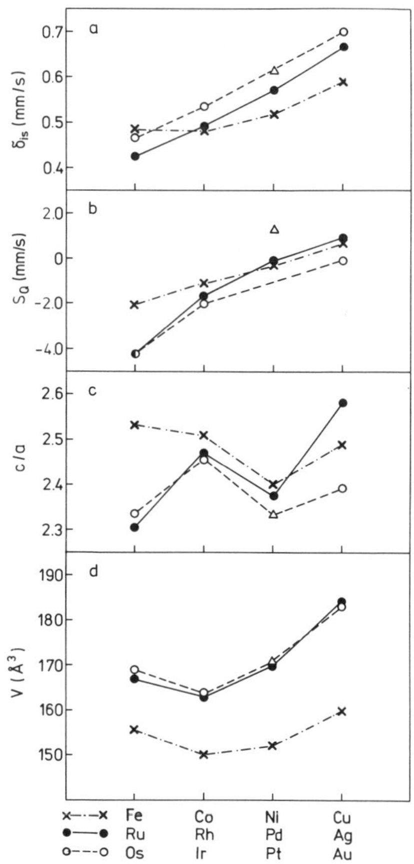

Figure 7. Results derived from X-ray diffraction and ^{155}Gd Mössbauer spectra for compounds GdT_2Si_2 (T = transition metal) versus the atomic number, Z_{nd}, for the 3*d*, 4*d*, and 5*d* series of transition metals: (a) isomer shifts, (b) quadrupole splittings, (c) axial ratios *c*/*a*, (d) volumes of the unit cell. (From Ref. 59.) The crystal structure of $GdPt_2Si_2$ differs from that of the other compounds.[59] Consequently, results obtained for $GdPt_2Si_2$ are marked by triangles rather than by circles. Lines are guides to the eye.

number Z_{nd} are very similar for the series with $3d$, $4d$, and $5d$ transition metals (Fig. 7a,b). Volume changes, shown in Fig. 7d, account at most for a small fraction of the observed shifts. Furthermore, results for S_Q are in no way correlated with changes of the axial ratio c/a (Fig. 7c). According to calculations of Coehoorn *et al.*, the main cause for the observed variation of S_Q with Z_{nd} is the variation of the $6p$ contribution to V_{zz} which progressively increases from negative values at the beginning of each series to positive values at the end.[49] Values of the isomer shifts, calculated with a generalized version of the model of Miedema and van der Woude for a few silicides,[121] are in reasonable agreement with experimental data.

In another group of ternary silicides, of general formula $R_2Fe_3Si_5$, compounds with nonmagnetic rare earths, R = Sc, Y, Lu, and for R = Tm, become superconductors at low temperatures.[144] In $Gd_2Fe_3Si_6$, the Gd moments order antiferromagnetically below $T_N = 8.6$ K. The rare-earth site in these compounds has very low symmetry, and Mössbauer spectra of ^{155}Gd yield a large value for η, in the range $0.7 \leq \eta \leq 1$. From results for V_{zz} and η, estimates for crystal field parameters B_2^m were derived using the approach described in Section 3.3.[144]

5.5. Ternary Borides

The discussion of results for two classes of ternary borides will be deferred to later sections:

- Compounds RRh_4B_4, with a fascinating interplay of superconductivity and magnetism, will be treated in the section devoted to superconductors.
- Results for compounds of composition $R_2T_{14}B$, materials of interest for permanent magnets, will be presented in the next section.

In a study of structures and magnetic properties of the series $R_3Ni_7B_2$ (R from Sm to Lu), two different structures, one for R = Sm, Eu, another one for compounds with heavier rare earths, were found.[145] In $Gd_3Ni_7B_2$ ($T_N = 37$ K), Gd occupies two inequivalent sites. For one of the sites [2(c) in Wyckoff notation], the quadrupole splitting seemed extraordinarily large for a metallic compound: $S_Q(2c) = (5.22 \pm 0.14)$ mm/s. Later, in other ternary borides to be mentioned below, even larger quadrupole splittings were found. For the other, 4(f), site, the quadrupole splitting is rather modest; $S_Q(4f) = (1.08 \pm 0.14)$ mm/s. Also, the magnetic hyperfine field is quite different at the two sites:

$$B_{hf}(2c) = (21.5 \pm 0.5)\mathrm{T}$$

$$B_{hf}(4f) = (11.5 \pm 0.5)\mathrm{T}$$

The direction of $\mathbf{B}_{hf}$ is perpendicular to the hexagonal *c* axis.

Some ternary borides RCo_2B_2 have the $ThCr_2Si_2$ structure like the silicides and germanides discussed in the preceding section. In$GdCo_2B_2$, the isomer shift, $\delta_{IS} = (0.34 \pm 0.01)$ mm/s, is smaller than in the corresponding silicide; the value of the quadrupole splitting, $S_Q = -(4.14 \pm 0.04)$ mm/s, however, is much larger.[146] The magnetic hyperfine field has nearly the same value as in the silicide, and its direction is also nearly perpendicular to the tetragonal symmetry axis.

Another type of boride, also with Co, is derived from the hexagonal compound $GdCo_5$ by successive substitution of B for Co atoms in the planes in which Gd atoms are located.[147] This leads to a series of ordered compounds of compositions $GdCo_4B$, $Gd_3Co_{11}B_4$, $Gd_2Co_7B_3$, and in $GdCo_3B_2$. Different sites of Gd are distinguished by the presence or absence of B atoms in the plane of the Gd site and in adjacent planes. The quadrupole splitting increases drastically with increasing B content in the local environment, up to $S_Q = (7.12 \pm 0.04)$ mm/s in $GdCo_3B_2$. Point-charge calculations of V_{zz} with the empirical values for atomic charges: $q_{Gd} = +1$, $q_{Co} \times -1/6$, $q_B = -1/3$ reproduce all experimental results for the different sites in the compounds under discussion, and also the field gradients derived from ^{155}Gd Mössbauer spectra in the compounds $Gd_2Co_{14}B$ and $Gd_2Fe_{14}B$.[147] Results of an earlier investigation[148] of $GdCo_3B_2$ are in good agreement with those reported by Smit *et al.*

Tetragonal compounds $R_{1+\varepsilon}Fe_4B_4$ are distinguished by a structural peculiarity: They consist of two interpenetrating chainlike lattices containing rare earths and Fe + B atoms, respectively, with slightly different lattice spacings along the *c* axis. The ratio $c_{Fe}/c_R = 1 + \varepsilon$ varies between 1.06 and 1.18, depending on the size of the rare-earth atoms.[149]

Hyperfine interactions at ^{155}Gd nuclei in $Gd_{1+\varepsilon}Fe_4B_4$ and in alloys $(R_{1-x}Gd_x)_{1+\varepsilon}Fe_4B_4$ with R = Nd, Sm have been determined by Mössbauer spectroscopy.[150–153] The quadrupole splitting, $S_Q = (8.34 \pm 0.04)$ mm/s, is even larger than that reported for $GdCo_3B_2$. As expected for the different spacings of Gd and Fe–B chains, a distribution of field gradients is indicated by broadening of the lines. The spectra are well described by the assumption of a Gaussian distribution of splittings with an rms width of 0.4 mm/s. In alloys with Nd and Sm, the average splitting is practically unchanged, but the width of the distribution increases with increasing Nd or Sm concentration.

The magnetic structure of $Gd_{1+\varepsilon}Fe_4B_4$ seems to be rather complex. The spectra are well described by a unique value of the hyperfine field at all probe nuclei, but a wide range of angles between the EFG *z* axis and the direction of $\mathbf{B}_{hf}$ must be assumed, corresponding to a nearly isotropic distribution of field directions. A low concentration of Nd is sufficient to induce a field direction perpendicular to the tetragonal symmetry axis, whereas the field direction is parallel to the tetragonal axis when Sm is substituted for Gd.

5.6. Materials for Permanent Magnets

Studies of materials of interest for permanent magnets by Mössbauer spectroscopy have been reviewed quite recently, along with a discussion of properties which particularly qualify a material as suitable for permanent magnets.[154] Consequently, in this section, the emphasis will rest on newer developments. Since the discovery of remarkable properties of a phase in Nd–Fe–B samples,[155,156] a phase which soon was identified as the tetragonal compound $Nd_2Fe_{14}B$,[157,158] searches for materials with improved properties for permanent magnets were enormously intensified.

Because of the inherent magnetic isotropy of Gd moments, compounds containing Gd are not useful for permanent magnets themselves. However, Gd is a very useful probe for the extraction of information on local properties at rare-earth sites in compounds. Except for the occupation of the 4*f* shell, differences between the electronic structure of rare-earth atoms are marginal, and many results obtained for Gd can be transferred to other rare earths.

Mössbauer spectra of powder samples $Gd_2Fe_{14}B$ consist of the superposition of two components corresponding to two inequivalent rare-earth sites in this compound. These spectra are rather unstructured, and values of hyperfine interaction parameters deduced from the spectra by different groups differ considerably.[159,160] Spectra of single crystals have a lot more structure (Fig. 8), and their analysis clarified the situation.[58] Moreover, the differences between hyperfine fields derived from single-crystal spectra in a zero field and in an applied field of 8 T revealed the hyperfine field to be positive (in the sense

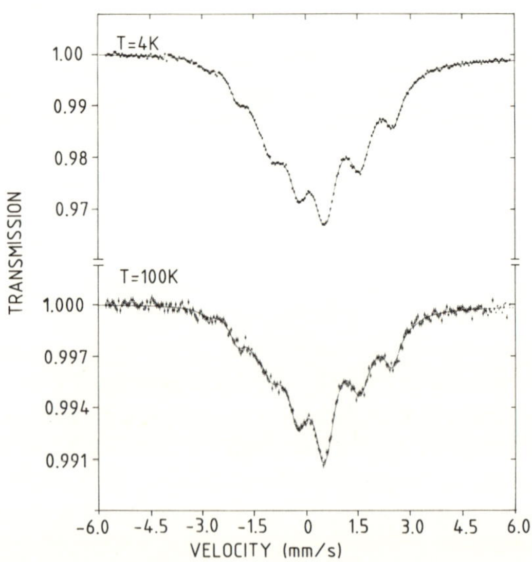

Figure 8. Mössbauer spectra of ^{155}Gd in a $Gd_2Fe_{14}B$ single-crystal absorber at the temperatures indicated. (From Ref. 78.)

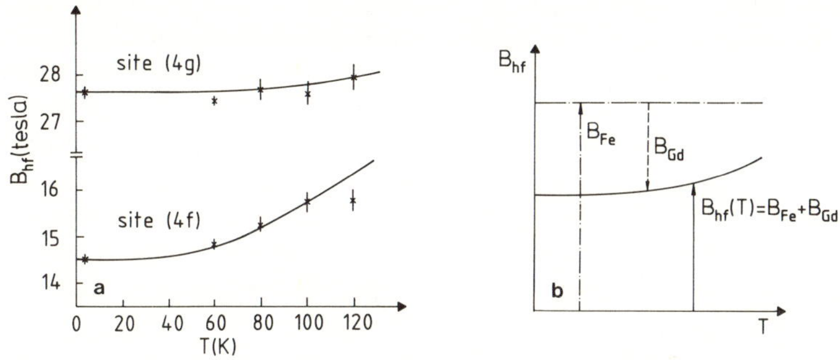

Figure 9. (a) Temperature dependence of the hyperfine fields B_{hf} at the two Gd sites in $Gd_2Fe_{14}B$. (b) A schematic presentation of the basis for the unusual *increase* of B_{hf} in $Gd_2Fe_{14}B$ with increasing temperature, as discussed in the text. (From Ref. 78.)

defined in Section 3.4) at both Gd sites. Measurements at several temperatures between 4 and 120 K yield an apparently paradoxical *increase* of the hyperfine fields with increasing temperature (Fig. 9a).[78] This is easily explained by a weak temperature dependence of Fe moments and consequently of the large positive hyperfine field B_N^{Fe} transferred to Gd sites, and a somewhat stronger temperature dependence of Gd moments and of their negative contribution B_{Gd} to the hyperfine field, as shown schematically in Fig. 9b. This interpretation is in accordance with conclusions derived from a comparison of the temperature dependences of the bulk magnetization of $Y_2Fe_{14}B$ and of $Gd_2Fe_{14}B$.[160]

Based on this interpretation of the data, estimates for the strength of the Gd–Fe exchange interaction can be derived from the experimental results, although with a high degree of uncertainty, since several unknown details are involved in the derivation.[78] In particular, it has not been possible to obtain a reliable estimate for B_N^{Fe} from hyperfine fields at Y and Lu nuclei measured in the analogous compounds by NMR,[161] because polarized neutron diffraction data have shown that Fe moments at sites close to rear-earth atoms are larger in compounds with magnetic rare earths than in the Y compound.[162] Thus, differences between transferred fields B_N^{Fe}, depending on rare-earth moments beyond the differences due to variations of the hyperfine coupling constant with Z,[163] must be assumed.

Values of crystal-field parameters A_2^m, estimated[58] by conversion from V_{zz} and η (see Section 3.3), are larger than values derived from the analysis of high-field magnetization data.[164] The sign is in both cases the same. The discrepancy is not too surprising in the case of metallic compounds, as pointed out in Section 3.3.

Mössbauer spectra of ^{155}Gd in $Gd_2Co_{14}B$ were registered with an absorber made with powder which had been aligned in a magnetic field.[165] As in the binary Gd–Co compounds (Section 5.2), the hyperfine fields at both Gd sites have small values. Spectra in an applied field showed both fields to be negative. Result for V_{zz} and η are quite similar to those for $Gd_2Fe_{14}B$.

One major drawback of $Nd_2Fe_{14}B$ for its use as a permanent-magnet material is the rather low Curie temperature ($T_C < 600$ K). In consequence, the magnetic anisotropy which is very strong at low temperatures, rapidly weakens above 100°C. This is a serious limitation for the use of this compound in high-power motors, for example. Thus, many efforts were devoted to the search for other compounds with properties similar to those of $Nd_2Fe_{14}B$, but with a higher Curie temperature. Some improvements were achieved by partial substitution of Co for Fe in $Nd_2Fe_{14}B$, but at the expense of a reduced anisotropy.

Besides attempts to achieve higher Curie temperatures in compounds derived from or related to $Nd_2Fe_{14}B$, other types of compounds were studied as well. Examples are the ternary compounds of composition $RFe_{12-x}M_x$ (M = Ti, W, V, Cr, Mo, Si) with the tetragonal $ThMn_{12}$ structure.[166] They brought no substantial improvement. The Curie temperatures are rather low, and from Mössbauer spectra of ^{155}Gd in several of these compounds, small values of A_2^0 were deduced, implying a weak magnetic anisotropy.[167]

More promising results were obtained with ternary compounds derived from R_2T_{17} by introduction of interstitial atoms. Carbon interstitials raise the Curie temperatures which are quite low (<500 K) in the pure R_2Fe_{17} compounds. In $Gd_2Fe_{17}C_x$, T_C increases from 475 K for $x = 0$ to a value near 600 K for $x = 1.4$.[168] By neutron diffraction, the sites of C interstitials were determined to be in close vicinity to rare-earth atoms.[169] Mössbauer spectra of ^{155}Gd in $Gd_2T_{17}C_x$ (T = Fe, Co) are superpositions of up to three components, ascribed to Gd sites with zero, one, and two C atoms, respectively, in the vicinity.[168,170] The component for the site associated with one neighbor is distinguished by a very large quadrupole splitting. In the Fe compound, B_{hf} is strongly reduced by C neighbors. Surprisingly, in $Gd_2Ni_{17}C_x$, the spectra do not change with C content. For $x = 0.$ 0.6, and 1.2, practically identical spectra were obtained.[170]

It seems that the carbon uptake in compounds $R_2T_{17}C_x$ is limited to a maximum value $x \approx 1.4$. In addition, the Curie temperature shows a tendency to saturate for the largest values of x achieved.[168] An exception is $Gd_2Mn_{17}C_x$ with highest C concentrations x close to 3.[171] However, with magnetic ordering temperatures below 20 K, these materials are of no interest for permanent magnets.

A real breakthrough was achieved by the discovery that rather large amounts of nitrogen can be introduced into R_2Fe_{17} and into $R_2Fe_{17}C_x$, leading to increases of T_C up to about 750 K. For R = Sm, the magnetization and the anisotropy field are similar to those of $Nd_2Fe_{14}B$.[172] Thus, nitrided compounds $Sm_2Fe_{17}N_x$ with $x \approx 3$ are extraordinarily promising new materials for permanent magnets.

Mössbauer spectra of ^{155}Gd in $Gd_2Fe_{17}N_x$ with $x \approx 3$ consist of a

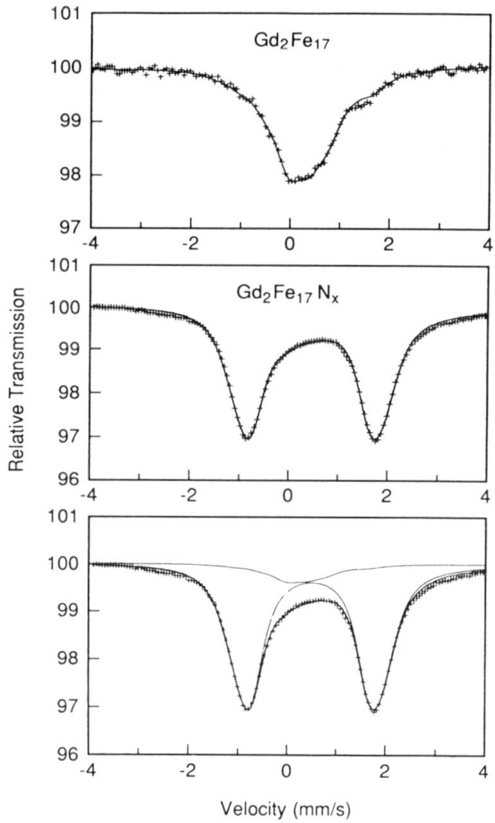

Figure 10. Mössbauer spectra of ^{155}Gd in Gd_2Fe_{17} and in $Gd_2Fe_{17}N_x$ ($x \approx 3$). The bottom panel shows the decomposition of the $Gd_2Fe_{17}N_x$ spectrum into that for the nitride with $x = 3$ and a small amount of unnitrided Gd_2Fe_{17}. (With permission from Ref. 173.)

component with very large quadrupole splitting and a rather small magnetic hyperfine field, ascribed to the stoichiometric nitride with $x = 3$, and a weak component corresponding to that of unnitrided Gd_2Fe_{17} (Fig. 10).[173] The large quadrupole splitting implies a very large value for the crystal field parameter A_2^0 and consequently a strong magnetic anisotropy. The small value of the magnetic hyperfine field indicates that the transferred field B_N^{Fe} is strongly reduced in the presence of N atoms near the Gd sites.

6. SOLIDS WITH IONIC OR COVALENT BONDS

Several ionic and covalently bonded solids were among the large number of compounds studied by Cashion and co-workers.[36] The results provided the basis for establishing an isomer shift scale for ^{155}Gd and for the comparison with shift scales reported for other rare-earth nuclei.[37] Quadrupole splittings measured for some ionic compounds, in conjunction with calculated splittings, yielded an estimate for the Sternheimer factor, γ_∞, as discussed in Section 3.3.[52]

Beyond this initial effort, only a few studies of ionic and covalently bonded solids by ^{155}Gd Mössbauer spectroscopy have been reported although this technique can yield information which contributes substantially to our understanding of the properties of the materials. In particular, the determination of magnetic structures by neutron diffraction is very difficult in the case of Gd compounds because of the gigantic absorption cross section (~50,000 b) of Gd with natural isotopic composition. Mössbauer spectroscopy with ^{155}Gd, combined with results of other experimental techniques, is well suited as an alternative. This has been demonstrated by Cook and Cashion[71,174,175] in investigations of $GdAlO_3$ and of $GdVO_4$ by ^{155}Gd Mössbauer spectroscopy. For $GdAlO_3$, a compound with an orthorhombic crystal structure, results derived from Mössbauer spectra of single-crystal absorbers, in combination with data obtained by several other techniques, allowed the complete determination of the noncollinear magnetic structure in spite of its complexity. In addition, information on the variation of the canting angle with temperature, on exchange coupling constants, and on crystal-field parameters was obtained.[174] In an analogous study of $GdVO_4$, a simple collinear antiferromagnet, values of exchange interaction energies and of crystal-field parameters were derived from the experimental results.[175] In contrast to the highly ionic character of the aluminate, contributions to crystal fields and to the EFG due to covalent overlap of electronic charges are important in the vanadate.

Mössbauer spectra of ^{155}Gd in compounds $Gd_2M_2O_7$ (M = Sn, Ti, Ru, Ir) with cubic pyrochlore structure have been observed by several groups.[19,21,22,36,176,177] The interest in these compounds was primarily concentrated on the strong Goldanskii–Karyagin effect,[20] the anisotropy of the recoilless fraction. Since it is difficult to determine the anisotropy quantitatively with the 86.5-keV transition of ^{155}Gd, other resonances in Gd nuclei were used: the 105.3-keV transition in ^{155}Gd,[21,177] the 123-keV transition in ^{154}Gd,[177] and the 89-keV transition in ^{156}Gd.[19,22] Furthermore, the quadrupole splitting was found to decrease linearly with increasing lattice parameter *a*.[177] Point-charge calculations of the EFG in pyrochlore compounds yielded results which are quite well correlated with experimental data derived from Gd, Sn, and Fe Mössbauer spectra.[178] Thus, estimates of the Sternheimer factors, γ_∞, for these nuclei were obtained from a comparison of the experimental splittings with the calculated values of the EFG.

In general, bonding in sulfides tends to have stronger covalent character than in oxides since the electronegativity of sulfur is much smaller than that of oxygen. Nevertheless, the isomer shift reported for Gd_2S_3 (~0.5 mm/s) has nearly the same value as the shift in Gd_2O_3.[36] Presumably, contributions of electrons with 6*s* and 5*d* (or 6*p*) character involved in covalent bonds nearly cancel each other. In contrast, in GdS with a formal Gd valence 2+, a significantly larger shift $\delta_{IS} = 0.68$ mm/s points to an increased occupation of 5*d* states at the Gd atoms.[179] In spite of this difference, the magnetic hyperfine fields have nearly the same value in both sulfides: in Gd_2S_3, $B_{hf} = -28\,T$; in GdS, $B_{hf} = -29.6\,T$.[36,179]

Ternary sulfides of general formula AR_2S_4 where A is an alkaline earth or a divalent transition metal and R is a rare earth have various different crystal structures. Only for $MnGd_2S_4$, Mössbauer spectra of ^{155}Gd have been reported.[180] This compound has the body-centered cubic Th_3P_4 structure. The temperature dependence of the magnetic susceptibility closely follows a Curie–Weiss law. The value of the Curie–Weiss constant is that expected for the free-ion moments of both Mn^{2+} and Gd^{3+}. However, no indication for magnetic order was found above 4.2 K. In accordance with this, the ^{155}Gd Mössbauer spectrum measured at 4.2 K shows no magnetic hyperfine splitting. Charge transfer leads to a somewhat reduced shift, $\delta_{IS} = 0.44\,mm/s$, compared with the shift in Gd_2S_3.[180]

7. SUPERCONDUCTORS

7.1. Magnetism and Superconductivity in Classical Superconductors

A magnetic field destroys the superconducting state when its strength exceeds a critical value. Thus, it is not surprising that magnetic impurities in a superconductor reduce its transition temperature, T_c, to the superconducting state. The first experimental confirmation was found in a study of the effects of rare-earth impurities in La.[181] A comparison of the rates of depression induced by the different rare earths revealed that the *spin* of the impurities, not their magnetic moment, determines the rate of depression. This observation suggested pair breaking due to exchange scattering of conduction electrons by impurity spins as the basic mechanism for the depression of T_c.[182] The theory of Abrikosov and Gorkov, based on this mechanism,[183] describes very well the dependence of the T_c depression on impurity concentration x in many superconductors.[184] However, this is not true for all systems. As discussed by Maple, qualitatively different shapes of $T_c(\mathbf{x})$ curves are found, depending both on the host material and on the stability of the impurity moments.[184]

Only a few months after publication of the data for rare-earth doped La, Matthias and co-workers reported results for pseudobinary alloys $(Ce_{1-x}R_x)Ru_2$, R = Pr, Gd, showing a much weaker effect of the magnetic atoms on T_c than expected. The data even gave indications for the coexistence of superconductivity and ferromagnetism in a narrow region of impurity concentrations.[185] Matthias and co-workers emphasized that it is fairly easy to detect a transition to superconductivity occurring below the magnetic ordering temperature by bulk methods, whereas the detection of a magnetic ordering temperature below T_c requires application of a strong magnetic field which destroys the superconducting state. Thus, for studies of materials in which superconductivity and magnetism coexist, nuclear techniques such as neutron scattering, spin-echo NMR, and Mössbauer spectroscopy[186] are very useful techniques.

In the case of $(Ce_{1-x}Gd_x)Ru_2$, neutron scattering is difficult because Gd nuclei strongly absorb neutrons. Very high transverse relaxation rates rendered NMR signals in zero field undetectable in the interesting concentration range.[187] However, evidence for the coexistence of magnetic order and superconductivity for the Gd concentration $x = 0.105$ was obtained by NMR measurements in small applied fields (much smaller than the critical field which destroys superconductivity in this compound).[187] Further support for the occurrence of magnetic ordering in the superconducting state was derived from measurements of the specific heat.[188]

In Mössbauer spectra of ^{57}Fe in ^{57}Co-doped alloys $(Ce_{1-x}R_x)Ru_2$, R = Gd, Ho, at low temperatures, a magnetic and a nonmagnetic component are superimposed.[189] The relative intensity of the magnetic component is proportional to the concentration x of the magnetic rare earth. Hence, Erickson *et al.* ascribe this component to ^{57}Fe nuclei which are nearest neighbors of a magnetic rare earth.[189]

Direct evidence for the ordering of Gd moments was obtained in an investigation of alloys $(Ce_{1-x}Gd_x)Ru_2$, $0.05 \leq x \leq 0.5$, by ^{155}Gd Mössbauer spectroscopy.[190] The temperature dependence of the magnetic hyperfine field corresponds to a Brillouin curve only for the highest Gd concentration, $x = 0.5$. Deviations of $B_{hf}(T)$ from the Brilloiun curve, evolving progressively with decreasing x, are similar to those frequently observed in magnetic glasses.[95] In the glasses, they result from distributions of exchange constants. In the case of the crystalline alloys, they were interpreted as an indication for short-range ferromagnetic order with correlation lengths, ξ_m, which decrease with decreasing Gd concentration, x.[190] This interpretation is supported by results of neutron scattering experiments which yielded $\xi_m \sim 15$ Å in $(Ce_{0.8}Tb_{0.2})Ru_2$,[191] $\xi_m \sim 80$ Å in $(Ce_{0.73}Ho_{0.27})Ru_2$.[192]

Experimental evidence for short-range magnetic correlations in the superconducting state was also derived from ^{151}Eu Mössbauer spectra in dilute *La*–Eu alloys.[193] These findings are in accordance with theoretical arguments by Anderson and Suhl that ferromagnetic correlations in coexistence with superconductivity must be restricted to extensions which are smaller than the superconducting coherence length.[194]

In these alloy systems, magnetic atoms are distributed randomly over certain lattice sites. In other classes of materials in which magnetic order and superconductivity can coexist, regular lattice sites are fully occupied by magnetic rare-earth atoms, R. Examples are the ternary borides, RRh_4B_4, and the Chevrel-phase compounds, RMo_6X_8, X = S, Se.

In the ternary Rh-borides of rare earths with a partially filled $4f$ shell, long-range magnetic order has been found below a magnetic critical temperature, T_M.[195] In the compounds with Nd, Sm, and Tm, the order is antiferromagnetic, and T_M is below the transition temperature, T_c, to superconductivity. In zero applied field, the occurrence of antiferromagnetic order does not affect the superconducting state. However, in $ErRh_4B_4$, the transition to superconductivity, at $T_c = 8.7$ K, is followed by a first-order transition, at $T_{c2} \sim 0.9$ K, from the superconducting to a normal state, with ferromagnetic order of the Er moments.[196] In borides with Gd, Tb, Dy, and Ho, higher transition temperatures, T_M, (~6 to 12 K) to ferromagnetic order indicate stronger magnetic exchange interactions which obviously preclude the formation of Cooper pairs. No transition to the superconducting state is found in these compounds.

Several experimental results concerning magnetic properties of compounds RRh_4B_4 indicate strong influences of crystal fields acting on the 4*f* state of the rear-earth atoms.[197] For the tetragonal symmetry at the rare-earth site, a dominant role of the second-order crystal-field parameter B_2^0 is expected. A strong quadrupole splitting in the Mössbauer spectrum of ^{155}Gd in $GdRh_4B_4$ indeed points to large values of B_2^0.[198] Conversion from the ^{155}Gd quadrupole splitting to B_2^0, by the approach described in Section 3.3, results in values and signs of B_2^0 which agree with qualitative trends in the experimental data, if this term is assumed to dominate crystal-field effects.[197,198] Deviations in quantitative details show that other terms in the crystal field are not entirely negligible.

The magnetic hyperfine splitting in $GdRh_4B_4$ below $T_M = 5.8$ K indicates a complex noncollinear magnetic structure, as the direction of the hyperfine field is parallel to the tetragonal *c* axis at only about 70% of the Gd nuclei. At the other nuclei, an angle of (50 ± 10) degrees between the field direction and the *c* axis was deduced from the Mössbauer spectrum.[199] Alternatively, the canting can be caused by structural defects in the vicinity of these nuclei as Noakes *et al.* point out.[199] In alloys $(Gd_{1-x}R_x)Rh_4B_4$, R = Ho, Er, the strong anisotropy of the solute atoms forces the moments to align either parallel to the *c* axis for very low Ho concentrations, or perpendicular to the *c* axis for very low Er concentrations.[199]

In nearly all Chevrel-phase compounds RMo_6S_8, a transition to antiferromagnetic order occurs at a temperature, T_M, below the superconducting transition temperature, T_c, without affecting the superconducting state.[200] Exceptions are $EuMo_6S_8$, with no transition to superconductivity, and the compounds with Sm and Tm with no magnetic ordering. For the selenides, fewer experimental results have been reported. Data for RMo_6Se_8, R = Gd, Tb, and Er, indicate similar behavior as in the sulfides.[200] In $GdMo_6S_8$, the two transition temperatures are very close; $T_c = 1.1$ K, $T_M = 0.85$ K. Due to difficulties with the preparation of pure-phase materials, only one study of $GdMo_6S_8$ by ^{155}Gd Mössbauer spectroscopy has been carried out so far.[201] The quadrupole splitting is larger than in $GdRh_4B_4$. As in the case of ternary borides, estimates of the crystal-field parameter, B_2^0, led to predictions which agree qualitatively with experimental data.[201] No results concerning magnetic hyperfine interactions below T_M are available.

7.2. High-T_c Cuprates and Related Compounds

The months after Bednorz and Müller reported their observation[3] of a transition to superconductivity near 30 K in a mixed oxide of La, Ba, and Cu were a hectic time, full of excitement, for many solid state physicists and chemists. An account of the explosive development during this period has been given by Ginsberg.[202] The emphatic search for new compounds with higher transition temperatures culminated in 1988 in the discovery of $Tl_2Ba_2Ca_2Cu_3O_{10}$ with $T_c \approx 120$ K.[203] The search still goes on, but the emphasis has shifted to thorough studies of the known compounds, aiming, on one hand, at the preparation of materials and devices suitable for technical applications, and on the other hand, at an understanding of the physical properties and processes which might be the cause of the high transition temperatures. The nature of the underlying mechanism is still in dispute[204] in spite of an overwhelming amount of experimental information about these materials.

Nearly all high-T_c compounds have a layered structure with CuO_2 planes. Examples of these structures are shown in Fig. 11. Superconducting compounds are derived from insulating compounds by appropriate doping of mobile current carriers. In most cases, the current carriers are holes located in the CuO_2 planes, mainly in the oxygen p bands.[205] Examples are

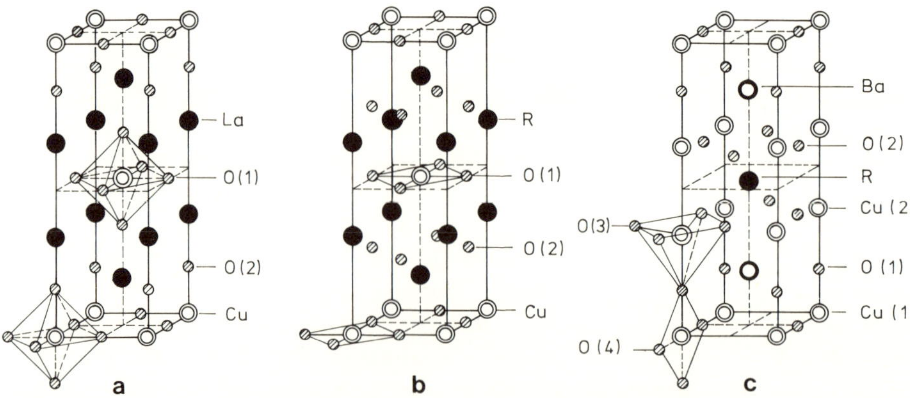

Figure 11. Crystal structures of some high-T_c superconducting cuprates. (a) High-temperature tetragonal structure of La_2CuO_4 (T structure). (b) Tetragonal structure of R_2CuO_4, R = Pr, Nd, Sm, Eu, and Gd (T′ structure). The difference between T′ and T structures lies in the positions of the O(2) atoms. (c) Orthorhombic structure of $RBa_2Cu_3O_7$, R = Y, rare earth. By heat treatment in a reducing atmosphere, O(4) atoms are removed from the Cu(1)–O(4) chains. The compound $RBa_2Cu_3O_6$ thus obtained is an insulator with antiferromagnetic ordering of the Cu(2) moments below $T_N^{Cu} \approx 430$ K.

$(La_{1-x}Sr_x)_2CuO_4$, doped through replacement of some La^{3+} by Sr^{2+} and $YBa_2Cu_3O_{6+y}$, doped through introduction of O^{2-} beyond $y \sim 0.5$. In the class of compounds $(R_{1-x}Ce_x)_2CuO_4$, R = Pr, Nd, Sm, Eu, however, replacement of trivalent rare earths by Ce ions with a valence close to 3.5+ introduces mobile electrons in the 3*d* bands of copper.[206]

In high-T_c cuprates, coexistence of superconductivity and magnetism is a common phenomenon. In the insulating parent compounds, La_2CuO_4, $YBa_2Cu_3O_6$, and R_2CuO_4, antiferromagnetic order of the copper moments occurs at high Néel temperatures, T_N^{Cu}, between about 250 and 400 K. At low doping levels, the Néel temperature decreases rapidly with increasing dopant concentration. However, magnetic fluctuations of copper moments have been observed at high dopant concentrations, even in the superconducting state.[207] Furthermore, many of the doped high-T_c cuprates contain magnetic rare earths as structural constituents. The rare-earth moments have no adverse effect on the superconducting state, and antiferromagnetic order of the rare-earth moments is found in coexistence with superconductivity. Thus, for example, the compounds $RBa_2Cu_3o_7$, which are isostructural with $YBa_2Cu_3O_7$, are formed with all rare earths, R, and with the exception of the compounds with Ce, Pr, and Tb,[208] the superconducting transition temperature is near or above 90 K.[209] Ordering temperatures, T_N^R, of the rare-earth moments are quite low and on the order of 1 K.

Nuclear methods have been employed in many investigations of high-T_c materials. Magnetic structures have been determined by neutron diffraction,[210–216] and inelastic neutron scattering has been applied to observations of spin fluctuations and magnetic excitations.[207] From measurements of internal fields by muon spin rotation, information on the magnetic penetration depth has been obtained.[217] Valuable clues concerning the electron structure and fluctuations of Cu moments have been derived from the temperature dependence of relaxation rates and Knight shifts in NMR and NQR spectroscopy.[9,10,218]

Investigations by Mössbauer spectroscopy can be classified according to the sites of the probe nuclei in the cuprates:

1. Mössbauer spectroscopy with nuclei of atoms which are substituted in Cu sites, mainly ^{57}Fe, ^{119}Sn, and ^{67}Zn
2. Mössbauer spectroscopy with rare-earth nuclei in sites which are intrinsically occupied by rare-earth atoms.

Mössbauer spectroscopy with ^{155}Gd, to be discussed here, clearly belongs to the second class. Other rare-earth nuclei employed in studies of high-T_c cuprates by Mössbauer spectroscopy are (only a few references can be given here): ^{151}Eu,[219–224] ^{161}Dy,[221,225,226] ^{166}Er,[227] ^{169}Tm,[226] and ^{170}Yb.[228]

When it became known that the transition temperature to superconductivity in $GdBa_2Cu_3O_7$ was above 90 K, several groups more or less simultaneously turned to studies of this compound by ^{155}Gd Mössbauer spectroscopy.[73,74,81,221,229–232] On essential points, the results reported by all groups are in mutual agreement, with some minor discrepancies resulting from different approaches in the analysis of the spectra. All groups obtain isomer shifts $\delta_{IS} \approx 0.5$ mm/s, close to the shift in Gd_2O_3, and field gradients $V_{zz} \approx -6 \times 10^{21}$ V/m^2, derived from the observed quadrupole splitting with the value $Q_0 = 1.3b$ for the ground-state quadrupole moment. Wortmann *et al.*[73,221] and Smit *et al.*[229] quote smaller values for V_{zz} as they have used the old value $Q_0 = 1.59b$. After correction for the different moment values, their results agree well with those of the other groups. For the asymmetry parameter, η, values in the range $0.5 < \eta < 0.6$ are quoted by most groups. Wortmann *et al.*[73,221] obtain $\eta \approx 0.4$. This is a consequence of the Lorentz approximation used by them in the analysis of the spectra as demonstrated in Table 3.

Magnetic ordering of Gd moments at $T_N^{Gd} = 2.25$ K is indicated by a sharp peak in the specific heat.[74,233] This is confirmed by the magnetic hyperfine splitting in the ^{155}Gd Mössbauer spectra below 2.25 K. The direction of the hyperfine field was reported to be parallel to the crystallographic *c* axis, in agreement with the direction of the Gd moments deduced from neutron diffraction.[210] This result is of some importance as it proves exchange interactions between Gd moments to be effective in the superconducting state. For purely dipolar interactions, the moment direction is perpendicular to the *c* axis in the configuration of lowest energy.[74] Only Alp *et al.*[230] deduced a field direction perpendicular to the *c* axis. In their spectra, the signal-to-noise ratio is lower by an order of magnitude because the depth of the absorption dip is much smaller than that in the spectra of other groups. This is an illuminating example for the importance of choosing the appropriate absorber thickness, emphasized in Section 4.1.

Spectra of the oxygen-reduced, nonsuperconducting compounds, $GdBa_2Cu_3O_{6+\delta}$ with $\delta < 0.5$, gave similar results as the spectra of $GdBa_2Cu_3O_7$.[74,81,221,229] The isomer shift is slightly more positive, by about 0.01 mm/s, and the absolute value of V_{zz} is smaller ($\approx -5.3 \times 10^{21}$ V/m^2). Wortmann *et al.* report $\eta \approx 0$, as expected for the tetragonal crystal structure of $GdBa_2Cu_3O_6$. The other groups obtain $\eta \approx 0.3$. In this case, the difference may well be a consequence of different degrees of reduction, that is, of different values of the oxygen excess, δ, in the formula given above. Also, in the reduced compound, the ordering temperature, T_N^{Gd}, is very close to that in $GdBa_2Cu_3O_7$,[74,221] and the moment direction is parallel to the *c* axis, in agreement with the result of neutron diffraction.[211,212]

Both for the superconducting and for the reduced compound, the highly two-dimensional character of the magnetic interactions is indicated by the following observations:[74,221] (1) considerable line broadening due to slow magnetic relaxation over an extended temperature range above T_N^{Gd}, and (2) the temperature dependence of the magnetic hyperfine field below T_N^{Gd} is well described by $B_{hf} \propto (1 - T/T_N^{Gd})^\beta$, with $0.09 \leq \beta \leq 0.15$. Muon spin rotation

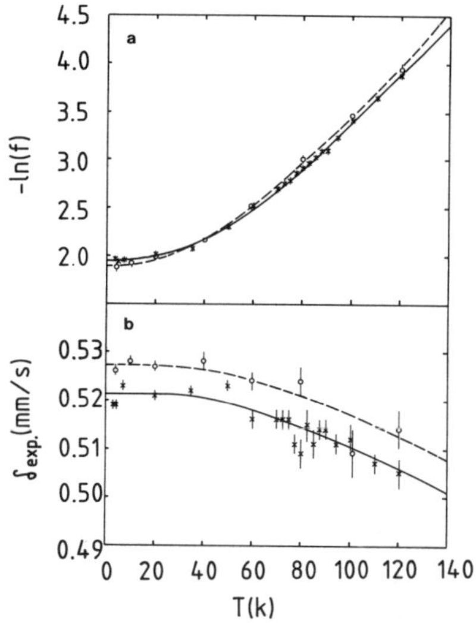

Figure 12. Temperature dependence (a) of the recoilless fraction f and (b) of the center shift, δ_{exp}, derived from Mössbauer spectra of ^{155}Gd in $GdBa_2Cu_3O_7$ (crosses, solid lines), and in $GdBa_2Cu_3O_{6+\delta}$, $\delta < 0.5$ (circles, dashed lines). The lines result from fits of $\ln f$ in terms of the Debye plus Einstein model described in Section 4.2. (From Ref. 81.)

experiments yielded a temperature dependence for the hyperfine field at the muon sites in good agreement with that observed in ^{155}Gd Mössbauer spectra.[232]

The temperature dependence of the recoilless fraction, $f(T)$, derived from Mössbauer spectra of both $GdBa_2Cu_3O_7$ and $GdBa_2Cu_3O_{6+\delta}$, over the temperature range up to 120 K (Fig. 12), was analyzed in terms of the Debye plus Einstein model described in Section 4.2.[81] The Debye temperature in the reduced compound is lower than that in $GdBa_2Cu_3O_7$. This indicates softening of acoustic vibrations resulting from missing bonds due to the removal of oxygen atoms from the Cu(1)–O chains in the reduced compound. The frequency of optical vibrations, corresponding to the Einstein temperature, however, is lower in $GdBa_2Cu_3O_7$ than in the oxygen-deficient material. This is presumably caused by the screening of the ionic potentials by the conduction electrons, a screening which results in the softening of optical vibrations. The opposite sign of the difference between Einstein and Debye temperatures in the two samples is reflected in the crossing of the $f(T)$ curves obtained for the two compounds (Fig. 12a). Recent studies of two compounds, $HoBa_2Cu_3O_{6+\delta}$ with $\delta \approx 0.5$ and 1, by neutron diffraction in the temperature range from 67 to 300 K yield differences between the mean square amplitudes of Ho vibrations in the two samples,[234] differences which are in good agreement with the results derived from ^{155}Gd Mössbauer spectroscopy.

In contrast to results obtained by Mössbauer spectroscopy with ^{151}Eu,[220,222] Mössbauer spectra of ^{155}Gd give no indication of anomalies in the temperature dependence of δ_{IS} or V_{zz} near T_c. Because of the low sensitivity of ^{155}Gd Mössbauer spectra to changes of the electronic density, a small discontinuity in δ_{IS} as reported by Wortmann *et al.*[220] would hardly be detectable in the case of ^{155}Gd. On the other hand, ^{155}Gd is the more sensitive probe for changes in V_{zz}. An anomaly in the temperature dependence of V_{zz}, as reported by Jin *et al.*,[222] would certainly be observable with ^{155}Gd, unless the anomaly is entirely associated with the 4*f* contribution to V_{zz}, which is negligible in the case of ^{155}Gd.

Supplementary information on the magnetic hyperfine field below T_N^{Gd} has been derived from Mössbauer spectra of ^{151}Eu in $Eu_{0.1}Gd_{0.9}Ba_2Cu_3O_7$.[219] In these compounds, Eu is in the nonmagnetic trivalent state. Thus, the observed magnetic hyperfine field, which saturates near 3T at zero temperature, can be identified with the transferred hyperfine field, B_N, defined in Eq. (10), Section 3.4. In the case of ^{155}Gd, a value of B_N close to that determined for ^{151}Eu can be assumed.

When impurities are substituted on copper sites, the transition temperature, T_c, decreases quite rapidly with increasing impurity concentration, but different rates of T_c-depression for a given impurity have been reported by different groups.[235] Results for the iron-doped samples $GdBa_2(Cu_{1-x}Fe_x)O_7$, prepared under a variety of conditions and preparation parameters,[236] suggest that relatively high values of T_c are obtained when the iron atoms on the Cu(1) sites segregate into linear chains along the (110) directions. The existence of linear iron clusters aligned along the (110) directions has been inferred from neutron diffraction studies.[237]

Results of a systematic investigation of iron-doped compounds, with iron concentrations up to $x = 0.25$, by ^{155}Gd Mössbauer spectroscopy show that variations of the distribution of iron atoms over the Cu(1) sites do not influence the hyperfine interactions of the Gd nuclei.[238] The isomer shift is completely insensitive to the presence of iron impurities. The absolute value of V_{zz} slowly decreases with increasing iron concentration (Fig. 13a). The asymmetry parameter drops rapidly for $x \leq 0.06$, but it does not vanish throughout the range of iron concentrations studied (Fig. 13b), although X-ray diffraction patterns indicate tetragonal symmetry for all samples with $x > 0.03$. Obviously, oxygen order in the Cu(1)–O chains is maintained, but the size of the twin domains becomes so small that this order escapes observation by X-ray diffraction.

An interesting system is the series of mixed compounds $(Gd_{1-x}Pr_x)Ba_2Cu_3O_7$. No transition to superconductivity occurs in the pure Pr compound ($x = 1$).[239] Initially, the cause was suspected to be a tetravalent state of Pr and the filling of mobile holes in the oxygen *p* bands by the additional electrons introduced into the system. However, results of studies by several spectroscopic methods indicate that Pr is trivalent,[240] and the number of holes in the oxygen *p* bands is not reduced.[241] Obviously, the unique properties of $PrBa_2Cu_3O_7$, as compared with those of the compounds with other rare earths, are due to strong hybridization of the Pr 4*f* states with the oxygen *p* states.[241] The hybridization furthermore leads to strong exchange

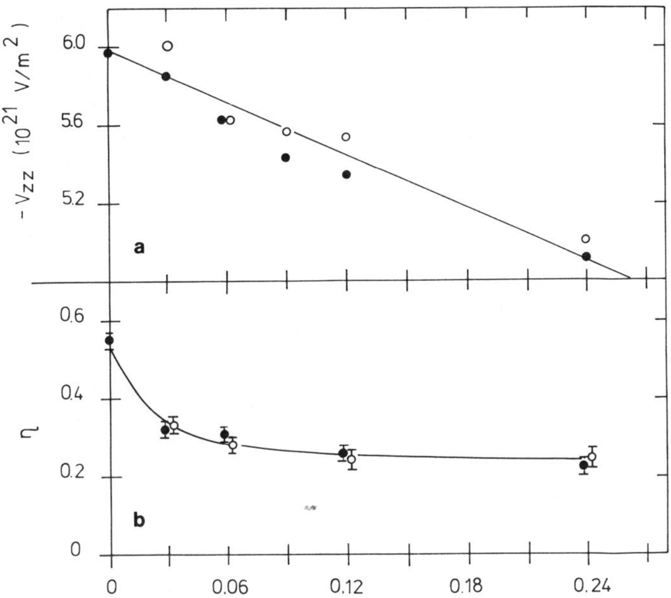

Figure 13. Electric field gradient, V_{zz}, and asymmetry parameter, η, for ^{155}Gd in $GdBa_2(Cu_{1-x}Fe_x)_3O_7$, as a function of the iron concentration, x. Lines are guides to the eye.

interactions between Pr moments, evidenced by a remarkably high ordering temperature, $T_N^{Pr} \approx 17$ K.[213]

Substitution of Pr for Gd in $(Gd_{1-x}Pr_x)Ba_2Cu_3O_7$ rapidly depresses T_c, and no transition to superconductivity is found for Pr concentrations $x \geq 0.45$.[242] The quadrupole splitting in ^{155}Gd Mössbauer spectra decreases slowly with increasing Pr content. In contrast, the quadrupole splitting increases rapidly at low concentrations when Ca^{2+} is substituted for Gd^{3+}. Tomkowicz *et al.* interpret these results as further evidence for a Pr valence near 3+.[242]

The ordering of the Pr moments in $Pr_{0.925}Gd_{0.075}Ba_2Cu_3O_{7-\delta}$ was monitored by observation of the magnetic hyperfine splitting of the ^{155}Gd Mössbauer spectra.[243] For an oxygen-rich sample ($\delta \approx 0.1$), line broadening due to magnetic splitting was found to set in near 16 K, in good agreement with the Néel temperature, T_N^{Pr}, reported in the literature.[213] In a reduced sample ($\delta \approx 0.9$), T_N^{Pr} is shifted to a much lower temperature near 7.5 K.[243]

Mössbauer spectra of ^{155}Gd in $GdBa_2Cu_4O_8$ ($T_c \approx 74$ K), a compound whose structure is closely related to that of $GdBa_2Cu_3O_7$,[244] yielded results which are quite similar to those for ^{155}Gd in $GdBa_2Cu_3O_7$.[245] The electric field gradient, $V_{zz} = (-6.34 \pm 0.01) \times 10^{21}$ V/m^2, has a slightly larger value. However, the asymmetry parameter, $\eta = 0.37 \pm 0.07$, is smaller than in

$GdBa_2Cu_3O_7$, in accordance with a smaller difference between the values of the lattice parameters a and b.[245]

Several compounds of different stoichiometric composition are formed by combined oxides of Bi, Sr, Ca, and Cu.[246] At low temperatures, these oxides are superconductors. The compound with the highest transition temperature in this system, $T_c \approx 120$ K, has the composition $Bi_2Sr_2Ca_2Cu_3O_{10}$.[202] A study of effects on bulk properties, due to doping other elements into these compounds, gave results which indicated that Gd can be substituted both in Ca and in Bi sites.[247] In a subsequent investigation of compounds of nominal composition $Bi_{4-x}Gd_xSr_3Ca_3Cu_4O_z$, with $x = 0.3$, 0.5, and $Bi_2Sr_2Ca_{1-y}Gd_yCu_2O_z$, with $y = 0.25$, 0.5, by ^{155}Gd Mössbauer spectroscopy, the spectra of all samples were found to be dominated by a component with a quadrupole splitting corresponding to a field gradient, $|V_{zz}| \approx 10^{22}$ V/m^2.[248] A second component of small relative intensity (≈6%), but with a large quadrupole splitting, $|V_{zz}| \approx 2 \times 10^{22}$ V/m^2, appeared only in the spectrum of $Bi_{3.5}Gd_{0.5}Sr_2Ca_3Cu_4O_z$. The conclusion from these results is a strong preference of Gd to occupy the Ca sites in the Bi-based cuprates.[248]

Preliminary results of an ongoing investigation of mixed compounds $(Gd_{1-x}R_x)_2CuO_4$, R = Pr, Nd, Sm, and Eu, all with the T′ structure (Fig. 11b), can be reported. All compounds R_2CuO_4, except for that with R = Gd, become superconductors at low temperatures when doped with some Ce on the rare-earth sites. The transition temperature for the doped materials, $R_{1.85}Ce_{0.15}CuO_4$, is in the range between 20 and 25 K for R = Pr, Nd, Sm, and ≈13 K for R = Eu.[249] Apparently, magnetic moments of the rare earths have no influence on the transition to superconductivity:[250] trivalent Eu is nonmagnetic, and the crystal-field ground state of Pr is a nonmagnetic singlet; in contrast, Sm moments in Sm_2CuO_4 order at a surprisingly high temperature, $T_N^{Sm} = 6$ K. In the case of Nd_2CuO_4, a maximum in the specific heat, C_p, at 1.65 K was initially interpreted as an indication of a Néel temperature T_N^{Nd}.[250] However, there is now ample experimental evidence that the C_p maximum is a Schottky anomaly, and that the magnetic order of Nd moments is induced by Nd–Cu exchange coupling.[60,214,215] In Gd_2CuO_4, displacements of O(1) atoms from the sites in the ideal structure, in a direction perpendicular to the c axis and to the Cu–O(1) bonds, lead to an orthorhombic distortion of the crystal structure.[60] A consequence of this distortion is weak ferromagnetism caused by antisymmetric exchange interactions[251] between the Cu moments.[252]

The different magnetic characters of the compounds are reflected in the concentration dependence of the magnetic hyperfine field, B_{hf}, at the ^{155}Gd nuclei in the mixtures $(Gd_{1-x}R_x)_2CuO_4$. For R = Pr, Nd, and Eu, the hyperfine field, measured at 1.6 K, decreases monotonically with increasing x. For $x > 0.5$, $B_{hf}(1.6\,K)$ saturates at a constant value near 7T in the Nd series,[60] and near 3T in the mixtures with Pr and Eu. The cause of the saturation is possibly an exchange coupling between the Gd and Cu moments. In $(Gd_{1-x}Sm_x)_2CuO_4$, in contrast, both the ordering temperature of the rare-earth moments and B_{hf} go through a minimum at the Sm concentration $x = 0.25$.

Changes of the lattice parameters, *a* and *c*, with concentration *x* correspond approximately to the differences in the ionic radii of the rare earths. The variations in the unit cell volume and the axial ratio $c/3a$, which are relevant for the electric hyperfine interactions, are displayed in Fig. 14a and b, respectively. However, the changes of the quadrupole splitting, S_Q (Fig. 14c), and of the isomer shift, δ_{IS} (Fig. 14d), with *x* do not correspond to the trends in the structural parameters. Nearly identical results for S_Q and δ_{IS} are obtained in mixtures with Pr, Nd, and Sm, in spite of considerable differences of the structural parameters. In the series with Eu, changes with *x* are much smaller.

Thus, the electric hyperfine interactions at the ^{155}Gd nuclei and their variation with concentration in the mixed series are *not* determined primarily

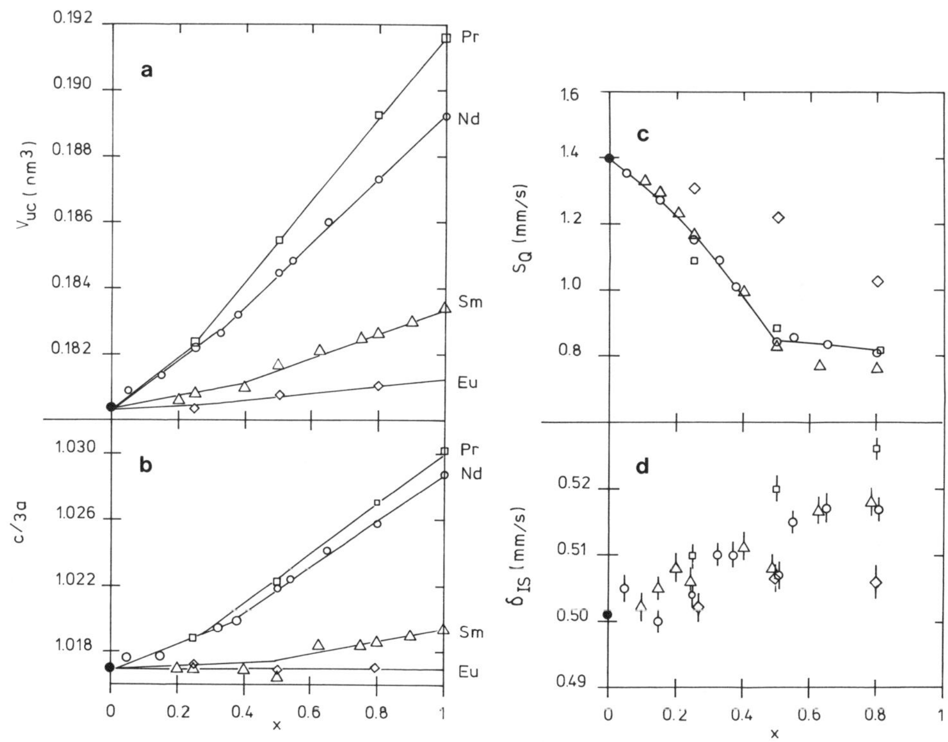

Figure 14. Concentration dependence of the structural parameters and the electric hyperfine interactions at the ^{155}Gd nuclei in mixed compounds $(Gd_{1-x}R_x)_2CuO_4$, R = Pr (□), Nd (○), Sm (△), and Eu (◇). Black circles mark the results for Gd_2CuO_4. (a) Volume of the unit cell; (b) axial ratio $c/3a$; (c) ground-state quadrupole splitting, S_Q, defined in Eq. (3); (d) isomer shift. Lines are guides to the eye.

by structural parameters. Rather, the dominant cause for the electric hyperfine interactions must be sought in the local electronic charge distribution near the rare-earth atoms, which appears to be quite similar in the compounds with Pr, Nd, and Sm. This could be of some interest as the data show a correlation with the transition temperatures in the Ce-doped materials. The explanation of these experimental results will be a challenge for theoretical studies of electronic charge distributions and the resulting hyperfine interactions and crystal fields in these compounds.

8. OUTLOOK

After having reported about studies of solid materials by Mössbauer spectroscopy with ^{155}Gd over a period of almost exactly 30 years, I want to consider some of the directions one can anticipate for future work.

In view of the intense activities in the fields of materials for permanent magnets and of high-temperature superconductors, one can be sure of the emergence of new compounds for which information on properties on the atomic scale, derived from Mössbauer spectra, will be of interest. Some new areas which have up to now been out of reach of this technique are easily identified. These include:

1. Artificial multilayer-superstructures. Thin films of Gd/Y and Gd/(other rare earth) multilayers have been made and studied by various techniques.[253] Since the film thickness hardly ever exceeds 1 μm, a large number of such films must be stacked together for absorbers whose thickness is sufficient to yield Mössbauer spectra of good quality. Because of the high stability of the Gd 4f moments, results which differ significantly from those obtained with ^{57}Fe in multilayer structures[254] may be expected.
2. Quasicrystals. Apparently, no crystalline compounds containing rare earths have been produced to date. However, there is no compelling reason which would preclude the possibility of their existence. Quasicrystals of Pd–U–Si alloys have been made,[255] and, in view of the chemical similarity of U and rare earths, attempts to combine rare earths with Pd and Si seem a promising perspective. Mössbauer spectroscopy with ^{155}Gd and with other rare-earth nuclei would certainly yield valuable information on the local symmetry at the rare-earth sites and on the resulting crystal fields. Possibly, some of the questions which are left unanswered[256] by Mössbauer spectra of ^{57}Fe may be resolved by Mössbauer spectroscopy with rare-earth nuclei.
3. Fullerenes, hollow all-carbon cage molecules, C_{2n}, $n \geq 16$, which contain rare earths. Fullerenes enclosing La have been produced,[257] although in very small quantities. However, progress in the new branch

of fullerene chemistry has been very rapid during the last few years, and one may expect that studies of rare earths enclosed in fullerenes by Mössbauer spectroscopy will be realized in the near future.

It is remarkable that only 10 years ago, no one would have predicted any of the topics listed here since they concern materials which became known only during the course of this time span. The same holds for $Nd_2Fe_{14}B$-type compounds and for high-T_c superconducting cuprates, the foremost topics of the present chapter. Thus, there is a high probability that a similar review, written 10 years from now, will report on studies of materials which at present are beyond our imagination.

ACKNOWLEDGMENTS. I want to use this occasion to express my warm thanks to Felix E. Obenshain, who taught me all of the skills needed by a Mössbauer spectroscopist, to Jörg Fink, who has installed the highly efficient Mössbauer laboratory at Kernforschungszentrum Karlsruhe, and to Krzysztof Tomala, whose contributions to several aspects of our operations, notably to the programs employed in analyzing the Mössbauer spectra, have been of great value.

REFERENCES

1. C. L. Herzenberg, L. Meyer-Schützmeister, L. L. Lee, Jr., and S. S. Hanna, *Bull. Am. Phys. Soc.* **7**, 39 (1962).
2. R. L. Mössbauer, *Z. Phys.* **151**, 124 (1958); *Naturwissenschaften* **45**, 538 (1958); *Z. Naturforsch.* **14a**, 211 (1959).
3. J. G. Bednorz and K. A. Müller, *Z. Phys. B Condens. Matter* **64**, 189 (1986).
4. D. B. Prowse, A. Vas, and J. D. Cashion, *J. Phys. D* **6**, 646 (1973).
5. S. Ofer, I. Nowik, and S. G. Cohen, in *Chemical Applications of Mössbauer Spectroscopy,* V. I. Goldanskii and R. H. Herber, eds., p. 427, Academic Press, New York (1968).
6. S. P. Taneja and C. W. Kimball, in *Advances in Mössbauer Spectroscopy,* B. V. Thosar and P. K. Iyengar, eds., p. 814, Elsevier, Amsterdam (1983).
7. E. R. Bauminger, G. M. Kalvius, and I. Nowik, in *Mössbauer Isomer Shifts,* G. K. Shenoy and F. E. Wagner, eds., p. 661, North-Holland, Amsterdam (1978).
8. E. Dormann, in *Handbook on the Physics and Chemistry of Rare Earths,* K. A. Gschneidner, Jr., and L. Eyring, eds., Vol. 14, p. 63, North-Holland, Amsterdam (1991).
9. K.-I. Kumagai, in *Electronic Properties and Mechanicms of High T_c Superconductors,* T. Oguchi, K. Kadowaki, and T. Sasaki, eds., p. 109, North-Holland, Amsterdam (1992).
10. M. Takigawa, in *Electronic Properties and Mechanisms of High T_c Superconductors,* T. Oguchi, K. Kadowaki, and T. Sasaki, eds., p. 123, North-Holland, Amsterdam (1992).
11. J. Fink and P. Kienle, *Phys. Lett.* **17**, 326 (1965).
12. J. Fink, *Z. Phys.* **207**, 225 (1967).
13. K. E. Rehm, W. Henning, and P. Kienle, *Phys. Rev. Lett.* **22**, 790 (1969).
14. J. Eck, Y. K. Lee, E. T. Ritter, R. R. Stevens, Jr., and J. C. Walker, *Phys. Rev. Lett.* **17**, 120 (1966).

15. R. R. Stevens, Jr., Y. K. Lee, and J. C. Walker, *Phys. Lett.* **21**, 401 (1966); R. R. Stevens, Jr., J. S. Eck, E. T. Ritter, Y. K. Lee, and J. C. Walker, *Phys. Rev.* **158**, 1118 (1967).
16. P. B. Russell, G. D. Sprouse, G. L. Latshaw, S. S. Hanna, and G. M. Kalvius, *Phys. Lett.* **32B**, 35 (1970).
17. R. Sielemann, M. Menningen, B. Keck, G. Vogl, and Y. Yoshida, *Hyp. Interact.* **56**, 1643 (1990); S. Laubach, P. Schwalbach, M. Hartick, E. Kankeleit, B. Keck, and R. Sielemann, *Hyp. Interact.* **53**, 75 (1990).
18. W. Henning, *Z. Phys.* **217**, 438 (1968).
19. H. Armon, E. R. Bauminger, A. Diamant, I. Nowik, and S. Ofer, *Nucl. Phys.* **A233**, 385 (1974).
20. V. I. Goldanskii, G. M. Gorodinskii, S. V. Karyagin, L. A. Korytko, L. M. Krizhanskii, E. F. Makarov, I. P. Suzdalev, and V. V. Khrapov, *Proc. Acad. Sci. USSR Phys. Chem. Sect.* **147**, 766 (1963); S. V. Karyagin, *Dokl. Akad. Nauk SSSR* **148**, 1102 (1963).
21. H. Armon, E. R., Bauminger, and S. Ofer, *Phys. Lett.* **43B**, 380 (1973).
22. H. Armon, E. R. Bauminger, A. Diamant, I. Nowik, and S. Ofer, *Solid State Commun.* **15**, 543 (1974).
23. J. Göring and R. Herbig, *Z. Naturforsch.* **72a**, 1684 (1972).
24. H. Prange and P. Kienle, *Phys. Lett.* **23**, 681 (1966).
25. H. Prange, *Z. Phys.* **212**, 415 (1968).
26. J. Göring, *Z. Phys.* **251**, 185 (1972).
27. H. Blumberg, B. Persson, and M. Bent, *Phys. Rev.* **170**, 1076 (1968).
28. P. J. Unsworth, *J. Phys. B* **2**, 122 (1969).
29. Y. Tanaka, D. B. Laubacher, R. M. Steffen, E. B. Shera, H. D. Wohlfahrt, and M. V. Hoehn, *Phys. Lett.* **108B**, 8 (1982).
30. K. Tomala, in *Hyperfine Interactions,* R. Kulessa and K. Krolas, eds., p. 71, Jagiellonian University, Cracow (1979).
31. C. Sauer, E. Matthias, and R. L. Mössbauer, *Phys. Rev. Lett.* **21**, 961 (1968).
32. G. T. Trammel and J. P. Hannon, *Phys. Rev.* **180**, 337 (1969).
33. Y. M. Kagan, A. M. Afanas'ev, and V. K. Voitovetskii, *JETP Lett.* **9**, 91 (1969).
34. R. A. Meyer and J. W. T. Meadows, *Nucl. Phys.* **A132**, 177 (1969).
35. W. Henning, G. Baehre, and P. Kienle, *Phys. Lett.* **31B**, 203 (1970).
36. J. D. Cashion, D. B. Prowse, and A. Vas, *J. Phys. C* **6**, 2611 (1973).
37. J. D. Cashion, M. A. Coulthard, and D. B. Prowse, *J. Phys. C* **7**, 3620 (1974).
38. J. D. Cashion, M. A. Coulthard, and D. B. Prowse, *J. Phys. C* **8**, 1267 (1975).
39. G. K. Shenoy and B. D. Dunlap, in *Mössbauer Isomer Shifts,* G. K. Shenoy and F. E. Wagner, eds., p. 869, North-Holland, Amsterdam (1978).
40. J. G. Stevens and W. L. Gettys, in *Mössbauer Isomer Shifts,* G. K. Shenoy and F. E. Wagner, eds., p. 901, North-Holland, Amsterdam, (1978).
41. T. E. Katila, V. K. Typpi, G. K. Shenoy, and L. Niinistö, *Solid State Commun.* **11**, 1147 (1972).
42. H. A. Buskes and J. D. Cashion, *J. Phys. (Paris) Colloq.* **35** (C6), 221 (1974).
43. G. M. Kalvius, U. F. Klein, and G. Wortmann, *J. Phys. (Paris) Colloq.* **35** (C6), 139 (1974).
44. J. C. Love, F. E. Obenshain, and G. Czjzek, *Phys. Rev. B* **3**, 2827 (1971).
45. W. L. Gettys and J. G. Stevens, in *CRC Handbook of Spectroscopy,* J. W. Robinson, ed., Vol. III, p. 514, CRC Press, Boca Raton, Fla. (1981).
46. B. D. Dunlap, in *Mössbauer Effect Data Index,* J. G. Stevens and V. E. Stevens, eds., p. 25, Plenum Press, New york (1970).
47. K. D. Sen, P. C. Schmidt, and A. Weiss, *Z. Naturforsch.* **41a**, 37 (1986).
48. E. N. Kaufmann and R. J. Vianden, *Rev. Mod. Phys.* **51**, 161 (1979).
49. R. Coehoorn, K. H. J. Buschow, M. W. Dirken, and R. C. Thiel, *Phys. Rev. B* **42**, 4645 (1990).
50. K. Schwarz, C. Ambrosch-Draxl, and P. Blaha, *Phys. Rev. B* **42**, 2051 (1990).
51. C. Ambrosch-Draxl, P. Blaha, and K. Schwarz, *Phys. Rev. B* **44**, 5141 (1991).

52. W. A. Barton and J. D. Cashion, *J. Phys. C* **12**, 2897 (1979).
53. R. E. Watson and A. J. Freeman, *Phys. Rev.* **135**, A1209 (1964).
54. A. Abragam and B. Bleaney, *Electron Paramagnetic Resonance of Transition Ions,* Clarendon Press, Oxford (1970).
55. R. P. Gupta and S. K. Sen, *Phys. Rev. A* **7**, 850 (1973).
56. R. G. Barnes, R. L. Mössbauer, E. Kankeleit, and J. M. Poindexter, *Phys. Rev.* **136**, A175 (1964).
57. J. Pelzl, S. Hüfner, and S. Scheller, *Z. Phys.* **231**, 377 (1970).
58. M. Bogé, G. Czjzek, D. Givord, C. Jeandey, H. S. Li, and J. L. Oddou, *J. Phys. F* **16**, L67 (1986).
59. G. Czjzek, V. Oestreich, H. Schmidt, K. Łątka, and K. Tomala, *J. Magn. Magn. Mater.* **79**, 42 (1989).
60. P. Adelmann, R. Ahrens, G. Czjzek, G. Roth, H. Schmidt, and C. Steinleitner, *Phys. Rev. B* **46**, 3619 (1992).
61. A. T. Boothroyd, S. M. Doyle, D. McK. Paul, and R. Osborn, *Phys. Rev. B* **45**, 10075 (1992).
62. J. Van Dongen Torman, R. Jagannathan, and J. M. Trooster, *Hyp. Interact.* **1**, 135 (1975).
63. R. E. Watson and A. J. Freeman, in *Hyperfine Interactions,* A. J. Freeman and R. B. Frankel, eds., p. 53, Academic Press, New York (1967).
64. R. E. Gegenwarth, J. L. Budnick, S. Skalski, and J. H. Wernick, *Phys. Rev. Lett.* **18**, 9 (1967).
65. G. Longworth, in *Advances in Mössbauer Spectroscopy,* B. V. Thosar and P. K. Iyengar, eds., p. 122, Elsevier, Amsterdam (1983).
66. T. E. Cranshaw, in *Mössbauer Spectroscopy Applied to Inorganic Chemistry,* G. J. Long, ed., Vol. 1, p. 27, Plenum Press, New York (1984).
67. A. E. Balabanov, N. N. Delyagin, and H. el Sayes, *Sov. J. Nucl. Phys.* **3**, 150 (1966).
68. E. R. Bauminger, D. Froindlich, A. Mustachi, I. Nowik, S. Ofer, and S. Samuelov, *Phys. Lett.* **30B**, 531 (1969).
69. G. J. Long, T. E. Cranshaw, and G. Longworth, *Mössbauer Eff. Ref. Data J.* **6**, 42 (1983).
70. G. K. Shenoy, J. M. Friedt, H. Maletta, and S. L. Ruby, in *Mössbauer Effect Methodology,* I. J. Gruverman, C. W. Seidel, and D. K. Dieterly, eds., Vol. 9, p. 277, Plenum Press, New York (1974).
71. D. C. Cook and J. D. Cashion, *Hyp. Interact.* **5**, 479 (1978).
72. D. B. Prowse and J. D. Cashion, *Phys. Status Solidi A* **22**, 631 (1974).
73. G. Wortmann, C. T. Simmons, and G. Kaindl, *Solid State Commun.* **64**, 1057 (1987); G. Wortmann, A. Kolodziejczyk, C. T. Simmons, and G. Kaindl, *Physica C* **153–155**, 1547 (1988).
74. C. Meyer, H. J. Bornemann, H. Schmidt, R. Ahrens, D. Ewert, B. Renker, and G. Czjzek, *J. Phys. F* **17**, L345 (1987).
75. P. C. M. Gubbens, A. M. van der Kraan, and K. H. J. Buschow, *Hyp. Interact.* **53**, 37 (1990).
76. G. Czjzek and W. G. Berger, *Phys. Rev. B* **1**, 957 (1970).
77. R. M. Housley, U. Gonser, and R. W. Grant, *Phys. Rev. Lett.* **20**, 1279 (1968); R. M. Housley, R. W. Grant, and U. Gonser, *Phys. Rev.* **178**, 514 (1969).
78. G. Czjzek, H. J. Bornemann, and H. S. Li, *J. Magn. Magn. Mater.* **80**, 23 (1989).
79. U. Gonser and H. Fischer, in *Mössbauer Spectroscopy II,* U. Gonser, ed., p. 99, Springer, Berlin (1981).
80. J. K. Srivastava, S. C. Bhargava, P. K. Iyengar, and B. V. Thosar, in *Advances in Mössbauer Spectroscopy,* B. V. Thosar and P. K. Iyengar, eds., p. 1, Elsevier, Amsterdam (1983).
81. H. J. Bornemann, G. Czjzek, D. Ewert, C. Meyer, and B. Renker, *J. Phys. F* **17**, L337 (1987).

82. R. M. Housley and F. Hess, *Phys. Rev.* **146**, 517 (1966).
83. R. Brusetti, J. M. D. Coey, G. Czjzek, J. Fink, F. Gompf, and H. Schmidt, *J. Phys. F* **10**, 33 (1980).
84. F. J. Shore, C. A. Reynolds, V. L. Sailor, and G. Brunhart, *Phys. Rev.* **138**, B1361 (1965).
85. B. Persson, H. Blumberg, and M. Bent, *Phys. Lett.* **27A**, 189 (1968).
86. H. Zmora, M. Blau, and S. Ofer, *Phys. Lett.* **28A**, 668 (1969).
87. Y. Seiwa, T. Tsuda, A. Hirai, and C. W. Searle, *Phys. Lett.* **43A**, 23 (1973); J. Durand and C. Robert, *J. Phys. F* **3**, L90 (1973).
88. E. R. Bauminger, A. Diamant, I. Felner, I. Nowik, and S. Ofer, *Phys. Rev. Lett.* **34**, 962 (1975).
89. J. W. Cable and E. O. Wollan, *Phys. Rev.* **165**, 733 (1968).
90. P.-A. Lindgård, *Phys. Rev. B* **14**, 4074 (1976); **16**, 2168 (1977).
91. E. A. Görlich, K. Łątka, R. Kmieć, and K. Tomala, *J. Phys. (Paris) Colloq.* **40** (C2), 654 (1979).
92. I. Nowik, in *Proceedings International Conference on Mössbauer Spectroscopy,* A. Z. Hrynkiewicz and J. A. Sawicki, eds., Vol. 2, p. 83, Jagiellonian University, Cracow (1975).
93. T.-W. E. Tsang, K. A. Gschneidner, Jr., F. A. Schmidt, and D. K. Thome, *Phys. Rev. B* **31**, 235 (1985).
94. D. Wohlleben, *Phys. Rev. Lett.* **21**, 1343 (1968).
95. K. Moorjani and J. M. D. Coey, *Magnetic Glasses,* Elsevier, Amsterdam (1984).
96. G. Czjzek, *Nucl. Instrum. Methods* **199**, 37 (1982), *Hyp. Interact.* **24–26**, 667 (1985).
97. G. S. Cargill III, *Solid State Phys.* **30**, 227 (1975).
98. P. H. Gaskell, *Nucl. Instrum. Methods* **199**, 45 (1982).
99. G. Czjzek, J. Fink, F. Götz, H. Schmidt, J. M. D. Coey, J.-P. Rebouillat, and A. Liénard, *Phys. Rev. B* **23**, 2513 (1981).
100. M. Maurer, J. M. Friedt, and J. P. Sanchez, *Nucl. Instrum. Methods* **199**, 219 (1982); J. M. Friedt, M. Maurer, J. P. Sanchez, and J. Durand, *J. Phys. F* **12**, 821 (1982).
101. C. Jeandey, J. L. Oddou, B. Boucher, R. Tourbot, and G. Czjzek, *J. Magn. Magn. Mater.* **89**, 335 (1990).
102. K. N. R. Taylor, *Adv. Phys.* **20**, 335 (1971).
103. K. H. J. Buschow, *Rep. Prog. Phys.* **40**, 1179 (1977).
104. K. H. J. Buschow, *Rep. Prog. Phys.* **42**, 1373 (1979).
105. K. H. J. Buschow, in *Ferromagnetic Materials,* E. P. Wohlfarth, ed., Vol. 1, p. 297, North-Holland, Amsterdam (1980).
106. F. J. van Steenwijk, H. T. Lefever, R. C. Thiel, and K. H. J. Buschow, *Physica* **92B**, 52 (1977).
107. K. Tomala, G. Czjzek, J. Fink, and H. Schmidt, *Solid State Commun.* **24**, 857 (1977).
108. M. Seh, S. Cohen, and I. Nowik, *J. Phys. (Paris) Colloq.* **41** (C1), 201 (1980).
109. E. Dormann, *J. Magn. Magn. Mater.* **6**, 87 (1977).
110. S. K. Arif, J. W. Ross, and M. A. H. McCausland, *Physica* **86–88B**, 158 (1977).
111. O. Prakash, M. A. Chaudhry, J. W. Ross, and M. A. H. McCausland, *J. Magn. Magn. Mater.* **36**, 271 (1983).
112. I. Felner and I. Nowik, *J. Phys. Chem. Solids* **39**, 951 (1978).
113. I. Felner and I. Nowik, *J. Phys. Chem. Solids* **40**, 1035 (1979).
114. I. Felner, M. Seh, M. Rakavy, and I. Nowik, *J. Phys. Chem. Solids* **42**, 369 (1981).
115. I. Felner, M. Seh, and I. Nowik, *J. Phys. Chem. Solids* **42**, 1091 (1981).
116. I. Felner, I. Nowik, and M. Seh, *J. Magn. Magn. Mater.* **38**, 172 (1983).
117. J. W. Ross and J. Sigalas, *J. Phys. F* **5**, 1973 (1975).
118. A. R. Miedema and F. van der Woude, *Physica* **100B**, 145 (1980).
119. H. de Graaf, R. C. Thiel, and K. H. J. Buschow, *J. Phys. F* **12**, 2079 (1982).
120. J. W. C. de Vries, R. C. Thiel, and K. H. J. Buschow, *J. Phys. F* **15**, 1413 (1985).
121. J. W. C. de Vries, R. C. Thiel, and K. H. J. Buschow, *J. Less-Common Met.* **111**, 313 (1985).
122. A. R. Miedema, P. F. de Châtel, and F. R. de Boer, *Physica* **100B**, 1 (1980).

123. K. H. J. Buschow, in *Handbook on the Physics and Chemistry of Rare Earths,* K. A. Gschneidner, Jr., and L. Eyring, eds., Vol. 6, p. 1, North-Holland, Amsterdam (1984).
124. R. Wordel, F. J. Litterst, and F. E. Wagner, *J. Phys. F* **15**, 2525 (1985).
125. U. Potzel, R. Raab, J. Völkl, H. Wipf, A. Magerl, D. Salomon, and G. Wortmann, *J. Less-Common. Met.* **101**, 343 (1984).
126. S. J. Lyle, P. T. Walsh, A. D. Witts, and J. W. Ross, *J. Chem. Soc. Dalton Trans.* **1975**, 1406 (1975); S. J. Lyle and P. T. Walsh, *J. Chem. Soc. Dalton Trans.* **1978**, 601 (1978).
127. G. Wortmann, W. Thal, A. Mustachi, and I. Nowik, *J. Less-Common Met.* **125**, 105 (1986).
128. J. W. C. de Vries, R. C. Thiel, and K. H. J. Buschow, *Physica* **128B**, 183 (1985).
129. H. de Graaf, R. C. Thiel, and K. H. J. Buschow, *J. Phys. F* **12**, 1239 (1982).
130. I. Jacob, E. R. Bauminger, D. Davidov, I. Felner, S. Ofer, and D. Shaltiel, *J. Magn. Magn. Mater.* **15–18**, 1269 (1980).
131. E. R. Bauminger, D. Davidov, I. Felner, I. Nowik, S. Ofer, and D. Shaltiel, *Physica* **86–88B**, 201 (1977).
132. A. Szytula and J. Leciejewicz, in *Handbook on the Physics and Chemistry of Rare Earths,* K. A. Gschneidner, Jr., and L. Eyring, eds., Vol. 12, p. 133, North-Holland, Amsterdam (1989).
133. F. Steglich, J. Aarts, C. Bredl, W. Lieke, D. Meschede, W. Franse, and J. Schäfer, *Phys. Rev. Lett.* **43**, 1892 (1979).
134. I. Nowik, I. Felner, and M. Seh, *J. Magn. Magn. Mater.* **15–18**, 1215 (1980).
135. D. R. Noakes, A. M. Umarji, and G. K. Shenoy, *J. Magn. Magn. Mater.* **39**, 309 (1983).
136. I. Felner and I. Nowik, *J. Phys. Chem. Solids* **45**, 419 (1984).
137. K. Łątka, *Acta Phys. Pol.* **A72**, 73 (1987).
138. M. W. Dirken, R. C. Thiel, and K. H. J. Buschow, *J. Less-Common Met.* **147**, 97 (1989).
139. I. Felner and I. Nowik, *J. Phys. Chem. Solids* **39**, 763 (1978).
140. J. P. Sanchez, K. Tomala, and A. Szytula, *Solid State Commun.* **78**, 419 (1991).
141. I. Felner and I. Nowik, *Solid State Commun.* **47**, 831 (1983).
142. D. Gignoux, P. Morin, and D. Schmitt, *J. Magn. Magn. Mater.* **102**, 33 (1991).
143. J.-M. Mignot, J. Flouquet, R. Haen, F. Lapierre, L. Puech, and J. Voiron, *J. Magn. Magn. Mater.* **76–77**, 97 (1988); D. M. Edwards, *Physica B* **169**, 271 (1991).
144. D. R. Noakes, G. K. Shenoy, D. Niarchos, A. M. Umarji, and A. T. Aldred, *Phys. Rev. B* **27**, 4317 (1983).
145. I. Felner, *J. Phys. Chem. Solids* **44**, 43 (1983).
146. I. Felner, *Solid State Commun.* **52**, 191 (1984).
147. H. H. A. Smit, R. C. Thiel, and K. H. J. Buschow, *J. Phys. F* **18**, 295 (1988).
148. S. K. Malik, A. M. Umarji, and G. K. Shenoy, *J. Appl. Phys.* **57**, 3252 (1985).
149. D. Niarchos, G. Zouganelis, A. Kostikas, and A. Simopoulos, *Solid State Commun.* **59**, 389 (1986).
150. H. R. Rechenberg, M. Bogé, C. Jeandey, J. L. Oddou, J. P. Sanchez, and P. Tenaud, *Solid State Commun.* **64**, 277 (1987).
151. M. Bogé, D. Givord, C. Jeandey, J. L. Oddou, H. R. Rechenberg, and P. Tenaud, *Hyp. Interact.* **40**, 373 (1988).
152. J. L. Oddou, M. Bogé, C. Jeandey, and H. R. Rechenberg, *J. Phys. (Paris) Colloq.* **49** (C8), 429 (1988).
153. J. L. Oddou, C. Jeandey, and H. R. Rechenberg, *J. Magn. Magn. Mater.* **87**, 135 (1990).
154. O. A. Pringle and G. J. Long, in *Mössbauer Spectroscopy Applied to Inorganic Chemistry,* G. J. Long and F. Grandjean, eds., Vol. 3, p. 13, Plenum Press, New York (1989).
155. J. J. Croat, J. F. Herbst, R. W. Lee, and F. E. Pinkerton, *J. Appl. Phys.* **55**, 2078 (1984).
156. M. Sagawa, S. Fujimura, M. Togawa, H. Yamamoto, and Y. Matsuura, *J. Appl. Phys.* **55**, 2063 (1984).
157. J. F. Herbst, J. J. Croat, F. E. Pinkerton, and W. B. Yelon, *Phys. Rev. B* **29**, 4176 (1984).
158. D. Givord, H. S. Li, and J. M. Moreau, *Solid State Commun.* **50**, 497 (1984).
159. K. H. J. Buschow, J. W. C. de Vries, and R. C. Thiel, *Physica* **132B**, 13 (1985); *J. Phys. F* **15**, L93 (1985).

160. M. Bogé, J. M. D. Coey, G. Czjzek, D. Givord, C. Jeandey, H. S. Li, and J. L. Oddou, *Solid State Commun.* **55**, 295 (1985).
161. Y. Berthier, M. Bogé, G. Czjzek, D. Givord, C. Jeandey, H. S. Li, and J. L. Oddou, *J. Magn. Magn. Mater.* **54–57**, 589 (1986).
162. D. Givord, H. S. Li, and F. Tasset, *J. Appl. Phys.* **57**, 4100 (1985).
163. I. A. Campbell, *J. Phys. C* **2**, 1338 (1969).
164. R. J. Radwański and J. J. M. Franse, *J. Magn. Magn. Mater.* **74**, 43 (1988).
165. H. H. A. Smit, R. C. Thiel, and K. H. J. Buschow, *Physica* **145B**, 329 (1987).
166. D. B. de Mooij and K. H. J. Buschow, *Philips J. Res.* **42**, 246 (1987).
167. M. W. Dirken, R. C. Thiel, and K. H. J. Buschow, *J. Less-Common Met.* **146**, L15 (1989).
168. M. W. Dirken, R. C. Thiel, L. J. de Jongh, T. H. Jacobs, and K. H. J. Buschow, *J. Less-Common Met.* **155**, 339 (1989); T. H. Jacobs, M. W. Dirken, R. C. Thiel, L. J. de Jongh, and K. H. J. Buschow, *J. Magn. Magn. Mater.* **83**, 293 (1990).
169. R. B. Helmholdt and K. H. J. Buschow, *J. Less-Common Met.* **155**, 15 (1989).
170. M. W. Dirken, R. C. Thiel, T. H. Jacobs, and K. H. J. Buschow, *Hyp. Interact.* **54**, 787 (1990).
171. M. W. Dirken, R. C. Thiel, T. H. Jacobs, and K. H. J. Buschow, *J. Less-Common Met.* **168**, 269 (1991).
172. J. M. D. Coey and H. Sun, *J. Magn. Magn. Mater.* **87**, L251 (1990).
173. M. W. Dirken, R. C. Thiel, R. Coehoorn, T. H. Jacobs, and K. H. J. Buschow, *J. Magn. Magn. Mater.* **94**, L15 (1991).
174. D. C. Cook and J. D. Cashion, *J. Phys. C* **9**, L79 (1976); **13**, 4199 (1980).
175. D. C. Cook and J. D. Cashion, *J. Phys. C* **12**, 605 (1979).
176. R. Kmieć, K. Łątka, T. Matlak, K. Ruebenbauer, and K. Tomala, *Phys. Status Solidi B* **68**, K125 (1975).
177. E. R. Bauminger, A. Diamant, I. Felner, I. Nowik, A. Mustachi, and S. Ofer, *J. Phys. (Paris) Colloq.* **37** (C6), 49 (1976).
178. Y. Calage and J. Pannetier, *J. Phys. Chem. Solids* **38**, 711 (1977).
179. J. P. Sanchez, J. M. Friedt, K. Tomala, and F. Holtzberg, *Phys. Lett.* **111A**, 83 (1985).
180. L. Ben-Dor, I. Shilo, and I. Felner, *J. Solid State Chem.* **24**, 401 (1978).
181. B. T. Matthias, H. Suhl, and E. Corenzwit, *Phys. Rev. Lett.* **1**, 92 (1958).
182. H. Suhl and B. T. Matthias, *Phys. Rev.* **114**, 977 (1959).
183. A. A. Abrikosov and L. P. Gorkov, *Sov. Phys. JETP* **12**, 1243 (1961).
184. M. B. Maple, in *Magnetism,* H. Suhl, ed., Vol. V, p. 289, Academic Press, New York (1973).
185. B. T. Matthias, H. Suhl, and E. Corenzwit, *Phys. Rev. Lett.* **1**, 449 (1958).
186. G. K. Shenoy, in *Advances in Mössbauer Spectroscopy,* B. V. Thosar and P. K. Iyengar, eds., p. 561, Elsevier, Amsterdam (1983).
187. K.-L. Kumagai and K. Asayama, *J. Phys. Soc. Jpn.* **39**, 543 (1975).
188. M. Peter, P. Donzé, Ø. Fischer, A. Junod, J. Ortelli, A. Treyvaud, E. Walker, M. Wilhelm, and B. Hillenbrand, *Helv. Phys. Acta* **44**, 345 (1971); G. Rupp, *J. Phys. F* **3**, 1403 (1973).
189. D. J. Erickson, C. E. Olson, and R. D. Taylor, in *Mössbauer Effect Methodology,* I. J. Gruverman and C. W. Seidel, eds., Vol. 8, p. 73, Plenum Press, New York (1973). J. O. Willis, D. J. Erickson, C. E. Olsen, and R. D. Taylor, *Phys. Rev. B* **21**, 79 (1980).
190. K. Ruebenbauer, J. Fink, H. Schmidt, G. Czjzek, and K. Tomala, *Phys. Status Solidi B* **84**, 611 (1977).
191. S. Roth, K. Ibel, and W. Just, *J. Phys. C* **6**, 3465 (1973).
192. J. W. Lynn, D. E. Moncton, L. Passell, and W. Thomlinson, *Phys. Rev. B* **21**, 70 (1980).
193. P. Steiner, D. Gumprecht, and S. Hüfner, *Phys. Rev. Lett.* **30**, 1132 (1973); P. Steiner and D. Gumprecht, *Solid State Commun.* **22**, 501 (1977).
194. P. W. Anderson and H. Suhl, *Phys. Rev.* **116**, 898 (1959).
195. M. B. Maple, in *Ternary Superconductors,* G. K. Shenoy, B. D. Dunlap, and F. Y. Fradin, eds., p. 131, North-Holland, Amsterdam (1981).

196. W. A. Fertig, D. C. Johnston, L. E. DeLong, R. W. McCallum, M. B. Maple, and B. T. Matthias, *Phys. Rev. Lett.* **38**, 987 (1977).
197. G. K. Shenoy, G. W. Crabtree, D. Niarchos, F. Behroozi, B. D. Dunlap, D. Hinks, and D. R. Noakes, in *Crystalline Electric Field Effects in f-Electron Magnetism,* R. P. Guertin, W. Suski, and Z. Zolnierek, eds., p. 431, Plenum Press, New York (1982).
198. G. K. Shenoy, D. R. Noakes, and D. G. Hinks, *Solid State Commun.* **42**, 411 (1982).
199. D. R. Noakes, A. M. Umarji, G. K. Shenoy, and D. G. Hinks, *J. Appl. Phys.* **55**, 1847 (1984).
200. M. Ishikawa, in *Ternary Superconductors,* G. K. Shenoy, B. D. Dunlap, and F. Y. Fradin, eds., p. 43, North-Holland, Amsterdam (1981).
201. D. R. Noakes, G. K. Shenoy, and D. G. Hinks, *Phys. Rev. B* **31**, 5712 (1985).
202. D. M. Ginsberg, in *Physical Properties of High Temperatures Superconductors,* D. M. Ginsberg, ed., Vol. I, p. 1, World Scientific, Singapore (1989).
203. Z. Z. Sheng and A. M. Hermann, *Nature* **332**, 55 (1988); R. M. Hazen, L. W. Finger, R. J. Angel, C. T. Prewitt, N. L. Ross, C. G. Hadidiacos, P. J. Heaney, D. R. Veblen, Z. Z. Sheng, A. El Ali, and A. M. Hermann, *Phys. Rev. Lett.* **60**, 1657 (1988).
204. V. J. Emery, *Physica B* **169**, 17 (1991); D.-H. Lee, *Physica B* **169**, 37 (1991).
205. N. Nücker, J. Fink, J. C. Fuggle, P. J. Durham, and W. M. Temmerman, *Phys. Rev. B* **37**, 5158 (1988).
206. M. Alexander, H. Romberg, N. Nücker, P. Adelmann, J. Fink, J. T. Markert, M. B. Maple, S. Uchida, H. Takagi, Y. Tokura, A. C. W. P. James, and D. W. Murphy, *Phys. Rev. B* **43**, 333 (1991).
207. J. M. Tranquada and G. Shirane, in *Dynamics of Magnetic Fluctuations in High-Temperature Superconductors,* G. Reiter, P. Horsch, and G. C. Psaltakis, eds., p. 1, Plenum Press, New York (1991), and subsequent articles in the same volume.
208. K. N. Yang, B. W. Lee, M. B. Maple, and S. S. Laderman, *Appl. Phys. A* **46**, 229 (1988).
209. P. H. Hor, R. L. Meng, Y. Q. Wang, L. Gao, Z. Huang, J. Bechtold, K. Forster, and C. W. Wu, *Phys. Rev. Lett.* **58**, 1891 (1987).
210. D. McK. Paul, H. A. Mook, A. W. Hewat, B. C. Sales, L. A. Boatner, J. R. Thompson, and M. Mostoller, *Phys. Rev. B* **37**, 2341 (1988).
211. T. Chattopadhyay, H. Maletta, W. Wirges, K. Fischer, and P. J. Brown, *Phys. Rev. B* **38**, 838 (1988).
212. H. A. Mook, D. McK. Paul, B. C. Sales, L. A. Boatner, and L. Cussen, *Phys. Rev. B* **38**, 12008 (1988).
213. W. H. Li, J. W. Lynn, S. Skanthakumar, T. W. Clinton, A. Kebede, C.-S. Lee, J. E. Crow, and T. Mihalisin, *Phys. Rev. B* **40**, 5300 (1989).
214. J. W. Lynn, I. W. Sumarlin, S. Skanthakumar, W.-H. Li, R. N. Shelton, J. L. Peng, Z. Fisk, and S.-W. Cheong, *Phys. Rev. B* **41**, 2569 (1990).
215. M. Matsuda, K. Yamada, K. Kakurai, H. Kadowaki, T. R. Thurston, Y. Endoh, Y. Hikada, R. J. Birgeneau, M. A. Kastner, P. M. Gehring, A. M. Moudden, and G. Shirane, *Phys. Rev. B* **42**, 10098 (1990).
216. T. Chattopadhyay, P. J. Brown, A. A. Stepanov, P. Wyder, J. Voiron, A. I. Zvyagin, S. N. Barilo, D. I. Zhigunov, and I. Zobkalo, *Phys. Rev. B* **44**, 9486 (1991).
217. Y. J. Uemura, G. M. Luke, B. J. Sternlieb, I. P. Le, J. H. Brewer, R. Kadono, R. F. Kiefl, S. R. Kreitzman, T. Riseman, C. L. Seaman, J. J. Neumeier, Y. Dalichaouch, M. B. Maple, G. Saito, and H. Yamochi, in *Dynamics of Magnetic Fluctuations in High-Temperature Superconductors,* G. Reiter, P. Horsch, and G. C. Psaltakis, eds., p. 127, Plenum Press, New York (1991), and subsequent articles in the same volume.
218. M. Takigawa, in *Dynamics of Magnetic Fluctuations in High-Temperature Superconductors,* G. Reiter, P. Horsch, and G. C. Psaltakis, eds., p. 61, Plenum Press, New York (1991), and subsequent articles in the same volume.
219. R. D. Taylor, J. O. Willis, and Z. Fisk, *Hyp. Interact.* **42**, 1257 (1988).
220. G. Wortmann, S. Blumenröder, A. Freimuth, and D. Riegel, *Phys. Lett. A* **126**, 434 (1988).

221. G. Wortmann, A. Kolodziejczyk, M. Bergold, G. Stadermann, C. T. Simmons, and G. Kaindl, *Hyp. Interact.* **50**, 555 (1989).
222. M. Z. Jin, X. W. Liu, M. L. Liu, and Y. Q. Jia, *Phys. Lett. A* **148**, 217 (1990).
223. E. R. Bauminger, I. Felner, Y. Lehavi, and I. Nowik, *Hyp. Interact.* **55**, 1199 (1990).
224. M. Taniwaki, H. Itoh, and M. Kusuhashi, *Physica C* **185–189**, 1123 (1991); *Hyp. Interact.* **70**, 1189 (1992).
225. J. A. Hodges, P. Imbert, J. B. Marimon da Cunha, and J. P. Sanchez, *Hyp. Interact.* **50**, 569 (1989).
226. M. Bergold, G. Wortmann, and G. A. Stewart, *Hyp. Interact.* **55**, 1205 (1990).
227. J. A. Hodges, P. Bonville, P. Imbert, and G. Jéhanno, *Physica C* **184**, 283 (1991).
228. J. A. Hodges, P. Bonville, P. Imbert, and G. Jéhanno, *Physica C* **184**, 259 (1991); J. A. Hodges, P. Bonville, P. Imbert, G. Jéhanno, and P. Debray, *Physica C* **184**, 270 (1991).
229. H. H. A. Smit, M. W. Dirken, R. C. Thiel, and L. J. de Jongh, *Solid State Commun.* **64**, 695 (1987); J. van den Berg, C. J. van der Beek, P. H. Kes, J. A. Mydosh, G. J. Nieuwenhuys, and L. J. de Jongh, *Solid State Commun.* **64**, 699 (1987).
230. E. E. Alp, L. Soderholm, G. K. Shenoy, D. G. Hinks, D. W. Capone II, K. Zhang, and B. D. Dunlap, *Phys. Rev. B* **36**, 8910 (1987).
231. J. D. Cashion, J. R. Fraser, A. C. McGrath, R. H. Mair, and R. Driver, *Hyp. Interact.* **42**, 1253 (1988).
232. P. Dalmas de Réotier, P. Vulliet, A. Yaouanc, O. Hartmann, E. Karlsson, R. Wäppling, P. Chaudouët, S. Garçon, J. P. Sénateur, F. Weiss, and T. O. Niinikoski, *Physica C* **153–155**, 1541 (1988); P. Dalmas de Réotier, P. Vulliet, A. Yaouanc, P. Chaudouët, S. Garçon, J. P. Sénateur, F. Weiss, L. Asch, and G. M. Kalvius, *Physica C* **153–155**, 1543 (1988).
233. J. O. Willis, Z. Fisk, J. D. Thompson, S.-W. Cheong, R. M. Aikin, J. L. Smith, and E. Zirngiebel, *J. Magn. Magn. Mater.* **67**, L139 (1987).
234. P. Schweiss, W. Reichardt, M. Braden, G. Collin, G. Heger, H. Claus, and A. Erb, (to be published).
235. T. J. Kistenmacher, *Phys. Rev. B* **38**, 8862 (1988).
236. H. J. Bornemann, R. Kmieć, G. Czjzek, P. Adelmann, and A. Seidel, *Physica C* **159**, 173 (1989).
237. P. Bordet, J. L. Hodeau, P. Strobel, M. Marezio, and A. Santoro, *Solid State Commun.* **66**, 435 (1988); B. D. Dunlap, J. D. Jorgensen, C. Segre, A. E. Dwight, J. L. Matykiewicz, H. Lee, W. Peng, and C. W. Kimball, *Physica C* **158**, 379 (1989).
238. H. J. Bornemann, R. Kmieć, and G. Czjzek, *Physica B* **165–166**, 1323 (1990).
239. M. B. Maple, Y. Dalichaouch, J. M. Ferreira, R. R. Hake, B. W. Lee, J. J. Neumeier, M. S. Torikachvili, K. N. Yang, H. Zhou, R. P. Guertin, and M. V. Kuric, *Physica B* **148**, 155 (1987).
240. S. Horn, J. Cai, S. A. Shaheen, Y. Jeon, M. Croft, C. L. Chang, and M. L. den Boer, *Phys. Rev. B* **36**, 3895 (1987); J.-S. Kang, J. W. Allen, Z.-X. Shen, W. P. Ellis, J. J. Yeh, B. W. Lee, M. B. Maple, W. E. Spicer, and I. Lindau, *J. Less-Common Met.* **148**, 121 (1989).
241. J. Fink, N. Nücker, H. Romberg, M. Alexander, M. B. Maple, J. J. Neumeier, and J. W. Allen, *Phys. Rev. B* **42**, 4823 (1990).
242. Z. Tomkowicz, K. Łątka, A. Szytula, A. Bajorek, M. Balanda, R. Kmieć, K. Kruk, and A. Zygmunt, *Physica C* **174**, 71 (1991).
243. G. Wortmann and I. Felner, *Solid State Commun.* **75**, 981 (1990).
244. J. Karpinski, E. Kaldis, E. Jilek, S. Rusiecki, and B. Bucher, *Nature* **336**, 660 (1988).
245. H. J. Bornemann, D. E. Morris, C. Steinleitner, and G. Czjzek, *Phys. Rev. B* **44**, 12567 (1991).
246. R. M. Hazen, in *Physical Properties of High Temperature Superconductors*, D. M. Ginsberg, ed., p. 121, World Scientific, Singapore (1990).
247. E. V. Sampathkumaran, N. Ikeda, K. Kohn, and R. Vijayaraghavan, *J. Phys. F* **18**, L163 (1988).
248. E. V. Sampathkumaran, G. Wortmann, and G. Kaindl, *Bull. Mater. Sci.* **14**, 703 (1991).

249. M. B. Maple, N. Y. Ayoub, J. Beille, T. Bjørnholm, Y. Dalichaouch, E. A. Early, S. Ghamaty, B. W. Lee, J. T. Markert, J. J. Neumeier, G. Nieva, L. M. Paulius, I. K. Schuller, C. L. Seaman, and P. K. Tsai, in *Transport Properties of Superconductors,* R. Nicolsky, ed., p. 536, World Scientific, Singapore (1990).
250. S. Ghamaty, B. W. Lee, J. T. Markert, E. A. Early, T. Bjørnholm, C. L. Seaman, and M. B. Maple, *Physica C* **160**, 217 (1989).
251. I. Dzyaloshinsky, *J. Phys. Chem. Solids* **4**, 241 (1958); T. Moriya, *Phys. Rev.* **120**, 91 (1960).
252. Z. Fisk, J. D. Thompson, S.-W. Cheong, C. Rettori, S. B. Oseroff, D. C. Vier, S. Schultz, M. Tovar, and R. Zysler, in *Studies of High Temperature Superconductors,* A. Narlikar, ed., Vol. 8, p. 69, Nova Science Publishers, New York (1991).
253. J. Kwo, M. Hong, D. B. McWhan, Y. Yafet, R. M. Fleming, F. J. DiSalvo, J. V. Waszczak, C. F. Majkrzak, D. Gibbs, A. I. Goldmann, P. Boni, J. Bohr, H. Grimm, C. L. Chien, and J. W. Cable, *J. Phys.* (*Paris*) *Colloq.* **49** (C8), 1651 (1988).
254. C. L. Chien, *Hyp. Interact.* **53**, 3 (1990).
255. S. J. Poon, A. J. Drehmann, and K. R. Lawless, *Phys. Rev. Lett.* **55**, 2324 (1985).
256. R. D. Werkmann, P. J. Schurer, and F. van der Woude, *Hyp. Interact.* **53**, 241 (1990).
257. R. E. Smalley, in *Fullerenes,* G. S. Hammond and V. J. Kuck, eds., p. 141, American Chemical Society, Washington, D.C. (1992).

Author Index

Italicized numbers indicate reference numbers wherein the specified author can be found.

Subject Index